BIOFERTILIZERS AND BIOPESTICIDES IN SUSTAINABLE AGRICULTURE

BIOFERTILIZERS AND BIOPESTICIDES IN SUSTAINABLE AGRICULTURE

Edited by
B. D. Kaushik
Deepak Kumar
Md. Shamim

Apple Academic Press Inc.	Apple Academic Press Inc.
3333 Mistwell Crescent	1265 Goldenrod Circle NE
Oakville, ON L6L 0A2	Palm Bay, Florida 32905
Canada	USA

© 2020 by Apple Academic Press, Inc.

First issued in paperback 2021

Exclusive worldwide distribution by CRC Press, a member of Taylor & Francis Group

No claim to original U.S. Government works

ISBN 13: 978-1-77463-466-0 (pbk)
ISBN 13: 978-1-77188-793-9 (hbk)

Library and Archives Canada Cataloguing in Publication

Title: Biofertilizers and biopesticides in sustainable agriculture / edited by B.D. Kaushik, Deepak Kumar, Md. Shamim.

Names: Kaushik, B. D., editor. | Kumar, Deepak (Research scientist), editor. | Shamim, Md., 1985-editor.

Description: Includes bibliographical references and index.

Identifiers: Canadiana (print) 20190141727 | Canadiana (ebook) 20190141735 | ISBN 9781771887939 (hardcover) | ISBN 9780429059384 (ebook)

Subjects: LCSH: Biofertilizers. | LCSH: Natural pesticides. | LCSH: Soil management. | LCSH: Field crops—Diseases and pests—Control. | LCSH: Sustainable agriculture.

Classification: LCC S654.5 .B56 2019 | DDC 631.8/6—dc23

CIP data on file with US Library of Congress

Apple Academic Press also publishes its books in a variety of electronic formats. Some content that appears in print may not be available in electronic format. For information about Apple Academic Press products, visit our website at **www.appleacademicpress.com** and the CRC Press website at **www.crcpress.com**

About the Editors

B. D. Kaushik, PhD, has worked in different capacities before retiring in 2007 as Head, Division of Microbiology at the Indian Agricultural Research Institute, New Delhi, India. He has expertise in biological nitrogen fixation through cyanobacteria, amelioration of salt-affected soils through cyanobacteria, and Nif gene organization and nitrogen fixation by non-heterocystous cyanobacteria. He developed technologies on indoor production technology for BGA biofertilizer and amelioration of salt-affected soils. One of the technologies on improved soil-based BGA biofertilizer was transferred by the institute to a private entrepreneur. He has filed two patents on different technologies. Dr. Kaushik extensively has visited many countries in his work, including Germany, England, USA, Bangladesh, Nepal, and many others. He was expert to Government of Mauritius, Ministry of Agriculture & Food Mauritius. At national level he served on different committees in Indian Council of Agricultural Research, Council of Scientific and Industrial Research, Indian Institutes of Technology, and various universities. Dr. Kaushik has published more than 160 papers, including 83 research papers in refereed journals and seven books. For his academic contribution, Dr. Kaushik was bestowed with several awards, including a Life Time Achievement Award from KIA Chennai. He is a Fellow of the National Academy of Agricultural Sciences and the Association of Microbiologists of India. He was President of the Association of Microbiologists of India, and currently he is President of the Indian Phycological Society. During his professional career, he acted as a research guide for over 20 MSc and PhD students. He obtained his master's degree and PhD in microbiology from the Indian Agricultural Research Institute, New Delhi, India.

Deepak Kumar, PhD, is presently working as Research Scientist cum Quality In-charge at the Department of Scientific & Industrial Research (DSIR), Government of India approved lab of Shri Ram Solvent Extraction Pvt. Ltd., Jaspur, Uttarakhand, India. He was involved in quality production of biofertilizers and biopesticides from last 9 years. He has participated in many national and international conferences. He received a Young Scientist Award in International Conference on Advances in Agricultural and Applied Sciences for Promoting Food Security (AAPS–2017) organized by Society for Agriculture Innovation and Development (SAID) at Battisputli, Nepal. He has also completed and guided different indigenous and externally funded (DST, UCB, and UCS&T) research projects. He is a member of several research associations, such as Association of Microbiologist of India (AMI), Food Scientists & Nutritionist Association of India (FSNAI), National Environmental Association of India (NEA), and the Indian Phytopathological Society of India (IPSI). He has done his early research work on soybean meal hydrolysates based organic fertilizer production through microbial fermentation and development of potent liquid formulations of indigenous PGPR and biopesticides. He earned his MSc (Biotechnology) from C.C.S. University, Meerut, India; his MBA (HR) from Punjab Technical University, Jalandhar, India; and his PhD (Ag Biotechnology) from N. D. University of Agriculture & Technology, Faizabad, Uttar Pradesh, India.

Md. Shamim, PhD, is presently working as Assistant Professor-cum-Jr. Scientist in the Department of Molecular Biology and Genetic Engineering of Dr. Kalam Agricultural College, Kishanganj (Bihar Agricultural University), India. He is the author or coauthor of 30 peer-reviewed journal articles, 15 book chapters, and two conference papers. He has three authored books along with one edited and one practical book to his credit. He is an editorial board member of several national and international journals. Recently, Dr. Shamim received the Young Faculty Award 2016 from the Venus International Foundation, Chennai, India. Before joining Bihar Agricultural University, Sabour, Dr. Shamim worked at the Indian Agricultural Research Institute, New Delhi, where he was engaged in heat-responsive gene regulation in wheat. Dr. Shamim also has working experience at the Indian Institute of Pulses Research, Kanpur, India, on molecular and phylogeny analysis of several *Fusarium* fungi and has also done research at the Biochemistry Department of Dr. Ram Manohar Lohia Institute on plant protease inhibitor isolation and their characterization. He is a member of the soil microbiology core research group at Bihar Agricultural University (BAU), where he helps with providing appropriate direction and assisting with prioritizing the research work on PGPRs. His research has focused on biotic stress management in rice, especially in yellow stem borer management by isolating protease inhibitor from jackfruit seeds, and he is also doing work in sheath blight resistance mechanism in wild rice, cultivated rice, and other hosts. Dr. Shamim acquired his master's degree (Biotechnology) and PhD (Agricultural Biotechnology) from Narendra Deva University of Agriculture and Technology, Kumarganj, Faizabad, India, with specialization in biotic stress management in rice through molecular and proteomics tools.

Contents

Contributors

Israr Ahmad
Division of Crop Improvement and Biotechnology, ICAR-CISH, Rehmankhera, Lucknow 226016, Uttar Pradesh, India

Md. Shamsher Ahmad
Department of Post Harvest Technology, Bihar Agricultural University, Sabour, Bhagalpur 813210, Bihar, India

Salman Ahmad
Integral Institute of Agricultural Science and Technology, Integral University, Dasauli, Lucknow 226021, Uttar Pradesh, India. E-mail: salmanamd@iul.ac.in

Md. Arshad Anwer
Department of Plant Pathology, Bihar Agricultural University, Sabour, Bhagalpur 813210, Bihar India

Kumari Apurva
Department of Forestry, Forest Research Institute (Indian Council of Forestry Research and Education), P.O. New Forest, Dehradun 248006, Uttrakhand, India. E-mail: apurvashahishiats@gmail.com
Department of Molecular Biology and Genetic Engineering, Dr. Kalam Agricultural College, Kishanganj, Bihar Agricultural University, Sabour, Bhagalpur 813210, Bihar, India

Kahakashan Arzoo
Department of Plant Pathology, G. B. Pant University of Agriculture and Technology, Pantnagar 263145, Uttarakhand, India

Rabia Basri
Department of Plant Protection, Faculty of Agricultural Science, Aligarh Muslim University, Aligarh 202001, Uttar Pradesh, India

Tara Singh Bisht
Department of Biotechnology, Bhimtal Campus, Kumaun University, Nainital 263136 Uttarakhand, India

Asheesh Chaurasiya
Department of Agronomy, Bihar Agricultural University, Sabour, Bhagalpur 813210, Bihar, India

Arpita Das
Department of Genetics and Plant Breeding, Bidhan Chandra Krishi Viswavidyalaya, Mohanpur, Nadia 741252, West Bengal, India

Erayya
Department of Plant Pathology, Bihar Agricultural University, Sabour, Bhagalpur 813210, Bihar, India. E-mail: erayyapath@gmail.com

Faria Fatima
Integral Institute of Agricultural Science and Technology, Integral University, Dasauli, Lucknow 226021, Uttar Pradesh, India

Parveen Fatima
Department of Agricultural Biotechnology, Sardar Vallabhbhai University of Agriculture and Technology, Modipuram, Meerut 250110, Uttar Pradesh, India. E-mail: parveenfrizvi@gmail.com

Arun Kumar Gupta
Department of Agricultural Biotechnology, Sardar Vallabhbhai University of Agriculture and Technology, Modipuram, Meerut 250110, Uttar Pradesh, India

Gora Chand Hazra
Directorate of Research, Bidhan Chandra Krishi Viswavidyalaya, Kalyani, Nadia 741235, West Bengal, India

Neelakanth S. Hiremani
ICAR-Central Institute for Cotton Research, Shankar Nagar, Nagpur 440010, Maharashtra, India

V. B. Jha
Department of Plant Breeding and Genetics, Dr. Kalam Agricultural College, Kishanganj, Bihar Agricultural University, Sabour, Bhagalpur 813210, Bihar, India

Vijay Kumar Jha
Department of Botany, Patna University, Patna 800005, Bihar, India

Mayank Kaashyap
The Pangenomics Group, School of Science, RMIT University, Melbourne 3083, Victoria, Australia.
E-mail: mayankkaashyap@gmail.com

B. D. Kaushik
Research and Development Unit, Shri Ram Solvent Extractions Pvt. Ltd. Jaspur, U.S. Nagar 244712, Uttarakhand, India. E-mail: bdkaushik@hotmail.com

Nadeem Khan
Integral Institute of Agricultural Science and Technology, Integral University, Dasauli, Lucknow 226021, Uttar Pradesh, India

Avinash Kumar
Department of Plant Breeding and Genetics, Dr. Rajendra Prasad Central Agricultural University, Pusa, Samastipur 848125, Bihar, India

Deepak Kumar
Research and Development Unit, Shri Ram Solvent Extractions Pvt. Ltd., U. S. Nagar, Jaspur 244712, Uttarakhand, India

Mahesh Kumar
Department of Molecular Biology and Genetic Engineering, Dr. Kalam Agricultural College, Kishanganj, Bihar Agricultural University, Sabour, Bhagalpur 813210, Bihar, India.
E-mail: maheshkumara2z@gmail.com

Mayank Kumar
Amity Institute of Biotechnology, Amity University, Mumbai 410206, India

Pankaj Kumar
Department of Agricultural Biotechnology and Molecular Biology,
Dr. Rajendra Prasad Central Agricultural University, Pusa, Samastipur 848125, Bihar, India.
E-mail: pankajcocbiotech@gmail.com
Department of Molecular Biology and Genetic Engineering, Bihar Agricultural University, Sabour, Bhagalpur 813210, Bihar, India

Pushpendra Kumar
Department of Agricultural Biotechnology, Sardar Vallabhbhai University of Agriculture and Technology, Modipuram, Meerut 250110, Uttar Pradesh, India

Ranjeet Ranjan Kumar
Division of Biochemistry, ICAR-Indian Agricultural Research Institute, Pusa 110012, New Delhi, India

Ravi Ranjan Kumar
Department of Molecular Biology and Genetic Engineering, Bihar Agricultural University, Sabour, Bhagalpur 813210, Bihar, India

Sanjeev Kumar
Department of Plant Pathology, Bihar Agricultural University, Sabour, Bhagalpur 813210, Bihar, India

Santosh Kumar
Department of Plant Pathology, Bihar Agricultural University, Sabour, Bhagalpur 813210, Bihar, India

Vinod Kumar
Department of Molecular Biology and Genetic Engineering, Bihar Agricultural University, Sabour, Bhagalpur 813210, Bihar, India

Rima Kumari
Department of Agricultural Biotechnology and Molecular Biology, Dr. Rajendra Prasad Central Agricultural University, Pusa, Samastipur 848125, Bihar, India

Abha Kumari
Department of Horticulture (Fruit & Fruit Technology), Bihar Agricultural University, Sabour, Bhagalpur 813210, Bihar, India

Pratibha Laad
School of Life Sciences, Devi Ahilya Vishwavidylaya, Indore 452001, Madhya Pradesh, India

Hemant S. Maheshwari
ICAR-Indian Institute of Soybean Research, Khandwa Road, Indore 452001, Madhya Pradesh, India

Rashmi Maurya
Council of Scientific and Industrial Research-National Botanical Research Institute, Rana Pratap Marg, Lucknow 226001, Uttar Pradesh, India

Alka Mishra
Council of Scientific and Industrial Research-National Botanical Research Institute, Rana Pratap Marg, Lucknow 226001, Uttar Pradesh, India

Anurag Mishra
Department of Agricultural Biotechnology, Sardar Vallabhbhai University of Agriculture and Technology, Modipuram, Meerut 250110, Uttar Pradesh, India.
E-mail: anuragmishraa@gmail.com

Naga Teja Natra
Department of Plant Pathology, Irrigated Agriculture Research and Extension Center, Washington State University, 24106 N. Bunn Road, Prosser, WA 99350, USA.
E-mail: tejanatra@gmail.com

Md. Abu Nayyer
Integral Institute of Agricultural Science and Technology, Integral University, Dasauli, Lucknow 226021, Uttar Pradesh, India

Akanksha Nigam
Department of Microbiology and Molecular Genetics, IMRIC, The Hebrew University Hadassah Medical School, Jerusalem, 91120, Israel

Rambalak Nirala
Department of Plant Breeding and Genetics, Bihar Agricultural University, Sabour, Bhagalpur 813210, Bihar, India

Hari Om
Department of Agronomy, Dr. Kalam Agricultural College, Kishanganj, Bihar Agricultural University, Sabour, Bhagalpur 813210, Bihar, India

Dhaneshwar Padhan
Directorate of Research, AICRP on Integrated Farming System,
Bidhan Chandra Krishi Viswavidyalaya, Kalyani, Nadia 741235, West Bengal, India

Awadhesh Kumar Pal
Department of Biochemistry and Crop Physiology, Bihar Agricultural University, Sabour, Bhagalpur 813210, Bihar, India

Rachna Pande
ICAR-Central Institute for Cotton Research, Shankar Nagar, Nagpur 440010, Maharashtra, India

Veena Pande
Department of Biotechnology, Bhimtal Campus, Kumaun University, Nainital 263136 Uttarakhand, India. E-mail: veena_kumaun@yahoo.co.in

Manuraj Pandey
Molecular Mechanism and Biomarkers Group, International Agency for Research on Cancer, 150 Cours Albert Thomas, 69008 Lyon, France

Pramila Pandey
Krishi Vigyan Kendra, Haidergarh, Barabanki, 227301, Uttar Pradesh, India

Santosh Kumar Patel
Department of Agronomy, Mata Gujri College, Fatehgarh Sahib 140406, Punjab, India

Sajal Pati
Directorate of Research, Bidhan Chandra Krishi Viswavidyalaya, Kalyani, Nadia 741235, West Bengal, India

Parthendu Poddar
Department of Agronomy, Uttar Banga Krishi Viswavisyalaya, Pundibari, Coochbehar 736165, West Bengal, India

Bishun Deo Prasad
Department of Molecular Biology and Genetic Engineering, Bihar Agricultural University, Sabour, Bhagalpur 813210, Bihar, India. E-mail: bdprasadbau@gmail.com

Shivam Priya
Laboratory of Developmental Biology, Institute of Dental Science,
Hebrew University Hadessah Ein Kerem, P.O. Box 12272, Jerusalem 91120, Israel

Tushar Ranjan
Department of Molecular Biology and Genetic Engineering, Bihar Agricultural University, Sabour, Bhagalpur 813210, Bihar, India.

Bholanath Saha
Department of Soil Science Agricultural Chemistry, Dr. Kalam Agricultural College, Kishanganj, Bihar Agricultural University, Sabour, Bhagalpur 813210, Bihar, India.
E-mail: bnsaha1@gmail.com

Sushanta Saha
Directorate of Research, AICRP on Integrated Farming System, Bidhan Chandra Krishi Viswavidyalaya, Kalyani, Nadia 741235, West Bengal, India

Md. Shamim
Department of Molecular Biology and Genetic Engineering, Dr. Kalam Agricultural College, Kishanganj, Bihar Agricultural University, Sabour, Bhagalpur 813210, Bihar, India.
E-mail: shamimnduat@gmail.com

Mahaveer P. Sharma
ICAR-Indian Institute of Soybean Research, Khandwa Road, Indore 452001, Madhya Pradesh, India.
E-mail: mahaveer620@gmail.com

Preeti Sharma
Department of Microbiology, Shri R.L.T. College of Science, Akola 444001, Maharashtra, India

Mohd Haris Siddiqui
Integral Institute of Agricultural Science and Technology, Integral University, Dasauli, Lucknow 226021, Uttar Pradesh, India

Mohammed Wasim Siddiqui
Department of Post Harvest Technology, Bihar Agricultural University, Sabour, Bhagalpur, 813219, Bihar, India

Saba Siddiqui
Integral Institute of Agricultural Science and Technology, Integral University, Lucknow 226021, Uttar Pradesh, India

Chandrakant Singh
Wheat Research Station, Jungadh Agricultural University, Junagadh 362001, Gujarat, India

Hemant Kumar Singh
Krishi Vigyan Kendra, Hawai Adda Road Khagra, Kishanganj 855107, Bihar, India

K. N. Singh
Department of Plant Molecular Biology and Genetic Engineering, N. D. University of Agriculture and Technology, Kumarganj, Faizabad 224229, Uttar Pradesh, India

M. K. Singh
Research and Development Unit, Shri Ram Solvent Extractions Pvt. Ltd., Jaspur, U.S. Nagar 244712, Uttarakhand, India

Deepti Srivastava
Integral Institute of Agricultural Science and Technology, Integral University, Dasauli, Lucknow 226021, Uttar Pradesh, India. E-mail: deeptifzd@gmail.com

Vishwa Vijay Thakur
Department of Molecular Biology and Genetic Engineering, Dr. Kalam Agricultural College, Kishanganj, Bihar Agricultural University, Sabour, Bhagalpur 813210, Bihar, India

Pooja Verma
Indian Council of Agricultural Research-Central Institute for Cotton Research, Shankar Nagar, Nagpur 440010 Maharashtra, India. E-mail: poojaverma1906@gmail.com

Dharmendra Kumar Verma
Department of Soil Science and Agricultural Chemistry, Bihar Agricultural University, Sabour, Bhagalpur 813210, Bihar, India. E-mail: dkvermabhu@gmail.com

Abbreviations

ACC	1-aminocyclopropane-1-carboxylate
ADH	alcohol dehydrogenase
AGS	Advanced Genetic Sciences
AM	arbuscular mycorrhiza
AMF	arbuscular mycorrhizal fungi
ATP	adenosine triphosphate
BCA	biological control agent
BCAs	biocontrol agents
BNF	biological nitrogen fixation
Bt.	*Bacillus thuringiensis*
CA	carbonic anhydrase
CAMERA	Community Cyber Infrastructure for Advanced Microbial Ecology Research and Analysis
CMV	cucumber mosaic virus
Cry	crystal
DD	dichloropropene-dichloropropane
DDT	dichlorodiphenyltrichloroethane
DE	delayed early
ECC	environment controlled chambers
EDB	ethylene dibromide
EDTA	ethylenediaminetetraacetic acid
EFFs	environment friendly fertilizers
EPA	Environmental Protection Agency
FACS	florescence-activated cell sorting
FAO	Food and Agricultural Organization
FIFRA	Federal Insecticide, Fungicide and Rodenticide Act
FVW	fruit and vegetable wastes
GA	gluconic acid
GABA	gamma-aminobutyric acid
GDD	growing degree-days
GDH	glucose dehydrogenase
GEMs	genetically engineered microorganisms
GM	genetically modified
GMMs	genetically modified microorganisms

GRAS	generally recognized as safe
GVs	granulovirus
HCN	hydrogen cyanide
IAA	indole-3-acetic acid
IBMA	International Biocontrol Manufacturer's Association
IE	immediate early
IFC	immunofluorescence colony
IMG/M	Integrated Microbial Genomes and Metagenomes
IOBC	International Organization for Biological Control
IPM	integrated pest management
ISR	induced systemic resistance
JH	juvenile hormones
LMW	low-molecular-weight
LPS	lipopolysaccharides
MBr	methyl bromide
MEGAN	MEta Genome ANalyzer
MG-RAST	rapid annotation using subsystems technology for metagenomes
MH	moulting hormones
MPS	mineral phosphate solubilization
MSW	municipal solid waste
NFB	nitrogen-fixing bacteria
NPV	nuclear polyhedrosis virus
NS	non-symbiotic
OFM	oriental fruit moth
OMRI	organic materials review institute
Pdop	biochar polydopamine
PGPM	plant growth promoting microbes
PGPR	plant growth promoting rhizobacteria
PIPs	plant incorporated protectants
PPN	plant parasitic nematodes
PQQ	quinine-4.5-dione
PSB	phosphorous-solubilizing bio-fertilizers
PSM	phosphate solubilizing microorganisms
pUC	plasmid University of California
RAB	rice-associated bacteria
RNA	ribonucleic acid
ROS	reactive oxygen species
SCAR	sequence characterized amplified region
TMV	tobacco mosaic virus
ToMoV	tomato mottle virus

USEPA	US Environmental Protection Agency
UV	ultraviolet
VAM	vesicular arbuscular mycorrhiza
ZSB	zinc-solubilizing bacteria

Preface

The Green Revolution, or the so-called seed fertilizer revolution that was achieved during, 1970s through the use of high-yielding varieties, subsidized NPK fertilizers, and other agrochemicals, with assured facilities of modern crop production. The spectacular boost in the use of agrochemicals to attain optimum yield has become an integral component of present-day agricultural practices. The frequent and imbalanced application of such chemicals is not only expensive but also pollutes the environment at a faster rate and makes the soils unsuitable for cultivation. In addition, soil degradation, disturbances in composition and functional properties of soil microbial communities, and, consequently, loss of soil fertility following various soil management practices have further compounded the agronomic problems. In recent times, much interest has been generated in exploiting microbial strategies to facilitate plant growth and development through inoculating seed/soil to provide nutrients like phosphate, nitrogen, and other phyto-compounds, and in some cases, they have been commercialized for different crops.

In addition, microbes have also attracted worldwide attention due to their role in disease management and remediation of polluted soils. Thus, the microbial communities in general are the potential tools for sustainable production of crops and the trend for the future. Scientific researchers, however, involve multidisciplinary approaches to understand the complexity and practical utility of the wide spectrum of microbes for the benefit of crops. Substantial amounts of research work has been done to highlight the role of microbes in crop improvement, but very little attempt is made to organize such findings in a way that could substantially help scientists and farmers.

This book presents strategies for the management of soils and crop diseases, and explores means of integrating various approaches to achieve the desired levels of crop yield under both conventional and neglected soils through the use of biopesticides and other botanicals as well as biomolecules. This book also presents a broad and updated view of the nitrogen-fixing, phosphate-solubilizing, and sulphur-transforming microbes for nutrition of crops vis-à-vis the role of metal-tolerant microbes in providing protection to plants grown in metal-contaminated soils. The preparation and application of biofertilizers, utilization of household waste materials, and use of genetically modified microorganisms (GMOs) in plant growth and development are also well discussed. The book further describes the functional diversity of plant growth promotion activity of rhizobacteria, use of vesicular arbuscular mycorrhizal fungi, and utilization of microbial resources for availability of micronutrients (viz., Fe and Zn) as well as secondary nutrients (viz. sulfur).

Special attention is paid to highlight the role of fertile soil in crop improvement and to understand the impact of various management practices on variability of microbes and their function.

This edited book will be very useful not only for students, teachers, and researchers but also for those interested in microbiology, biotechnology, physiology of plant growth and development, phytoprotection, agronomy, and environmental sciences.

We sincerely wish to acknowledge our colleague authors who participated in this endeavor from different countries and who have assisted in the development of *Biofertilizers and Biopesticides in Sustainable Agriculture* by providing the recent information and comprehensive scripts without which it would have been extremely difficult to complete this herculean task. We crave to recognize all the theme specialists who were involved in the book and cooperated with their auxiliary, expensive assistance to create this book a success.

Recognitions are due to our research team members, especially to Dr. Mohammad Wasim Siddiqui, Dr. Mahesh Kumar, Dr. Erayya, Dr. Tushar Ranjan, and Kumari Apoorva, who generously supported the compilation and completion of this assignment. I extend my sincere thanks to Dr. K. N. Singh and her colleagues for their valuable support to facilitate the completion of this volume.

We also invite suggestions and healthy criticism from the readers of this book so that the information on the subject it contains can be improved in future.

Finally, we acknowledge Almighty God, who provided all the inspirations, insights, positive thoughts, and channels to complete this book project.

India, January, 2018

—**B. D. Kaushik**
Deepak Kumar
Md. Shamim

CHAPTER 1

General Introduction of Bio-Inputs Versus Chemical Inputs in Agriculture and Ill Effects

DEEPTI SRIVASTAVA[1,*], RASHMI MAURYA[2], NADEEM KHAN[1], MD. ABU NAYYER[1], ALKA MISHRA[2], FARIA FATIMA[1], SALMAN AHMAD[1], SABA SIDDIQUI[1], and MOHD HARIS SIDDIQUI[1]

[1]*Integral Institute of Agricultural Science and Technology, Integral University, Dasauli, Lucknow 226021, Uttar Pradesh, India*

[2]*Council of Scientific and Industrial Research-National Botanical Research Institute, Rana Pratap Marg, Lucknow 226001, Uttar Pradesh, India*

Corresponding author. E-mail: deeptifzd@gmail.com

ABSTRACT

To fulfil the requirement of increasing population and to enhance the productivity, various inputs are used in agriculture and among them fertilizer is frequently used. Fertilizer is a material that supplies nutrients to the plant, which are essential for the plant growth and development. Fertilizer is broadly classified into two major categories, that is, chemical and bio-fertilizers/organic fertilizers. Chemical fertilizers are widely adopted because of their quick action and cost effectiveness. Despite the associated benefits of chemical fertilizers, their long-term use can cause impairment of environment as they pollute soil and water due to the leaching of harmful chemicals, chemical burn to the crops, and acidification of soil that leads to mineral depletion. In comparison with chemical fertilizer; bio-fertilizers are economic and environment friendly. Bio-fertilizer adds nutrients to the natural processes of nitrogen fixation, phosphorus solubilization, and stimulates plant growth through the synthesis of growth promoting substances. In the present chapter, we discussed the general introduction of chemical and bio-inputs used in agriculture along with their associated benefits and harmful effects.

1.1 INTRODUCTION

World's population is increasing day by day due to which demand for food is also increasing. But threats in agricultural production such as loss of soil fertility, lack of nutrients, and pests may result in low yield. If these problems are not addressed, there would not be enough food to fulfill the rising needs of the people. Any agricultural revolution depends on the adaption of modern agricultural inputs (Hazell and Wood, 2008). These agricultural inputs include agricultural machinery, knowledge of agricultural practices, irrigation, improved seed, feedstuffs, fertilizers, permitted plant protection products as well as cleaning agents and additives used in food, etc. Among these inputs, fertilizer supplies nutrients to the soil. The term "fertilizer" can be defined as any material, synthetic or natural, that is applied to soil or to plant tissues to supply one or more plant nutrients essential for plant growth. Fertilizers are meant for increasing plant growth by the following two different manners: (1) chemical addition for providing nutrients and (2) enhancing soil's efficiency by modification of aeration and water retention capacity. The use of manure and composts as fertilizers is probably as old as agriculture. Fertilizer can be grouped into two categories, that is, chemical fertilizer and bio-fertilizer. Chemical fertilizer has significantly improved the crop production with appropriate use for different soil type and crop. There is worldwide concern regarding the effect of modern farming practices on quality of soil and water, and most of the recent researchers has focused on management of reducing leaching and nutrient runoff (Gillingham and Thorrold, 2000; Ledgard et al., 2000) and improving the health of soil quality (Liebig and Doran, 1999). Increased use of fertilizers and pesticides adversely affects the soil, environment, as well as human health. To overcome these problems, there is a need to adopt sustainable agriculture. Sustainable agriculture refers to an agricultural system that is economically viable, ecologically sound, and socially adaptable. The major aspect of sustainable agriculture is to adapt future potential change and attainment of sustainability. It aims to reduce the utilization of inputs from non-renewable sources and replace them with those from renewable resources and ensures that the basic nutritional requirement of current and future generations are made in both quality and quantity terms. For example, use of bio-inputs such as fertilizers and biopesticides in sustainable agriculture can certainly help us to cope with the problems faced due to the detrimental effect of conventional agriculture (Raja, 2013; Ahmad et al., 2015).The relevant and important technological components of sustainable agriculture are reduced use of synthetic chemical inputs, use of organic manures, biological pest control, soil and water conservation practices, crop rotations, biological nitrogen fixation, etc. (Rao, 2002). In the present chapter, we described different types of bio-inputs and chemical inputs along with their associated benefits and ill effects (Fig. 1.1).

1.2 CHEMICAL FERTILIZER

Chemical fertilizers are widely accepted due to instant manufacturing, fast delivery, definite action, and cost-effective availability. Chemical fertilizer has significantly improved the crop production with appropriate use for different soil type and crop. A chemical fertilizer is defined as "any inorganic material of complete or partial synthetic origin that is added to the soil to sustain plant growth." In India, diverse geographical areas are associated with diverse soil deficiencies that depend on soil texture, monsoon, and crop type. The problem was aggravated by monotonous irrigation strategy that depletes soil

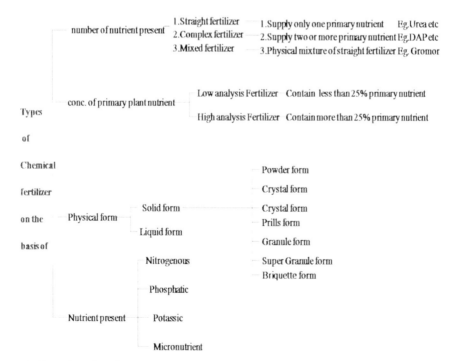

FIGURE 1.1   Classification of chemical fertilizers.

for the specific nutrient type. Plant nutrients have been categorized into major, minor, and traces elements on the basis of plant requirement for proper growth and development. Nitrogen, potassium, and phosphorus are considered under major nutrient; calcium, magnesium, and sulfur are considered under minor; and iron, manganese, zinc, copper, boron, and molybdenum are considered as trace elements. Usually, fertilizers having nitrogen, phosphorus, and potassium, collectively termed as NPK fertilizers, are provided in different ratio. Among the fertilizers, nitrogen, potassium, and phosphorus are the three main elements or macronutrients most rapidly removed from the soil by crops. Nitrogen helps in leaf growth; potassium aids sturdy stem growth, water movement, and promotion of flowering and quality fruiting; and phosphorus helps in storage and distribution of energy throughout the plant.

1.3 TYPES OF CHEMICAL FERTILIZER

1.3.1 NITROGENOUS FERTILIZERS

Nitrogen is a component of protein, DNA, and enzymes, and it regulates vital biological processes including photosynthesis, growth, and development. Most nitrogen on earth is present in inert gaseous form and is not available to plant. Thus, nitrogenous fertilizer can be given in the form of nitrate, ammonium, amide, and ammonical nitrate that can be utilized by plants. These break down to form either nitrate or amide that could be taken up by plants. Some nitrogenous fertilizers, such as nitrate, leach down into the soil and remain unavailable to plants; however, some, like ammonical and amide form, are resistant to leach down. The leaching of fertilizer further leads to water table contamination or soil acidity. The most commonly used nitrogenous fertilizer is urea, which is water soluble, resistant to leach down, and can be easily degraded by soil micro bacteria. The global potential nitrogen balance (i.e., the difference between N potentially available for fertilizers and N fertilizer demand) as a percentage of N fertilizer demand is increasing during the period, from 3.7% in 2014 to 5.4% in 2015, and then 6.9% in 2016, 8.8% in 2017, and reached 9.5% in 2018. The global potential balance of phosphorous is expected to rise from 2,700,000 tonnes in 2014 to 3,700,000 tonnes in 2018 or from 6.4% of total demand to 8.5% (Anonymous, 2017). Plant height, the number of branches, and number of bolls per plant were increased with the recommended dosage of fertilizer along with 2% potassium nitrate (Channakeshava et al., 2013). Foliar application of potassium nitrate at 4% improved the fruit set in olive cv. "Picual" (Hegazi et al., 2011). Foliar application of

1% potassium nitrate also increases the number of cobs per plant, cob length, number of grains per row, number of grains per cob, and 100-grain weight in maize (Singh et al., 2017). However, in rice seedlings, exogenous application of potassium nitrate increases the salt stress (Siringam et al., 2013). A spray of calcium nitrate was effective in olive seedlings and resulted in the increase in plant height, stem diameter, root length, and lateral shoot length but the foliar application of the lower rate of calcium nitrate resulted in high root number (Laila et al., 2011). Similarly, application of calcium nitrate and mixed application of calcium chloride with calcium chloride increased the plant height and tuber yield in varieties of potato, namely, Shenkola and Gera (Seifu and Deneke, 2017). The dry weight of tomato plants increased with treatment with 150 mg of ammonium sulfate but on increasing the conc. above 150 mg, plant dry weight reduces (Hzhbryan and Kazemi, 2014). In soybean, leaf burning was observed after the foliar application due to the toxic amount of urea (Bremer, 1992). In rice, nitrogen accumulation increases in Sulfur-coated urea treated plot of rice (Kiran et al., 2010).

1.3.2 PHOSPHATE FERTILIZERS

Phosphorus is a major nutrient that is rapidly removed by crops. Phosphorous is water soluble. Phosphatic fertilizer is of different types such as single superphosphate, double superphosphate, triple superphosphate, rock phosphate, bone meal, ammonium phosphate, etc. It was observed that rock phosphate and superphosphate have increased the grain yield, dry matter, and crude protein content (Lukiwati, 2002). Application of phosphatic fertilizer decreases the salinity and increases the growth of corn and nutritional status of the plant (Siahpoush et al., 2015).

1.3.3 POTASSIUM FERTILIZERS

Potassium is the third major plant nutrient after nitrogen and phosphorous. It is known to regulate photosynthesis, protein synthesis, water, and nutrient transport. Although its deficiency is less crucial as compared to nitrogen and phosphate and soil replenish it in a faster way. The available chemical forms are potassium sulfate and muriate of potash. As part of various compounds, potassium sulfate affects the sucrose concentration in sugarcane. It is observed that sugarcane quality is not affected by potassium alone but is also influenced by chloride and sulfate. In pineapple, biomass accumulation of pineapple plants was impaired by the chlorine addition with potassium chloride. When potassium chloride is added in pineapple, its chloride damages the biomass accumulation. However, potassium sulfate showed better results than potassium chloride (Teixeira et al., 2011).

1.3.4   ZINC FERTILIZER

Zinc is one of the necessary micronutrients and an important component of several enzymes and proteins, and is required by the plants in small quantities. However, it is vital for plant development, as it plays a significant part in a wide range of processes. Although it is toxic when applied in excess amounts (Broadley et al., 2007). Zn in soil is found in three major parts: (1) water-soluble Zn (including Zn^{2+} and soluble organic fractions); (2) adsorbed and exchangeable Zn in the colloidal fraction; and (3) insoluble Zn complexes and minerals (Alloway, 1995; Barber, 1995). The chief function of Zn is to activate enzymes that are responsible for the production of certain proteins. Zinc is essential for growth regulation and stem elongation as the component of auxins. The deficiency symptoms are expressed as altering pattern of interveinal chlorosis of the young leaves and necrotic spots may be formed on leaf margins and tips. These new leaves are smaller in size and often distorted or cupped upward. Foliar application of zinc fertilizer under drought stress on wheat plants affected physiological response and yield. Zinc application increases the antioxidants content (Ma et al., 2017). Zinc application on maize increases the chlorophyll content in leaves and yield (Liu et al., 2016).

1.3.5   CALCIUM FERTILIZER

Calcium (Ca) is an essential plant nutrient. Ca^{2+} is a divalent cation, and is needed for structural roles in the cell membranes and cell wall, as a counter-cation for inorganic and organic anions in the vacuole, and as an intracellular messenger in the cytosol (Marschner, 1995; White and Broadley, 2003). Its deficiency is unusual in plants. Calcium is taken up by roots from the soil and is transported to the shoot via the xylem. It may cross the root either through the cytoplasm of cells linked by plasmodesmata (the symplast) or through the spaces between cells (the apoplast). The relative involvements of the apoplastic and symplastic pathways to the delivery of Ca to the xylem are unknown (White, 2001; White and Broadley, 2003). Ca in soils originates from the decomposition of bedrock and minerals that contain these elements (Tisdale and Nelson, 1975; Cole et al., 2016). Ca weathers comparatively rapidly and become unavailable to plants via leaching in highly weathered soils (Pilbeam and Morley, 2007).

1.3.6 MAGNESIUM FERTILIZER

Magnesium is an essential element for chlorophyll production. High concentration of potassium in soil reduces the uptake of magnesium. Magnesium deficiency causes interveinal reddening of old leaves. Calcium, potassium, and magnesium can be applied when plant is in dormant stage so that nutrients move into the root zone and are available to plants when plant growth begins again. Magnesium is applied as dolomite.

1.3.7 SULFUR FERTILIZERS

Sulfur is another essential plant nutrient and all crops require sulfur. These are mobile in soil and prone to leaching, particularly in sandy soil. Sulfur can be applied in the form of ammonium sulfate, potassium sulfate, etc. Sulfur is not mobile in plant so continuous supply of sulfur is needed for crop maturity. Deficiency of sulfur causes reduction in yield. In sunflower, application of sulfur (between 60 and 80 kg/ha) increased the yield of sunflower as well as uptake of N, P, K, and S (Nasreen and Haq, 2002).

1.3.8 MICRONUTRIENT FERTILIZER

Micronutrient fertilizer includes iron, zinc, copper, and boron. Among these, iron sulfate (19% Fe) and iron EDTA (5–14% Fe), zinc EDTA (14%), copper chelate (13% Cu), and sodium borate (11%) are used in foliar spray, but zinc sulfate (35% Zn) is broadcasted and can also be applied on foliage. Copper sulfate (25% Cu) is applied on foliage.

1.3.9 PESTICIDE

According to the Food and Agricultural Organization (FAO), pesticide is "any substance or mixture of substances intended for preventing, destroying, or controlling any pest, including vectors of human or animal disease, causing loss in production, interference in processing, storage, transport, or marketing of food, agricultural commodities, wood and wood products or animal feedstuffs, or substances that may be administered to animals for the control of insects or other pests in or on their bodies." Pesticide includes insecticides, fungicides, herbicides, rodenticides, molluscicides, and nematicides. Herbicides are used for killing weeds, fungicides for fungus, rodenticides for rats and rodents, molluscicides for mollusk, and nematicides for nematodes. Ideally pesticide must kill only targeted organism and should not affect non-target organism but unfortunately this is not followed. Because of the harmful effect of pesticide on non-target organism, use of pesticide has controversy. Pesticides production started in 1952 in India with the establishment of a plant for the production of BHC near Calcutta, and India is now the second largest manufacturer of pesticides in Asia after China and ranks twelfth globally (Mathur, 1999). Production of the technical grade of pesticide has increased over the years such as from 5000 metric tons in 1958 to 102,240 metric tons in 1998.

1.4   EFFECT OF CHEMICAL INPUTS

1.4.1   *EFFECT OF CHEMICAL INPUTS ON SOIL*

During the last decades, there has been an increase in the spread of agro-chemical residues in the environment that cause contamination (Carson, 1962; EEA 2013). However, chemical inputs, such as fertilizers and pesti-cides, have advantages and disadvantages with their use. Fertilizers and pesticides both increase yield but they also cause water pollution as when erosion and water runoff occurs. Excessive use of chemical fertilizer and pesticide causes acidification and neutralization of soil that leads to change in the pH of soil. Change in pH results in the reduction of enzymatic activity of microbes and microbes have to enter into encysting phase or will die. Excessive use of ammonium sulfate hinders the nitrogen fixation and kills earthworms. Gypsum attracts water to itself; therefore, frequent use of gypsum takes away water from plant roots as well as from soil microbes (Ohio Environmental Protection Agency, 2001). Similarly, urea is converted into anhydrous ammonia and carbon dioxide by bacteria where anhydrous ammonia is highly toxic. Disproportionate use of chemical inputs also causes air pollution such as NO, N_2O, and NO_2 emissions.

Nowadays, pesticide residue has become a serious problem for food such as in fruits, vegetables, etc. (Battu et al., 2004). Due to the excessive use of chemical fertilizers, productive soils converted into unproductive soils due to which land degradation occurs (Singh et al., 2017). Singh et al. also studied the effect of long-term use of chemical fertilizer on wheat and maize and found the decrease in bulk density of soil; he also found that the continuous use of urea causes acidification (Singh et al., 2017). The continuous use of chemical fertilizer leads to the problem in protein synthesis in leaves due to which crops develop pathological condition and causes problem in human and animal's consumption (Talukdar et al., 2003). These chemical inputs also affect human health, for example, pesticides can cause skin cancer (Yong, 1994).

1.5 ENVIRONMENT-FRIENDLY FERTILIZERS

Chemical fertilizers cause negative impact on environment that can be improved by fertilization application methods, increasing water-use efficiency via irrigation methods, and with the use of environment friendly fertilizers (EFFs; Lü et al., 2016). EFFs reduce the negative effects of chemical fertilizers by controlling the release of nutrients into soil. EFFs are covered with environmental friendly material that degrade easily into the soil and produce carbon dioxide, methane, water, and inorganic compounds (Naz and Sulaiman, 2016). EFFs can also be developed by covering with micro or nano materials (Zhou et al., 2015a, Pereira et al., 2015). Mostly used materials are natural materials, such as chitosan, sodium alginate, starch and its derivatives, cellulose and their derivatives, lignin, agricultural residues, and biochar polydopamine (Pdop; Chen et al., 2018). These materials are biodegradable and easily available (Schneider et al., 2016). These EFFs decrease the nitrogen oxide and dinitrogen emission, increases soil organic matter, adjust soil pH, and improve water retention and water holding capacity of soil (Chen et al., 2018). Despite of various benefits of EFFs, their use is limited due to high cost and difficult production process. More study is required to know the nutrient release behavior of EFFs under different environmental conditions, as they have various benefits over chemical fertilizers.

1.6 BIOINPUTS

Nowadays, the management of soil mainly depends on the use of chemical inputs such as chemical fertilizers, pesticides, etc., which not only cause a serious threat to soil ecology and its fertility but also have a negative impact on human health and environment as well as increase agricultural cost and wastage of energy. These are challenge for sustainable agriculture. To avoid these issues, use of bio-inputs such as bio-fertilizers and biopesticide (use of beneficial microorganisms) has become an important alternative in agriculture sector because of their vital role in increasing soil fertility by acting as harmless inputs and safeguarding soil health and also the quality of crop products. Application of different kinds of rhizospheric microbes belonging to several taxa such as bacteria, cyanobacteria, fungi, algae, and protozoa kingdoms that live in moiety with rhizospheric region as plant growth promoting microbes (PGPM) opened a new way for plant growth and its productivity, nutrient and hormonal balance, and plant defense and plant resistant against various abiotic and biotic stress. PGPMs are soil microbes that assist rooting by direct or indirect methods (Mayak et al., 1999). They play various important roles in maintaining soil health and plant productivity. These microbes colonize in the rhizospheric region and protect plants from pathogens by producing secondary metabolites in the form of antibiotics suppressing harmful bacteria, siderophores production, releasing phytohormone, atmospheric nitrogen fixation, phosphate solubilization, and by producing biologically active substances that have a positive influence on plant growth and development (Arshad and Frankenberger, 1998). Apart from this, these PGPM can also clean environment by detoxifying pollutants like heavy metals and pesticides. Several types of research are still going on to understand the diversity and importance of soil PGPM communities and their roles in the betterment of agricultural productivity. Bio-fertilizers (microbes or their products) improve nutrient availability to plants by facilitating nutrient uptake and by increasing primary nutrient availability in the rhizosphere by different methods, like nitrogen fixation, solubilization, mineralization, and by phytohormone production (Bhardwaj et al., 2014; Arora et al., 2011). They can be used as a substitute when applied complementary or as the substitute for chemical fertilizers. The plant–PGPM cooperation plays an important role in enhancing growth and health of plants (Table 1.1). Soil microorganisms belonging to different taxa such as protozoa, bacteria, and fungi that colonize the rhizosphere or the plant tissues as well as PGPM can be used as bio-fertilizer (Vessey, 2003; Lucy et al., 2004; Smith and Read, 2008). Thus, the utilization of these bio-fertilizers will help in sustainable agricultural productivity in reducing problems associated with the use of chemicals fertilizers.

1.6.1 TYPES OF BIO-FERTILIZERS

1. Nitrogen-fixing bio-fertilizers (*Rhizobium, Bradyrhizobium, Azospirillum,* and *Azotobacter*)
2. Phosphorous-solubilising bio-fertilizers or PSB (*Bacillus, Pseudomonas, Aspergillus, Penicillium, Fusarium, Trichoderma, Mucor, Ovularopsis, Tritirachium,* and *Candida*)
3. Potash-solubilising bio-fertilizers (*Aspergillus* spp. (*A. fumigates, A. niger, A. terreus*) Ectomycorrhizal fungi)
4. Zinc-solubilizing bio-fertilizers
5. Phosphate-mobilizing bio-fertilizers (*Mycorrhiza*)
6. Plant-growth-promoting bio-fertilizers (*Pseudomonas*)
7. Enriched compost bio-fertilizers = Cellulolytic fungal cultures (*Chaetomium bostrychodes, C. olivaceum, Humicola fuscoatra, Aspergillus flavus, A. nidulans, A. niger, A. ochraceus, Fusarium solani,* and *F. oxysporum*).

1.6.2 NITROGEN-FIXING BIO-FERTILIZERS

Nitrogen played an important role in plant growth and productivity. It is not available to the plants due to high losses by emission or leaching. Nitrogen fixing bio-fertilizers have the capability to fix atmospheric nitrogen and thus make it available to plants by two mechanisms—either by symbiotic or by non-symbiotic. Symbiotic nitrogen fixation is a symbiotic relationship between rhizospheric microbes and plant. Symbiotic bacterial species such as *Rhizobium, Sinorhizobium, Bradyrhizobium,* and *Mesorhizobium* interact with leguminous plants, while *Frankia* with non-leguminous trees and shrubs (Zahran, 2001). Non-symbiotic nitrogen fixation is carried out by free-living diazotrophs that belong to different genera, such as *Azotobacter, Pseudomonas,* cyanobacteria (*Anabaena, Nostoc*), *Azospirillum, Burkholderia, Enterobacter,* and *Gluconacetobacter* (Bhattacharyya and Jha, 2012). Non-symbiotic free-living N-fixing bacteria species showed enhanced N uptake of plants (Bardi and Malusà, 2012; Lucy et al., 2004; Okon and Labandera-Gonzalez, 1994).

Rhizobia, the best-known N_2-fixing bacteria symbionts of legume plants, are able to provide up to 90% of the N requirements of the host through atmospheric N_2 fixation (Franche et al., 2009), but they can also behave as plant growth promoting rhizobacteria (PGPR) with non-legumes such as maize, wheat, rice, and canola (Hayat et al., 2010; Yanni et al., 2001), which can derive nitrogen from biological nitrogen fixation in 7–58% range in cereals (Baldani et al., 2000; Malik et al., 2002) and up to 60–80% in sugarcane (Boddey et al., 1991). Cyanobacteria (Anabaena, Aulosira, and Nostoc), as free-living or in symbiosis with Azolla—a small free floating freshwater fern—were found to fix N and to release it for rice uptake in the range of 30–40 up to 70–110 kg N ha^{-1} (Wagner, 1997). Arbuscular mycorrhizal fungi (AMF) may supply more than 50% of plant N requirements (Govindarajulu et al., 2005; Leigh et al., 2009; Subramanian and Charest, 1999). AM fungi can take up nitrogen in both inorganic (either ammonium or nitrate) and organic form (Hawkins et al., 2000).

1.6.3 PHOSPHATE SOLUBILIZATION

Phosphorus (P) is the most important key element involved in physiological processes such as respiration, energy transfer, photosynthesis, signal transduction, and macromolecular biosynthesis (Khan et al., 2010). P is present in the soil in both inorganic and organic forms but in insoluble, immobilized, and in precipitated form. PGPMs can directly solubilize and mineralize inorganic phosphorus or facilitate the mobility of organic phosphorus (Richardson and Simpson, 2011) by secreting various organic acids (e.g., fumaric acids, carboxylic acid), which lowers the pH in the rhizospheric region and thus releasing the bound forms of phosphate like $Ca_3(PO_4)_2$ in the calcareous soils (Table 1.1). Some bacterial species of PSB belongs to genera *Bacillus*, *Pseudomonas*, and *Staphylococcus*, *whereas Aspergillus, Fusarium, Trichoderma, Penicillium, Phichia norvegensis, Cryptococcus albidus var. albidus, Cryptococcus luteollus, Rhodotrula aurantiaca B, Cryptococcus albidus var. diffluens, Candida etchellsii* constitute the fungal species. Some cyanobacteria such as *Anabaena variabilis, Westiellopsis prolific* (Mahesh et al., 2011) showed the presence of organic acids, extracellular compounds, or enzymes that might promotes phosphate solubilization. *Arbuscular mycorrhizal* are the major group that contribute uptake of phosphorous by solubilization of inorganic P forms (Cavagnaro et al., 2005; Smith and Read, 2008; Tawaraya et al., 2006) and by hydrolyzation of organic P (Richardson et al., 2009). Several PGPR are very effective in solubilizing P from the highly insoluble tricalcium phosphate, hydroxyl apatite, and rock phosphate (Rodríguez and Fraga, 1999; Owen et al., 2015).

TABLE 1.1 Different PGPR and Their Mechanism of Fixation Used in Crop Production.

PGPR	PGPR mechanisms	Crops	References
Azoarcus	Nitrogen fixation	Rice	Hurek et al. (1998)
Azobacter	Cytokinin synthesis	Cucumber	Aloni et al. (2006)
Azospirillum	Nitrogen fixation	Sugar cane	Felipe et al. (2006); Tejera et al. (2005); Sahoo et al. (2014); Breg et al. (1980)
Bacillus	Auxin synthesis, Cytokinin synthesis, Gibberelin synthesis, Potassium solubilization, Antibiotic production, Siderophore production	Potato, cucumber, peeper, peeper alfalfa, maize	Ahmed and Hasnain (2010); Sokolova et al. (2011); Joo et al. (2005); Han et al. (2005, 2006)
Paenibacillus	Indole acetic acid, Synthesis	Pine	Bent et al. (2001)
Pseudomonas	Chitinase and glucanases production, ACC deaminase synthesis, Antibiotic production	Several crops, mung beans, wheat, cotton, maize	Arora et al. (2008); Ahmad et al. (2013); Shaharoona et al. (2008); Yao et al. (2010); Mazzola et al. (1995)
Rhizobium	Indole acetic acid, Synthesis, Nitrogen fixation, Siderophore production	Pepper, tomato, lettuce, carrot, rice, tomato, pepper, lettuce	Yanni et al. (2001); Fraile et al. (2012); Felix et al. (2013)
Streptomyces	Indole acetic acid, Synthesis, Siderophore production	Indian lilac	Verma et al. (2011)
Fungi	Phosphate solubilization	Haricot bean, faba bean, cabbage, tomato, sugarcane	Richardson and Simpson, (2011)
Fusarium Oxysporum	Antibiotic production	Tomato	Lafontaine et al. (1996)
Serratia marcescens	Chitinase expression	Vegetables and melons	Ordentlich et al. (1988)
Cercospora	Phytohormones production	Leguminous plant	Assante et al. (1977)

1.6.4 *POTASH SOLUBILIZING BIO-FERTILIZERS*

A wide array of bacterial genera (e.g., Pseudomonas, Acidothiobacillus, Burkholderia, Bacillus, and Paenibacillus) are able to liberate potassium from minerals such as illite, mica, biotite, muscovite, and orthoclases (Bennett et al., 1998, 2001; Liu et al., 2012) and can increase K availability up to 15% (Supanjani et al., 2006). Thus, the role of PGPMs as bio-fertilizer can be commercialized by doing some genetic modifications which could further contribute to sustainable development of agriculture. Therefore, the selected microbial strains can be used practically for the development of plant growth promoting or biocontrol inoculants, together with other plant growth promoting microbes (EvaLaslo et al., 2012).

1.6.5 EFFECT OF BIO-FERTILIZERS

Bio-fertilizer enriches the soil and is well-suited for long-term sustainability. Chemical fertilizers are used to increase the crop productivity and thus added in soil to cover the deficiency of inorganic nutrients such as N and P, which results in environmental pollution and is also costly. Increased use of chemical fertilizer also damages the soil texture. Thus, bio-fertilizers opened a new alternative to overcome these problems. Microbial activity have vital role in agriculture because they are very important in the movement and availability of minerals essential for plant growth and ultimately lower the requirement of synthetic fertilizers and increases the crop productivity by 20–30% (Kennedy et al., 2004). Bio-fertilizers bind to atmospheric nitrogen so that they are directly available to the plants. They also boost the phosphorous content of the soil by solubilization and liberating unavailable phosphorous. Bio-fertilizers also increase beneficial association with bacteria such as bean plants when co-inoculated with *Rhizobium etli* and *R.tropici* and *Azospirillum brasilense* had more nodules than plants inoculated only with one of the two *Rhizobia* (Burdman et al., 1996). *Azolla* helps in improving low nitrogen use efficiency in intensive rice cropping system (Yao et al., 2017). Their continuous use leads to the increase in quality of the soil. Therefore, the use of bio-fertilizer is both economical and environment friendly (Sahoo et al., 2012). Thus, bio-fertilizers reduce the requirement of chemical fertilizers. Bio-fertilizers improve root proliferation due to the release of growth promoting hormones. They help in increasing the crop yield by 10–25%. The main reasons derive from the unpredictability of results, problems to identify and track inoculated strains in the field, the poor understanding of the interrelationships between microorganisms and plants, and the technology of production. Although biofertilizers are more beneficial than chemical fertilizers, care should be taken in using biofertilizers, such as right combination of biofertilizers should be used as well as use according to crop. Despite the associated benefits of biofertilizers, their use is limited because of the unavailability of proper and competent strains, short shelf life, lack of awareness among farmers, etc.

1.7 BIOPESTICIDE

Bio-inputs also include biopesticides. It is a non-chemical approach for the management of pests, diseases, and vectors, which can be included in Integrated Pest management programs. Biopesticides are derived from natural products, which may be a plant, animal, or microbial origin. Being less toxic in nature, biopesticides are preferred over conventional pesticides. Biopesticide affects only target pests and closely related organism whereas broad-spectrum conventional pesticides may affect non-target organisms like mammals, birds, and beneficial insects. In addition to these properties, biopesticides are effective in small quantities and decompose easily that result in less exposure to pollution problems. They can be used against agricultural pests (Isman, 2006, Ahmad et al., 2015), malarial vectors (Blanford et al., 2011), production of compost through solid state fermentation (Ballardo et al., 2017), and are harmless to organisms (Kim et al., 2018). The biopesticides may be of plant origin—like neem beneficial and azadirachtins, nicotine and other alkaloids, rotenone and rotenoids, pyrethrum, pyrethrins and pyrethroids, and essential oils (Fournier, 1988; Regnault-Roger and Philogène, 2008; Isman, 2006; Ahmad et al., 2015)—microbial origin—like viral (Nuclear Polyhedrosis Virus), bacterial (*Bacillus thuringiensis*), and fungal (*Beauveria bassiana*)—entomopathogenic nematodes, or insect pheromones that are applied for mating disruption, monitoring, or lure and kill strategies (Copping and Menn, 2000). Since the biopesticides are biosynthesized, they are enzymatically biodegradable and in general, they have short half-lives. The prospect of new sources of active substances within the plant biodiversity for being included in the pest management formulations will contribute to broadening the range of available biopesticides of plant origin to be incorporated in the integrated pest management programs of different crops.

1.8 NANO FERTILIZER

With the advent of nanotechnology, the agriculture sector has improved. Nanotechnology plays an important role in crop production. In the agriculture sector, nano-biosensors (to detect moisture content), nano-fertilizers (for nutrient management), nano-herbicides (for killing weeds), and nano-pesticides (for pest management) are frequently used. Among these, nanofertilizers are modified forms of traditional fertilizers and developed by chemical, physical, mechanical, or biological methods. Nano-particles are made from bulk materials and their physical and chemical properties differed than bulky materials at nano scale (Nel et al., 2006). Nanoparticles have more surface area than traditional fertilizers, targeted delivery system with controlled release, and sorption capacity. All these characteristics increase the rate of uptake of nutrient by plants due to which nutrient loss is minimized and thereby nutrient-use efficiency is enhanced (Solanki et al., 2015). In agriculture, nano-materials are used in two forms (1) carbon-based single and multiwalled carbon nanotube and (2) metal-based aluminium, zinc, gold, etc. Single and multiwalled carbon nanotubes are used as nanosensors and plant regulator to enhance plant growth (Khoda-kovskaya et al., 2012).

1.8.1   EFFECT OF NANO-FERTILIZERS

Nano-fertilizers are more advantageous than conventional fertilizers because they are nontoxic, less harmful, and efficiently used by crops due to which environmental protection cost is minimized (Naderi and Abedi, 2012). In wheat grains, protein content increases after the application of iron nano-fertilizers (Tabrizi et al., 2009). In saffron, nano-fertilizers improve the saffron yield by affecting the flowering. Mother corm weight, dry saffron yield, flower number increased after the application of Fe, P, and K nano-fertilizers (Amirnia et al., 2014). Better pH, moisture, and CEC were observed after the application of nano-fertilizers (Rajonee et al., 2017). When potassium nano-fertilizer and nitrogen were applied on tomato, better effect was observed and this suggested that combined application of both types had a synergistic effect on growth and yield of tomato (fruit per plant, fruit diameter, fruit weight) (Ajirloo et al., 2015). Despite the benefits of nano-fertilizers, it does not hit the market too much because of problems associated with it. It affects the soil, plant life, aquatic life, and, most importantly, human life. Toxicity of nano-fertilizers depends on the size and composition of nano-particles. Nano-fertilizers can change, contaminate soil, and can affect soil microbes, earthworm, etc. (Du et al., 2011; Mura et al., 2013).

1.9   CONCLUSION

Chemical fertilizers are less expensive, easily available, and provide a known nutrient composition and strength and could be customized, if required. They are available in water-soluble form thus providing an immediate supplement to crop as compared to various traditional practices that allow slow release and biodegradation of manure. The soil health can be assessed and deficiencies could be supplemented by known and calculated composition of chemical fertilizers that could not be feasible with traditional manure approach or organic manure strategy. Production of chemical fertilizer heavily depends on the extraction and mining of mineral resources, such as phosphate rocks lead to scarcity of non-renewable resources. Phosphatic fertilizer production is associated with loss of vegetation, draining excess water, discharge of greenhouse gases, and radioactive hazards. Chemical fertilizers have adverse effect on soil health at biotic and compositional level. It does not support the growth of soil microbiota that is crucial for any soil for adequate productivity. Soil composition and structure is another important parameter that was not supported by the application of chemical fertilizers. Soil acidification is another alarming situation caused due to imbalanced use of nitrogen, phosphorous, and sulfur fertilizers, which leach down and raise the H+ ion concentration leading to increase in soil acidity. The excess use of chemical fertilizers run off the soil surface and accumulate to water bodies resulting into the overgrowth of aquatic vegetation known as eutrophication. The issue could be magnified as the deposited elements enter food cycles (trophic zone) leading to biomagnification. The accumulation of these elements disturbs the environmental ecological balance. Imbalanced use of chemical fertilizers could lead to several deficiencies or excess in soil composition resulting in too low productivity or toxicity to crop. Excess application of chemical fertilizers may leach down to water table and contaminate the drinking water leading to adverse health effects.

KEYWORDS

- **bio-fertilizer**
- **biological nitrogen fixation**
- **water conservation**
- **micronutrient**
- **agricultural practices**

REFERENCES

Ahmad, S.; Ansari, M. S.; Moraiet, M. A. Demographic Changes in *Helicoverpa Armigera* after Exposure to Neemazal (1% EC azadirachtin). **2013,** *50*, 30–36.

Ahmad, S.; Ansari, M. S; Muslim, M. Toxic Effects of Neem Based Insecticides on the Fitness of *Helicoverpa armigera* (Hübner). *Crop Protection* **2015,** *68*, 72–78.

Ajirloo, A. R.; Shaaban, M.; Motlagh Z. R. Effect of K Nano-fertilizer and N Bio-fertilizer on Yield and Yield Components of Tomato (*Lycopersicon esculentum* L.). *Int. J. Adv. Biol. Biom. Res.* **2015,** *3* (1), 138–143.

Amer, M. M.; Swelim, M. A.; Bouthaina, F.Abd El-Ghany; Omar, A. M. Effect of N_2 Fixing Bacteria and Actinomycetes as Biofertilizers on Growth and Yield of Cucumbers in Sandy Soil in Egypt. *Egyptian J. Desert Res.* **2002,** *52*, 113–126.

Amirnia, R.; Bayat, M.; Tajbakhsh, M. Effects of Nano Fertilizer Application and Maternal Corm Weight on Flowering of Some Saffron (*Crocus sativus* l.) Ecotypes. *Turkish J. Field Crops* **2014,** *19* (2), 158–168.

Anjuman, A. R.; Zaman, S.; Huq, S. M. I. Preparation, Characterization and Evaluation of Efficacy of Phosphorus and Potassium Incorporated Nano Fertilizer. *Adv. Nanoparticl.* **2017,** *6*, 62–74.

Anonymous. Food and Agriculture Organization, World Fertilizer Trends and Outlook to 2018. 2017.

Aune, J. B. Conventional, Organic and Conservation Agriculture: Production and Environmental Impact. In *Agroecology and Strategies for Climate Change;* Springer: Netherlands, 2012; pp 149–165.

Azcón-Aguilar, C.; Barea, J. M. Nutrient Cycling in the Mycorrhizosphere. *J. Soil Sci. Plant Nutr.* **2015,** *25*, 372–396.

Ballardo, C.; Barrena, Y.; Artola, A.; Sánchez, A. A Novel Strategy for Producing Compost with Enhanced Biopesticide Properties Through Solid-state Fermentation of Bio Waste and Inoculation with *Bacillus thuringiensis*. *Waste Manage.* (*in Press*). **2017.**

Blanford, S.; Shi, W.; Christian, R.; Marden, J. H.; Koekemoer, L. L.; Brooke, B. D.; Coetzee, M.; Read, A. F.; Thomas, M. B. Lethal and Pre-Lethal Effects of a Fungal Biopesticide Contribute to Substantial and Rapid Control of Malaria Vectors. *PLoS One* **2011,** *6* (8), e6591.

Bremner, J. M. Recent Research on Problems in the Use of Urea as a Nitrogen Fertilizer. *Fert. Res.* **1995,** *42* (1–3), 321–329.

Burdman, S.; Volpin, H.; Kigel, J.; Kapulnik, Y.; Okon, Y. Promotion of *Nod* Gene Inducers and Nodulation in Common Bean (*Phaseolus Vulgaris*) Roots Inoculated with *Azospirillum brasilense* Cd. *Appl. Environ. Microbiol.* **1996,** *62*, 3030–3033.

Chang, D. C. N. Effect of Three Glomus Endomycorrhizal Fungi on the Growth of Citrus Rootstocks. *Proc. Int. Soc. Citricult.* **1987,** *1*, 173–176.

Channakeshava, S.; Goroji, P. T.; Doreswamy, C.; Naresh, N. T. Assessment of Foliar Spray of Potassium Nitrate on Growth and Yield of Cotton. *Karnataka J. Agric. Sci.* **2013,** *26* (2), 316–317.

Chen, J.; Lu, S.; Zhang, Z.; Zhao, X.; Li, X.; Ning, P.; Liu, M. Environmentally Friendly Fertilizers: A Review of Materials Used and Their Effects on the Environment. *Sci. Total Environ* **2018,** *613–614*, 829–839.

Cherr, C. M.; Scholberg, J. M. S.; McSorley, R. Green Manure Approaches to Crop Production: A Synthesis. *Agron. J.* **2006,** *98*, 302–319.

Choate, J. Phosphorus Availability in Biosolids Amended Soils. M.S. Thesis, Oregon State University, 2004. http://hdl.handle.net/1957/20969.

Copping, L. G.; Menn, J. J. *Biopesticides*: A Review of Their Action, Applications and Efficacy. *Pest. Manage. Sci.* **2000**, *56*, 651–676.

Dongyun, M.; Dexiang, S.; Chenyang, W.; Huina, D.; Haixia, Q.; Junfeng, H.; Xin, H.; Yingxin, X.; Tiancai, G. Physiological Responses and Yield of Wheat Plants in Zinc-mediated Alleviation of Drought Stress. *Front Plant Sci.* 2017, *8*, 860.

Fournier, J.; Chimie des Pesticides. Cultures et Techniques, Paris, Nantes et Agence de Cooperation Culturelle et Tech-´nique, **1988**, 351.

Gabriele, B. Plant–Microbe Interactions Promoting Plant Growth and Health: Perspectives for Controlled Use of Microorganisms in Agriculture. *Appl. Microbiol. Biotechnol.* **2009**, *84*, 11–18.

Hazell, P.; Wood, S. Drivers of Change in Global Agriculture. *Phil. Trans. Royal Soc. B Biol. Sci.* **2008**, *363* (1491), 495–515.

Hegazi, E. S.; Samira, M. M.; El-Sonbaty, M. R.; Abd El-Naby, S. K. M.; El-Sharony, T. F. Effect of Potassium Nitrate on Vegetative Growth, Nutritional Status, Yield and Fruit Quality of Olive cv. "Picual." *J. Horticult. Sci. Ornament. Plants* **2011**, *3* (3), 252–258.

Horton, T. R.; Bruns, T. D.; Parker, V. T. Ectomycorrhizal Fungi Associated with Arctostaphylos Contribute to *Pseudotsuga Menziesii* Establishment. *Can. J. Bot.* **1999**, *77*, 93–102.

Huber, D. M.; Graham, R. D. The Role of Nutrition in Crop Resistance and Tolerance to Disease. In *Mineral Nutrition of Crops Fundamental Mechanisms and Implications;* Rengel, Z., Ed.; Food Product Press, New York, 1999; pp 205–226.

Hzhbryan, M. Kazemi, S. Effects of Ammonium Sulphate on the Growth and Yield of Different Tomato (*Lecopersicon Esculentum*) Plant in the City Jahrom. *J. Novel Appl. Sci.* **2014**, *3-1*, 62–66.

Isman, M. B. Botanical Insecticides, Deterrents, and Repellents in Modern Agriculture and An Increasingly Regulated World. *Ann. Rev. Entomol.* **2006**, *51*, 45–66.

James, S. C. Seaweed Extracts Stimuli in Plant Science and Agriculture. *J. Appl. Phycol.* **2011**, *23*, 371–393.

Jyoti, S. K. Role of Bio-fertilizers and Biopesticides for Sustainable Agriculture. *J. Bio. Innov.* **2013**, *2*, 73–78.

Kennedy, I. R.; Choudhury, A. T. M. A.; Kecskés, M. L. Non-symbiotic Bacterial Diazotrophs in Crop-farming Systems: Can Their Potential for Plant Growth Promotion Be Better Exploited? *Soil Biol. Biochem.* **2004**, *36*, 1229–1244.

Khajuria, A.; Yamamoto, Y.; Morioka, T. Estimation of Municipal Solid Waste Generation and Landfill Area in Asian Developing Countries. *J. Environ. Biol.* **2010**, *31*, 649–654.

Khan, M. S.; Zaidi, A.; Ahemad, M.; Oves, M.; Wani, P. A. Plant Growth Promotion by Phosphate Solubilizing Fungi—Current Perspective. *Arch. Agron. Soil Sci.* **2010**, *56*, 73–98.

Kim, S. A.; Ahn, H. G.; Ha, P. J.; Lim, U. T.; Lee, J. H. Toxicities of 26 Pesticides Against 10 Biological Control Species. *J. Asia-Pacific Entomol.* **2018**, *21,* 1–8.

Kiran, J. K.; Khanif, Y. M.; Amminuddin, H.; Anuar, A. R. Effects of Controlled Release Urea on the Yield and Nitrogen Nutrition of Flooded Rice. *Comm. Soil Sci. Plant Anal.* **2010**, *41*, 811–819.

Krull, E.; Skjemstad, J.; Baldock, J. *Functions of Soil Organic Matter and the Effect on Soil Properties: A Literature Review*; Report for GRDC and CRC for Greenhouse Accounting: CSIRO Land and Water Client Report. Adelaide: CSIRO Land and Water, 2004.

Laila, F.; Hagag, M. F.; Shahin, M.; El-Migeed, M. M. M. Effect of Rates and Methods of Calcium Nitrate Application on Vegetative Growth of Dolcy Olive Seedlings. *Am.-Eur. J. Agric. Environ. Sci.* **2011**, *11* (6), 802–806.

Liu, H.; Gan, W.; Rengel, Z.; Zhao, P. Effects of Zinc Fertilizer Rate and Application Method on Photosynthetic Characteristics and Grain Yield of Summer Maize. Effects of Zinc Fertilizer Rate and Application Method on Photosynthetic Characteristics and Grain Yield of Summer Maize. 2016, *16* (2), 550–562.

Mazid, M.; Khan, T. A. Future of Bio-fertilizers in Indian Agriculture: An Overview. *Int. J. Agricul. Food Res.* **2015**, *3* (3), 10–23.

Morin, C.; Samson, J.; Dessureault, M. Protection of Black Spruce Seedlings Against Cylindrocladium Root Rot with Ectomycorrhizal Fungi. *Can. J. Bot.* **1999**, *77*, 169–174.

Morrissey, J. P.; Dow, J. M.; Mark, L.; O'Gara, F. Are Microbes at the Root of a Solution to World Food Production? *EMBO Rep.* **2004**, *5*, 922–926.

NAAS. Nanotechnology in Agriculture, Scope and Current Relevance. National Academy of Agricultural Sciences, New Delhi, 2013.

Naderi, M. R.; Abedi, A. Application of Nanotechnology in Agriculture and Refinement of Environmental Pollutants. *J. Nanotech.* **2012,** *11* (1), 18–26.

Naz, M. Y.; Sulaiman, S. A. Slow Release Coating Remedy for Nitrogen Loss from Conventional Urea: A Review. *J. Cont. Release* **2016**, *225*, 109–120.

Nel, A.; Xia, T. M.; Li, N. Toxic Potential of Materials at the Nanolevel. *Science*. **2006**, *311*, 622–627.

Pathma, J.; Pathma, N. S. Microbial Diversity of Vermicompost Bacteria that Exhibit Useful Agricultural Traits and Waste Management Potential. 2012, *1*, 26.

Pereira, E. I.; Cruz, Da.; Solomon, C. C.; Cavigelli, A. E. A.; Ribeiro, M. A. Novel Slow Release Nano-composite Nitrogen Fertilizers: The Impact of Polymers on Nanocomposite Properties and Function. *Ind. Eng. Chem. Res.* **2015**, *54*, 3717–3725.

Pierre, M. J.; Bhople, B. S.; Kumar, A.; Erneste, H.; Emmanuel, B.; Singh, Y. N. Contribution of Arbuscular Mycorrhizal Fungi (AM Fungi) and *Rhizobium* Inoculation on Crop Growth and Chemical Properties of Rhizospheric Soils in High Plants. *IOSR J. Agri. Vet. Sci.* **2014,** *7*, 45–55.

Rao, N. H. Sustainable Agriculture: Critical Challenges Facing the Structure and Function of Agricultural Research and Education in India. National Workshop on Agricultural Policy: Redesigning R&D to Achieve the Objectives, 2002.

Regnault-Roger, C.; Philogène, B. J. R. Past and Current Prospects for the Use of Botanicals and Plant Allelochemicals in Integrated Pest Management. *Pharma. Biol.* **2008**, *46*, 41–52.

Richardson, A. E.; Simpson, R. J. Soil Microorganisms Mediating Phosphorus Availability. *Plant Physiol.* **2011**, *156*, 989–996.

Ryglewicz, P. T.; Anderson, C. P. Mycorrhizae Alter Quality and Quantity of Carbon Below Ground. *Nature* **1994**, *369*, 58–60.

Sahu, D.; Priyadarshani, I.; Rath, B. Cyanobacteria—As Potential Bio-fertilizer. *CIB Tech. J. Microbiol.* **2012**, *1* (2-3), 20–26.

Sánchez-Monederoa, M. A.; Mondini, C.; Nobilic, M.de.; Leitab, L. Land Application of Biosolids. Soil Response to Different Stabilization Degree of the Treated Organic Matter. *Waste Manag.* **2004**, *24* (4), 325–332.

Schmutterer, H. Properties and Potential of Natural Pesticides from the Neem Tree, Azadirachta Indica. *Ann. Rev. Entomol.* **1990**, *35*, 271–297.

Schneider, T.; Deladino, L.; Zaritzky, N. Yerba Mate (*Ilex paraguariensis*) Waste and Alginate as a Matrix for the Encapsulation of N Fertilizer. *ACS Sustain.Chem. Eng.* **2016**, *4*, 2449–2458.

Shukla, K. H. Sustainable Development in Agricultural Sector in India. *Bus. Manage. Rev.* **2015**, *5* (4), 220–222.

Singh, H.; Singh, M.; Kang, J. S. Effect of Potassium Nitrate on Yield and Yield Attributes of Spring Maize (*Zea mays* L.) under Different Dates of Planting. *Int. J. Curr. Microbiol. App. Sci.* **2017**, *6* (3), 1581–1590.

Sinha, R. K.; Herat, S.; Valani, D.; Chauhan, K.; Vermiculture and Sustainable Agriculture. *Am-Euras. J. Agric. Environ. Sci.* **2009**, *5*, 1–55.

Siringam, K.; Juntawong, N.; Cha-um, S.; Kirdmanee, C. Exogenous Application of Potassium Nitrate to Alleviate Salt Stress in Rice Seedlings. *J. Plant Nutr.* **2013**, *36*, 607–616.

Solanki, P.; Bhargava, A.; Chhipa, H.; Jain, N.; Panwar, J. Nano-fertilizers and Their Smart Delivery System. In *Nanotechnologies in Food and Agriculture;* Rai, M., Ribeiro, C., Mattoso, L., Eds.; Springer: New York, 2015; pp 81–102.

Somaye Soltani, S.; Kazem, H.; Nosratolah, N. The Effect of Diammonium Phosphate Fertilization on Salinity Tolerance of Maize (*Zea mays* L.). *J. Soil Environ.* **2015**, *1* (1), 18–27.

Stanley, M. R.; Koide, R. T.; Shumway, D. L. Mycorrhizal Symbiosis Increases Growth, Reproduction and Recruitment of Abutilon Theophrasti Medic. In the field. *Oecologia* **1993**, *94*, 30–35.

Teixeira, L. J.; Quaggio, J. A.; Cantarella, H.; Mellis, V. E. Potassium Fertilization for Pineapple: Effects on Plant Growth and Fruit Yield. *Rev. Bras. Frutic. Jaboticabal.* **2011**, *33* (2), 618–626.

Venkatashwarlu, B. Role of Bio-fertilizers in Organic Farming: Organic Farming in Rain Fed Agriculture. Central Institute for Dry Land Agriculture, Hyderabad. 2008; pp 85–95.

Vitale, J. D.; Penn, C.; Park, S.; Payne, J.; Hattey, J.; Warren, J. Animal Manure as Alternatives to Commercial Fertilizers in the Southern High Plains of the United States. In *How Oklahoma Can Manage Animal Waste, Integrated Waste Management;* Kumar, S., Ed., InTech: Rijeka, Croatia, 2011; Vol. 2, pp 143–164.

Seifu, W., Deneke, Y. S. Effect of Calcium Chloride and Calcium Nitrate on Potato (*Solanum tuberosum* L.) Growth and Yield. *J. Hortic.* **2017**, *4*, 3.

Yao, Y.; Zhang, M.; Tian, Y.; Zhao, M.; Zeng, K.; Zhang, B.; Zhao, M.; Yin, B. Azolla Biofertilizer for Improving Low Nitrogen Use Efficiency in an Intensive Rice Cropping System. *Field Crop. Res.* **2016**, *216*, 158–164.

Zhen, Z.; Liu, H.; Wang, N.; Guo, L.; Meng, J.; Ding, N.; Wu, G.; Jiang, G. Effects of Manure Compost Application on Soil Microbial Community Diversity and Soil Microenvironments in a Temperate Cropland in China. *PLoS One* **2014**, *9* (10), e108555.

CHAPTER 2

Bacterial Biopesticides and Their Use in Agricultural Production

KUMARI APURVA[1,*], CHANDRAKANT SINGH[2], VINOD KUMAR[3], and V. B. JHA[4]

[1]*Department of Forestry, Forest Research Institute (Indian Council of Forestry Research & Education), P.O. New Forest, Dehradun 248006, Uttarakhand, India*

[2]*Wheat Research Station, Junagadh Agricultural University, Junagadh 362001, Gujarat, India*

[3]*Department of Molecular Biology and Genetic Engineering, Bihar Agricultural University, Sabour, Bhagalpur 813210 Bihar, India*

[4]*Department of Plant Breeding and Genetics, Dr. Kalam Agricultural College, Kishanganj, Bihar Agricultural University, Sabour, Bhagalpur 813210, Bihar, India*

Corresponding author. E-mail: apurvashahishiats@gmail.com

ABSTRACT

Biopesticides are naturally occurring substances (biochemicals), microbes and plants, and environmentally safe pesticides, and amidst its numerous types, bacterial biopesticides are used extensively and is of prodigious importance. The inclusive use of biopesticides is substantial in prospective aids to agriculture and public health programmes; hence, usage of such bacterial biopesticides can be beneficial in such programmes. Biopesticides are key components of integrated pest management (IPM) programs, which are gaining consideration by means of reduced synthetic chemical product extent used to control several plant pest, diseases, and safeguard stored crop produce. A huge number of bacterial-derived products are introduced into market among which quite a few have already played a vital role. Bacterial pesticides have different mechanisms to control pests, pathogens, and weeds where they might act as competitors or inducers of host resistance in plant or by inhibiting growth, feeding, development, or reproduction of a pest or pathogen. The aim of this chapter is to provide a synopsis on use of bacterial-derived biopesticides for pest management and to deliberate the current development and application of their various types. This chapter encloses thorough classification of *Bacillus thuringiensis*, *Bacillus subtilis*, and *Bacillus sphaericus* based biopesticide along with their insecticidal, mosquitocidal, nematicidal, and antimicrobial activities. This section divulges great potential for auxiliary manipulation of bacterial-derived biopesticides in fortification of plants. India has a vast potential for bacterial biopesticides but its adoption by farmers in India needs education for maximum benefit. The stress on organic farming and on residue-free commodities would certainly warrant increased adoption of bacterial biopesticides by the farmers. Being target pest specific, it is presumed to be relatively safe to non-target organism including humans.

2.1 INTRODUCTION

In developing countries like India, agriculture plays a vital role as it not only makes sure whether the food requirements of growing population is satisfied or not but also is important in improvement of the country's economy; then again this sector has been facing many problems like the destructive activities of numerous pests as fungi, weeds, and insects from time immemorial, leading to radical decrease in yields. Insect–pests are frequently being introduced in new areas either naturally or inadvertently, or, in some cases, organisms that are intentionally introduced become pests. Global trade has stemmed in increased numbers of invasive non-native pest species being introduced to new areas. Controlling these invasive species presents a supreme challenge worldwide. Agriculture and forests makes an important resource to sustain global economic, environmental, and social system; this is the reason global challenge is to secure high end quality yields and to make agricultural produce environmentally compatible. Protection of plants by means of chemicals acquires the leading place as regards to its total volume of application in integrated pest management (IPM) and diseases of plants. But pesticides cause toxicity to humans and animals. The Green Revolution (GR) technology adopted in between 1960 and 2000 has increased wide varieties of agricultural crop yield/ha which increased 12–13% food supply in developing countries. Southeast Asia and India were the first developing countries to show the impact of GR on varieties of rice yields. Inputs like fertilizers, pesticides helped a lot in this regard. But in spite of this fact, food insecurity and poverty still prevails prominently in our country. Use of chemical biopesticides and fertilizers have caused negative impact on environment by affecting soil fertility, water hardness, development of insect resistance, genetic variation in plants, increase in toxic residue through food chain, and animal feed, thus increasing health problems and many more. This has made it essential to introduce measures which can harness foresaid challenges. Use of biopesticides (bacterial, viral, fungal) and biofertilizers can play a major role in dealing with these challenges in a sustainable way (Suman, 2010).

2.2 PESTICIDES AND ENVIRONMENTAL SAFETY

Biopesticides are natural constituent that is nothing but biochemical pesticides which by nontoxic mechanisms controls the pest; it can be living organisms (natural enemies) or their products (phytochemicals, microbial products) or byproducts (semiochemicals) which can be used for the management of pests that are harmful to plants. They pose less threat to the environment and to human health. Biopesticides are pathogenic for the pest of interest; among all, the most commonly used biopesticides are living organisms, these include biofungicides (Trichoderma), bioherbicides (*Phytophthora*), and bioinsecticides (*Bacillus thuringiensis*, Bt). Also, there are few plant products which can now be used as a major biopesticide source (Salma et al., 2011). Plants naturally on genetic modification produce substances that act as plant-incorporated protectants. Such examples are the incorporation of Bt gene, protease inhibitor, lectines, chitinase, etc., into the plant genome so that the transgenic plant synthesizes its own substance that destroys the target pest. The use of bacterial biopesticides is considerable in agriculture and public health programs and has potential advantage.

2.3 WHAT ARE BIOPESTICIDES?

Biopesticides are particular type of pesticides derived from natural materials as animals, plants, bacteria, and certain minerals. For example, canola oil and baking soda have pesticidal properties and are considered biopesticides. In 2014, 430 registered biopesticide active ingredients and 1320 active Product registrations were identified. Biopesticides fall into three major classes:

1. The active component of microbial pesticides consists of microorganism (e.g., a bacterium, fungus, virus, or protozoan). Microbial pesticides can control several kinds of pests, although each have separate active ingredient which is relatively specific for its target pest (Desai, 1997). For example, there are fungi that control certain weeds, and other fungi that kill specific insects (Desai, 1997). Subspecies and strains of Bt are the most extensively used microbial pesticides. Each strain of this bacterium produces a different mix of proteins, which specifically kills one or a few related species of insect larvae. Although some Bt's control moth larvae found on plants, other are specific for larvae of flies and mosquitoes. The target insect species are determined by whether the particular Bt produces a protein that can bind to a larval gut receptor, thereby causing the insect larvae to starve (Kalra and Khanuja, 2007).

2. Plant-incorporated protectants are pesticidal substances that plants produce from genetic material that has been introduced in the plants. For example, scientists can introduce the gene for the Bt pesticidal protein into the plant's genetic material, then the plant, instead of the Bt bacterium will manufacture the substance that destroys the pest. The protein and its genetic material, except the plant itself, are regulated by EPA (Thakore, 1997).

3. Biochemical pesticides are naturally occurring elements that control pests by nontoxic mechanisms. Conventional pesticides, by contrast, are generally synthetic materials that directly kill or inactivate the pest. Biochemical pesticides include substances, such as insect sex pheromones that interfere with mating as well as various scented plant extracts that attract insect pests to traps. As it is sometimes difficult to determine whether a substance meets the criteria for classification as a biochemical pesticide, EPA has established a special committee to make such decisions.

2.4 BIOPESTICIDES IN INDIA

Biopesticides represent only 2.89% (as on 2005) of the overall pesticide market in India and is expected to increase significantly in coming years. In India, so far, only 12 types of biopesticides have been registered under the Insecticide Act, 1968 (Thakore, 2006). Neem-based pesticides, Bt, NPV, and Trichoderma are the major biopesticides produced and used in India, whereas more than 190 synthetics are registered for use as chemical pesticides. Most of the biopesticides find use in public health, except a few that are used in agriculture. Besides, (1) transgenic plants and (2) beneficial organisms called bioagents are used for pest management in India (Kalra and Khanuja, 2007).

Some success stories about successful utilization of biopesticides and biocontrol agents in Indian agriculture include (Kalra and Khanuja, 2007):

- Control of diamondback moths by Bt,
- Control of mango hoppers and mealy bugs and coffee pod borer by *Beauveria,*
- Control of *Helicoverpa* on cotton, pigeon-pea, and tomato by Bt,
- Control of white fly on cotton by Neem products.

2.5 MARKET TRENDS OF BIOPESTICIDES

Biopesticides are used globally for controlling insects, pests, and diseases. Bioinsecticides, biofungicides, and bionematicides are rapidly growing market segments and are expected to boost the demand for biopesticides in near future. Globally, there are 175 registered biopesticide active ingredients, out of which 700 products are available in the market (Hajeck, 1994). The global market for biopesticides was valued at US$1.3 billion in 2011, and it is expected to extent US$3.2 billion by 2017. Increasing demand for residue-free crop produce is one of the key drivers of the biopesticide market along with growing organic food market and its easier registration than chemical pesticides. North America dominated the global biopesticide market and accounted for about 40% of the global biopesticide demand in 2011. The US biopesticides market is valued at around $205 million and expected to increase to approximately $300 million by 2020 (Hoy Myths, 1999). European market is estimated close to $200 million, and due to the stringent pesticide regulations and increasing demand from organic producers, it is expected to be the fastest growing market. As China and India are accepting more biopesticides, Asian market holds a good opportunity and demand for biopesticides.

2.6 BACTERIAL BIOPESTICIDES

Bacterial biopesticides are the type of microbial pesticides that works in different ways. More often than not, they are used as insecticides, even though they can be used to control the growth of plant pathogenic bacteria and fungi. As an insecticide, they are highly specific to individual pest species. To be effective, they should come into contact with the target vermin and might be required to be ingested (Table 2.1). Bacterial pathogens used as a biopesticides are spore-forming, rod-shaped bacteria belonging to genus *Bacillus.* Its habitat is mostly soil, and several insecticidal strain has been obtained from soil samples. The genetic biodiversity of this genus is very wide, ranging from sea water to soil and some time also in very extreme climate such as hot springs (Harwood and Wipat, 1996). Because of several precious attributes, the strain of this genus could be one of the major sources of potential microbial biopesticides (Ongena and Jacques, 2008). The endotoxins produced by bacteria disrupt the digestive system of targeted pests which is highly specific to particular pests (O'Brien et al., 2009). To recognize that bacterial biopesticides when used to control pathogenic microorganisms or parasite, the bacterial biopesticide colonizes on the plant and group out the pathogenic species.

TABLE 2.1 Bacterial Biopesticides.

S. no.	Name of the bacterial biopesticide	Registration act with years
1.	*Bacillus thuringiensis var. israelensis*	Registered under Insecticides Act, 1968
2.	*Bacillus thuringiensis var. galleriae*	Registered under Insecticides Act, 1968
3.	*Bacillus sphaericus*	Registered under Insecticides Act, 1968
4.	*Pseudomonas fluoresens*	Registered under Insecticides Act, 1968

Source: Adapted with permission from Gupta et al. (2010).

The members of the genus *Bacillus* are often considered as an important source of wide range of biologically active molecules, some of which are potential inhibitor for fungal growth (Schallmey et al., 2004). The strain of Bt covered approximately 90 % of the biopesticides market in the United States and is considered as most widely used biopesticides (Chattopadhyay et al., 2004). Since its discovery in 1901, Bt has been widely used to manage insect pests in agriculture, forestry, and medicine (Mazid and Kalita, 2011). Its principal characteristic is the synthesis, during sporulation, of crystalline inclusions containing proteins known as δ-endotoxins or Cry proteins, which have insecticidal properties. To date, over 100 Bt-based bioinsecticides, biopesticides, and biofungicides have been developed. Microbial pesticides containing Bt var. *kurstaki* kill the caterpillar stage of a wide array of butterflies and moths. In addition, the genes that code for the insecticidal crystal proteins have been successfully transferred into different crop plants including cotton, tomato, brinjal, etc., that lead to significant economic benefits. It is identified that due to their high specificity and safety in the environment, Bt and Cry proteins are efficient, safe, and sustainable alternatives to chemical pesticides for the control of insect pests (Roy et al., 2007; Kumar, 2012).

2.6.1   ADVANTAGES OF USING BACTERIAL BIOPESTICIDES

- Bacterial biopesticides are usually inherently less toxic than conventional pesticides.

- Bacterial biopesticides generally affect only the target pest and closely related organisms, in contrast to broad spectrum, conventional pesticides that may affect organisms as different as birds, insects, and mammals.
- Biopesticides often are effective in very small quantities and often decompose quickly, thereby resulting in lower exposures and largely avoiding the pollution problems caused by conventional pesticides (Gupta and Dikshit, 2010).
- When used as a component of IPM programs, biopesticides can greatly decrease the use of conventional pesticides, while crop yields remain high. To use biopesticides effectively (and safely), however, users need to know a great deal about managing pests and must carefully follow all label directions.

2.6.2   Bacillus thuringiensis

Bt is an aerobic, Gram-positive, spore-forming soil bacterium that shows unusual ability to produce endogenous different kinds of crystal protein inclusions during its sporulation. Bt (commonly known as "Bt") is an insecticidal bacterium having market over the world for control of many important plant pests—mainly caterpillars of the Lepidoptera (butterflies and moths) also mosquito larvae, and simulid blackflies that is the vector of river blindness in Africa. The commercial Bt products are powder containing a mixture of dried spores and toxin crystals. They are applied to leaves or other environments where the insect larvae feed. The toxin genes have also been genetically engineered into several crop plants. The method of use, mode of action, and host range of this biocontrol agent may vary within other Bacillus insecticidal species (Meadows et al., 1993).

The *Bacillus* species, Bt, has developed many molecular mechanisms to produce pesticidal toxins; most of these toxins are coded by several *cry* genes. Since its discovery in 1901 as a microbial insecticide, Bt has been widely used to control insect, pests significant in agriculture, forestry, and medicine. Its principal characteristic is the synthesis of a crystalline inclusion during sporulation, containing proteins known as endotoxins or Cry proteins, which have insecticidal properties (Schnepf et al., 1998). The crystal protein also called as δ-endotoxins or insecticidal crystal proteins inclusions are composed of one or more crystal (Cry) and cytolytic (Cyt) toxins. Some of these proteins are highly toxic to certain insects but they are harmless to most other organisms including vertebrates and beneficial insect. Since their insecticidal potential has been discovered, it has been produced commercially and accepted as a source of environment friendly biopesticide all over the world.

There are different strains of Bt, each strain produces a different mix of proteins, and specifically kills one or a few related species of insect larvae. There are reports on moth larvae control by Bt toxins that are reported on plants, other Bt's are specific for larvae of flies and mosquitoes. The target insect species are determined by whether the particular Bt produces a protein that can bind to a larval gut receptor, thus causing the insect larvae to starve. The most widely used strains of Bt have started against three genera of mosquitos: *Culex, Culiseta,* and *Aedes* and their study have shown that Bt spores can survive both on the ground as well as in animals. What's more, wind, rain, and animals can carry them to nearby areas. In the splashing rain drops, they can even "hop" from the ground up onto leaves—another means of transport. Bt bacteria are also known to be able to easily transfer their toxicity genes to other bacteria in the application area as it was seen that when the bacteria were sprayed on cabbage plants, it was found that all the cabbage white butterfly larvae were killed. In addition, although, the field study revealed that the bacteria are able to survive for a considerable time. After spraying, so far the majority of the spores were found to be present in the top 2 cm of the soil (National Environmental Research Institute of Denmark). "Their toxic effects disappeared after a few days, but half of the bacteria still survived as spores 120 days later, and one-fifth of them were still alive after a year. They existed in a dormant state, however, and did not produce toxins, although the spores are able to germinate later and produce insecticide again," explain microbiologists Bjarne Munk Hansen and Jens Chr. Pedersen of the National Environmental Research Institute. Until now, it was generally believed that the majority of Bt bacteria disappear rapidly after they have been sprayed. "It was thought that when the toxic effect disappeared, the bacteria had also disappeared. What in fact happens, although, is that the bacteria convert to a dormant stage and become spores," concluded the two scientists. In the present era of transgenic technology, insecticidal toxins of Bt assume considerable significance in the production of insect-resistant crops such as cotton, maize, potato, rice, etc.

Insects can be infected with many species of bacteria but those belonging to the genus Bacillus, as already mentioned, are most widely used as pesticides. Bt has developed many molecular mechanisms to produce *Cry* genes (Schnepf et al., 1998). Since its discovery in 1901, over 100 Bt-based bioinsecticides have been developed, which are mostly used against lepidopteran, dipteran, and coleopteran larvae (Roh et al., 2007). In addition, the genes that code for the insecticidal crystal proteins have been successfully transferred into different crops plants by means of transgenic technology which has led to significant economic benefits. Because of their high specificity and their safety in the environment, Bt and Cry protein toxins are efficient, safe, and sustainable alternatives to chemical pesticides for the control of insect pests. The toxicity of the Cry proteins have traditionally been explained by the formation of transmembrane pores or ion channels that lead to osmotic cell lysis (Zhang et al., 2006). In addition to this, Cry toxin monomers also seem to promote cell death in insect cells by mechanism of adenylyl cyclase/ PKA signaling pathway. However, despite this entomopathogenic potential, controversy has arisen regarding the pathogenic lifestyle of Bt. Recent reports claim that Btrequires the cooperation of commensal bacteria within the insect gut to be fully pathogenic (Broderick et al., 2009; Baum and Malvar, 1995).

The first developed Bt insecticidal agent is a mixture of its spores and toxin. As a pesticide, Bt accounts for more than 90% of total share of today's bioinsecticides market and has been used as biopesticides for several decades. The discovery of the strain Bt serovar *israelensis* made possible efficient microbiological control of Diptera Nematocera vectors of diseases, such as mosquitoes (Culicidae) and black flies (Buss and Park-Brown, 2002). In global market, variety of products is available for control of caterpillars (var. *kurstaki, entomocidus, galleriae,* and *aizawai*), mosquito and blackfly larvae (var. *israeliensis*), and beetle larvae (var. *tenebrionis*). Actively growing cells lack the crystalline inclusions and hence nontoxic to insects. The Bt preparations remain stable without any disintegration over years even in the presence of UV sun rays. As the insect feeds on the foliage, the crystals too are eaten up. These are then hydrolyzed in the insect's midgut to produce an active endotoxin, these active toxins binds to receptor sites on gut epithelial cells and creates imbalance in the ionic make-up of the cell. This is seen by swelling and bursting of the cells due to osmotic shock.

Subsequent symptoms are paralysis of the insect's mouthparts and gut. So, obviously the feeding process is inhibited. (Milner, 1994; Lambert et al., 1992). Also, a relatively new mechanism of action of Cry toxins have been proposed which involves the activation of Mg^{2+}-dependent signal cascade pathway that is triggered by the interaction of the monomeric 3-domain Cry toxin with the primary receptor, the cadherin protein Bt-R1 (Schnepf et al., 1998). This triggering of the Mg^{2+}-dependent pathway has a knock-on effect and initiates a series of cytological events that include membrane blebbing, appearance of nuclear ghosts, and cell swelling followed by cell lysis. The Mg^{2+}-dependent signal cascade pathway activation by Cry toxins is observed to be analogous to similar effect imposed by other pore-forming toxins on their host cells when they are applied at subnanomolar concentration (Porta et al., 2011; Nelson et al., 1999). Although the two mechanisms of action seem to differ, with series of downstream events following on from toxin binding to receptors on target cell membranes, there is a degree of commonality in that initially the crystals have to be solubilized in vivo or in vitro, and activated by proteases before and/or after binding to receptors such as cadherin (Aronson et al., 1991; Soberón et al., 2007).

2.6.3   Bt POTENTIAL TO AGRICULTURE

Bt and its products have been formulated into various forms for application as biological control agents. Such formulations could be solid (powdery or granulated) or liquid. Presently, there are over 400 of Bt-based formulations that have been registered in the market and most of them contain insecticidal proteins and viable spores although the spores are inactivated in some products (Table 2.2). Formulated Bt products are applied directly in the form of sprays. An alternative, and highly successful, method for delivering the toxins to the target insect has been to express the toxin-encoding genes in transgenic plants (Jogen and Kalita, 2011; Ali et al., 2010).

2.6.4   Bacillus sphaericus (Bs)

Entomopathogenic bacteria, namely Bt, have been known from the early 1900s but the control of dipteran species has been established only since the discovery of Btserovar *israelensis* (Bti) in 1977 and a highly toxic strain of *B. sphaericus* (Bs) was also isolated and used entomopathogenic bacteria (George and Crickmore, 2012). Bs belonging to genus Bacillus is an aerobic

TABLE 2.2 Example of Bacterial Biopesticide Products for Control of Insect Pest.

Sr. no.	Active agent bacteria	Manufacturer	Product example	Market
1.	*Bacillus thuringiensis* var. kurstaki	Novo, Abbott, Sandoz	Biobit, Dipel, Delfin	Vegetables and forestry
2.	*Bacillus thuringiensis* var. aizawai	Sandoz	Certan	Apiculture
3.	*Bacillus thuringiensis* var. israelensis	Novo, Abbott, Cyanamid	Acrobe, Skectal, Vectobac	Mosquito control
4.	*Bacillus thuringiensis* var. tenebrionis	Sandoz, Novo	Trident, Novodor	Vegetables
5.	*Bacillus thuringiensis* var. conjugates	Ecogen, Geigy	Ciba-Foil, Agree, Cutlass	Vegetables, forestry
6.	*Pseudomonas fluorescens* (Bt toxin)	Mycogen	MVP, M-Trak	Vegetables
7.	*Bacillus papilliae*	Fairfax	Doom	Turf
8.	*Serratia entomophila*	Monsanto	Invade	Turf

bacterium and widely used as biopestisides. Bs, like Bt, is a naturally occurring soil bacterium and have potential mosquito larvicidal properties. Since the isolation of highly larvicidal strains of this bacterium, it has become an alternative agent for biological control of mosquitoes. Some bacteria like *Psorophora*, and few other members of the genus *Aedes. Ae. aegypti* and *Ae. albopictus* are insensitive to Bs strain. The first reported Bs active against mosquito larvae was isolated from Moribund area in Argentina to mosquito larvae of *Culiseta incidens* in 1965 (Myers and Yousten, 1980). Strain 2362, isolated from *Simulium* in Nigeria, is nottoxic to black flies, but it is regarded as the most promising isolate for field use against mosquitoes. Pasteurization of the soils makes the medium selective for Bs. The efficacy of strain 2362 against field populations of mosquitoes from the genera *Culex* has been demonstrated. Since the 1960s, when a strain of Bs was discovered to have larvicidal activity against mosquito species, a large number of other mosquitocidal Bs strains have been described. The larvicidal activity of this first isolate was so low that its use in mosquito control would not have been considered indeed. But only after isolation in Indonesia from dead mosquito, larvae of strain 1593 which exhibited a much higher mosquitocidal activity against *Culex quinquefasciatus* was potential of Bs as a biological control agent for some species of mosquitoes, and used as insecticide in the field as part of vector control programmes (Kellen et al., 1965). It has terminally located spherical spores. One of the phenotypic characters examined was pathogenicity of some of them to mosquito larvae. A protoxin produced during sporulation as in the case of BT causes fatal cellular alterations when ingested by larvae of some dipteran species. This bacterium has been used to control *Culex* and *Anopheles* populations in various countries replacing chemical larvicides with certain advantages. They include reduction in cost and selectivity to the target populations. The toxic activity of the Bs strains increased at the time of sporulation, it is logical to look for parasporal (solid protein crystal) inclusions in this bacterium. Since filter sterilized culture supernatants had been shown to be nontoxic, all of the toxin must be retained on or within the cells themselves. In the cells fractioned in the process of sporulation, the cell walls gave more toxic character there than the cyto-plasmic part. On the other hand, the mature spores isolated from the cells were more toxic than the cell wall fraction, thus it appeared that some toxin may be located in several parts of the cell but that the spore contains the highest concentration of the toxin (Charles et al., 1993; Brownbridge and Margalit, 1987; Klein et al., 2002).

2.7   USE OF BACTERIAL BIOPESTICIDES

Plant pathogenic fungi and oomycetes are major threats in crops and plant production. Therefore, the control of fungal diseases by *Bacillus*-based biopesticides represents an interesting opportunity for agricultural biotech-nology. Indeed, several commercial products based on various *Bacillus* species such as *B. amyloliquefaciens*, *B. licheniformis*, *B. pumilus*, and *B. subtilis* have been marketed as biofungicides (Fravel, 2005). These *Bacillus*-based products have been developed especially for the control of fungal diseases (Tables 2.3 and 2.4). A large number of reports have described the practical effects of several *Bacillus* species against diseases elicited by oomycetes and fungal pathogens. Some examples are the suppression of root diseases (such as avocado root rot, tomato damping off, and wheat take-all), foliar diseases (such as cucurbit and strawberry powdery mildews), and post-harvest diseases (such as green, grey, and blue molds) (Cazorla et al., 2007; Pertot et al., 2008; Arrebola et al., 2010). Certain strains of *B. subtilis* are being used against a range of plant pathogens that cause damping off and soft rots (Kloepper et al., 2004; Haas and Defago 2005; Berg, 2009). Due to their catabolic versatility and excellent root-colonizing capability, pseudomonas is also being investigated extensively for the use in biocontrol of pathogens in agriculture (Ganeshan and Kumar, 2006). They are known to enhance plant growth and yield, reduce severity of many diseases, and are considered to

TABLE 2.3 Examples of Bacterial Biopesticide Products for Control of Fungi and Bacterial Diseases and Weeds.

Sr. no.	Active agent	Manufacturer	Product example	Market
1.	Fungal disease *Streptomyces griseoviridus*	Kemira Oy	Mycostop	Protected horticulture
2.	Bacterial diseases a. *Agrobacterium radiobacter* b. *Pseudomonas fluroscencens*	Burns phillips	Conquer	Mushrooms
3.	Weeds a. *Xanthomonas* sp. b. *Pseudomonas* sp./chemical herbicide	Mycogen Crop Genetics International	Field trials	Turf Arable crops

TABLE 2.4 Examples of Bacterial Biopesticides and Their Modes of Action.

Sr. no.	Bacterial species	Mechanism of action
1.	*Agrobacterium radiobacter,* Strain K84	Competitive inhibition of pathogenic bacterial species (*Agrobacteriumc tumefaciens*)
2.	*Bacillus licheniformis* strain SB3086	Production of antifungal enzyme
3.	*Bacillus subtilis* GBO3	Competitive inhibition of pathogenic fungal species
4.	*Bacillus thuringiensis* subspecies *israelensis*	Production of insecticidal crystal (Cry) proteins
5.	*Pseudomonas chlororaphis* strain 63-28	Competitive inhibition of pathogenic fungal species
6.	*Pseudomonas syringae* A506	Prevents frost damage by competitively inhibiting growth of bacteria that promote ice formation
7.	*Streptomyces griseoviridus* strain *K61*	Competitive inhibition of pathogenic metabolites

be among the most prolific PGPRs (Hoffland et al., 1996; Wei et al., 1996). Several species of *Pseudomonas* are being used for designing biopesticides that include *P. fluorescence, P. aeruginosa, P. syringae*, etc. Certain strains of *Pseudomonas aureofaciens* are being used against a range of plant pathogens including damping off and soft rots (Kloepper et al., 2004; Haas and Defago, 2005; Berg, 2009). The cell suspensions of pseudomonas are immobilized on certain carriers and are prepared as formulations for easy application, storage, commercialization, and field use. In India, *P. fluorescens* biopesticide is effectively being used against late blight of potato; it is available commercially under diverse brand names such as Krishi bio rahat, Krishi bio nidan, Mona, etc. Virulent cells of bacterial antagonist *P. fluorescens* are taken to prepare a biopesticide formulation that is effective against phytopathogens *Ralstonia solanacearum* (Bora and Deka, 2007; Chakravarty and Kalita, 2011). *P. syringae* strains ESC-10 and ESC-11 were initially registered (licensed for sale and distribution) in 1995; at the end of April 2000, there were three end products containing ESC-10 and two end products containing ESC-11 in the USA (Bull et al., 1997). An attractive role of fluorescent pseudomonas in biological control of fungal plant pathogens has been illustrated against *Aspergillus, Alternaria, Fusarium, Macrophomina, Pythium, Sclerotinia,* and *Rhizoctonia* (Dunne et al., 1998; Gupta et al., 2001). Several commercially available biopesticides in USA that are developed from *Pseudomonas* and are effective against fungal phytopathogens are spotless, At-Eze, Bio-Save 10LP, and Bio-Save 11LP (Vargas, 1999; Nakkeeran et al., 2005; Khalil et al., 2013). Bioformulation, biopesticides, and bioinoculants developed from fluorescent pseudomonas can serve multifaceted functions of plant growth promotion, bioremediation, and disease management (Arora et al., 2008, 2013; Khare and Arora, 2011; Tewari and Arora 2013). Certain other bacterial strains like that of *Agrobacterium radiobacter* are also used to control pests such as *Agrobacterium tumefaciens*.

2.8 FUTURE OPPORTUNITIES

Specialists believe that there would be a greater progress in biopesticides sector and more shift toward the use of ecofriendly agricultural products (Nicholson, 2007). Due to its rich biodiversity, India offers plenty of scope in terms of sources for natural biological control organisms as well as natural plant-based pesticides. The traditionally rich knowledge available with the highly diverse indigenous communities in India may provide valuable clues for developing such newer and effective biopesticides (bacterial, viral, and fungal). More stress on organic farming and on residue-free commodities would certainly warrant an increased adoption of biopesticides by the farmers (Baker et al., 1991). Increased adoption further depends on:

1. Concrete evidences of efficacy of biopesticides in controlling crop damage and resulting increase in crop yield.
2. Availability of high quality products at affordable prices.
3. Strengthening of supply chain management in order to enhance the usage of biopesticides.

In this regard, an efficient delivery system from the place of production (factory) to place of utilization (farm) of biopesticides is quite essential (Salom et al., 1995). The National Farmer Policy, 2007, strongly recommends the promotion of biopesticides for increasing agricultural production and sustaining the health of farmers and environment. It also includes a clause that biopesticides would be treated at par with chemical pesticides in terms of support and promotion. Further research and development of biological pest control methods must be given priority and people in general and agriculturists in particular must be educated about the handling and usage of such control measures. All this will lead to a general understanding about the benefits of biopesticides as green alternative. However, the present day need is about IPM, INM, ICM, and GAP and by practicing these, the quality of life and health will be assured. Governments are likely to continue imposing stringent safety criteria on conventional chemical pesticides, and this will result in rarer products in the market. This will generate a real opportunity for biopesticide companies to help fill the gap, although there will also be major challenges for them, most of which are small and medium enterprises with limited resources for R&D, product registration, and promotion. Perhaps the biggest advances in biopesticide development will come through exploiting knowledge of the genomes of pests and their natural competitors. Herniou et al., 2003; Wang et al., 2005; Muthumeenakshi et al. 2007 reported that researchers are already using molecular-based technologies to reconstruct the evolution of microbial natural antagonists and pull apart the molecular basis for their pathogenicity (Herniou et al., 2003; Wang et al., 2005; Muthumeenakshi et al. 2007); to understand how weeds compete with crop plants and develop herbicide resistance (Tranel and Horvath, 2009); and to identify and characterize the receptor proteins used by insects to detect semiochemicals (Palletire and Leal, 2009). This information will give us new insights into the ecological interactions of pests and biopesticides and lead to new possibilities for improving biopesticide efficacy, for example, through strain improvement of microbial natural enemies (Aiuchi et al., 2008). As the genomes of more and more pests are getting sequenced, the use of techniques such as RNA interference for pest management is also likely to be put into commercial practice (Bum et al., 2007).

2.9 CONCLUSION

Global warming and climatic changes have made us aware of environmental pollution and various health jeopardies to life. Increasing population, urbanization, and industrialization especially in developing countries is diverting the focus of governments from the severity of pollution hazards. For developing countries, it is imperative to use pesticides to control famine and communicable diseases like malaria which means to be expedient to accept a reasonable degree of risk. Pesticides are often considered a quick, easy, and less expensive solution for controlling weeds and insect pests in urban landscapes which have contaminated almost every part of our environment and nontarget organisms ranging from beneficial soil microorganisms, to insects, plants, fish, and birds. Pesticide residues are found in soil, air, in surface and ground water across the countries and contribute to the problem. The best way to reduce pesticide contamination (and the harm it causes) in our environment is to use safer, non-chemical pest control (including weed control) methods (i.e., biopesticides and other alternatives) in agriculture to develop the ecofriendly biosphere in soil, air, and water. Biopesticides are highly specific, affects only the targeted pest or closely related pests, and do not cause any harm to humans or beneficial organisms while chemical pesticides are broad spectrum and are known to affect nontarget organisms including predators and parasites as well as humans. Our efforts should include critical use of insecticides, pesticides, fungicides, chemicals, and investigations of biopesticides as safe alternatives as far as possible. By achieving pecuniary benefits, social and environmental benefits should not be neglected at any of the level and the ultimate benefit that should be focused on global level is the safe existence of human on the earth for longer period of time. This is high time for every individual, communities, institutions, and governments in the world to understand plus to plan the economic development in one hand and socioenvironmental benefits in another hand for sustainable livelihood on the earth. There is thus every reason to develop health education packages based on knowledge, aptitude, and practices and to disseminate them within the community in order to avail alternate natural resources and minimize human exposure to the controlled and justified use of pesticides for safe human life on the earth. It is very likely that in future their role will be more significant in agriculture and forestry. Biopesticides clearly have a potential role to play in development of future IPM strategies hopefully, more rational approach will be gradually adopted toward biopesticides in the near future and short-term profits from chemical pesticides will not determine the fate of biopesticides.

KEYWORDS

- **biopesticides**
- ***Bacillus* sp.**
- ***Pseudomonas* sp.**
- **larvicidal potency**
- **nematicidal activity**
- **plant protection**

REFERENCES

Aiuchi, D.; Inami, K.; Kuramochi, K.; Koike, M.; Sugimoto, M.; Tani, M.; Shinya, R. A New Method for Producing Hybrid Strains of the Entomopathogenic Fungus *Verticillium Lecanii* (*Lecanicillium* spp.) Through Protoplast Fusion by Using Nitrate Non-utilizing (Nit) Mutants. *Micol. Aplicada Int.* **2008**, *20*, 1–16.

Ali, S.; Zafar, Y.; Ali, G. M.; Nazir, F. *Bacillus thuringiensis* and its Application in Agriculture. *Afr. J. Biotechnol.* **2010**, *9*(14), 2022–2031.

Aronson, A. I.; Han, E. S.; McGaughey, W.; Johnson, D. The Solubility of Inclusion Proteins from *Bacillus Thuringiensis* is Dependent upon Protoxin Composition and is a Factor in Toxicity to Insects. *Appl. Environ. Microbiol.* **1991**, *57*(4), 981–986.

Arora, N. K.; Khare, E.; Naraian, R.; Maheshwari, D. K. Sawdust as a Superior Carrier for Production of Multipurpose Bioinoculant Using Plant Growth Promoting Rhizobial and Pseudomonad Strains and Their Impact on Productivity of *Trifolium repens*e. *Curr. Sci.* **2008**, *95*, 90–94.

Arora, N. K.; Tewari, S.; Singh, R. Multifaceted Plant-associated Microbes and Their Mechanisms Diminish the Concept of Direct and Indirect Pgprs. In *Plant Microbe Symbiosis Fundamentals and Advances*; Arora N. K., Ed.; Springer: India, 2013, pp 411–449.

Arrebola, E.; Jacobs, R.; Korsten, L. Iturin A is the Principal Inhibitor in the Biocontrol Activity of *Bacillus Amyloliquefaciens* PPCB004 Against Postharvest Fungal Pathogens. *J. Appl. Microbiol.* **2010**, *108*, 386–395.

Baker, T. C.; Staten, R. T.; Flint, H. M.; Ridgeway, R. L.; Silverstein, R. M.; Dekker, M. Use of Pink Bollworm Pheromone in the South-western United States. In *Behaviour Modifying Chemicals for Insect Management*. New York, NY. 1991, 417-436.

Baum, J. A., et al. Control of Coleopteran Insect Pests Through RNA Interference. *Nat. Biotech.* **2007**, *25*, 1322–1326.

Baum, J. A.; Malvar, T. Regulation of Insecticidal Crystal Protein Production in *Bacillus thuringiensis*. *Mol. Microbiol.* **1995**, *18* (1), 1–12.

Berg, G. Plant–microbe Interactions Promoting Plant Growth and Health: Perspectives for Controlled Use of Microorganisms in Agriculture. *Appl. Microbiol. Biotechnol.* **2009**, *84* (1), 11–18.

Bora, L. C.; Deka, S. N. Wilt Disease Suppression and Disease Enhancement in (*Lycopersicon esculentum*) by Application of *Pseudomonas fluorescens* Based biopesticide (Biofor-Pf) in Assam. *Indian. J. Agr. Sci.* **2007,** *77*, 490–494.

Broderick, N. A.; Raffa, K. F.; J. Handelsman. Midgut Bacteria Required for *Bacillus Thuringiensis* Insecticidal Activity. *Proc. Natl. Acad. Sci.* USA. **2006,** *103*, 15196–15199.

Broderick, N. A.; Robinson, C. J.; McMahon, M. D.; Holt, J.; Handelsman, J.; Raffa, K. F. Contributions of Gut Bacteria to *Bacillus Thuringiensis*-induced Mortality Vary Across a Range of Lepidoptera. BMC *Biol.* **2009,** *7*, 11.

Brownbridge, M.; Margalit, J. Mosquitoe Active Strains of *Bacillus Sphaericus* Isolated from Soil and Mud Samples Collected in Israel. *J. Invertebr. Pathol.* **1987,** *50*, 106–112.

Bull, C. T.; Stack, J. P.; Smilanick, J. L. *Pseudomonas syringae* Strains ESC-10 and ESC-11 Survive in Wounds on Citrus and Control Green and Blue Molds of Citrus. *Biol. Control.* **1997,** *8*, 81–88.

Buss, E. A.; Park-Brown, S. G. *Natural Products for Insect Pest Management*, University of Florida, 2002.

Cazorla, F. M.; Romero, D.; Garcia, A. P.; Lugtenberg, B. J. J.; Vicente, A.; Bloemberg, G. Isolation and Characterization of Antagonistic *Bacillus subtilis* Strains from the Avocado Rhizoplane Displaying Biocontrol Activity. *J. App. Microbiol.* **2007,** *103*, 1950–1959.

Chakravarty, G.; Kalita, M. C. Management of Bacterial Wilt of Brinjal by *P. Fluorescens* Based Bioformulation. *ARPN J. Agri. Biol. Sci.* **2011,** *6* (3), 1–11.

Chandler, D.; Davidson, G.; Grant, W. P.; Greaves, J.; Tatchell, G. M. Microbial Biopesticides for Integrated Crop Management: An Assessment of Environmental and Regulatory Sustainability. *Trends Food Sci. Technol.* **2008,** *19*, 275–283.

Charles, J.-F., Hamon, S.; Baumann, P. Inclusion Bodies and Crystals of *Bacillus Sphaericus* Mosquitocidal Proteins Expressed in Various Bacterial Hosts. *Res. Microbiol.* **1993,** *144*, 411–416.

Chattopadhyay, A.; Bhatnagar, N. B.; Bhatnagar, R. Bacterial Insecticidal Toxins. *Crit. Rev. Microbiol.* **2004,** *30*, 33–54.

Dean, D. H. *Bacillus thuringiensis* and Its Pesticidal Crystal Proteins. *Microbiol. Mol. Biol. Rev.* **1998,** *62*, 775–806.

Desai, S. T. Chemical Industry in the Post-Independence Era: A Finance Analysis Point of View. *Chem. Bus.* **1997,** *11* (1), 25–28.

Dunne, C.; Moënne-Loccoz, Y.; McCarthy, J.; Higgins, P.; Powell, J.; Dowling, D. N.; O'Gara, F. Combining Proteolytic and Phloroglucinol-producing Bacteria for Improved Biocontrol of Pythium-mediated Damping Off of Sugar Beet. *Plant Pathol.* **1998,** *47*, 299–307.

Fravel, D. R. Commercialization and Implementation of Biocontrol. *Ann. Rev. Phytopathol.* **2005,** *43*, 337–359.

Ganeshan, G.; Kumar, M. A. *Pseudomonas fluorescens*, a Potential Bacterial Antagonist to Control Plant Diseases. *J. Plant. Interact.* **2006,** *1*, 123–134.

George, Z.; Crickmore, N. 2 *Bacillus thuringiensis* Applications in Agriculture. In: E. Sansinenea. *Bacillus thuringiensis biotechnology Chapter*. Springer: Verlag Science+Business Media B.V.: Falmer, Brighton, UK: University of Sussex. 2012; pp 19–39.

Gupta, C. P.; Dubey, R. C.; Kang, S. C.; Maheshwari, D. K. Antibiosis Mediated Necrotrophic Effect of *Pseudomonas* GRC2 Against Two Fungal Pathogens. *Curr. Sci.* **2001,** *81*, 91–94.

Gupta, S.; Dikshit, A. K. Biopesticides: An Ecofriendly Approach for Pest Control. *J. Biopesticides* **2010,** *3* (1), 186–188.

Haas, D.; De'fago, G. Biological Control of Soil Borne Pathogens by Fluorescent Pseudomonads. *Nat. Rev. Microbiol.* **2005,** *3*, 307–319.

Hajeck, A. E.; Leger, St. Interactions Between Fungal Pathogens and Insect Hosts. *Ann. Rev. Entomol.* **1994,** *39*, 293–322.

Harwood, C. R.; Wipat, A. Sequencing and Functional Analysis of the Genome of *Bacillus subtilis* Strain 168. *FEBS Lett.* **1996,** *389,* 84–87.

Herniou, E. A.; Olszewski, J. A.; Cory, J. S.; O'Reilly, D. R. The Genome Sequence and Evolution of Baculoviruses. *Ann. Rev. Entomol.* **2003,** *48*, 211–234.

Hoffland, E.; Hakulinen, J.; van Pelt, J. A. Comparison of Systemic Resistance Induced by Avirulent and Non-pathogenic *Pseudomonas* species. *Phytopathology* **1996,** *86*, 757–762.

Hoy Myths, M. A. Models and Mitigation of Resistance to Pesticides. In *Insecticide Resistance: From Mechanisms to Management;* CABI Publishing: New York, 1999; pp 111–119.

Kalra, A.; Khanuja, S. P. S. Research and Development Priorities for Biopesticide and Biofertiliser Products for Sustainable Agriculture in India. In. *Business Potential for Agricultural Biotechnology,* Asian Productivity Organisation, 2007, pp 96–102.

Kellen, W. R.; Clark, T. B.; Lindegren, J. E.; Ho, B. C.; Rogoff, M. H.; Singer, S. *Bacillus sphaericus* Neide as a Pathogen of Mosquitoes. *J. Invertebr. Pathol.* **1965,** *7*, 442–448.

Khalil, I. A. I. M.; Appanna, V.; Rick, D. P.; Ronald, J. H.; Lucie, G.; Tharcisse, B.; Kelvin, L.; René, P.; Kathy, A. D.; Ian, K. M.; Sharon, L. I. L.; Kithsiri, E. J. Efficacy of Bio-save 10LP and Bio-save 11LP (*Pseudomonas syringae)* for Management of Potato Diseases in Storage. *Biol. Control.* **2013,** *64*, 315–322.

Klein, D.; Uspensky, I.; Braun, S. Tightly Bound Binary Toxin in the Cell Wall of *Bacillus sphaericus*. *Appl. Environ. Microbiol.* **2002,** *68* (7), 3300–3307.

Kloepper, J. W.; Ryu, C. M.; Zhang, S. Induced Systemic Resistance and Promotion of Plant by *Bacillus* spp. *Phytopathology* **2004,** *94*, 1259–1266.

Kumar, S. Biopesticides: A Need for Food and Environmental Safety. *J. Biofert. Biopest.* **2012,** *3*, 1–3.

Kumar, S.; Chandra, A.; Pandey, K. C. *Bacillus thuringiensis* (*Bt*) Transgenic Crop: An Environmentally Friendly Insect–Pest Management Strategy. *J. Environ. Biol.* **2008,** *29*, 641–653.

Lambert, B.; Hofte, H.; Annys, K.; Jansens, S.; Soetaert, P.; Peferoen, M. Novel *Bacillus thuringiensis* Insecticidal Crystal Protein with a Silent Activity Against Coleopteran Larvae. *Appl. Environ. Microbiol.* **1992,** *58* (8), 2536–2542.

Mazid, S.; Kalita, J. C. A Review on the Use of Biopesticides in Insect Pest Management. *Int. J. Sci. Adv. Technol.* **2011,** *1*, 169–178.

Meadows, M. P. *Bacillus Thuringiensis* in the Environment: Ecology and Risk Assessment. In *Bacillus thuringiensis: an Environmental Biopesticide: Theory and Practice*; Entwistle, P. F.; Cory, J. S.; Bailey, M. J.; Higgs, S., Eds, John Wiley: Chichester, USA, 1993, pp 193–220.

Milner, R. J. History of *Bacillus thuringiensis*. *Agric. Ecosyst. Environ.* **1994,** *49* (1), 9–13.

Muthumeenakshi, S.; Sreenivasaprasad, S.; Rogers, C. W.; Challen, M. P.; Whipps, J. M. Analysis of cDNA Transcripts from Coniothyrium Minitans Reveals a Diverse Array of Genes Involved in Key Processes During Sclerotial Mycoparasitism. *Fungal Genet. Biol.* **2007,** *44*, 1262–1284.

Myers, P. S.; Yousten, A. A. Localization of a Mosquito-larval Toxin of *Bacillus sphaericus* 1593. *Appl. Environ. Microbiol.* **1980,** *39*, 1205–1211.

Nakkeeran, S.; Dilantha Fernando, W. G.; Zaki, A. Plant Growth Promoting Rhizobacteria Formulations and Its Scope in Commercialization for the Management of Pests and Diseases. In *PGPR: Biocontrol and Biofertilization*; Siddiqui, Z. A., Ed., Springer: Dordrecht, **2005,** pp 257–296.

Nelson, K. L.; Brodsky, R. A.; Buckley, J. T. Channels Formed by Subnanomolar Concentrations of the Toxin Aerolysin Trigger Apoptosis of T Lymphomas. *Cell Microbiol.* **1999,** *1* (1), 69–74.

Nicholson, G. M. Fighting the Global Pest Problem: Preface to the Special Toxicon Issue on Insecticidal Toxins and Their Potential for Insect Pest Control. *Toxiconomy* **2007,** *49,* 413–422.

O'Brien, K. P.; Franjevic, S.; Jones, J. Green Chemistry and Sustainable Agriculture: The Role of Biopesticides. *Advancing Green Chem.* 2009. http://advancinggreenchemistry. org/ wp-content/uploads/Green-Chemand-Sus.- Ag.-the-Role-of-Biopesticides.pdf.

Ongena, M.; Jacques, P. *Bacillus* Lipopeptides: Versatile Weapons for Plant Disease Biocontrol. *Trends Microbiol.* **2008,** *16* (3), 115–125.

Pelletier, J.; Leal, W. S. Genome Analysis and Expression Patterns of Odorant-binding Proteins From the Southern House Mosquito Culex Pipiens Quinquefasciatus. *PLoS One* **2009,** *4,* e6237.

Pertot I.; Gobbin, D.; De Luca, F.; Prodorutti, D. Methods of Assessing the Incidence of Armillaria Root Rot Across Viticultural Areas and the Pathogen's Genetic Diversity and Spatial–Temporal pattern in Northern Italy. *Crop Prot.* **2008,** *27,* 1061–1070.

Porta, H.; Cancino-Rodezno, A.; Soberón, M.; Bravo, A. Role of MAPK p38 in the Cellular Responses to Pore-forming Toxins. *Peptides* **2011,** *32* (3), 601–606.

Roh, J. Y.; Choi, J. Y.; Li, M. S.; Jin, B. R. Je, Y. H. *Bacillus thuringiensis* as a Specific, Safe, and Effective Tool for Insect Pest Control. *J. Microbiol. Biotechnol.* **2007,** *17,* 547–559.

Roy, A.; Moktan, B.; Sarkar, P. K. Characteristics of *Bacillus cereus* Isolates from Legume-based Indian Fermented Foods. *Food Contr.* **2007,** *18,* 1555–1564.

Salom, S. M.; Grossman, D. M.; McClellan, Q. C.; Payne, T. L. Effect of an Inhibitor Based Suppression Tactic on Abundance and Distribution of Southern Pine Beetle (Coleoptera: scolytidae) and its Natural Enemies. *J. Econ. Entomol.* **1995,** *88,* 1703–1716.

Schallmey, M.; Singh, A.; Ward, O. P. Developments in the Use of *Bacillus* Species for Industrial Production. *Can. J. Microbiol.* **2004,** *50,* 1–17.

Schnepf, E.; Crickmore, N.; Van Rie, J.; Lereclus, D.; Baum, J.; Feitelson, J.; Zeigler, D. R.; Dean, D. H. *Bacillus thuringiensis* and Its Pesticidal Crystal Proteins. *Microbiol. Mol. Biol. Rev.* **1998,** *62,* 775–806.

Soberón, M.; Pardo-López, L.; López, I.; Gómez, I.; Tabashnik, B. E.; Bravo A. Engineering Modified Bt Toxins to Counter Insect Resistance. Science **2007,** *318* (5856), 1640–1642.

Thakore, Y. The Biopesticide Market for Global Agricultural Use. *Ind. Biotechnol.* **2006,** *2,* 194–208.

Tranel, P. J.; Horvath, D. P. Molecular Biology and Genomics: New Tools for Weed Science. *Bioscience* **2009,** *59,* 207–215.

Wang, C.; St Leger, R. J. Developmental and Transcriptional Responses to Host and Non-host Cuticles by the Specific Locust Pathogen Metarhizium anisopliaesf. acridum. *Eukaryot. Cell* **2005,** *4,* 937–947.

Zhang, X.; Candas, M.; Griko, N. B.; Taussig, R.; Bulla, L. A., Jr. A Mechanism of Cell Death Involving an Adenylyl Cyclase/PKA Signaling Pathway Is Induced by the Cry1Ab Toxin of *Bacillus thuringiensis, Proc. Natl. Acad. Sci. USA* **2006,** *103,* 9897–9902.

CHAPTER 3

Fungal Biopesticides and Their Uses for Control of Insect Pest and Diseases

DEEPAK KUMAR[1,*], M. K. SINGH[1], HEMANT KUMAR SINGH[2], and K. N. SINGH[3]

[1]Research and Development Unit, Shri Ram Solvent Extractions Pvt. Ltd. Jaspur, U.S. Nagar 244712, Uttarakhand, India

[2]Krishi Vigyan Kendra, Hawai Adda Road Khagra, Kishanganj 855107, Bihar, India

[3]Department of Plant Molecular Biology and Genetic Engineering, Narenda Deva University of Agriculture and Technology, Kumarganj, Faizabad 224229, Uttar Pradesh, India

*Corresponding author. E-mail: deepak_rajora84@rediffmail.com

ABSTRACT

Pest control by chemical pesticides is extensively used in most of the countries of the world; however, chemicals are regarded as ecologically unsafe. The growing demands for dropping chemical pesticides in agriculture and augmented resistance to insecticides have afforded huge impetus to the development of another form of insect-pest control. Therefore, there is an increased social pressure to substitute them progressively with biopesticides which are harmless to humans and nontarget organisms. By the arrival of greener approach of developing and using biopesticides, the condition is slowly changing but in fact can shift far more quickly in this track which will be sustainable and ecofriendly. This leads to increased development of compounds based on the models of naturally occurring toxins of biological origin, having various biological activities. Biopesticides comprise a broad range of microbial pesticides, biomolecules derived from microorganisms and other normal sources. The biopesticides also processes by genetic modification of plants to express genes encoding insecticidal toxins. Fungal biopesticides are potentially the most adaptable biological control means owing to their broad host range. These fungi comprise a various group of more than 90 genera with approximately 750 species. Fungal biopesticides are naturally occurring organisms which are professed as less damaging to the environment. Although biopesticides are gradually replacing the chemical pesticides, absolute global look at the scenario shows that the former and particularly the industries based on them are silent in an insecure situation in comparison to the chemicals which rule the agriculture. The book chapter revises us regarding the recent progress in the field of fungal biopesticides of insect pests and their probable mechanism of achievement to further enhance our understanding about the natural control of insect pests.

3.1 INTRODUCTION

Agriculture has been facing the vicious activities of many pests like fungi, weeds, and insects from time immemorial, leading to drastic decline in yields. Pests are regularly being introduced to fresh areas either naturally or accidentally, or, in a few cases, organisms that are deliberately introduced become pests. Global employment has resulted in increased numbers of insidious nonnative pest species being introduced to fresh areas. Controlling these invasive species presents an unparalleled challenge worldwide. Although many years of successful control by conventional agrochemical insecticides, a number of features are bullying the effectiveness and continued use of these agents. These contain the development of insecticide resistance and use-cancellation or deregistration of a little insecticide due to human health and environmental disquiet. Therefore, an ecofriendly substitute is the need of the present scenario. Advances in pest control approaches symbolize one method to generate elevated quality and greater quantity of agricultural products. Therefore, there is a requirement to develop biopesticides which are efficient, biodegradable, and do not put down any harmful effect on environment. Their shield against pests is a main concern and due to the poor impact of chemical insecticides, use of biopesticides is escalating.

3.2 CONCEPT OF BIOPESTICIDES

According to the US Environmental Protection Agency (USEPA), biopesticides are pesticides prepared from natural resources such as animals, plants, bacteria, and minerals. Biopesticides also include live organisms that obliterate agricultural pests. The EPA divides biopesticides into three main classes on the basis of their type of active ingredient used, namely, biochemical, plant-incorporated protectants, and microbial pesticides (USEPA, 2008). Biochemical pesticides are chemicals either isolated from natural resource or manufactured to have the similar structure and function as the naturally occurring chemicals. Biochemical pesticides are differed from conventional pesticides together by their construction (source) and mode of action (method by which they kill or control pests) (O'Brien et al., 2009). At a global level, there is a discrepancy in understanding the term biopesticide as aforementioned functioning definition of the term biopesticide given by USEPA is not pursued in the whole world and that is why International Biocontrol Manufacturer's Association (IBMA) and the International Organization for Biological Control (IOBC 2008) endorse to use the term biocontrol agents (BCAs) instead of biopesticide (Guillon, 2003). IBMA categorize BCAs into four groups: (1) macrobials, (2) microbials, (3) natural products, and (4) semiochemicals (insect behavior-modifying agents). Among all the BCAs, the mainly important products are microbials (41%), followed by macrobials (33%), and, ultimately, other natural products (26%) (Guillon, 2003). This chapter spotlights on microbe-based biopesticides (specifically fungal biopesticides).

Biopesticides or biological pesticides stand on pathogenic microorganisms specific to a target pest and present an ecologically thud and effective solution to pest problems. They cause less threat to the environment and to human health. The most frequently used biopesticides are living organisms, which are pathogenic for the pest of concern. These include biofungicides (*Trichoderma*), bioherbicides (*Phytopthora*), and bioinsecticides (*Beauvarria bassiana, Metarhizium anisopliae*). The impending benefits to agriculture and public health programs throughout the use of biopesticides are substantial. The interest in biopesticides is based on the rewards associated with such products which are:

- inherently fewer harmful and fewer environmental load,
- designed to concern only one specific pest or, in a few cases, some target organisms,
- often effective in very minute quantities and often decay quickly, thereby resulting in lesser exposures and largely avoiding the pollution problems and
- when used as a component of integrated pest management (IPM) programs, biopesticides can contribute very much.

3.2.1   FUNGAL BIOPESTICIDES

The fungal pathogens participate in a major position in the development
of diseases on various imperative field and horticultural crops, resulting in
huge plant yield losses (Khandelwal et al., 2012). Intensified use of fungicides has effected in amassing of toxic compounds potentially hazardous to
human and environment and also in the build-up of resistance in the pathogens. Fungal biopesticides can be used to manage insects and plant diseases
including other fungi, bacteria, nematodes, and weeds. The means of action
are varied and depends on together the pesticidal fungus and the target pest.
One benefit of fungal biopesticides in contrast with many of the bacterial
and all of the viral biopesticides is that they do not require to be eaten to
be effective. However, they are living organisms that often need a narrow
range of situations including moist soil and cool temperatures to flourish.
BCAs like *Trichoderma* are acclaimed as efficient, ecofriendly, and cheap,
nullifying the ill effects of chemicals (Table 3.1). Therefore, of late, these
BCAs are recognized to act against an array of important soil-borne plant
pathogens causing severe diseases of crops (Bailey and Gilligan, 2004).
Fungal biopesticides used against plant pathogens comprise *T. harzianum*,
which is an antagonist of *Rhizoctonia, Pythium, Fusarium*, and other soil-
borne pathogens (Harman, 2005). *Trichoderma* is a fungal antagonist that
rises into the main tissue of a disease-causing fungus and exudes enzymes
that degrade the cell walls of the other fungus and then eat the contents of the
cells of the target fungus and multiplies its own spores. *Trichoderma* is one of
the general fungal BCAs being used worldwide for appropriate management
of a variety of foliar and soil-borne plant pathogens like *Ceratobasidium,
Fusarium, Rhizoctonia, Macrophomina, Sclerotium, Pythium*, and *Phytophthora* spp. (Dominguesa et al., 2000; Anand and Reddy, 2009). *Trichoderma
viride* has shown to be very promising against soil-borne plant parasitic fungi
(Khandelwal et al., 2012). A precise strain *Muscodor albus* QST 20799 is a
naturally occurring fungus initially isolated from the bark of a cinnamon tree
in Honduras. *M. albus* strain is reported to generate a number of volatiles,
mainly alcohols, acids, and esters, which inhibit and kill definite bacteria and
other organisms that generate soil-borne and postharvest diseases in the plants.
Products containing QST 20799 can be also used in fields, greenhouses, and
warehouses (USEPA, 2008) for disease control. Other two fungi namely; *B.
bassiana* (Balsamo) Vuillemin and *Metarhizium anisopliae* (Metchnikoff)
Sorokin are naturally occurring entomopathogenic fungi that infect sucking
pests including *Nezara viridula* (L) (green vegetable bug) and *Creontiades*
sp. (green and brown mirids) (Sosa-Goméz and Moscardi, 1998). Fungi have

TABLE 3.1 Fungal Biopesticides Developed or Being Developed for the Biological Control of Pests.

Product	Fungus	Target pests	Producer companies and country
BIO 1020	*Metarhizium anisopliae*	Vine weevil	Licensed to Taensa, USA
Bio-Blast	*Metarhizium anisopliae*	Termites	EcoScience, USA
Biogreen	*Metarhizium anisopliae*	Scarab larvae on pasture	Bio-care Technology, Australia
Bio-Path	*Metarhizium anisopliae*	Cockroaches	EcoScience, USA
Boverin	*Beauveria bassiana*	Colorado beetle	former USSR
Conidia	*Beauveria bassiana*	Coffee berry borer	Live Systems Technology, Colombia
Corn Guard	*Beauveria bassiana*	European corn borer	Mycotech, USA
Engerlingspilz	*Beauveria brongniartii*	Cockchafers	Andermatt, Switzerland
Jas Bassi	*Beauveria bassiana*	Colorado beetle	Shri Ram Solvent Ext. Pvt., India
Jas Meta	*Metarhizium anisopliae*	Sugarcane spittle bug, termites	Shri Ram Solvent Ext. Pvt., India
Jas Verti	*Verticillium lecanii*	Whitefly and thrips	Shri Ram Solvent Ext. Pvt., India
Laginex	*Lagenidium giganteum*	Mosquito larvae	AgraQuest, USA
Metaquino	*Metarhizium anisopliae*	Spittle bugs	Brazil
Mycotal	*Verticillium lecanii*	Whitefly and thrips	Koppert, The Netherlands
Mycotrol WP	*Beauveria bassiana*	Whitefly, aphids, thrips	Mycotech, USA
Naturalis-L	*Beauveria bassiana*	Cotton pests including bollworms	Troy Biosciences, USA
Ostrinil	*Beauveria bassiana*	Corn borer	Natural Plant Protection (NPP), France
Pae-Sin	*Paecilomyces fumosoroseus*	Whitefly	Agrobionsa, Mexico
PFR-97	*Paecilomyces fumosoroseus*	Whitefly	ECO-tek, USA
Procol	*Beauveria bassiana*	Army worm	Probioagro, Venezuela
Schweizer Beauveria	*Beauveria brongniartii*	Cockchafers	Eric Schweizer, Switzerland
Vertalec	*Verticillium lecanii*	Aphids	Koppert, The Netherlands

Source: Adapted and modified from Butt et al. (2001) and Rai et al. (2014).

the unique ability to attack insects by penetrating through the cuticle making them ideal for the control of sucking pests. *B. bassiana* is currently registered in the USA as Mycotrol ES® (Mycotech, Butte) and Naturalis L® (Troy Biosciences). These biopesticedes products are registered against sucking pests such as whitefly, aphids, thrips, mealybugs, leafhoppers, and weevils. Studies also show that *B. bassiana* is virulent against *Lygus hesperus* Knight (Hemiptera: Miridae), a major pest of alfalfa and cotton in the United States (Noma and Strickler, 2000).

3.2.2   Beauvaria spp.

The genus *Beauveria* includes at least 49 species of which approximately 22 are regarded as pathogenic (Kirk, 2003). *Beauveria bassiana*, a white muscardine fungus, is one of the most historically central fungus commonly used in this genus and originally recognized as *Tritirachium shiotae,* renamed after the Italian lawyer and scientist Agostino Bassi who first implicated it as the causative agent of a white (later yellowish or occasionally reddish) muscardine disease in domestic silkworms (Furlong and Pell, 2005; Zimmermann, 2007). *Beauveria bassiana* is a fungus that grows naturally in soils all through the world and acts as a pathogen on various insect species belonging to the entomopathogenic fungi (Sandhu et al., 2004; Jain et al., 2008). When the microscopic spores of the fungus get in touch with the body of an insect host, they germinate, penetrate the cuticle, and grow inside, killing the insect within a matter of days. Later, a white mold appears from the cadaver and produces fresh spores. A typical isolate of *B. bassiana* can attack a broad range of insects; various isolates differ in their host range. An attractive attribute of *Beauveria* sp. is the high host specificity of several isolates. Hosts of agricultural and forest implication include the Colorado potato beetle, the codling moth, and several genera of termites, American bollworm, *Helicoverpa armigera* (Thakur et al., 2010). *Beauveria bassiana* can easily be isolated from insect cadavers or from soil in forested region by using media (Beilharz et al., 1982), as well as by baiting soil with insects.

3.2.3   Metarhizium spp.

A fungus *Metarhizium anisopliae,* firstly known under the name *Entomphthora anisopliae*, was first described near Odessa in Ukraine from infected larvae of the wheat cockchafer *Anisopliae austriaca* in 1879, and later on, *Cleonus punctiventis* by Metschnikoff. It was shortly renamed as *M. anisopliae* by Sorokin in 1883 (Tulloch, 1976). *Metarhizium* causes a disease known as "green muscardine" in insect hosts as the green color of its conidial cells. When these mitotic (asexual) spores (called conidia) of the fungus come into touch with the body of an insect host, they germinate and the hyphae that emerge penetrate the cuticle. The fungus then expands within the body and ultimately killing the insect after a few days; this lethal outcome is very likely aided by the creation of insecticidal cyclic peptides (destruxins). If the ambient moisture is high enough, a white mould then develops on the cadaver that soon revolves green as spores are produced. Several insects living near the soil have advanced natural defences against entomopathogenic fungi like *M. anisopliae*. The genus *Metarhizium* is pathogenic to a bulky number of insect species, numerous of which are agricultural and forest insects (Ferron, 1978). The taxonomy of *Metarhizium* is intricate. Using the morphological characteristics of conidia and conidiogenous cells, Tulloch (1976) documented *M. flavoviride* and *M. anisopliae* of which the latter was further subdivided into two variants, namely, *majus* (short conidia up to 9 µm) and *anisopliae* (long conidia up to 18 µm). Currently, *M. anisopliae* consists of four varieties or genetic groups (Driver et al., 2000).

3.2.4 *Verticillium lecanii*

Verticillium lecanii is a broadly distributed fungus, which can cause huge epizootic in tropical and subtropical regions, as well as in warm and humid environments (Nunez et al., 2008). It was accounted by Kim et al. (2002) that *V. lecanii* was an effective BCAagainst *Trialeurodes vaporariorum* in South Korean greenhouses. The conidia (spores) of *V. lecanii* are slimy and connect to the cuticle of insects. The fungus infects to the insects by producing hyphae from germinating spores that enter the insect's integument; the fungus then obliterates the internal contents and the insect dies. The fungus eventually rises out through the cuticle and sporulates on the outside of the insect body. Infected insects emerge as white to yellowish cottony particles. Diseased insects typically appear in 7 days. However, due to environmental conditions, there may be a few considerable lag time from infection to death of insects. The fungus *V. lecanii* works best at temperatures of 15–25°C and a relative humidity of 85–90%. This fungus attacks nymphs and adults and sticks to the leaf underside by means of a filamentous mycelium (Nunez et al., 2008). In 1970s, *V. lecanii* was developed to manage whitefly and several aphids species, including the green peach aphids (*Myzus persicae*) for use in the greenhouse chrysanthemums (Hamlen et al., 1979).

3.2.5   *Paecilomyces* spp.

Fungus *Paecilomyces fumosoroseus* is one of the most important natural enemies of whiteflies worldwide, and causes the sickness called "yellow muscardine" (Nunez et al., 2008). Tough epizootic prospective against *Bemisia* and *Trialeurodes* spp. in both greenhouse and open field environments has been reported by *Paecilomyces fumosoroseus*. The aptitude of this fungus to grow broadly over the leaf surface under humid conditions is a characteristic that certainly enhances its ability to extend rapidly through whitefly populations (Wraight et al., 2000). Kim et al. (2002) accounted that *P. fumosoroseus* is best for managing the nymphs of whitefly. These fungi cover the whitefly's body with mycelial threads and stick them to the bottom of the leaves. The nymphs show a "feathery" aspect and are surrounded by mycelia and conidia (Nunez et al., 2008). *Paecilomyces lilacinus* has been characterized as aggressive and Dunn et al. (1982) stated, "The fungal egg parasites as group appear to be more promising to investigate as potential biological control agents of nematodes." Research results utilizing this fungus have been contradicting and erratic. In one experiment, the fungus caused a 71% decline of root galls and 90% decline in egg masses on root-knot nematode infected corn (Ibrahim et al., 1987).

3.2.6   *Nomuraea* spp.

Nomuraea rileyi, an extra potential entomopathogenic fungi, is a dimorphic hyphomycete that can cause epizootic death in various plant insects. It has been demonstrated that several insect species belonging to Lepidoptera including *Spodoptera litura* and few belonging to *Coleoptera* are susceptible to *N. rileyi* (Ignoff, 1981). The host specificity of *N. rileyi* and its ecofriendly nature persuade its use in insect pest management. Its mode of infection and development have been reported for several insect hosts such as *Trichoplusiani, Heliothis zea, Plathypena scabra, Bombyx mori, Pseudoplusia includes,* and *Anticarsia gemmatalis.*

3.3 PATHOGENICITY AND MODE OF ACTION OF ENTOMOPATHOGENIC FUNGAL BIOPESTICIDES

Insect pathogenic fungi are diverse in pathogenicity than bacteria and viruses in that they infect insects by breaching the host cuticle. The cuticle is poised of chitin fibrils embedded in a matrix of proteins, lipids, pigments, and N-acylcatecholamines (Richard et al., 2010). They exude extracellular enzymes proteases, chitinases, and lipases to humiliate the major constituents of the cuticle (i.e., protein, chitin, and lipids) and allow hyphal infiltration (Wang et al., 2005; Cho et al., 2006). The successfulness of infection was unswervingly proportional to emission of exo-enzymes (Khachatourians, 1996). It is considered that both mechanical force and enzymatic action are engaged in the penetration of fungus to the hemocoel of the insect. There are a large number of toxic compounds accounted in the filtrate of entomopathogenic fungi such as small secondary metabolites, cyclic peptides, and macromolecular proteins. *B. bassiana* is reported to secrete low-molecular weight cyclic peptides and cyclosporins A and C with insecticidal properties such as beauvericin, enniatins, bassianolide (Roberts, 1981; Vey et al., 2001). Unlike other biopesticides such as bacteria and viruses, entomopathogenic fungi do not have to be ingested to cause infection, making them valuable as BCAs. Although little information suggest a mode of infection through the siphon tips or gut of insect larvae (Lacey et al., 1988; Goettel and Inglis, 1997), entomopathogenic fungi generally infect or penetrate their targets percutaneously (Charnley, 1989). This can occur by adhesion of spores to the insect integument, particularly the intersegmental folds, or by simple tarsal contact (Clarkson and Charnley, 1996).

Normally, germinated conidia manufacture an appressorium, which then forms a contagion peg (St. Leger et al., 1991). The penetration of conidia throughout the host`s cuticle engage both mechanical pressure and enzymatic degradation (Bidochka et al., 1995; Clarkson and Charnley, 1996). Enzymatic degradation occupies the production of numerous and diverse amounts of cuticle degrading enzymes which differ according to the species and strains of the fungi. These enzymes will reveal changeable levels of pathogenicity toward their hosts. Subsequent successful penetration of the cuticle, the fungus then produces blastospores or hyphae bodies, which are inactively allocated in the hemolymph and the fat body (Hajek and Leger, 1994). So as to kill their host, fungal pathogens discharge a broad variety of secondary metabolic compounds, commonly called toxins, inside the insect host, mainly in the hemocoel. Two fungi; *M. anisopliae* and *B. bassiana,* secrete large amounts of single extracellular protease called chymoelastase protease or Pr1 to degrade the host cuticle (St. Leger et al., 1992 and 1996). The endomoprotease Pr1, which is the major enzyme secreted by *Metarhizium* throughout the degradation process of the cuticle, also differs in terms of biochemistry among strains. However, the cuticle degrading enzymes are manufactured in a sequential manner with the proteolytic enzymes and esterases first followed by chitinases, in other words proteins surrounding the cuticle must be degraded before the actions of chitinases begins (Leger et al., 1996).

Time to fatality of an infected insect differs from 2 to 15 days post-infection depending on the fungal strain and species, however, extra particularly on the characteristics of the host (Boucias and Pendland, 1998) required. When the infection procedure followed by the death of the host is complete, the fungus changes back to its hyphal mode. Under relatively humid conditions, the fungus consequently grows out of the cadaver surface to produce new, external, infective conidial saprophytic growth (Jianzhong et al., 2003; Mitsuaka, 2004). Under very dry conditions, the fungus may persevere in the hyphal stage inside the cadaver where the conidia are produced inside the body of host (Hong et al., 1997). Under positive conditions, sporulated cadavers can infect other individuals from the same target species through horizontal transmission (Meadow et al., 2000; Quesada-Moraga et al., 2004).

3.4   BIOFUNGICIDE OR MYCOFUNGICIDE

A range of fungal species can be used as biological control means and may offer effective activity against a variety of pathogenic microorganisms; these are known as biofungicides or mycofungicides (Table 3.2). Examples of these biofungicides or mycofungicides are *Trichoderma harzianum*— a species with biocontrol potential against *Botrytis cineria, Fusarium, Pythium,* and *Rhizoctonia* (Khetan, 2001); *Ampelomyces quisqualis*—a hyperparasite of powdery mildew (Liang et al., 2007; Viterbo et al., 2007); *Chaetomium globosum* and *C. cupreum*, having biocontrol activity against root rot disease caused by *Fusarium, Phytophthora,* and *Pythium* (Soytong et al., 2001); *Gliocladium virens* effective biocontrol of soil-borne pathogens (Viterbo et al., 2007), respectively. An efficient BCA should be genetically constant, effective at low concentrations, easy to mass produce in culture on inexpensive media, and also be very effective against a wide range of pathogens (Wraight et al., 2001; Irtwange, 2006).

TABLE 3.2 Fungal Biofungicides/Mycofungicides Developed or Being Developed for the Biological Control of Diseases.

Product	Fungus	Target pests	Producer companies and country
AQ10 Biofungicide	*Ampelomyces quisqualis*	Powdery mildews	Ecogen Inc.USA
Aspire	*Candida oleophila*	*Botrytis* spp., *Penicillium* spp.	Ecogen Inc.USA
Binab T	*Trichoderma harzianum,*	Fungi causing Wilt	Bio-Innovation, Sweden
Biofox C	*Fusarium oxysporium*	*Fusarium oxysporium, F. moniliforme*	SIAPA, Italy
Bio-Trek, RootShield	*Trichoderma harzianum*	*Rhizoctonia solani, Sclerotium rolfsii, Pythium*	BioWorks (= TGT Inc) Geneva, USA
Cotans WG	*Coniothyrium minitans*	*Sclerotinia* species	Prophyta, Germany. KONI, Germany
Fusaclean	*Fusarium oxysporium*	*Fusarium oxysporium*	Natural Plant Protection, France
Ketomium®	*Chaetomium sp.*	*Botrytis cinerea, Didymella applanata, Fusarium oxysporum and Rhizoctonia solani*	Thailand
Neemoderma	*Trichoderma harzianum, Trichoderma viride*	Wide range of fungal diseases	Shri Ram Solvent Ext. Pvt., India
Polygandron, Polyversum	*Pythium oligandrum*	*Pythium ultimum*	Plant Protection Institute, Slovak Republic
Primastop	*Gliocladium catenulatum*	Several plant diseases	Kemira, Agro Oy, Finland
SoilGard (= GlioGard)	*Gliocladium virens*	Several plant diseases, Damping off & root pathogens	ThermoTrilogy, USA
T-22 and T-22HB	*Trichoderma harzianum*	*Rhizoctonia solani, Sclerotium rolfsii, Pythium*	BioWorks (= TGT Inc) Geneva, USA
Trichoderma 2000	*Trichoderma harzianum*	*Rhizoctonia solani, Sclerotium rolfsii, Pythium*	Mycontrol (EfA1)Ltd, Israel
Trichodex	*Trichoderma harzianum*	Fungal diseases; e.g., *Botrytis cinerea*	Makhteshim-Agan, Several European companies e.g., DeCeuster, Belgium.
Trichodowels, Trichoject,	*Trichoderma viride*	*Chondrostereum purpureum,* & other soil & foliar pathogens	Agrimms Biologicals, New Zealand
Trichopel	*Trichoderma harzianum*	Wide range of fungal diseases	Agrimm Technologies Ltd, New Zealand
YIELDPLUS	*Cryptococcus albidus*	*Botrytis* spp., *Penicillium* spp.	Anchor Yeast, S. Africa

Source: Adapted and modified from Butt et al. (2001) and Soytong et al. (2001).

3.4.1   TRICHODERMA

Trichoderma species are frequent in soil and root ecosystems of several plants and they are easily isolated from soil, decaying wood, and other organic material (Howell, 2003; Zeilinger and Omann, 2007). *Trichoderma* species have been used as biological control means against a wide range of pathogenic fungi, for example, *Rhizoctonia* spp., *Pythium* spp., *Botrytis cinerea*, and *Fusarium* spp. *Phytophthora palmivora, P. parasitica,* and different species can be used, for example, *T. harzianum, T. viride, T. virens* (Sunantapongsuk et al., 2006; Zeilinger and Omann, 2007). Among them, *Trichoderma harzianum* is accounted to be most widely used as an efficient BCA (Szekeres et al., 2004; Abdel-Fattah et al., 2007). The biocontrol mechanism in *Trichoderma* is a combination of mechanisms (Benítez et al., 2004; Zeilinger and Omann, 2007). The main mechanism is mycoparasitism and antibiosis (Vinale et al., 2008). Mycoparasitism relies on the recognition, binding, and enzymatic disturbance of the host fungus cell wall (Woo and Lorito, 2007). *Trichoderma* species have been very fruitfully used as mycofungicides as they are fast growing, have high reproductive capability. *Trichoderma* also inhibit a broad spectrum of fungal diseases and have a diversity of control mechanisms. *Trichoderma* have excellent competitors in the rhizosphere, and capacity to modify the rhizosphere are tolerant or resistance to soil fungicides. This fungus also have the ability to survive under unfavorable conditions, are efficient in utilizing soil nutrients, have strong aggressiveness against phytopathogenic fungi, and also promote plant growth (Benítez et al., 2004; Vinale et al., 2006). Their capability to colonize and produce in association with plant roots is known as rhizosphere competence.

3.4.2   CHAETOMIUM

Chaetomium species are usually found in soil and organic compost (Soytong et al., 2001). The application of *Chaetomim* as a biological control means for controlling of plant pathogens has been firstly commenced in about 1954 when Martin Tviet and M. B. Moor found *C. globosum* and *C. cochliodes* occurring on oat seeds and that these taxa provided some control of *Helminthosporium victoriae* (Tviet and Moor, 1954). *Chaetomium* species have been accounted to be impending antagonists of various plant pathogens, especially soil-borne and seed-borne pathogens (Aggarwal et al., 2004; Park et al., 2005). Numerous species of *Chaetomium* with probable BCAs capacity suppress the growth of bacteria and fungi through competition (for substrate and nutrients), mycoparasitism, antibiosis, or various combinations of these (Marwah et al., 2007; Zhang and Yang, 2007). *Chaetomium globosum* and *C. cupreum* have been extensively considered and successfully applied to control root rot disease of citrus, black pepper, strawberry, and have been reported to minimize damping off disease of sugar beet (Soytong et al., 2001; Tomilova and Shternshis, 2006). These taxa have been prepared in the form of powder and pellets as Ketomium®, a broad spectrum mycofungicide. Ketomium® has been also registered as a biofertilizer for mortifying organic matter and for inducing plant immunity and stimulating plant growth (Soytong et al., 2001).

3.4.3   GLIOCLADIUM

Gliocladium species are general soil saprobes and numerous species have been accounted to be parasites of several plant pathogens (Viterbo et al., 2007), for example, *Gliocladium catenulatum* parasities, *Sporidesmium sclerotiorum,* and *Fusarium* spp. It demolishes the fungal host by direct hyphal contact and forms pseudoappressoria (Punja and Utkhede, 2004; Viterbo et al., 2007). *Gliocladium catenulatum* (Strain JI446) has also been employed as a wettable powder named Primastop® by Kemira Agro Oy, Finland. This product can be used to soils, roots, and foliage to minimize the frequency of damping-off disease caused by *Pythium ultimum* and *Rhizoctonia solani* in the greenhouse (Paulitz and Belanger, 2001; Punja and Utkhede, 2004). *Gliocladium virens* produce antibiotic metabolites such as gliotoxin which have important activities including antibacterial, antifungal, antiviral, and antitumor. Recently, molecular confirmation indicates that *G. virens* is more strongly related to *Trichoderma* than those of *G. virens*. This chains suggest that this taxon should be referred to as *Trichoderma virens* (Hebber and Lumsden, 1999; Punja and Utkhede, 2004).

3.4.4   AMPELOMYCES

Ampelomyces quisqualis is the mycoparasitic anamorphic ascomycete that decreases the growth and kills powdery mildews. It can concern the pathogen through antibiosis and parasitism (Kiss, 2003; Viterbo et al., 2007). The fungus *A. quisqualis* was the initial organism accounted to be a hyperparasite of powdery mildew and it can be simply established in association with powdery mildew colonies (Paulitz and Belanger, 2001). Hyphae of *Ampelomyces* pierce the hyphae of powdery mildews and rise inside and then kill all the parasitized cells (Kiss, 2003). *Ampelomyces quisqualis* isolate M-10 has been prepared as AQ10 Biofungicide, developed by Ecogen, Inc, USA. This mycofungicide include conidia of *A. quisqualis* and formulated as water-dispersible granules for the control of powdery mildew of carrot, cucumber, and mango (Kiss, 2003; Viterbo et al., 2007).

3.4.5   OTHER FUNGI AS BIOFUNGICIDES

Coniothyrium minitans is an anamorphic coelomycete (Gong et al., 2007) which has been accounted to be a mycoparasite of *Sclerotinia* species, for example, *Sclerotinia minor*, *S. sclerotiorum*, *S. trifoliorum,* and *S. cepivorum* (Viterbo et al., 2007; Whipps et al., 2008). It has been used effectively to control disease in numerous crops including lettuce (Jones et al., 2004), oil oilseed rape (Li et al., 2006), peanut (Partridge et al., 2006), and alfalfa (Li et al., 2005). The utilization of nonpathogenic strains of *Fusarium oxysporum* to control Fusarium wilt has been accounted for several crops, but there has been small commercial production, due of lack of understanding of their genetics, biology, and ecology (Fravel et al., 2003; Kvas et al., 2009). Nonpathogenic *F. oxysporum* strain Fo47 was promoted as liquid formulation named as Fusaclean® by Natural Plant Products, Nogueres, France for soil less culture (Khetan, 2001; Paulitz and Belanger, 2001). *Pythium oligandrum* confirms its ability to control soil-borne pathogens both in the laboratory and in the field condition. *Pythium oligandrum* oospores have been used as seed treatments which minimize damping-off disease caused by *P. ultimum* in sugarbeet (Khetan, 2001). *Pythium oligandrum* has been formulated as a granular or powder product named as Polygangron® by Vyskumny Ustav of Slovak Republic (Khetan, 2001). This fungus has tortuous effects by controlling pathogens in the rhizosphere and/or direct effects by inducing plant resistance after application. It also provokes plants to respond more quickly and efficiently to pathogen infections and enhance phosphorus uptake (Le Floch et al., 2003). Several other fungi that can be employed as mycofungicides are *Aspergillus* and *Penicillium* species. *Aspergillus* species are useful against the white-rot basidiomycetes (Bruce and Highley, 1991). The fungal antagonists *Aureobasidium pullulans* and *Ulocladium atrum* have also been analyzed for the control of *Botrytis aclada* which causes onion neck rot in the field condition (Köhl et al., 1997).

3.5 MECHANISMS OF BIOLOGICAL CONTROL

Biological control may outcome from direct or indirect communications between the beneficial microorganisms and the pathogen. A direct communication may entail physical contact and synthesis of hydrolytic enzymes, toxic compounds, or antibiotics as well as competition. An indirect communication may consequence from induced resistance in the host plant, the use of organic soil amendments to advance the activity of antagonists against the pathogens (Benítez et al., 2004; Viterbo et al., 2007). The mechanisms of BCAs and reaction with the pathogen are several and complex interaction among the host and microorganisms. These mechanisms are prejudiced by soil type, temperature, pH, and moisture of the plant and soil environment and also by the existence of other microorganisms (Howell, 2003). There are four principle machinery of biological control in plants, namely, antibiosis, competition, mycoparasitism, or lysis and induced resistance (Fravel et al., 2003; Viterbo et al., 2007).

3.5.1 ANTIBIOSIS

Antibiosis is described as the inhibition or devastation of the microorganism by essence such as specific or nonspecific metabolites or by the production of antibiotics that restrain the growth of one more microorganism (Benítez et al., 2004; Haggag and Mohamed, 2007). Many BCAs agents produce several types of antibiotics (Lewis et al., 1989; Handelsman and Stabb, 1996). A little antibiotics have been reported to play function in disease restraint (Lewis et al., 1989) either impede spore germination (fungistasis) or kill the cells (antibiosis) (Benítez et al., 2004; Haggag and Mohamed, 2007). *Gliocladium* and *Trichoderma* species are well-known BCAs which create a wide range of antibiotics and restrain disease by various mechanisms (Whipps, 2001; Harman et al., 2004). Gliovirin metabolites fabricated by *Gliocladium virens* can kill *Pythium ultimum* by causing coagulation of the protoplasm of the cell (Whipps, 2001; Viterbo et al., 2007).

3.5.2 COMPETITION

There are competition which occurs between microorganisms when space and nutrients are a limiting factor from the host (Lewis et al., 1989; Viterbo et al., 2007). The rhizosphere of the plant and soil zone is a main concern where competition for space and nutrient occurs mostly (Viterbo et al., 2007). Competition between the BCA and the pathogen can outcome in displacement of the pathogen. BCAs can contend with other fungi for food and essential elements in the soil and around the rhizosphere (Chet et al., 1990; Irtwange, 2006) and can inclusive the space or change the rhizosphere by acidifying the soil, so that pathogens cannot nurture (Benítez et al., 2004). For example, *Trichoderma harzianum* T-35 control of *Fusarium* species on diverse crops arises via competition for nutrients and rhizosphere colonization (Viterbo et al., 2007).

3.5.3 MYCOPARASITISM

Mycoparasitism involves the complex process that includes the following steps: (1) the chemothophic growth of the antagonist to the host; (2) recognition of the host by mycoparasite; (3) attachment; (4) excretion of extracellular enzymes; and (5) lysis and exploitation of the host (Whipps, 2001; Benítez et al., 2004; Viterbo et al., 2007). BCAs are able to lyse hyphae of pathogens by releasing the lytic enzymes and this is an important and powerful tool for control of plant disease (Chet et al., 1990, Viterbo et al., 2007) such as chitinases, proteases, and β-1, 3 glucanases (Whipps, 2001). For example, β-1, 3 glucanases produced from *Chaetomium* sp. can degrade cell walls of plant pathogens including *Rhizoctonia solani*, *Gibberella zeae*, *Fusarium* sp. *Colletotrichum gloeosporioides*, and *Phoma* sp. (Sun et al., 2006). Proteases produced by *Trichoderma harzianum* T-39 are involved in the degradation of pathogen hyphal membranes and cell walls.

3.5.4 INDUCED RESISTANCE

Induced resistance reported in large number of plants in response to infestation by phytopathogens (Harman et al., 2004). Induced resistance of host plants can be restricted and/or systematic, depending on the nature, source, and amount of stimuli of pathogens (Pal and Gardener, 2006). Induced resistance by BCAs entails the similar suite of genes and gene products engaged in plant response known as systematic acquired resistance (SAR) (Handelsman and Stabb, 1996; Whipps, 2001). *Trichoderma* strains are proficient of creating interaction-induced metabolic changes in plants that enhance resistance to a broad range of plant pathogenic fungi (Harman et al., 2004).

3.6 BIOHERBICIDES

Phytopathogenic microorganisms or microbial phytotoxins are valuable source for biological weed management and applied in related ways to conventional herbicides (Goeden, 1999; Boyetchko et al., 2002; Boyetchko and Peng, 2004). Bioherbicides serves additional significant role as a complimentary component in flourishing integrated management strategies (Hoagland et al., 2007), and not as a replacement for chemical herbicides and other weed management tactics (Singh et al., 2006). There have been several microbial components under evaluation for their prospective as bioherbicides with horticultural crops, turf, and forest trees, including obligate fungal parasites, soil-borne fungal pathogens, nonphytopathogenic fungi, pathogenic and nonpathogenic bacteria, and nematodes (Kremer, 2005). DeVine (Encore Technologies, Plymouth, MN, USA) is one of the first herbicides registered with the active ingredient *Phytophthora palmivora*, which was developed to manage strangler vine (Morrenia odorata) on citrus in Florida (Charudattan, 2005, Table 3.3). Plant pathogens are used as BCAs and can cause brutal damage to target weed species. In order to become appropriate pathogens, they have to be mass produced and their pathogenicity tested on weeds in a variety of environmental conditions, followed by field efficacy and host range tests (Ayres and Paul, 1990). A range of phytotoxins produced by plant pathogens can obstruct with plant metabolism, sorting from delicate effects on gene expression to plant mortality (Walton, 1996). A few fungal pathogens are toxic to a broad range of weed species. The early mycoherbicides ("DeVine," "Collego" with the active ingredient *Colletotrichum gloeosporioides* f. sp. *aeschynomene*, "Biomal" with the active ingredient *Colletotrichum gloeosporioides*) had extremely virulent fungal plant pathogens that could be mass cultured to produce large quantities of inoculum for inundative application to the weed host. These fungi infect the aerial portion of weed hosts, effecting in noticeable disease symptoms (Charudattan, 2005).

The rust fungus *Puccinia canaliculata* is a foliar pathogen of yellow nutsedge (*Cyperus esculentus*), and it can be mass cultured on the weed host in little field plots or in greenhouse condition (Phatak et al., 1983). Application of fungal pathogen *Chonrotereum purpureum* to wounded branches or stumps of weedy tree species inhibited resprouting and corroded the woody tissues of the applied plant (Prasad, 1996). Weidemann et al. (1992) accounted that the fungal pathogen *Microsphaeropsis amaranthi* controlled few pigweed (Amaranthus) species, while *Phoma proboscis* controlled field bindweed (*Convolvulus arvensis*) and *Colletotrichum*

TABLE 3.3 Fungal Bioherbicides/Mycoherbicides Developed or Being Developed for the Biological Control of Weeds.

Sl. no.	Fungus	Product name	Target weed	Supplier or countries, where registered
1.	*Cercospora rodmanii*	"ABG 5003"	Water hyacinth (*Eichhornia crassipes*)	Abbott Labs, USA
2.	*Chondrosterium purpureum*	BioChon	Black cherry (*Prunus serotina*) in forestry	Koppert, The Netherlands
3.	*Colletotrichum gloeosporioides* f. sp. *Malvae*	Biomal	Mallow (*Malva pusilla*) in wheat and lentils	Canada
4.	*Alternaria cassiae*	Casst	sicklepod (*Cassia obtusifolia*) and coffee senna (*C. occidentalis*) in soybeans and peanuts	USA
5.	*Colletotrichum gloeosporioides* f. sp *aeschynomene*	Collego	Northern jointvetch (*Aeschynomene virginica*) in rice	Encore Technologies, USA
6.	*Phytophthora palmivora*	Devine	Milkweed vine (*Morrenia odorata*) in Florida citrus	Sumitomo, Valent, USA
7.	*Colletotrichum gloeosporioides* f. sp *cuscutae*	Luboa 2	*Cuscuta chinensis, C. australis* in soybeans	PR China
8.	*Colletotrichum coccodes*	Velgo	Velvetleaf (*Abutilon theophrasti*) in corn and soybeans	USA, Canada

capsici controlled morning glory (*Ipomoea spp.*). The naturally occurring fungus *Phoma macrostoma* has been considered as control agent for dandelion (*Taraxacum officinale*), Canada thistle (*Cirsium arvense*), chickweed (*Stellaria media*), and scentless chamomile (*Matricaria perforata*), and its outcome is comparable to the industry standard synthetic herbicide pendimethalin (Bailey and Derby, 2001). A study by Héraux et al. (2005) on *Trichoderma virens* (*Gliocladium virens*) reported that colonization of composted chicken manure significantly reduced the emergence and growth of redroot pigweed (*Amaranthus retroflexus*) and broadleaf weeds in fields of horticulture crops.

3.7 LIMITATION IN SUCCESSFUL UTILIZATION OF FUNGAL BIOPESTICIDES

Billions of dollars has been invested by the biopesticide companies for the development of a variety of microbial products so as to eliminate crop diseases. It is impossible to blatantly tell the market trends for biopesticides, and there is a substantial difference in both predication of global sales and selecting category of biopesticides. Agriculture market is observing an enhancement in demand for environment friendly, chemical residue-free organic products. Growth in several regions is however hindered due to vigorous established chemical pesticide markets. However, there is lack of awareness about benefits of biopesticides, and uneven efficiency of biopesticides. The lack of awareness, knowledge, and confidence in farmers is one of the major reasons for the lagging of these ecofriendly pest control options. Commercial biological control absorbing fungal biopesticides in the present global scenario is a hi-tech venture both in terms of safety and sustenance. The viability and virulence of fungal inoculum (conidia) after field application is the prerequisite threshold for their efficacy (Doust and Roberts, 1983). Diverse isolates of *B. bassiana* and *M. anisopliae* have been the most entrusted entomopathogens that have been greatly researched and find considerable effects on field tests and their commercial usage for insect pest management (Easwaramoorthy, 2003). Upon field application, the entomopathogens are exposed to a collection of abiotic stresses like temperature (Rangel et al., 2005a), UV radiations (Rangel et al., 2006a), humidity–osmolarity (Lazzarini, 2006), edaphic factors, and also on nutrient source (Shah, 2005) that negatively affect the field use of fungi as BCAs. The solar radiation, which includes visible light, UV radiations, infrared rays, and radio waves have been the dominant source in which all organisms are evolved and adapted. In biological context, the UV radiations acclaim a special mention in terms of their impact on life (Bjorn, 2006). Soil temperature is a major factor, which affects the success or failure in the establishment and production of fungal inoculum (Thomas and Jenkins, 1997). The entomopathogenic fungi have tolerance to the soil temperature and also have capacity to survive through thermoregulatory defence response of the host insect (Ouedraogo et al., 2003). It has been confirmed that high/low temperature changes the vegetative growth among isolates of entomopathogenic fungi (Ouedraogo et al., 2004). Dry heat exposure causes DNA breakage through base loss leading to depurination and this may cause mutation in several cases (Nicholson et al., 2000). Wet heat, that is, heat in conjunction with high humidity results in protein denaturation and membrane disorganization of the cell. It has been accounted that *M. anisopliae* has temperature tolerance upper limit as 37–40°C temperature (Thomas and Jenkins, 1997). *B. bassiana* on the other hand can survive up to a maximum temperature of 37°C temperature (Fargues et al., 1997). In fungi, the temperature range for germination and mycelial growth has been reported to be in similar pattern. Environmental factors also affect pathogenicity as well as mode of virulence of entomopathogenic fungi (Hasan, 2014).

In case of fungal bioherbicides, the efficacy of bioherbicides is the major limiting feature for their use, often due to environmental factors. There are needs of humidity for the establishment and spread of many foliar and stem fungal pathogens for weed control. There are also few necessary elements for the development of special formulations to ensure the effectiveness of agents applied in the field. An extended dew period is needed by few pathogens for infection on the aerial surfaces of target weeds (Auld et al., 2003). The bioherbicide application process should be judged for enhancing efficacy of the BCA. These include attention to spray droplet size, droplet retention and distribution, spray application volume, and the equipment used (Charudattan, 2001). Other factors, such as the spectrum of the bioherbicide, whether broad or targeted to specific species, the type of formulation, and if it involves amino acid-excreting strains, can significantly affect efficacy. Broad-spectrum and Abbas, 1994).

3.8 CONCLUSION

The relevance of fungal biopesticides in biological control is increasing largely because of superior environmental awareness, food safety concerns, and the breakdown of conventional chemicals due to an increasing number of insecticide-resistant species. In formative, whether the use of fungal biopesticides have been thriving in pest and disease management, it is necessary to regard as each case individually, and direct comparisons with chemical insecticides are usually unsuitable. Fungal biopesticides being component of an integrated advancement that can offer significant and selective insect and disease control. In the near future, we suppose to see synergistic combinations of microbial control agents with other technologies (in combination with semiochemicals, soft chemical pesticides, fungicides, other natural enemies, resistant plants, chemigation, remote sensing, etc.) that will augment the effectiveness and sustainability of integrated control strategies of plant disease.

KEYWORDS

- **fungal biopesticides**
- **pest control**
- **ecofriendly approach**
- **insect pest**
- **agriculture**

REFERENCES

Abdel-Fattah, M. G.; Shabana, M. Y.; Ismail, E. A.; Rashad, M. Y. *Trichoderma harzianum*: a Biocontrol Agent Against *Bipolaris oryzae*. *Mycopathologia* **2007**, *164*, 81–89.

Aggarwal, R.; Tewari, A. K.; Srivastava, K. D.; Singh, D. V. Role of Antibiosis in the Biological Control of Spot Blotch (*Cochliobolus sativus*) of Wheat by Chaetomium Globosum. *Mycopathologia* **2004**, *157*, 369–377.

Anand, S.; Reddy, J. Biocontrol Potential of *Trichoderma* Sp Against Plant Pathogens. *Inter. J. Agri. Sci.* **2009**, *2*, 30–39.

Auld, B. A.; Hethering, S. D.; Smith, H. E. Advances in Bioherbicide Formulation. *Weed Biol. Man.* **2003**, *3*, 61–67.

Ayres, P.; Paul, N. Weeding with Fungi. *New Sci.* **1990**, *732*, 36–39.

Bailey, K. L.; Derby, J. Fungal Isolates and Biological Control Compositions for the Control of Weeds. U.S. Patent 60/294,475, May 20, 2001.

Bailey, D. J.; Gilligan, C. A. Modeling and Analysis of Disease Induced Host Growth in the Epidemiology of Take All. *Phytopathology* **2004**, *94*, 535–540.

Benítez, T.; Rincón, M. A.; Limón, M. C.; Codón, C. A. Biocontrol Mechanisms of *Trichoderma* Strains. *Int. Microbiol.* **2004**, *7*, 249–260.

Bidochka, M. J.; St Leger, R. J.; Joshi, L.; Roberts, D. W. An Inner Cell Wall Protein (Cwp1) From Conidia of the Entomopathogenic Fungus *Beauveria bassiana*. *Microbiology* **1995,** *141*, 1075–1080.

Bjorn, L. O. Stratospheric Ozone, Ultraviolet Radiation, and Cryptogams. *Biol. Conservation* **2006,** *135*, 326–333.

Boucias, D. R.; Pendland, J. C. Entomopathogenic Fungi: Fungi Imperfecti. In *Principles Of Insect Pathology*; Boucias, D. R., Pendland, J. C., Eds; Kluwer Academic Publishers: Dordrecht, 1998, pp 321–359.

Boyetchko, S. M.; Rosskopf, E. N.; Caesar, A. J. Charudattan, R. Biological Weed Control with Pathogens: Search for Candidates to Applications. In *Applied Mycology and Biotechnology*; Khachatourians, G. G., Arora, D. K. Eds.; Elsevier: Amsterdam, 2002; pp 239–274, Vol. 2.

Boyetchko, S. M.; Peng, G. Challenges and Strategies for Development of Mycoherbicides. In *Fungal Biotechnology in Agricultural, Food, and Environmental Applications*; Arora, D. K., Ed., Marcel Dekker: NewYork, 2004; pp 11–121.

Boyette, C. D.; Abbas, H. K. Host Range Alteration of the Bioherbicidal Fungus Alternaria Crassa with Fruit Pectin and Plant Filtrates. *Weed Sci.* **1994,** *42*, 487–491.

Bruce, A.; Highley, L. T. Control of Growth of Wood Decay Basidiomycetes by *Trichoderma* Spp. and Other Potentially Antagonistic Fungi. *Forest Products J.* **1991,** *41*, 63–67.

Butt, T. M.; Jackson, C. W.; Magan, N., Eds. *Fungal Biological Control Agents: Progress, Problems and Potential.* CABI International: Wallingford, Oxon, UK, 2001.

Charnley, A. K. Mechanisms of Fungal Pathogenesis in Insects. In *The Biotechnology of Fungi for Improving Plant Growth*; Whipps, J. M., Lumsden, R. D., Eds.; Cambridge University: London, 1989; pp 85–125.

Charudattan, R. Biological Control of Weeds by Means of Plants Pathogens: Significance for Integrated Weed Management in Modern Agroecology. *Biocontrol* **2001,** *46*, 229–260.

Charudattan, R. Use of Plant Pathogens as Bioherbicides to Manage Weeds in Horticultural Crops. *Proc. Fla. State Hort. Soc.* **2005,** *118*, 208–214.

Chet, I.; Ordentlich, A.; Shapira, R.; Oppenheim, A. Mechanisms of Biocontrol of Soil-borne Plant Pathogens by *Rhizobacteria*. *Plant Soil* **1990,** *129*, 85–92.

Cho, E. M.; Boucias, D.; Keyhani, N. O. EST Analysis of Cdna Libraries from the Entomopathogenic Fungus *Beauveria* (Cordyceps) *Bassiana*. II. Fungal Cells Sporulating on Chitin and Producing Oosporein. *Microbiology* **2006,** *152*, 2855–2864.

Clarkson, J. M.; Charnley, A. K. New Insights into the Mechanisms of Fungal Pathogenesis in Insects. *Trends Microbiol.* **1996,** *4*, 197–203.

Dominguesa, F. C.; Queiroza, J. A.; Cabralb, J. M. S.; Fonsecab, L. P. The Influence of Culture Conditions on Mycelial Structure and Cellulose Production by *Trichoderma reesei* rut C-30. *Enz. Microb. Technol.* **2000,** *26*, 394–401.

Doust, R. A.; Roberts, D. W. Studies on the Prolonged Storage of *Metarhizium Anisopliae* Conidia: Effect of Temperature and Relative Humidity on Conidial Viability and Virulence Against Mosquitoes. *J. Invertebrate Pathol.* **1983,** *41*, 143–150.

Driver, F.; Milner, R. J.; Trueman, J. W. H. A Taxonomic Revision of *Metarhizium* Based on a Phylogenetic Analysis of RDNA Sequence Data. *Mycol. Res.* **2000,** *104*, 134–150.

Dumm, M. T.; Sayee, R. M.; Carrell, A.; Wergin, W. R. Colonization of Nematodes Eggs by *P. lilacinus* Samson of Observed with SEM. *SEM* **1982,** *3*, 1351–1357.

Easwaramoorthy, S. Entomopathogenic Fungi. In *Biopesticides and Bioagents in Integrated Pest Management of Agricultural Crops*; Srivastava, R. P. Ed., International Book Distributing Co.: India, 2003; pp 341–379.

Fargues, J.; Goettel, M. S.; Smits, N.; Ouedraogo, A.; Rougier, M. Effect of Temperature on Vegetative Growth of *Beauveria Bassiana* Isolates from Different Origins. *Mycologia* **1997**, *89*, 383–382.

Ferron, P. Biological Control of Insect Pests by Entomologenous Fungi. *Ann. Rev. Entomol.* **1978**, *23*, 409–442.

Fravel, D.; Olivain, C.; Alabouvette, C. *Fusarium oxysporum* and Its Biocontrol. *New Phytologist* **2003**, *157*, 493–502.

Furlong, M. J.; Pell, K. J. Interactions Between Entomopathogenic Fungi and Arthropod Natural Enemies. In *Insect–fungal Associations: Ecology and Evolution*; Vega, F. E.; Blackwell, M., Eds.; Oxford University Press: England, UK, 2005; pp 51–73.

Goeden, R. D. Projects on Biological Control of Russian Thistle and Milk Thistle in California: Failures That Contributed to the Science of Biological Weed Control. In *Abstracts of the 10th International Symposium on Biological Control of Weeds*; Spencer, N., Noweierski, R., Eds.; Montana State University: Bozeman, MT, USA, 1999; p 27.

Goettel, M. S.; Eilenberg, J.; Glare, T. R. Entomopathogenic Fungi and Their Role in Regulation of Insect Populations. In *Comprehensive Molecular Insect Science*; Gilbert, L. I., Iatrou, K., Gill, S., Eds.; Elsevier: Amsterdam, Netherlands, 2005; pp 361–406.

Guillon, M. L. Regulation of Biological Control Agents in Europe. In International Symposium on Biopesticides for Developing Countries, Roettger, U., Reinhold, M., Eds.; CATIE: Turrialba, 2003; pp 143–147.

Haggag, W. M.; Mohamed, H. A. A. Biotechnological Aspects of Microorganisms Used in Plant Biological Control. *Am. Eur. J. Sustainable Agric.* **2007**, *1*, 7–12.

Hajek, A. E.; St. Leger, R. J. Interactions Between Fungal Pathogens and Insects Hosts. *Ann. Rev. Entomol.* **1994**, *39*, 293–322.

Hamlen, R. A. Biological Control of Insects and Mites on European Greenhouse Crops: Research and Commercial Implementation. *Proc. Florida State Horticultural Soc.* **1979**, *92*, 367–368.

Handelsman, J.; Stabb, V. E. Biocontrol of Soilborne Plant Pathogens. *Plant Cell* **1996**, *8*, 1855–1869.

Harman, G. E. Overview of Mechanisms and Uses of *Trichoderma* spp. 648. *Phytopathology* **2005**, *96*, 190–194.

Harman, G. E.; Howell, C. R.; Viterbo, A.; Chet, I.; Lorito, M. *Trichoderma* Species Opportunistic, Avirulent Plant Symbionts. *Nat. Rev. Microbiol.* **2004**, *2*, 43–56.

Hasan, S. Entomopathogenic Fungi as Potent Agents of Biological Control. *Int. J. Eng. Technol. Res.* **2014**, *2*, 234–237.

Hebbar, P. K.; Lumsden, R. D. Biological Control of Seedling Diseases. In *Methods in Biotechnology Vol. 5: Biopesticides: Use and Delivery*; Frinklin, R. H., Julius, J. M., Eds.; Humana Press: New York, 1999; pp 103–116.

Héraux, F. M. G.; Hallett, S. G.; Ragothama, K. G.; Weller, S. C. Composted Chicken Manure as a Medium for the Production and Delivery of *Trichoderma virens* for Weed Control. *Hort. Sci.* **2005**, *40*, 1394–1397.

Hoagland, R. E.; Weaver, M. A.; Boyette, C. D. Myrothecium Verrucaria Fungus: A Bioherbicide and Strategies to Reduce Its Non-target Risks. *Allelopathy J.* **2007**, *19* (1), 179–192.

Hong, T. D.; Ellis, R. H.; Moore, D. Development of a Model to Predict the Effect of Temperature and Moisture on Fungal Spore Longevity. *Ann. Botany* **1997**, *79*, 121–128.

Howell, R. C. Mechanisms Employed by *Trichoderma* Species in the Biological Control of Plant Diseases: The History and Evolution of Current Concepts. *Plant Dis.* **2003**, *87*, 4–10.

Ibrahim, I. K. A.; Raza, M. A.; El-Saedy, M. A.; Ibrahim, A. A. M. Control of *Meloidogyne incognita* on Corn, Tomato and Okra with *P. lilacinus* and the Nematicide Aldecarb. *Nematologia Mediterria* **1987**, *15*, 265–268.

Ignoffo, C. M. The Fungus *Nomuraea rileyi* as a Microbial Insecticide. In *Microbial Control of Pests and Plant Diseases*; Burges, H. D., Ed., Academic Press: London, UK, 1981, 513–538.

IOBC. International Organization for Biological Control. *IOBC Newslet.* **2008**, *84*, 5–7.

Irtwange, V. S. Application of Biological Control Agents in Pre- and Postharvest Operations. Agricultural Engineering International: The CIGR Ejournal. *Invited Overview*. **2006**, *3*, 1–12.

Jain, N.; Rana, I. S.; Kanojiya, A.; Sandhu, S. S. Characterization of *Beaveria bassiana* Strains Based on Protease and Lipase Activity and Their Role in Pathogenicity. *J. Basic Appl. Mycol.* **2008,** *I-II,* 18–22.

Jianzhong, S.; Fuxa, J. R.; Henderson, G. Effects of Virulence, Sporulation, and Temperature on *Metarhizium anisopliae* and *Beauveria bassiana* Laboratory Transmission in *Coptotermes formosanus. J. Invertebrate Pathol.* **2003**, *84*, 38–46.

Jones, E. E.; Mead, A.; Whipps, J. M. Effect of Inoculum Type and Timing of Application of *Coniothyrium Minitans* on *Sclerotinia Sclerotio46 Rum*: Control of Sclerotinia Disease in Glasshouse Lettuce. *Plant Pathol.* **2004**, *53*, 611–620.

Khachatourians, G. G. Biochemistry and Molecular Biology of Entomopathogenic Fungi. In *Human and Animal Relationships*; Howard, D. H.; Miller, J. D., Eds, *Mycota* VI, Springer: Heidelberg, 1996, pp 331–363.

Khandelwal, M.; Datta, S.; Mehta, J.; Naruka, R.; Makhijani, K.; Sharma, G.; Kumar, R.; Chandra, S. Isolation, Characterization and Biomass Production of *Trichoderma Viride* Using Various Agro Products—A Biocontrol Agent. *Adv. Appl. Sci. Res.* **2012**, *3*, 3950–3955.

Khetan, S. K. *Microbial Pest Control.* Marcel Dekker Inc.: New York, Basel, 2001; p 300.

Kim, J. J.; Lee, M. H.; Yoon, C. S.; Kim, H. S.; Yoo, J. K.; Kim, K. C. Control of Cotton Aphid and Greenhouse Whitefly with a Fungal Pathogen. *J. Nat. Inst. Agric. Sci. Technol.* **2002**, 7–14.

Kirk, P. M. *Indexfungorum,* 2003. http://www.indexfungorum.org (accesses Oct 28, 2009).

Kiss, L. A Review of Fungal Antagonists of Powdery Mildews and Their Potential as Biocontrol Agents. *Pest Management Sci.* **2003**, *59*, 475–483.

Köhl, J.; Bélanger, R. R.; Fokkema, N. J. Interaction of Four Antagonistic Fungi with *Botrytis Aclada* in Dead Onion Leaves: A Comparative Microscopic and Ultrastructural Study. *Phytopathology* **1997**, *87*, 634–642.

Kremer, R. J. The Role of Bioherbicides in Weed Management. *Biopestic. Int.* **2005**, *1*, 127–141.

Kvas, M.; Marasas, W. F. O.; Wingfield, B. D.; Wingfield, M. J.; Steenkamp, E. T. Diversity and Evolution of *Fusarium* Species in the Gibberella Fujikuroi Complex. *Fungal Diversity* **2009**, *34*, 1–21.

Lacey, C. M.; Lacey, L. M.; Roberts, D. R. Route of Invasion and Histopathology of *Metrahizium anisopliae* in *Culex quinquefasciatus. J. Invertebrate Pathol.* **1988**, *52*, 108–118.

Lazzarini, G. M. J.; Rocha, L. F. N.; Luz, C. Impact of Moisture on In Vitro Germination of *Metarhizium anisopliae* and *Beauveria bassiana* and Their Activity on *Triatoma infestans. Mycol. Res.* **2006**, *100*, 485–492.

Le Floch, G.; Rey, P.; Benizri, E.; Benhamou, N.; Tirilly, Y. Impact of Auxin-compounds Produced by the Antagonistic Fungus *Pythium Oligandrum* or the Minor Pathogen *Pythium* Group F on Plant Growth. *Plant Soil* **2003**, *257*, 459–470.

Leger, R. J.; Joshi, L.; Bidochka, M. J.; Roberts, D. W. Construction of an Improved Mycoinsecticide Overexpressing a Toxic Protease. *Proc. Natl. Acad. Sci. U. S. Am.* **1996**, *93*, 6349–6354.

Lewis, K.; Whipps, J. M.; Cooke, R. C. Mechanisms of Biological Disease Control with Special Reference to the Case Study of Pytium Oligandrum as an Antagonist. In *Biotechnology of Fungi for Improving Plant Growth*; Whipps, J. M., Lumsden, R. D., Eds.; Cambridge University Press: Cambridge, UK, 1989; pp 191–217.

Li, G. Q.; Huang, H. C.; Acharya, S. N.; Erickson, R. S. Effectiveness of *Coniothyrium minitans* and *Trichoderma atroviride* in Suppression of Sclerotinia Blossom Blight of Alfalfa. *Plant Pathol.* **2005,** *54,* 204–211.

Li, G. Q.; Huang, H. C.; Miao, H. J.; Erickson, R. S.; Jiang, D. H.; Xiao, Y. N. Biological Control of Sclerotinia Diseases of Rapeseed by Aerial Applications of the Mycoparasite *Coniothyrium minitans. Eur. J. Plant Pathol.* **2006,** *114,* 345–355.

Liang, C.; Yang, J.; Kovács, G. M.; Szentiványi, O.; Li, B.; Xu, X. M.; Kiss, L. Genetic Diversity of *Ampelomyces Mycoparasites* Isolated From Different Powdery Mildew Species in China Inferred from Analyses of RDNA ITS Sequences. *Fungal Diversity* **2007,** *24,* 225–240.

Marwah, R. G.; Fatope, M. O.; Deadman, M. L.; Al-Maqbali, Y. M.; Husband, J. Musanahol: A New Aureonitol-related Metabolite from a *Chaetomium* sp. *Tetrahedron* **2007,** *63,* 8174–8180.

Meadow, R.; Vandberg, J. D.; Shelton, A. M. Exchange of Inoculum of *Beauveria bassiana* (Bals) Vuil. (Hyphomycetes) Between Adult Flies of the Cabbage Maggot *Delia radicum* L. (Diptera: Anthomyiidae). *Biocontrol Sci. Technol.* **2000,** *10,* :479–485.

Mitsuaka, S. Effect of Temperature on Growth of *Beauveria Bassiana* F-263, A Strain Highly Virulent to the Japanese Pine Sawyer, *Monochamus alternatus,* Especially Tolerance to High Temperatures. *Appl. Entomol. Zool.* **2004,** *39,* 469–475.

Nicholson, W. L.; Munakata, N.; Horneck, G.; Melosh, H. J.; Setlow, P. Resistance of *Bacillus* Endospores to Extreme Terrestrial and Extraterrestrial Environments. *Microbiol. Mol. Biol. Rev.* **2000,** *64,* 548–572.

Noma, T.; Strickler, K. Effects of *Beauveria bassiana* on Lygus *Hesperus* (Hemiptera: Miridae) Feeding and Oviposition. *Environ, Entmol.* **2000,** *29,* 394–402.

Nunez, E. J.; Iannacone; Omez, H. G. Effect of Two Entomopathogenic Fungi in Controlling *Aleurodicus cocois* (Curtis, 1846) (Hemiptera: Aleyrodidae). *Chilean J. Agric. Res.* **2008,** *68,* 21–30.

O'Brien, K. P.; Franjevic, S.; Jones, J. Green Chemistry and Sustainable Agriculture: The Role of Biopesticides. *Advancing Green Chem.* **2009.** http://advancinggreenchemistry.org/wp-content/uploads/Green-Chemand- Sus.-Ag.-the-Role-of-Biopesticides.pdf. (accessed May 7, 2016).

Ouedraogo, R. M.; Cusson, M.; Goettel, M. S.; Brodeur, J. Inhibition of Fungal Growth in Thermoregulating Locusts, *Locusta Migratoria*, Infected by the Fungus *Metarhizium anisopliae* var. *acridum. J. Invertebr. Pathol.* **2003,** *82,* 103–109.

Ouedraogo, A.; Fargues, J.; Goettel, M. S.; Lomer, C. J. Effect of Temperature on Vegetative Growth Among Isolates of *Metarhizium anisopliae* and *Metarhizium flavoviride. Mycopathologia* **2004,** *137,* 37–43.

Pal, K.; Gardener, B. M. Biological Control of Plant Pathogens. *Plant Health Instructor* **2006,** 1–25. doi: 10.1094/PHI-A-2006-1117-02. APSnet.

Park, J. H.; Choi, G. J.; Jang, S. K.; Lim, K. H.; Kim, T. H.; Cho, Y. K.; Kim, J. C. Antifungal Activity Against Plant Pathogenic Fungi of Chaetoviridins Isolated from *Chaetomium globosum. FEMS Microbiol. Lett.* **2005,** *252,* 309–313.

Partridge, D. E.; Sutton, T. B.; Jordan, D. L.; Curtis, V. L.; Bailey, J. E. Management of Sclerotinia Blight of Peanut with the Biological Control Agent *Coniothyrium minitans*. *Plant Dis.* **2006**, *90*, 957–963.

Paulitz, T. C.; Belanger, R. R. Biological Control in Greenhouse System. *Ann. Rev. Phytopathol.* **2001**, *39*, 103–133.

Phatak, S. C.; Summer, D. R.; Wells, H. D.; Bell, D. K.; Glaze, N. C. Biological Control of Yellow Nutsedge with the Indigenous Rust fungus *Puccinia canaliculata*. *Science* **1983**, *219*, 1446–1447.

Prasad, R. In *Development of Bioherbicides for Integrated Weed Management in Forestry*. Proceedings of the 2nd International Weed Control Congress; Brown, H., Ed., Department of Weed Control and Pesticide Ecology, Slagelse, Denmark, 25–28 June 1996; pp 1197–1203.

Punja, Z. K.; Utkhede, R. S. Biological Control of Fungal Diseases on Vegetable Crops with Fungi and Yeasts. In *Fungal Biotechnology in Agricultural, Food, and Environmental Applications*; Arora, D. K., Ed.; New York, Basel, 2004; pp 157–171.

Quesada-Moraga, E.; Santos-Quirós, R.; Valverde-García, P.; Santiago, Á. C. Virulence, Horizontal Transmission, and Sublethal Reproductive Effects of *Metarhizium Anisopliae* (Anamorphic Fungi) on the German Cockroach (Blattodea: Blattellidae). *J. Invertebr. Pathol.* **2004**, *87*, 51–58.

Rai, D.; Updhyay, V.; Mehra, P.; Rana, M.; Pandey, A. K. Potential of Entomopathogenic Fungi as Biopesticides. *Ind. J. Sci. Res. Tech.* **2014**, *2* (5), 7–13.

Rangel, D. E. N.; Braga, G. U. L.; Anderson, A. J.; Roberts, D. W. Variability in Conidial Thermo Tolerance of *Metarhizium Anisopliae* Isolates from Different Geographic Origins. *J. Invertebr. Pathol.* **2005a**, *88*, 116–125.

Rangel, D. E. N.; Butler, M. J.; Torabinejad, J.; Anderson, A. J.; Braga, G. U. L.; Day, A. W.; Roberts, D. W. Mutants and Isolates of *Metarhizium Anisopliae* are Diverse in Their Relationships Between Conidial Pigmentation and Stress Tolerance. *J. Invertebr. Pathol* **2006a**, *93*, 170–182.

Richard, J. S.; Neal, T. D.; Karl, J. K.; Michael, R. K. Model Reactions For Insect Cuticle Sclerotization: Participation of Amino Groups in the Cross-Linking of *Manduca Sexta* Cuticle Protein MsCP36. *Insect Biochem. Mol. Biol.* **2010**, *40*, 252–258.

Roberts, D. W. Toxins of Entomopathogenic Fungi. In *Microbial Control of Pests and Plant Diseases*; Burges, H. D., Ed.; Academic Press: New York, 1981; pp 441–464.

Sandhu, S. S.; Vikrant, P. Myco-insecticides: Control of Insect Pests. In *Microbial Diversity: Opportunities & Challenges*; Gautam, S. P.; Sandhu, S. S., Sharma, A., Pandey, A. K., Eds.; Indica Publishers: New Delhi, India, 2004.

Shah, F. A.; Wang, C. S.; Butt, T. M. Nutrition Influences Growth and Virulence of the Insect Pathogenic Fungus *Metarhizium anisopliae*. *Microbiol. Lett.* **2005**, *251*, 259–266.

Singh, H. P.; Batish, D. R.; Kohli, R. K. *Handbook of Sustainable Weed Management;* Food Products Press: Binghamton, NY, 2006.

Sosa-Gomez, D. R.; Moscardi, F. Laboratory and Field Studies on the Infection of Stink Bugs, Nezara viridula, Piezodorus guildinii, and Euschistus heros (Hemiptera: Pentatomidae) with Metarhizium anisopliae and *Beauveria bassianain*. *Brazil J. Invertebr. Pathol.* **1998**, *2*, 115–120.

Soytong, K.; Kanokmadhakul, S.; Kukongviriyapa, V.; Isobe, M. Application of *Chaetomium* Species (Ketomium®) As a New Broad Spectrum Biological Fungicide for Plant Disease Control: A Review Article. *Fungal Diversity* **2001**, *7*, 1–15.

St Leger, R. J.; Bidochka, M. J. Staples, R. C. Preparation Events During Infection of Host Cuticle by *Metarhizium anisopliae. J. Invertebrate Pathol.* **1991,** *58,* 168–179.

St Leger, R. J.; May, B.; Allee, L. L.; Frank, D. C.; Staples, R. C.; Roberts, D. W. Genetic Differences in Allozymes and in Formation of Infection Structures Among Isolates of the Entomopathogenic Fungus *Metarhizium anisopliae. J. Invertebrate Pathol.* **1992,** *60,* 89–101.

St Leger, R. J.; Joshi, L.; Bidochka, M. J.; Roberts, D. W. Construction of an Improved Mycoinsecticide Overexpressing a Toxic Protease. *Proc. Natl. Acad. Sci. U. S. A.* **1996,** *93,* 6349–6354.

Sun, H.; Yang, J.; Lin, C.; Huang, X.; Xing, R.; Zhang, K. Q. Purification and Properties of a B-1,3-Glucanase from *Chaetomium* Sp. that is Involved in Mycoparasitism. *Biotechnol. Lett.* **2006,** *28,* 131–135.

Sunantapongsuk, V.; Nakapraves, P.; Piriyaprin, S.; Manoch, L. In *Protease Production and Phosphate Solubilization from Potential Biological Control Agents Trichoderma Viride and Azomonas Agilis from Vetiver Rhizosphere,* International Workshop on Sustained Managament of Soil-Rhizosphere System for Efficient Crop Production and Fertilizer Use, Land Development Department, Bangkok, Thailand, 2006; pp 1–4.

Szekeres, A.; Kredics, L.; Antal, Z.; Kevei, F.; Manczinger, L. Isolation and Characterization of Protease Overproducing Mutants of *Trichoderma harzianum. Microbiol. Lett.* **2004,** *233,* 215–222.

Thakur, R.; Sandhu, S. S. Distribution, Occurrence and Natural Invertebrate Hosts of Indigenous Entomopathogenic Fungi of Central India. *Ind. J. Microbiol.* **2010,** *50,* 89–96.

Thomas, M. B.; Jenkins, N. E. Effect of Temperature on Growth of *Metarhizium Flavoviride* and Virulence to the Variegated Grasshopper, *Zonocerus variegatus. Mycol. Res.* **1997,** *101,* 1469–1474.

Tomilova, O. G.; Shternshis, M. V. The Effect of a Preparation from *Chaetomium* Fungi on the Growth of Phytopathogenic Fungi. *Appl. Biochem. Microbiol.* **2006,** *42,* 76–80.

Tulloch, M. The Genus *Metarhizium. Trans. Britannic Mycol. Soc.* **1976,** *66,* 407–411.

Tviet, M.; Moor, M. B. Isolates of Chaetomium that Protect Oats from *Helminthosporium victoriae. Phytopathology* **1954,** *44,* 686–689.

USEPA. What Are Biopesticides? 2008. http://www.epa.gov/pesticides/biopesticides/ whatare-biopesticides.htm. (accessed May 7, 2016).

Vey, A.; Hoagland, R.; Butt, T. M. Toxic Metabolites of Fungal Biocontrol Agents. In *Fungi as Biocontrol Agents*; Butt, T. M., Jackson, C. W, Magan, N., Eds.; CAB International: Wallingford, 2001; pp 311–345.

Vinale, F.; Marra, R.; Scala, F.; Ghisalbert, E. L.; Lorito, M.; Sivasithamparam, K. Major Secondary Metabolotes Produced by Two Commercial *Trichoderma* Strains Active Against Different Phytopathogens. *Lett. Appl. Microbiol.* **2006,** *43,* 143–148.

Vinale, F.; Sivasithamparam, K.; Ghisalberti, E. L.; Marra, R.; Woo, S. L.; Lorito, M. Trichoderma Plant Pathogen Interactions. *Soil Biol. Biochem.* **2008,** *40,* 1–10.

Viterbo, A.; Inbar, J.; Hadar, Y.; Chet, I. Plant Disease Biocontrol and Induced Resistance via Fungal Mycoparasites. In *Environmental and Microbial Relationships*, 2nd edn, *The Mycota IV*; Kubicek, C. P., Druzhinina, I. S., Eds.; Springer-Verlag: Berlin, Heidelberg, 2007; pp 127–146.

Walton, J. D. Host-selective Toxins: Agents of Compatibility. *Plant Cell* **1996,** *8,* 1723–1733.

Wang, C.; Hu, G.; St Leger, R. J. Differential Gene Expression by *Metarhizium Anisopliae* Growing in Root Exudate and Host (*Manduca Sexta*) Cuticle or Hemolymph Reveals Mechanisms of Physiological Adaptation. *Fungal Genetic Biol.* **2005,** *42,* 704–718.

Weidemann, G. J.; TeBeest, D. O.; Templeton, G. E. Fungal Plant Pathogens Used for Biological Weed Control. *Ark. Farming Res.* **1992**, *41*, 6–7.

Whipps, J. M. Microbial Interactions and Biocontrol in the Rhizosphere. *J. Exp. Botany* **2001**, *52*, 487–511.

Whipps, J. M.; Sreenivasaprasad, S.; Muthumeenakshi, S.; Rogers, C. W.; Challen, M. P. Use of *Coniothyrium minitans* as a Biocontrol Agent and Some Molecular Aspects of Sclerotial mycoparasitism. *Eur. J. Plant Pathol.* **2008**, *121*, 323–330.

Woo, L. S.; Lorito, M. Exploiting the Interactions Between Fungal Antagonists, Pathogens and the Plant for Biocontrol. In *Novel Biotechnologies for Biocontrol Agent Enhancement and Management*; Vurro, M., Gressel, J., Eds.; Springer: Germany, 2007; pp 107–130.

Wraight, S. P.; Carruthers, R. I.; Jaronski, S. T.; Bradley, C. A.; Garza, C. J.; Wraight, S. G. Evaluation of the Entomopathogenic Fungi *Beauveria Bassiana* and *Paecilomyces Fumosoroseus* for Microbial Control of the Silver Leaf Whitefly, *Bemisia argentifolii*. *Biol. Control* **2000**, *17*, 203–217.

Wraight, S. P.; Jackson, M. A.; de Kock, S. L. Production, Stabilization and Formulation of Fungi Biocontrol Agents. In *Fungi as Biocontrol Agents Progress, Problem and Potential;* Butt, T. M., Jackson, C. W., Magan, N. Eds.; CABi Publishing: Wallingford, UK, 2001; pp 253–288.

Zeilinger, S.; Omann, M. Trichoderma Biocontrol: Signal Tranduction Pathways Involved In Host Sensing and Mycoparasitism. *Gene Regulation Syst. Biol.* **2007**, *1*, 227–234.

Zhang, H. Y.; Yang, Q. Expressed Sequence Tags-based Indentification of Genes in the Biocontrol Agent *Chaetomium cupreum*. *Appl. Microbiol. Biotechnol.* **2007**, *74*, 650–658.

Zimmermann, G. Review on Safety of the Entomopathogenic Fungus *Beauveria bassiana* and *Beauveria brongniartii*. *Biocontrol Sci. Technol.* **2007**, *17*, 553–596.

CHAPTER 4

Viral Biopesticides: An Effective and Environment-Friendly Approach to Control Insects

NAGA TEJA NATRA[1,*], MAHESH KUMAR[2], KUMARI APURVA[2], and VISHWA VIJAY THAKUR[2]

[1]Department of Plant Pathology, Irrigated Agriculture Research & Extension Center, Washington State University, 24106 N. Bunn Road, Prosser, WA 99350, United States

[2]Department of Molecular Biology and Genetic Engineering, Dr. Kalam Agricultural College, Kishanganj, Bihar Agricultural University, Sabour, Bhagalpur 813210, Bihar, India

[*]Corresponding author. E-mail: tejanatra@gmail.com

ABSTRACT

Viruses are considered as one of the most potential groups of pathogenic microorganisms under consideration for control of biological insect pest. Approximately, more than 1200 virus–host associations have been discussed and majority of them are found in order *Lepidoptera* (83%), Hymenoptera (10%), Diptera (4%), and few examples from Coleoptera, Neuroptera, Orthoptera, etc. Biopesticides are gaining escalating significance as they are alternatives to chemical pesticides and are central component of many pest management programs including bacteria, fungi, entomopathogenic viruses, nematodes, and plant secondary metabolites. The virulence potential of several biopesticides (nuclear polyhedrosis virus (NPV), bacteria, and plant product) was successfully examined under laboratory conditions and the selected ones were also evaluated under field situation with huge promises. In recent years, there has been a large number of agreements between pesticide companies and bioproduct companies that advocate the commercial importance of viral-biopesticides. In comparison to chemical pesticides, viral-biopesticides are more effective, environment friendly and economical, because they have no detrimental residues detected, biodegradable and can be cheaper when locally produced. Three major classes of biopesticides are available such as microbial pesticides consisting of entomopathogenic bacteria (e.g., *Bacillus thuringiensis*), fungi (e.g., *Trichoderma* spp.), or viruses (e.g., Baculovirus) including their metabolites in some cases, entomopathogenic nematodes and protozoa also used. Biopesticides can be applied through various ways such as, augmentative releases, or through conserving existing field populations of natural pest control agents. The main aim of this chapter is to review the important and basic functions of major viral-biopesticides in the past and also discuss the future prospects for the development of new biopesticides.

4.1   INTRODUCTION

To manage the agricultural pest, use of biopesticides has gained importance in recent years because of increased pressure to diminish the use of agrochemicals and their residues within the environment and food. Chemical pest control products are extensively used within all countries of the world but they are regarded as ecologically unsafe. For that reason, there is an expanded social demand to replace them gradually with biopesticides (viral, bacterial, or fungal) which are eco-friendly. The harmful environmental impacts of the synthetic chemicals have forced the search for some other kind of alternative methods. This leads to development of compounds, which have naturally occurring toxins of biological origin. Biopesticides include a broad range of microbial pesticides, biochemicals derived from microorganisms and other natural sources, and processes to change the plants genetically so as to express specific genes, encoding insecticidal toxins.

Viruses of a few families are identified as biopesticides; however, highly specialized family Baculoviridae have been used as biopesticides, because they are harmless to natural world and their specificity is very narrow. However, some limitation of baculoviruses restricted its application, as well as bioinsecticides including slow killing action and technical complications for in vitro production at industrial level. For the broader application of baculoviruses as biopesticides, two approaches are enforced in future. In first approach, the improvements will be made at the level of diagnostics, in vitro production, and biopesticides formulations as these efforts are useful for countries where use of genetically modified organisms is restricted. In the second approach, the killing activity of baculoviruses could also be improved by genetic alteration of the baculoviruses genome with genes of another natural pathogen. It is assumed that the genetically modified baculoviruses will be gradually introduced in countries that have fewer issues related to genetically modified organisms.

4.1.1   ENTOMOPATHOGENIC VIRUSES

Several caterpillar pests can be naturally and successfully controlled by insect-specific viruses. Some naturally occurring strains, such as nuclear polyhedrosis virus (NPV) and granulosis virus are present in many insect populations at low levels. Epizootics will often devastate populations of some pests, particularly once insect numbers are high. Insect viruses should be eaten by an insect to cause infection yet may likewise spread from insect to insect amid mating or laying egg. In some cases, for instance, while looking for appropriate hosts for laying egg, useful insects like parasitoids could physically spread a virus through the pest population. No risk to people or natural life is postured by these insect viruses. Virus diseases of caterpillar pests possibly will cause circuitous mortality of some adored larval parasitoids, if the host insects infect before the parasitoids have completed development. Predators and mature parasitoids are not stanchly pretentious. There has been some limitation on undefeated exploitation of insect-pathogenic viruses, in order that NPV strains have solely been mass made in living insects, which is an expensive procedure. The virulence of various biopesticides is highly specific to one species or genus which limited its uses at broad spectrum and also hindered the further development of viral insecticide.

Caterpillar pests affect most of the crops and habitats, while they themselves may be affected by naturally occurring viruses. NPV has been found to be effective against alfalfa looper, corn earworm, imported cabbageworm, cabbage looper, cotton bollworm, cotton leafworm, tobacco budworm, armyworms, European corn borer, almond moth, spruce budworm, Douglas fir tussock moth, pine sawfly, and gypsy moth. Several caterpillar species, such as imported cabbageworm, cabbage looper, armyworm, fall webworm, and mosquitoes have been used as the source for the preparation of granulosis virus. Viruses invade an insect's body via the gut and disrupt components of an insect's physiology by replicating in the tissues, interfering with feeding, laying egg, and movement. Different viruses trigger varying symptoms, for example, NPV infected larvae may initially turn white and granular or very dark, while some mount on the crop canopy, stop feeding, weaken and hang from the upper leaves or stems (hence the common name "caterpillar wilt" or "tree top" disease). The granulosis virus affected larvae may turn milky and stop feeding.

Whichever be the case, the ultimate result is that all the body contents of the dead larvae are liquefied and the infectious viral particles are easily released through the ruptured cuticle. It usually takes about 3–8 days to die from a viral infection. Pest population can be very gravely depleted by a naturally occurring viral epizootic. Even though viral transmission through a population may take days or weeks, once suitable conditions are attained, it may lead to crumpling down of the entire population. In order to maintain tolerable levels of pest population, the combination of naturally occurring viruses and other natural enemies have been observed to be very effective in some cases, for example, in a study, about 28% of the imported cabbage-worm populations in Cole crops were destroyed by the virus and about 55% of the residual population were parasitized by several parasitoids. In the same study, cabbage looper population showed 40% viral infection. There has been successful application of the mass reared viruses in limited areas as microbial insecticides against pests.

Mashed water solution of infected caterpillars has been used as microbial insecticide by applying them to pest population. The best time to apply the insecticides is late afternoon as the viruses are adversely affected by the UV radiation. Viral susceptibility and the effectiveness of different strains of the same virus will be affected by the abundance of pest and the general fitness of the pest population. Frozen viruses are viable for many years. The recent researches are using genetic engineering in order to improve the performance of viral insecticides and to reduce the killing time as well.

4.1.2 BIOPESTICIDES

The general sources of biopesticides are the living organisms like plants, animals, and microbes, which, due to their nontoxic eco-friendly mode of actions, are very beneficial in managing and controlling serious plant—damaging insect pests. So, the crop-protection armory now has an added component in the form of biopesticides. The limitations of funds for research and development, shelf-life, persistence in the environment (in some instances also considered an advantage), and variable field performance are the major essentials that are affecting the future developments and accep-tance. With the incorporation of experienced knowledge of selection and marketing of new products, most of the problems will be overcome, thereby providing improved products in the global markets.

Even the developing countries are benefitting from biopesticides as it is providing them the opportunity to explore and develop their own natural biopesticides resources in crop protection. This, in return, will also help in conserving foreign cash reserves, improving safety to applicators and consumers, and protecting the environment.

Their by-products are mainly used for the effective control of pests deleterious to plants (Mazid et al., 2011). They are categorized into three different types known as plant-incorporated protectants, microbial pesticides, and biochemical pesticides. They do not have any noxious residue that may be matter of considerable concern for consumers, specifically for consumable products, such as fruits and vegetables. When they are utilized as a constituent of insect-pest management, the efficiency of biopesticides will be adequate that of conventional pesticides, significantly for crops like fruits, vegetables, nuts, and flowers. In blend with synthetic pesticide performance and environmental safety, biopesticides execute successfully with the elasticity of minimum application boundaries and with superior resistance management potential (Kumar, 2012; Senthil-Nathan, 2013). Biopesticides have been picking up consideration and interest among those worried with developing environmentally safe and harmless integrated crop management (ICM)-compatible approaches and tactics for pest control (Copping and Menn, 2000). Specifically, agriculturists' selection of biopesticides may take after the current pattern of "biologically based products" and the more powerful presentation of organically-based items with a wide range of biological conduct against key target organisms and also the creating acknowledgment that these agents can be used to replace synthetic chemical pesticides (Menn and Hall, 1999; Copping and Menn, 2000; Chandrasekaran et al., 2012; Senthil-Nathan, 2013).

4.2   RESEARCHERS VIEW ON VIRAL BIOPESTICIDES

Insecticides from microorganisms provided a golden opportunity to developing nations to research, and they have had to create common biopesticide assets in ensuring crops. The utilization of biopesticide projects would be needed to stop the event of resistance in target insect pests to synthetic chemical pesticides and toxins from biopesticides (Copping and Menn, 2000; Senthil-Nathan, 2006; Senthil-Nathan et al., 2009). In Comparison with chemical pesticides, biopesticides quenched equivalent regulative issues seen with chemical pesticides. Biopesticides are highly target specific, are benign to valuable insects, and do not cause air and water quality issues in the surroundings and furthermore agricultural crops may be re-entered soon after treatment.

In addition, the application of biopesticides has other several advantages, such as many target pests are highly susceptible to their effects (USEPA, 2006; Goettel et al., 2001) and biopesticides got from bacteria like *Bacillus thuringiensis* (Bt), a substantial cluster of fungi, viruses, protozoa, and some valuable nematodes have been formulated for greenhouse, field crop, orchard, and garden use (Butt et al., 2001; USEPA 2006; Grewal et al., 2005; Hom, 1996).

Baculoviruses are present in arthropods; mainly insects and it has double-stranded DNA as a genetic material.

Baculoviruses are usually profoundly pathogenic and have been routinely utilized in their natural form as biocontrol agents against numerous noxious insects (Moscardi, 1999). Even so, the application of baculoviruses is remains restricted within the field of agriculture where the limits for pest damage tend to be minimized.

4.2.1 BACULOVIRUS

Scientific advancements are leading to eco-friendly crop-pest control methods are being developed by using either biological agents or specifically designed synthetic antiinsect compounds. Baculoviruses are rod-shaped DNA viruses, which mostly replicates inside the cells. Occlusion bodies are formed in the nuclei of baculoviruses infected caterpillar cells by the multiplication and incorporation of the viral progeny into the protective polyhedron-shaped protein structures. These occlusion bodies are responsible for the death of caterpillars and contamination of the leaf surfaces. When the healthy caterpillars feed on the infected leaves, they consume the occlusion bodies, releasing the virus, thus, the cycle of infection and replication continues.

Natural baculovirus epidemics are responsible for eliminating populations of caterpillar's pests, such as corn earworm, cotton bollworm, tobacco budworm, and cabbage's nemesis—the diamondback. Keeping in view of the global market, mass-production of baculoviruses in insect cell cultures is the main focus of several companies as it will result in microbial contamination free biopesticides as well as save a lot of labor costs. Insect cell cultures particularly those from embryonic and nerve tissue serves various purposes. Along with acting as the growth media for baculoviruses, they can be used for the screening of natural or synthetic chemical compounds that have the possibility to be used as eco-friendly anti-insect compounds. Cell cultures may also reveal the reason behind the ability of certain chemicals to paralyze or kill a particular pest insect or hindering its eating pattern. Colored fluorescent protein markers have been introduced into the baculoviruses that enables the researchers to indentify the desired traits (such as, the ability to kill pests with low for baculoviruses production) in the recombinants. Among the eight insect lines developed from embryos and ovaries of *Helicoverpa* and *Heliothis* species member, one has the ability to produce approximately 10 times more of AcMNPV baculovirus than the other lines. The major advantage of this was the increased rate (about 2000 times greater) of infection in caterpillars infected by this virus rather than the ones exposed to either AcMNPV or to another baculovirus. Sensitivity of baculoviruses to the UVB rays of sunlight hinders their existence in normal environment as they damage their DNA, so the genes needed for virus establishment and reproduction is protected by a protein coat called a polyhedron. Hence, they have the potential to be used as better alternatives to the existing chemical insecticides.

As the genus NPV comprises the majority of baculoviruses used as biopesticides, all "baculovirus" or "virus" are generally referred as NPV. In cases when beneficial insects are being conserved to aid in an overall Environmental Protection Agency (EPA) program, or during the treatment of ecologically sensitive area, these viruses are potent candidate as they show no negative impacts on plants, mammals, birds, fish, or even on nontarget insects.

The high specificity of baculoviruses is sometimes considered as its weakness as well, such as in cases when growers may want one product to use against a variety of pests. Researchers are trying to use genetic engineering techniques to solve this problem.

4.2.1.1 APPEARANCE

Baculovirus killed insects have a distinctive shiny-oily appearance, are extremely delicate (may rupture to release fluid filled with infective virus particles), and are often seen hanging limply from vegetation. The attachment to foliage and rupturing on touch is very important in order to continue the virus-life cycle of infecting and replicating, as only then can they be consumed by healthy caterpillars.

Unlike other viruses, most baculoviruses can be easily seen with a light microscope with their polyhedral looking like clear, irregular crystals of salt, or sand when viewed at 400× or 1000× magnification.

4.2.1.2 BACULOVIRUS DIVERSITY

Baculoviruses have varied applications like from being used as a biopesticide in agriculture and forestry to being used as expression vectors in biotechnology. Baculoviruses are pathogenic to arthropods—predominantly holometabolous insects. Among the two generas of Baculoviridae (Van Regenmortel et al., 2000), NPV have virions embedded in a crystalline matrix of the protein polyhedron that vary in their configurations of enveloping nucleocapsids singly or in multiples (the occluded viruses are referred to as polyhedral) while the granuloviruses (GV) have one, or rarely two, virions embedded in a crystalline matrix of granulin with their nucleocapsid singly enveloped.

This virus has been used as a lethal weapon in their natural form against numerous notorious insect pests (Moscardi, 1999). The baculoviruses strain isolated from *Lepidoptera* which are the primary source where they have been isolated, only cause mortality in the larval stage (Cory, 2000). To initiate infection this viruses need to be ingested by the larvae. After intake, they enter the insect's body through the midgut and from that point they spread all through the body, in spite of the fact that in a few insects, disease can be restricted to the insects' midgut or the fat body. A typical feature of baculoviruses is that they are blocked, that is, the infection particles are surrounded in a protein matrix framework. The presence of occlusion bodies is very crucial for virus to survive outside the host (Cory, 2000).

The numbers of new recombinants baculoviruses now utilize insect-specific toxins. Genetic modification has been mostly carried out on those viruses for which complete information at molecular level is available (alfalfa looper, *Autographa californica* NPV [AcNPV] virus), which permits insertion of foreign genes in more precised way. Recently, the development of genetically engineered baculoviruses has expanded to other strains of industrial or regional interest (Cory, 2000; Popham et al., 1997). Baculovirus-infected insect larvae formed cascade of molecular and cellular appendages before death and after that development of enormous amounts of polyhedral occlusion body also known as rod shaped virions (Miller, 1997). A naturally occurring baculovirus (*Agrotis ipsilon* multiple NPV, family Baculoviridae, Agip MNPV) was showed the insecticidal properties for *A. ipsilon* in turf (Prater et al., 2006). To control caterpillar pests, a number of viral formulations are available in markets. For example, Certis has lately verified Madex, an increased-potency codling moth GV that similarly disturbs oriental fruit moth. Certis also deals Cyd-X, which also encompasses the codling moth GV and which can be an effectual means for codling moth management (Arthurs et al., 2005). Apart from Madex and Cyd-X, Certis markets Gemstar, which comprises *Heliothis zea* NPV, and Spod-X, which holds beet armyworm NPV. Gemstar is correspondingly listed for the control of lepidopteran pests, identical to the cotton bollworm and budworm, caterpillars that are mostly hazardous insect pests of corn, soybean, and other vegetables (Arthurs et al., 2005). Additionally, Certis has showed a celery looper (*Syngrapha falcifera*) NPV and an alfalfa looper (*Autographa californica*) NPV (USEPA, 2006).

To control lepidopteron larvae, various efforts have been made for genetic modification of baculoviruses. The main aim of present study was focused on to assess recent advancement made in the field of genetic engineering for improvement of pest control potency of this group of viruses (King, 1992; Maeda, 1989; Miller, 1988; O'Reilly, 1992; Summers and Smith, 1987).

4.2.1.3 HABITAT

Baculoviruses are cosmopolitan in nature. Rain and wind easily disperse baculoviruses from place to place. It is generally acknowledged by researchers that almost all products presently on the racks are "contaminated" by baculoviruses' particles. The results of tests conducted in conjunction indicate the presence of baculoviruses, this may be considered as both direct or indict evidence for the safety of these agents. With the exception of *Autographa californica* NPV and a few other viruses, most baculoviruses are species or genus specific. The viral genome responsible for controlling the host genome has been the main focus on all the genetic work being done to improve the baculoviruses-based pesticides.

4.2.1.4 BACULOVIRUS LIFE CYCLE

The life cycle of a baculovirus begins with the ingestion of NPV by the larva. In the midgut, the protective polyhedrin coat gets dissolved and the virions are released which begins to infect the midgut cells. Other tissues are infected by another viral phenotype called the budded virus or viral particles which are formed as a result of the viral replication within the nucleus of infected midgut cells. These budded viruses are coated with a viral protein-modified basal plasma membrane as they move through the cell membrane. Using the tracheal system of the insect as channel, different larval tissues are infected sequentially (Engelhard et al., 1994). However, there are arguments regarding it being the primary route of viral infection or whether hemocytes play this role. As the cycle continues, the progeny viruses become corked up within the nuclei of the infected cells. The wild-type baculoviruses infected larvae climb up the plant, succumb to the infection and hang from an elevated position where they are disseminated as the cadaver decomposes. The baculoviruses gene expression in the infected insect cells occurs in a temporally regulated cascade (Friesen and Miller, 1986). The transcription of the immediate early (IE) genes does not depend on production of other viral proteins unlike the other three temporal phases. This is because the host factors are responsible for the transcription of these genes. Their products are responsible for the regulation of the delayed early (DE) genes. There is simultaneous DNA replication and late-gene expression, encoding the structural proteins of the virus particles. The proteins required in the final infection stage and polyhedron morphogenesis (including the pl0 and polyhedrin proteins) are encoded by the very late genes.

4.3   VIRUS—AS A CONTROL AGENT OF INSECT PESTS

Steinhaus (1956), first recognized the potential of baculoviruses to be used as insect pest control. From then on, a number of baculoviruses are being used pest controls for like hymenopteran, lepidopteran, and coleopteran pests of crops such as coconuts, cotton, and cabbages, as well as pests of beehives. The work of Entwistle and Evans (1985) on the pest control on soybean crops in Brazil and on palm and coconut in the South Pacific by *Anticarsia gemmatalis* NPV and *Oryctes rhinoceros* baculovirus, respectively are very significant. The killing time of virus varies from days to weeks, affected by various factors like temperature, viral dose, insect age, and the particular host and virus species.

Throughout the infection, the insect continues to nourish. Henceforth, numerous pests presently controlled by baculoviruses are pests of crops that can tolerate some mutilation deprived of noteworthy economic loss. In accumulation, baculoviruses have been used effectively when the insects are small and it is conceivable to apply virus. For example, in Brazil the soybean crops are examined intensively for larvae for optimal timing of virus application before substantial loss has happened. Genetic engineering of the baculovirus to decrease the time taken by the virus to destroy the host insect will not prevent the use of the wild-type viruses in biocontrol but have potentials to yield viruses more competitive with classical insecticides. This expansion will considerably upsurge the prospective of baculoviruses for pest control, predominantly in row crop agriculture (Table 4.1).

TABLE 4.1 Showing Baculovirus Products.

Sr. no.	Baculovirus	Trade name	Manufacturer	Target pest	Crop
1.	*Spodoptera exigua* NPV	SPOD-X	Thermo trilogy	Beet army worm, *S. exigua*	Glasshouses in the Netherlands
2.	*S. littoralis* NPV	Sodopterin	NPP-calliope	Egyptian cotton leaf worm, *S. littoralis*	Cotton
3.	*Helicoverpa zea* NPV	Gemstar	Thermo trilogy	Cotton bollworm *H. zea,* Tobacco budworm, *Heliothis viriscence,* Febricius	Corn, cotton
4.	*Limentria dispar* NPV	Disparvirus	Canadian forest service	Gypsy moth larvae, *L. dispar*(L)	Forest
5.	*Anagrapha falcifera* (kirby) NPV Celery looper	AfNPV	Thermo trilogy	Broad-spectrum control of lepidopterous larvae	Many
	Anticarsia gemmatalis NPV	Polygon multigen	Androgen EMPRAPA	Velvet bean caterpillar *A. gemmatalis* Hubner	Soybean

4.4 DEVELOPING RECOMBINANT BACULOVIRUSES THROUGH GENETIC ENGINEERING

The insistence of genetic engineering of baculoviruses for its practice as insecticides is to associate the pathogenicity of the virus with the insecticidal achievement of a toxin, hormone, or enzyme. After infection of the insect larva with the recombinant baculovirus, the foreign protein is articulated. If this protein is lethal to the insect, the insect will die swiftly from their consequence, rather than from the viral infection itself. The recombinant method will perhaps be used to advance production, alter host range, and improve the efficacy of countless insect viruses as biopesticides. Though, the objective of the research revised here is to decrease the time from infection with the recombinant virus to demise of the insect such that feeding devastation is under the economic threshold. This goal demands an estimated lethal-time ratio (lethal time of test virus divided by lethal time of wild-type virus; Bonning and Hammock, 1993) of 0.4–0.5 for control of insect pests on various crops. Fall of the lethal time can also boost farmer or user approval of baculovirus insecticides. Two key baculovirus-expression systems have been established for the making of recombinant proteins for research and clinical use. These are grounded on the NPV imitative from the alfalfa looper, AcNPV, and an analogous virus from the silkworm, *Bombyx mori* (BmNPV). The sequences of the whole genomes of both AcNPV and BmNPV have now been determined (Ayers et al., 1994). Initial engineering work was carried out with BmNPV for high levels of protein production in larvae of *B. mori* is the only identified host for BmNPV, this method also delivered biological containment for the virus (Maeda et al., 1991). Current effort in developing the virus for insect control has focused on AcNPV. A range of freshly established techniques and transfer vectors significantly enable the engineering progression (Bishop, 1989; Davies, 1994; Luckow, 1994). Protein expression systems have also been recognized in other baculoviruses, such as *Helicoverpa zea* NPV (Corsaro, et al., 1989) and *Lymantria dispar* NPV (Yu et al., 1992). This research delivers the foundation for engineering of these viruses for practice as insect pest-control agents in the future. The circular genome of AcNPV is nearly 134 kilo base pairs (kbp; Ayers et al., 1994). Since direct manipulation of such a large piece of DNA is difficult, engineering of a baculovirus is commonly carried out in two steps. First, the foreign gene is integrated into a baculovirus-transfer vector. Furthermost, transfer vectors used are bacterial plasmid, University of California (pUC), derivatives, which encode an origin of replication for proliferation in *Escherichia coli* and an ampicillin-resistance gene. The pUC fragment is ligated to a small segment of DNA, from the viral genome. The foreign gene sequence is assimilated into a cloning site downstream of the promoter, designated to drive expression. For the subsequent step, the transfer vector is mixed with DNA from the parental virus. The engineered DNA is combined into the virus via homologous recombination events inside the nucleus of cultured insect cells. Contrasting genetic engineering in plants, which marks in a slightly random integration of new DNA into the genome, the baculovirus system permits the accurate insertion of foreign DNA without interference of other genes. No drug resistance markers are involved in the final clone, which eradicates some of the key objections upraised to recombinant organisms (Fox, 1995). Commercial kits and reagents, as well as numerous outstanding manuals are available for this work (King and Possee, 1992; O'Reilly et al., 1992). A number of freshly developed substitute methods for genetic engineering of baculoviruses have been reviewed in a different place (Davies, 1994). Initial exploration involved engineering of these viruses for use as protein-expression vectors rather than for insect control (Smith et al., 1983). The methodology involved swapping the gene encoding polyhedrin with the foreign gene of concern. Expression of the foreign gene was compelled by the polyhedrin promoter in a polyhedrin-negative virus. While these viruses can be influenced positively in cell culture for production of high levels of foreign protein, they drop a substitute approach to replacement of a viral gene with a foreign gene sequence is to duplicate a viral promoter. In this illustration, none of the viral genes are missing, and promoters of essential viral genes can be used for expression of foreign proteins. The level and timing of expression of a specific protein by a recombinant baculovirus is resolute in part by the promoter selected to drive transcription of the foreign gene sequence. Presently, the polyhedrin and p10 promoters are used most regularly for expression of recombinant proteins. Though, expression under the elementary protein promoter was higher in several occasions (Bonning et al., 1994; Sridhar et al., 1993). In the forthcoming, the practice of early promoters, hybrid promoters, and promoters from other species will upsurge, predominantly with the identification of peptides and proteins that interrupt insect biology at minor expression levels. Any improvement deliberated by the polyhedrin coat that defends the virus from inactivation by desiccation and UV light under field conditions. Replacement of the viral gene encoding the p10 protein (Vlak et al., 1990), which is elaborated in calyx attachment and nuclear lysis, also lead to reduced viral capability. The firmness of polyhedra produced by p10-negative viruses is significantly reduced (Williams et al., 1989).

4.4.1 ADVANTAGES OF RECOMBINANT BACULOVIRUS INSECTICIDES

Insect viruses are significant constituents of natural biological regulator, but this chapter focuses on their use as biological insecticides with reduced capacity to salvage in the field. The stress to discover innovative resources of pest control to lessen dependence on the synthetic chemical insecticides is growing. The number of compounds presently accessible to the cultivator is declining as effect of pest resistance (Roush and Tabashnik, 1990) and damaging effects of insecticides on both human health and the environment (Coats, 1994). In this respect, numerous benefits are related with the use of baculoviruses as insect-control agents. One of the most significant characteristics of baculoviruses for monitoring the pest is their host specificity. Several of them infect only a few species, frequently inside the same family. This feature makes them ideal for integration into integrated pest management (IPM) programs as one of the key concepts of IPM is to aim the pest species that surpasses the economic-damage threshold, while leaving the rest of the fauna uninterrupted (FAO, 1968a,b). Therefore, the entire potential of valuable organisms can be exploited. Viral pesticides can be functional using conventional techniques and do not generate the problems related with residues. They do not display cross resistance with chemical compounds. While insects have revealed resistance to baculoviruses in some circumstances (Fuxa, 1993), the resistance ratios are normally low, and in numerous cases, resistance is unbalanced in the lack of selection pressure. Deceptive adversely correlated resistance to chemical insecticides has even been noted (Fuxa and Richter, 1990; Mccutchen et al., 1995). The use of baculovirus insecticides in IPM programs with additional biological-control agents or with conventional chemical insecticides may lessen the probability of resistance evolving to the baculovirus. In some pest-management circumstances, the recombinant viruses will be used to supplement the action of conventional insecticides. In fact, some recombinant viruses can synergize and be synergized by classical insecticides, and this complementarity makes the viruses even more striking and operative as tools for IPM (Mccutchen et al., 1995). For the production of baculovirus insecticides, industry is following numerous strategies, counting both in vitro and in vivo technologies. Nevertheless, wild-type and recombinant viruses can be manufactured by cottage industries. With sensibly produced inoculant, local laboratories in developing countries could harvest these genetically cultured products with nominal technological input.

4.4.2   LIMITATIONS OF RECOMBINANT BACULOVIRUS INSECTICIDES

The enhanced insecticidal potency of genetically engineered viruses diminished their ability to recycle in the nature as only some polyhedra are produced compared with the wild-type virus. Apart from this confinement, the advantages and disadvantages of recombinant and wild-type baculoviruses are constant. In many cases, the attributes that are perceived beneficial really restrain the utility of the recombinant infections. For instance, the restricted host go permits a user to decrease one pest population with accuracy without any modification of biological-control agents and without the malicious impacts on nonfocused organisms related with utilization of established pesticides and even *B. thuringiensis*. However, this attribute also leads to limits the market size for the baculovirus and accordingly the improvement costs that can be invested in the innovation.

Cloning system that permits the recombinant technology to be functional to diverse viruses is missing even within the Baculoviridae family. Therefore, the use of genetically engineered baculoviruses for control of insect is currently restricted to a few pest species. The proportion of progress of these systems will have to be speed up. Nevertheless, of all the restrictions facing the use of genetically modified and wild-type viruses, the glitches linked with accurate application and low importunity are the most thoughtful. The much recompense of the recombinant viruses has appealed industrial investment in research to elucidate these problems. These technologies will be relevant to wild-type of viruses and other biological-control agents projected for augmentative release in pest management, in addition to recombinant viruses. Even with the abundant prospective of recombinant baculoviruses, they do not signify a solution or even a stand-alone technology for insect control. Somewhat, they will be used as added implements for application in IPM programs.

4.5 CONCLUSION

There are many difficulties that are inflated by the insect-pest frequently in the production of crops and forest produce by their activities. Among them there are mainly three key features of commercial difficulties provoked by them. The first one is concerning to the damage of crop produce, second is related to the health of human and pest, and the third one is related to the cost of attempt to prevent or control such production losses and health hazards. The chemical pesticides used to control them have created serious ecological problems such as pest evolved resistance, destroying of beneficial insects, beneficial soil micro fauna and natural predators, and pollution of environment has increased worldwide. As a result, increasing attention has been directed toward creation of biological agents, making them highly toxic to the target organism, possessing the desirable properties of a chemical pesticide, can be mass produced on an industrial scale and has a long shelf and minimal environmental impacts. Fortunately, many of insect are associated with pathogenic microbes that have been suggested as controlling agents of pests and developed as commercial products. entomopathogenic microbes (bacteria, fungi, viruses, nematodes, and protozoa) that cause natural epidemics in arthropods, mainly insects and mites are host specific and naturally widespread in the environment. These can be produced at commercial scale and applied as a pesticide, using sprays, dust, and drenches. To control pests like aphids, whiteflies, stem borers, leaf miners, locust, and grasshoppers, viral biopesticides agents can be used successfully. The effectiveness of a viral biopesticides agent relies upon two elements, first is its ability to kill and the second is to reproduce on pests, users can adopt such kind of biological agents that may give efficacy comparable with chemical insecticides. Technological advances in viral biopesticides production, formulations, quality control, application timing and conveyance, and especially in choosing ideal target habitats and target pests, have may broaden the efficacy gap among chemical and biocontrol agents. It is very likely that upcoming role of viral biopesticides will be more significant in integrated pest management strategies particularly in the field of agriculture and forestry. In general, the advantages from the achievements in utilizing viral biopesticides exceed by a wide margin the cash lost on disappointments which occasionally occur. Economic benefits, although vital, do not seem to be the sole advantages of biocontrol; it also reduces the exposure of harmful chemicals and greatly benefited to animal life, farmers, and their families.

KEYWORDS

- **viral-biopesticides**
- **baculoviruses**
- **expression vectors**
- **detection**
- **genetic modification**

REFERENCES

Arthurs, S. P.; Lacey, L. A.; Fritts, R. Jr. Optimizing Use of Codling Moth Granulovirus: Effects of Application Rate and Spraying Frequency on Control of Codling Moth Larvae in Pacific Northwest Apple Orchards. *J. Econ. Entomol.* **2005,** *98,* 1459–1468.

Ayers, M. D.; Howard, S. C. K.; Kuzio, J.; Lopez-Ferber, M.; Possee, R. D. The Complete Sequence of *Autographa californica* Nuclear Polyhedrosis Virus. *Virology* **1994,** *202,* 586–605.

Bishop, D. H. L. Genetically Engineered Viral Insecticides – A Progress Report 1986–1989. *Pestic. Sci.* **1989,** *27,* 173–189.

Bonning, B. C.; Hammock, B. D. Lethal Ratios: An Optimized Strategy for Presentation of Bioassay Data Generated from Genetically Engineered Baculoviruses. *J. Invert. Pathol.* **1993,** *62,* 196–197.

Bonning, B. C.; Roelvink, P. W.; Vlak, J. M.; Possee, R. D.; Hammock, B. D. Superior Expression of Juvenile Hormone Esterase and B-Galactosidease from the Basic Protein Promoter of *Autographa californica* Nuclear Polyhedrosis Virus Compared to the p¹⁰ Protein and Polyhedrin Promoters. *J. Gen. Virol.* **1994,** *75,* 1551–1556.

Butt, T. M.; Jackson, C. W.; Magan, N. *Fungi as Biocontrol Agents: Progress, Problems and Potential*; CAB International: Wallingford, 2001.

Chandrasekaran, R.; Revathi, K.; Nisha, S.; Kirubakaran, S. A.; Sathish-Narayanan, S.; Senthil-Nathan, S. Physiological Effect of Chitinase Purified from *Bacillus subtilis* Against the Tobacco Cutworm *Spodoptera litura. Fab. Pestic Biochem. Physiol.* **2012**, *104*, 65–71.

Coats, J. R. Risks from Natural Versus Synthetic Insecticides. *Annu. Rev. Eniomol.* **1994**, *39*, 489–515.

Copping, L. G.; Menn, J. J. Biopesticides: A Review of Their Action, Applications and Efficacy. *Pest. Manag. Sci.* **2000**, *56*, 651–676.

Corsaro, B. G.; DiRenzo, J.; Fraser, M. J. Transfection of Cloned *Heliothis zea* Nuclear Polyhedrosis Virus. *J. Virol. Methods.* **1989**, *25*, 283–291.

Cory, J. S. Assessing the Risks of Releasing Genetically Modified Virus Insecticides: Progress to Date. *Crop Prot.* **2000**, *19*, 779–785.

Cory, J. S.; Hirst, M. L.; Sterling, P. H.; Speight, M. R. Native Host Range Nucleopolyhedric Virus for Control of the Brown Tail Moth (Lepidoptera: Lymantriidae). *Environ. Entomol.* **2000**, *29*, 661–667.

Davies, A. H. Current Methods for Manipulating Baculovirus. *Biotechnology.* **1994**, *12*, 47–50.

Engelhard, E. K.; Kam-Morgan, L. N. W.; Washburn, J. O.; Volkman, L. E. The Insect Tracheal System: A Conduit for the Systemic Spread of *Autographa californica* M Nuclear Polyhedrosis Virus. *Proc. Natl. Acad. Sci. USA* **1994**, *91*, 3224–3227.

Entwistle, P. F; Evans, H. F. Viral Control. In *Comprehensive Insect Physiology Biochemistry and Physiology;* Gilbert, L. I., Kerkut, G. A., Eds.; Pergamon Press: Oxford, UK, 1985; pp 347–412.

Food and Agriculture Organisation of the United Nations (FAO). *Report on 1st Session of the FAO Panel of Experts on Integrated Pest Control*; Meeting Report: Rome, 1968a, PL/1967/M/7 FAO.

Food and Agriculture Organization of the United Nations (FAO). *Report on 1st Session of the FA0 Panel of Experts on Integrated Pest Control*; Meeting Report: Rome, 1968b PlJI967hW7 FAO.

Fox, J. L. EPA'S First Commercial Release Is Still Pending. *Biotechnology* **1995**, *13*, 114–115,

Friesen, P. D; Miller, L. K. The Regulation of Baculovirus Gene Expression. *Curr. Top. Microbiol. Immunol.* **1986**, *131*, 3149.

Fuxa, J. R. Insect Resistance to Viruses. In *Parasites and Pathogens of Insects Vol. 2;* Beckage, N. E., Thompson, S. N., Federici, B. A., Eds.; Academic Press: New York, 1993; pp 197–209.

Fuxa, J. R; Richter, A. R. Response of Nuclear Polyhedrosis Virus-Resistant *Spodoptera frugiperda* Larvae to Other Pathogens and to Chemical Insecticides. *J. Invert. Pathol.* **1990**, *155*, 272–277.

Goettel, M. S.; Hajek, A. E.; Siegel, J. P.; Evans, H. C. Safety of Fungal Biocontrol Agents. In *Fungi as Biocontrol Agents: Progress, Problems and Potential*; Butt, T. M., Jackson, C., Magan, N., Eds.; CAB International: Wallingford, 2001; pp 347–375.

Grewal, P. S.; Ehlers, R. U.; Shapiro-Ilan, D. I. *Nematodes as Biocontrol Agents*. CABI Publishing: Wallingford, 2005; p 505.

Hom, A. Microbials, IPM and the Consumer. *IPM Pract.* **1996**, *18*, 1–11.

King, L. A.; Possee, R. D. *The Baculovirus Expression System.* Chapman & Hall: London, 1992; p 229.

Korth, K. L.; Leavings, C. S. Baculovirus Expression of the Maize Mitochondrial Protein URF13 Confers Insecticidal Activity in Cell Cultures and Larvae. *Proc. Natl. Acad. Sci. USA* **1993**, *90*, 3388–3392.

Kumar, S. Biopesticides: A Need for Food and Environmental Safety. *J. Biofertil. Biopestic.* **2012**, *3*, 4.

Luckow, V. A. Insect Cell. Expression Technology. In *Principles and Practice of Protein Engineering*; Cleland, J. L., Craik, C. S., Eds.; Wiley & Sons: New York, 1994; pp 1–27.

Maeda, S. Expression of Foreign Genes in Insects Using Baculovirus Vectors. *Annu. Rev. Entomol.* **1989**, *34*, 351–372.

Maeda, S.; Volrath, S. L.; Hanzlik, T. N.; Harper, S. A.; Maddox, D. W et al. Insecticidal Effects of an Insect Specific Neurotoxin Expressed by a Recombinant Baculovirus. *Virology* **1991**, *184*, 777–780.

Mazid, S.; Kalida, J. C.; Rajkhowa, R. C. A Review on the Use of Biopesticides in Insect Pest Management. *Int. J. Sci. Adv. Technol.* **2011**, *1*, 169–178.

Mccutchen, B. F; Betana, M. D; Herrmann, R; Hammock, B. D. Interactions of Recombinant and Wild Type Baculoviruses with Classical Insecticides and Pyrethroid-Resistant Tobacco Budworm (Lepidoptera: Noctuidae). *J. Econ. Entomol.* **1995**, *90* (5), 1170–1180.

Menn, J. J.; Hall, F. R. Biopesticides: Present Status and Future Prospects. In *Biopesticides Use and Delivery*; Humana Press: Totowa, 1999; pp 1–10.

Miller, L. K. *The Baculovirus*. Plenum Press: New York, 1997; pp 7–32.

Miller, L. K. Baculoviruses as Gene Expression Vectors. *Annu. Rev. Microbiol.* **1998**, *42*, 177–199.

Moscardi, F. Assessment of the Application of Baculoviruses for Control of Lepidoptera. *Annu. Rev. Entomol.* **1999**, *44*, 257–289.

O'Reilly, D. R.; Miller, L. K.; Luckow, V. A. *Baculovirus Expression Vector—A Laboratory Manual*. Freeman: New York, 1992; p 347.

Popham, H. J. R.; Li, Y.; Miller, L. K. Genetic Improvement of *Helicoverpa zea* Nuclear Polyhedrosis Virus as a Biopesticide. *Biol. Control.* **1997**, *10*, 83–91.

Prater, C. A.; Redmond, C.; Barney, W.; Bonning, B. C.; Potter, D. A. Microbial Control of Black Cutworm (Lepidoptera: Noctuidae) in Turfgrass Using *Agrotis ipsilon* Multiple Nucleopolyhedrovirus. *J. Econ. Entomol.* **2006**, *99*, 1129–1137.

Roush, R. T.; Tabashnik, B. E. *Pesticide Resistance in Arthropoh.* Chapman & Hall: New York, 1990.

Senthil-Nathan, S. Effects of *Melia azedarach* on Nutritional Physiology and Enzyme Activities of the Rice Leaffolder *Cnaphalocrocis medinalis* (Guenée) (Lepidoptera: Pyralidae). *Pestic. Biochem. Physiol.* **2006**, *84*, 98–108.

Senthil-Nathan, S. Physiological and Biochemical Effect of Neem and Other *Meliaceae* Plants Secondary Metabolites Against Lepidopteran Insects. *Front. Physiol.* **2013**, *4*, 359.

Senthil-Nathan, S.; Choi, M. Y.; Paik, C. H.; Seo, H. Y.; Kalaivani, K. Toxicity and Physiological Effects of Neem Pesticides Applied to Rice on the *Nilaparvata lugens* Stål, the Brown Planthopper. *Ecotoxicol. Environ. Saf.* **2009**, *72*, 1707–1713.

Smith, G. E.; Summers, M. D.; Fraser, M. J. Production of Human β-Interferon in Insect Cell Infected with a Baculovirus Expression Vector. *Mol. Cell. Biol.* **1983**, *3*, 2156–2165.

Sridhar, P.; Panda, A. K.; Pal, R.; Talwar, G. P.; Hasnain, S. E. Temporal Nature of the Promoter and Not the Relative Strength Determines the Expression of an Extensively Processed Protein in a Baculovirus System. *FEES Lett.* **1993**, *315*, 282–286.

Steinhaus, E. A. Microbical Control: The Emergence of an Idea. *Hilgardia.* **1956**, *26*, 107–160.

Summers, M. D.; Smith, G. E. A Manual of Methods for Baculovirus Vectors and Insect Cell Culture Procedures. *Tex. AGR IC. Exp. Stn. Bull.* **1987**, *1555*, 1–57.

USEPA (United States Environmental Protection Agency). *New Biopesticide Active Ingredients,* 2006. New Biopesticide Active Ingredients-2006. http://www.epa.go/pesticides/biopesticide/ products lists/new_ai_2006.htm (accessed January 4, 2008).

Van Regenmortel, M. H. V; Fauquet, C. M; Bishop, D. H. L; Cartens, E. B; Estes, M. K; Lemon, S. M. et al. *Virus Taxonomy*. Seventh Report of the International Committee on Taxonomy of Viruses, New York: Academic, 2000.

Vlak, J. M.; Schouten, A.; Usmany, M.; Belsham, G. J.; Klinge-Roode, E. C et al. Expression of a Cauliflower Mosaic Virus Gene I Using a Baculovirus Vector Based on the p^{10} Gene and a Novel Selection Method. *Virology* **1990,** *178,* 312–320.

Williams, G. V.; Rohel, D. G.; Kuzio, J.; Faulkner, P. A Cytopathological Investigation of *Autographa californica* Nuclear Polyhedrosis Virus p^{10} Gene Function Using Insertion/ Deletion Mutants. *J. Gen. Virol.* **1989,** *70,* 187–202.

Yu, Z.; Podgwaite, J. D.; Wood, H. A. Genetic Engineering of a *Lymantria dispar* Nuclear Polyhedrosis Virus for Expression of Foreign Genes. *J. Gen. Virol.* **1992,** *73,* 1509–1514.

CHAPTER 5

Control of Insect Pest Through Biomolecules and Traps

ANURAG MISHRA[1,*], PREETI SHARMA[2], ARUN KUMAR GUPTA[1], PARVEEN FATIMA[1], and PUSHPENDRA KUMAR[1]

[1]Department of Agricultural Biotechnology, Sardar Vallabhbhai University of Agriculture and Technology, Modipuram, Meerut 250110, Uttar Pradesh, India

[2]Department of Microbiology, Shri R. L. T. College of Science, Akola 444001, Maharashtra, India

[*]Corresponding author. E-mail: anuragmishraa@hotmail.com

ABSTRACT

Agriculture is playing a very important role in Indian economy. Nowadays agricultural sector faces systematic annual losses due to pests and diseases. For controlling insect pests, many insecticides and pesticides are used in agriculture which are very harmful for humans. Long-term use of insecticides and pesticides in crops can produce resistance against insects and pests. Some plants have insecticidal properties; botanicals can be used as an alternative for insecticides and pesticides against insect pest and disease. They are less hazardous, biodegradable, and more effective than botanicals such as pyrethrins, rotenone, sabadilla, ryania, nicotine, linalool, and neem. Neem dust is used to control the stored grain pests and in toothpaste, soap, shampoo for nits, cosmetics, and cattle feed. Pheromones are one of the most powerful alternative approaches to control insect pests. Pheromones are natural substances that are produced by special glands in the abdomen of insects and it attracts the opposite gender of the same species. Insects produce pheromones for attracting the mate, for marking foraging routes, or to signal alarm to neighbors (e.g., aphids). Pheromones are used to control harmful insects and to attract insects to insecticide or chemosterilant baits, decrease the use of insecticidal practices. Light trap is used for monitoring insect populations with a view to provide early warning of the presence of pests, population dynamics of Lepidoptera and Coleoptera. Botanicals, pheromone trap, and light trap play a very important role in controlling insect pest in agriculture. It is an alternative approach against insecticides and pesticides for controlling the insect pests. It is less hazardous to human, easy to handle, and most effective in controlling insect pests.

5.1 INTRODUCTION

The use of toxic pesticides to control pest problems has become a common exercise in the world. However, pesticides are poisons and, unfortunately, they can harm more than just the "pests" at which they are targeted. They are also very toxic, and exposure to pesticides can cause a number of health effects to the other non-targeted individuals. They are connected to a variety of serious illnesses and diseases from respiratory problems to cancer in the living individuals. There are several plant-based natural products that has been developed for the insect control (Regnault-Roger et al., 2005). Utilization of manufactured insecticides sprays prompted various problems at the season of their introduction: intense and interminable harming of utensils, farmworkers, and even buyers; destruction of fish, fowls, and other wild animals; interruption of natural organic control and fertilization; broad groundwater contamination, possibly devastating human and natural well-being; and the advancement of protection from pesticides in insect populaces (Forget et al., 1993; Marco et al., 1987; National Research Council, 2000; Perry et al., 1998).

The act of utilizing plant extracts, or organic insecticides sprays as we now know them, in horticulture goes back to not less than two centuries in old China, Egypt, Greece, and India (Thacker 2002; Ware 1883). Indeed, even in Europe and North America, the recorded utilization of botanicals stretches out back over 150 years, significantly originating before disclosures of the real classes of synthetic chemical insecticides (e.g., organochlorines, organophosphates, carbamates, and pyrethroids) in the mid-1930s to 1950s. What is clear from history is that manufactured insecticides successfully consigned botanicals from an imperative part in agriculture to a basically inconsequential position in the commercial center among trim protectants. There are four major types of botanical products that are used for insect control at present (pyrethrum, rotenone, neem, and essential oils), along with three others with limited use (ryania, nicotine, and sabadilla). Some other additional plant extracts and oils (e.g., garlic oil and capsicum oleoresin) provincial are used (low volume) in different nations, yet these are not considered here.

The real advantages of herbal (botanicals) insecticides can be best acknowledged in creating nations, where farmers will be unable to bear the cost of synthetic chemical insecticides and the traditional utilization of plants and plant derivatives for insurance of stored items is for long time built up. Indeed, even where synthetic insecticides are moderate to cultivators (e.g., through government sponsorships), control proficiency and an absence of protective instruments result in a huge number of unintentional poisonings yearly (Forget et al., 1993). Some plant items could even be valuable in industrialized nations for the security of grain against insect pest (Fields et al., 2001).

There are numerous optional ways to deal with insect pest. Cultural methods, including crop cleanliness, utilization of safe assortments, and strategies to advance the exercises of common enemies and predators, demonstrate to minimize the danger from power of insect pest. Hence, holding insect pest bothers underneath harm edges, does not point decrease populaces to a position where hereditary change is incited, prompting the enlargement of resistant biotypes. Such biorational pesticides incorporate insecticides, for example, neem seed extracts, various juice of plant parts, organisms (Beaveria and Metarhizium), epidemic (atomic polyhedrosis infection), bacteria (*Bacillus thuringiensis*), and semiochemicals (Akhtar and Mandal, 2008; Mamun et al., 2008; Islam et al., 2008a; Islam, 2009; Islam and Becerra, 2011; Islam and Begum, 2011).

The insect pests are getting resistant to chemical pesticides as the regularity of spraying is increasingly expanding while their viability is regularly declining. Biological regulatory includes use of individually selecting organism to manage a particular insect pest. These selecting organisms might be a parasite or a predator, which kills harmful insects. Trial of field results showed that the use of organic substances does not have injurious effect on human health and it is also easily available at low cost. Biocontrol measures are easy, safe, sustainable, and attractive. Insects interact by signaling utilizing scents—pheromones, chemicals. With these, they find and recognize their mates. They are characteristic chemicals discharged in small scale as a vapor by for all intents and purposes every single known insect. Ever insect have their own unique fragrance. Indeed, sex gives us an accomplished means for observation and control in the insect world. A female insect releases a sex hormone (fragrance) in thousand millionth of a gram several times in a minute. Males of her type follow this sex hormone fragrance to mate with the female. It follows that if we can identify the pheromones then produce that scent. Further we manage the practices of controlling the males of that species. This is the mysterious occurrence of pheromone innovation (Islam, 2012). In 1959, isolation and recognition of insect pheromone was first done by a German scientist on the silkworm. From that point forward, hundreds, maybe a huge number of insect pheromones have been recognized by gradually refined gear. At present, we have a straight and clear perspective of the confinements and potential outcomes related with insect scents (pheromones) in IPM programs. Different sorts of traps are accessible industrially, while many others can be made by farmers cheaply at home. A pheromone baited draws inner side of the trap that attracts male moths inside the trap. Appropriate trap configuration is basic to murder the vermin once it enters the trap. The kind of trap to be used relies upon the decency of the objective insects. Different research works demonstrated the best traps in bother control which are winged traps, channel traps, and delta traps (Islam, 2012).

However, the agricultural sector faces systematic annual losses due to pests and diseases. The damage caused by insect pests is one of the primary factors leading to the reduced production of major crops. For controlling insect pests, many chemical-based pesticides are used in agriculture. Continuous use of pesticides may make the crops progressively susceptible to diseases and insect pests. Excessive use of chemical fertilizers and pesticides also amounts to hamper the productivity of farming lands.

5.2   WHAT ARE BOTANICALS?

Botanicals are the pesticides derived from the plant materials. Some plants have insecticidal properties; they are toxic to insects. Botanical insecticides are naturally occurring chemicals (insect toxins) extracted from plants or minerals. They are also called *natural insecticides*. They are less hazardous, biodegradable, low effective to humans, and more effective against insect-pest due to knock down, antifeedant, repellant, and broad spectrum properties. Botanicals use is an old age practice where people used plant extracts to control the insects. There are many such examples as use of neem dust to control the stored grain pest, use of various plant extracts like *Acorus sps, Melia sps, Acacia sps*, and so on. Products of plant extract containing insecticidal properties are considered as pesticides. However, these products must be registered for use by the Environmental Protection Agency (EPA) and used in accordance with the provisions of the Federal Insecticide, Fungicide and Rodenticide Act (FIFRA). Nicotine, rotenone, and pyrethrum were popular among the plant-based insecticides used for storage pests control and other pests in green houses (Schmutterer, 1981). In Philippines, farmers use a mixture of *Derris* roots, seeds of *Jatropha curcas* and *Barringtonia asiatica* to control *Leptocorisa acuta* on rice (Blauw, 1986; Stoll, 1988). In South-Eastern Nigeria, rural farmers mix chili pepper and wood ash of *Parkia* spp., *Elaeis guineensis*, *Eucalyptus* spp., or *Azadirachta indica* (A.Ju ss) to control *Podagrica*sp, on okra plants, *Abelmoschus esculentus,* Amadi, 1993, personal communication). The natives here also use the mixture of *Chromolaena odorata* L. and *Ocimum gratissimum* L. leaf extracts to repel termites, "tailor ants" and "soldier ants" around their houses.

5.3   IMPORTANCE OF BOTANICAL AND INSECT PEST MANAGEMENT

Botanical insecticides are less hazardous to environment and human health in comparison to synthetic chemical insecticides; it is good alternative of insecticides and pesticides for insect pest management. Bioactivity plant derivatives to arthropod pests continues to expand, yet only a handful of botanicals are currently used in agriculture in the industrialized world, and there are few prospects for commercial development of new botanical products. Pyrethrum and neem are well established commercially used botanicals by old days against insect pest. A number of plant derivatives or extracts are considered for use as insect antifeedants or repellents, some of plant derivatives are used as natural mosquito repellents, little commercial success has ensued for plant substances that modify arthropod behavior. Many commercially used botanicals are used in insect pest management as follows pyrethrins, rotenone, sabadilla, ryania, nicotine, d-limonene, linalool, and neem.

5.3.1   PYRETHRINS

Pyrethrins extracted from pyrethrum flowers (*Tanacetum cinerariifolium*) and are the most widely used botanical insecticide, economically most important class of compounds used in homes and organic agriculture against insect pest management (Casida, 1973). Pyrethrins are selective to target insects; it is neurotoxin that binds to voltage-gated sodium channels of neuronal cells causing the channels to remain open (Davies et al., 2007). Pyrethrins comprise a group of six closely related esters, named pyrethrin I and II, cinerin I and II, and jasmolin I and II. They are found in all aboveground parts of the pyrethrum plants, but predominantly in the ovaries of the flower heads (Brewer, 1973). On an average, the concentration of pyrethrins is about 0.1% (dry weight) in leaves and 1–2% (dry weight) in flowers (Baldwin et al., 1993). Assuming water content of 90%, pyrethrins account for around 0.01% of the fresh weight of leaves and 0.1–0.2% of the fresh weight of flowers. Pyrethrins are effective against a broad spectrum of insects, while their toxicity for mammals is very low, allowing their use as a preharvest spray (Casida and Quistad, 1995; Schoenig, 1995). Pyrethrins are widely regarded as better for the environment and can be harmless if used only in the field with localized sprays as UV exposure breaks them down into harmless compounds. Additionally, they have little lasting effect on plants, degrading naturally or being degraded by the cooking process (Vettorazzi, 1979).

5.3.2   ROTENONE

Rotenone has insecticidal property which is use against insect pest for centuries. Products containing rotenone is extracted from plant species of the genus *Derris* or *Lonchocarpus* (Leguminosae) with the majority from Cubé resin, a root extract of *Lonchocarpus utilis* and *Lonchocarpus urucu* (Isman, 2006). Although rotenone is the major constituent in Cubé resin and hence in rotenone products, the active substances deguelin, rotenolone, and tephrosin are also present. Rotenone-based products are approved for use as organic insecticides under many trade names and most of them are sold as blends containing both rotenone and pyrethrum extracts. Rotenone is a cytotoxic poison, when it gets in contact to the stomach of insects it inhibits electron chain transport system in mitochondria. It also acts as respiratory enzyme inhibitor between NAD^+ (enzyme involved in oxidation and reduction) and coenzyme Q (respiratory enzyme responsible for carrying electrons in electron transport chain system), resulting in the failure of the respiratory function (Ware and Whitecare, 2004). Rotenone was considered safer to be used as a biological agent against insect pest; however, in present studies that have related the rotenone chronic exposure to some appearance of Parkinson's disease in animals, it is no longer considered to be used as biological agent against insect pest. In addition, studies tell that rotenone causes tumors in rats.

5.3.3   SABADILLA

Sabadilla-based products are derived from the seeds of plants from the genus *Schoenocaulon* and are predominantly from the sabadilla lily (*Schoenocaulon officinale*). The activity of sabadilla preparations is primarily due to the alkaloids cevadine and veratridine which typically exist in a 2:1 ratio and are collectively referred to as veratrine. The mode of action of sabadilla alkaloids appears to be similar to that of the pyrethrins as they work on voltage-sensitive sodium channels. Sabadilla affects insect nerve cells causing loss of nerve function, paralysis, and death. The dust formulation of sabadilla is the least toxic of all registered botanical insecticides. However, pure extracts are very toxic if swallowed or absorbed through the skin and mucous membranes. It breaks down rapidly in sunlight and air leaving no harmful residues (Eileen and Sydney, 2002). Sabadilla is approved for use in the United States as an organic insecticide, as well as for other uses, by the Organic Materials Review Institute (OMRI). Sabadilla is a broad-spectrum contact poison but has some activity as a stomach poison. It is commonly used in organic fruit and vegetable production against squash bugs, harlequin bugs, thrips, caterpillars, leaf hoppers, and stink bugs. It is highly toxic to honeybees and hence should only be used in the evening after bees have returned to their hives. Formulations include baits, dusts, or sprays. Sabadilla is permitted by the EPA on certain vegetables, including squash, cucumbers, melons, beans, turnips, mustard, collards, cabbage, peanuts, and potatoes.

5.3.4 NICOTINE

Nicotine and the related alkaloids nornicotin and anabasin are located from aqueous extract from tobacco (*Nicotina* spp., *Solaneace*) and *A. Phylla* (*Chinopodeace*). They induced high insecticidal effects as they are synaptic poison and mimic the neurotransmitter acetylcholine. Therefore, they cause symptoms of poisoning similar to organophosphate and carbamate insecticides (Hayes, 1982). Nicotine is used mostly as a fumigant in greenhouse against soft-bodied pest. Nicotine fatty acids soaps are available in market with reduced bioavailability and toxicity to humans. The nicotinoids act on nervous system of insect pest leading to irreversible blockage of postsynaptic nicotinergic acetylcholine receptors (Casnova et al., 2002).

5.3.5 NEEM

Neem is a newer botanical insecticide. In India, it is well-known as botanical marvel that is, the gift of nature. It is an evergreen tree which is found in tropical and sub-tropical areas of Asia, Africa, America, and Australia. It works in variety of ways from killing all sucking and chewing insects, keeping insects at bay who refuse to eat the sprayed foliage and end up dying of starvation, as well as disrupting the sexual reproduction of insects so that their life cycle is both disrupted and ended.

Neem is perfectly safe to spray on vegetables and fruit crops as well as ornamental shrubs and plants. In India, it is also used in toothpaste, soap, shampoo for nits, cosmetics, and cattle feed. Neem is harmless to other beneficial insects like birds, bees, and ladybirds (Mehelhorn et al., 2011). Because neem is not toxic to humans and other animals, areas that are sprayed with neem are not areas to be avoided like other areas that are sprayed with synthetic pesticides (Mordue and Nisbet, 2000). Neem is being used to manufacture bio insecticide. They are environmental friendly and do not have any toxic effects on plants and soil. Neem insecticide is used to protect both food as well as cash crops like rice, pulses, cotton, oils seeds and so on. It is used in all crops, trees, plants, flowers, fruits, veggies round the home as well as organic and commercial growers. Active ingredient of neem is azadirachtin which acts as an insect repellent and insect-feeding inhibitor, thereby protecting the plants. This ingredient belongs to an organic molecule class called tetranortriterpenoids which is, similar in structure to insect hormones called "ecdysones" which control the process of meta-morphosis as the insects pass from larva to pupa to adult. The major parts/extracts of neem seed that are used for making neem insecticides are the neem seed kernels and the neem seed oil. Neem seed kernels extracts contain azadirachtin which in turn works by inhibiting the development of immature insects. Neem oil or the neem seed oil is extensively used to manufacture insecticides used for different crops. Neem oil enters the system of the pests and obstructs their proper working. Insects do not eat, mate, and lay eggs resulting in the breaking of their life cycle.

The efficacy of neem extract is dependent on the physiological action of aszadirechtin, a nor triterpinoid belonging to the limonoids. It is highly biologically active ingredient of neem. It acts as an insect growth regulator on larval insects like disruption of moulting, growth inhibitor, and malformation which may contribute to motility. This is attributed to a disruption of endo-crine events as the downregulation of hemolimph ecdysteroid level through the blockage of release of PTTH, prothoracicotropic hormone, from the brain carpous cardiacum complex are to a delay in the appearance of the last ecdys-teroid peek showing a complete moult inhibition. Neem products are also effective on allatropin and juvenile hormone titers (Mordue and Blackwell, 1993). It has an antifeedant property whose effect is highly variable among pest species and even those species which are initially deterred and are often capable of rapid desensitization to azadirachtin (Bomford and Isman, 1996).

5.3.6 *LIMONENE*

Limonene is a naturally occurring monoterpene which is found in citrus, other fruits, conifers, and spices. Limonene is the major component of oil recovered from citrus rind when fruits are juiced (Florida Chemical, 2004). It has a strong pleasant odor that is typically associated with the smell of oranges or lemons. Limonene is used in a variety of foods and beverages and is classed by the U.S. Food and Drug Administration as a Generally Recognized as Safe (GRAS) compound when used as a food additive or flavoring (EPA, 1994). Limonene is frequently included as an ingredient in cleaning solutions, particularly those that are designed to cut grease or remove wax or oil (Florida Chemical, 2004). The cost of technical grade d-limonene in truckload quantities generally varies from $0.60 to $1.00 per pound (Laurie Winget, personal communication). In 1994, limonene was registered as a pesticide active ingredient in 15 products: for use against ticks and bees, as an insecticide spray, as an outdoor dog and cat repellent, as a by repellent on tablecloths, as an insect repellent for use on humans, and as a mosquito larvicide (EPA, 1994). If a new pesticide product containing limonene as the active ingredient is developed for control of mealybugs and scales, registration of the product with EPA would be required. However, limonene is exempted from the requirement of a tolerance in food when it is used as an inert ingredient (or "occasionally active") as a solvent or fragrance in pesticide formulations (Code of Federal Regulations, 2003). At least one US patent (no. 5,653,991, published in 1997 by Robert L. Rod; USPTO 1997) refers to using various oil-based formulations of d-limonene, with or without a water carrier, against plant pests such as white bees. To my knowledge, the only product, currently in the market, intended for control of plant pests using limonene as the active ingredient is Orange Guard for Ornamental Plants (Orange Guard, Inc., Carmel Valley, CA). This product, which was originally developed for control of household pests and ants, contains 5.8% d-limonene. The label recommends a 1:4 Ð1:6 dilution of the product with water for use on plants (Orange Guard, Inc. 2005). The toxicity and neurotoxic effects of monoterpenoids (including d-limonene) are discussed by Coats et al. (1991) and the suitability of limonene for control of insect pests has been reviewed by Ibrahim et al. (2001). Several reports mention use of limonene for control of plant pests, including use on pine seedlings to reduce egg clusters of a notodontid moth (Tiberi et al., 1999) and the use of 1Ð 6% limonene solutions on carrot to repel a psyllid [Aaltonen et al., 2000, as cited in Ibrahim et al. (2001)]. Hummel Brunner and Isman (2001) used tobacco cutworms to test acute and sublethal effects of topically applied monoterpenoid essential oil compounds (including d-limonene). They found that mixtures of different monoterpenes produced a synergistic effect on mortality and they developed a proprietary monoterpene mixture containing 0.9% active ingredient for use against foliar feeding pests. Unlike soaps and oils, limonene evaporates quickly from leaves, leaving no residues that might cause a delayed phytotoxic response. Despite its advantages, phyto-toxicity concerns will limit the use of limonene on certain types of plants or plant parts. Limonene was phytotoxic to strawberries when used at concentrations exceeding 3%, and cabbage and carrot seedlings were damaged when concentrations exceeded 9% (Ibrahim et al., 2001). In my study, 1% limonene damaged moss, ferns, ginger, certain types of dracaena, and delicate powers. These developed water-soaked areas that later became necrotic. However, limonene solutions generally caused no damage to ornamentals with thick, waxy leaves, such as palms, cycads, and orchids.

5.3.7 PHEROMONES

Agriculture and environment are firmly intertwined. Agricultural food production system is totally dependent on the environment for usage of land, daylight duration, rainfall, insect pests, and diseases. Pest is one of the major problems for carrying out higher production in agriculture crops. Farmers are generally dependent on synthetic insecticides pesticides because they are easily available, highly advertised and popular, cheap, easy to use, and have quick results. However, use of insecticides also kills non-target arthropods, rare insects involved in pollination and predators such as spiders and ground beetles. Insecticide residues contaminate the watercourses, especially in rice cultivation, and affect the food and water we eat (Cork et al., 2003; 2005). Moreover, quite often the unplanned and unscientific use of insecticides has resulted in many problems, such as insects developing resistance, renaissance of once lesser pest into a major problem, and also food safety and environmental hazards. The pests are becoming resistant to almost all chemical pesticides as the frequency of spraying is gradually increasing while their efficacy is gradually decreasing. Use of pheromones is one among the most powerful alternative approaches to control insect pests. Since few decades, researchers are focusing on investigation and development of alternate strategies for the minimization of pest damage caused in field and also to reduce serious side effects like environment and health problems from the conventional use of traditional chemical pesticides for routine arthropod pest management.

Amazingly, diverse and effective uses of pheromones have been established since the discovery of first insect pheromone (Butenandt, 1959). Pheromones are an integral part of insect integrated pest management (IPM) systems programs since decades. Other behavior-modifying chemicals have had more limited success on commercial levels and it still remains to be seen whether other behavior-modifying chemicals such as host plant volatiles can become as widely used as pheromones for insect in comparable field situations. Pheromones have been described as substances produced by insects that have a specific effect on members of their own species (Bartell, 1977). They were originally defined as "substances which are secreted to outside by an individual and received by a second individual of the same species, in which they release a specific reaction, for example, a definite behavior or a developmental processs" (Karlson and Luscher, 1959). These chemical signals are secreted in minute quantities by many insects for communication with others of their kind. They can evoke many responses including aggregation, alarm, trail-following, defense, feeding, and reproduction. Pheromones, in small amounts, attract insects to the emitter. Pheromones are natural substances that are produced by special glands in the abdomen of insects and it attracts the opposite gender of the same species. Insects produce pheromones for attracting the mate (e.g., most moths), for marking foraging routes (e.g., ants) or to signal alarm to neighbors (e.g., aphids). The discovery and applications of insect pheromones is an active area of research and new progress continues to be made. Potentially, pheromones may be used to control certain harmful insects and to attract insects to insecticide or chemo sterilant baits, lower the number of insecticidal sprays, or to distract insects and disturb mating. Application of traps as a sampling tool to resolve need for and timing of control regulation can arrange the basis of IPM methods for these pests.

5.4 PHEROMONE TRAPS

Monitoring for pests and diseases is a fundamental first step in creating a proper integrated pest management (IPM) program. The use of pheromones for controlling insect's pest requires three items: a pheromone chemical, a trap, and a support to hang the trap in the field. Technically sex pheromones can be used in three principal ways: detection and monitoring, mass trapping, and mating disturbance. Insects produce pheromones for attracting the mate (e.g., most moths), for marking foraging routes (e.g., ants) or to signal alarm to neighbors (e.g., aphids). In case of pheromone traps, the lure slowly releases synthetic attractants that helps in detection of a single species of insect. The basic parts of a wing pheromone trap are: top section (nonsticky portion), bottom section (sticky portion), and lure (pheromone cap).

Frequently used pheromones are the "sex pheromones" which a female produces to attract a mate. Pheromones are most well-known for Lepidoptera (moths and butterflies) and these chemical messengers can be commercially produced by synthesizing and blending the appropriate chemicals. For pheromone trap use, the pheromone chemicals are commonly forced into a rubber "septa" (a small rubber cap), which can be placed in a sticky trap to attract male moths.

Different types of pheromone traps are easily commercially available, while others can be made by the farmers at a low cost at home. A pheromone bait attracts male moths toward the inner side of the trap and carries them inside the trap. Appropriate trap design is very tough to kill the pest once it enters the trap. The type of trap is used according the behavior of the particular insect. Various research reports show the delta traps, funnel traps, and winged traps are highly effective traps in pest control. Pheromone traps are very sensitive which means they attract insects that are less in number. They are often used to distinguish the presence of exotic pests, for observing, sampling, or to determine the first occurrence of a pest in an area. The potential of traps for observing pest populations also depends on suitable placement. Maximum of one trap per 10 acres should be fixed in the field's center or where prevailing winds will transfer the pheromone into the planting.

Pheromones are chemicals for species specific communication. Most often, these sex pheromones are produced by females to attract a mate and are most well-known for adult Lepidoptera. Commercially, it is produced by synthesizing and blending the appropriate chemicals. The sex pheromones are loaded into dispensers which can be placed in traps of various designs for deployment in agriculture, horticulture, forestry, and storage. Pheromone traps are the most popular and widely used tools for pest detection and population monitoring. Pheromone traps have been exploited for three useful applications: (1) monitoring; (2) mass trapping; and (3) mating disruption. The most important and widespread practical applications of sex pheromones in pest management have been reviewed recently (Witzgall et al., 2010). Population monitoring relates trap captures to the abundance of, or to the damage caused by, an insect species. The numbers caught over time have been used for initiating field scouting for egg laying, and assessing the need for timing of control measures based on action thresholds (Wall et al., 1987; Gurrero and Reddy, 2001). However, traps do not always accurately indicate the overall pest pressure for use as thresholds for action, as trap catches are influenced by the efficacy of the lure, the dispenser (Arn et al., 1997), the trap design (Fadamiro, 2004; Spear-O'Mara and Allen, 2007), and the trap location (Reardon et al., 2006; Gallardo et al., 2009). Pheromone traps are the most effective and sensitive enough to detect low-density populations. They are, therefore, handy tools for tracking invasive species in the establishment phase (El Sayed et al., 2006; Liebhold and Tobin, 2008) or for population monitoring to determine the extent of an outbreak area and the effectiveness of eradication campaigns (Cannon et al., 2004).

The timing of adult male caught in the trap indicates the start of the pest flight activity in the area. This information is important for some pests, as it is used as the biofix date for accumulation of heat units above a base temperature in phenology models or sustained first flight for others (Knutson and Muegge, 2010). Sex pheromone traps are useful for monitoring difficult pests that evade early detection of economic damage when a trap catch is used to calculate: (1) growing degree-days (GDD) for onset and completion of moth emergence (Spear-O'Mara and Allen 2007; Knutson and Muegge, 2010); (2) starting dates of egg hatch (Isaacs and van Timmeren, 2009); and (3) onset of first larval damage (Knutson and Muegge, 2010). A linear relationship between male caught in sex pheromone traps and GDD is possible after appropriate transformation of variables (Gallardo et al., 2009) and in some cases variability is better explained by including other variables related to density of host plants or suitable plant parts (Spear-O'Mara and Allen, 2007). Validation of the degree day model is done by comparing the timing of predicted and observed phenological events through field scouting and damage assessments, and estimating the prediction accuracy and error (Knutson and Muegge, 2010). Monitoring through a network of sites is most useful for studying spatial distributions of pests, early detection of infestations, and identification of hot-spot locations to initiate appropriate management interventions on a spatial scale. Monitoring at the regional level improves the reliability of population monitoring for implementation of appropriate area-wide IPM systems (Ayalew et al., 2008).

Moth captured in a network of pheromone trap sites established across the Canadian prairies, when used in conjunction with back ward trajectories provided by meteorological services, were helpful in providing early detection of diamondback moth infestations (Hopkinson and Soroka, 2010). Peak trap captures are often correlated with associated weather to identify positive or negative influences of weather parameters on moth activity and pest build-up (Gwadi et al., 2006; Reardon et al., 2006; Monobrullah et al., 2007; Prasad et al., 2008). However, trap catches and weather may not necessarily serve as predictors of the future abundance of certain species in cropping regions (Baker et al., 2010).

5.5 LIGHT TRAPS

Insect attraction to light has been exploited for monitoring insect populations with a view to provide early warning of the presence of pests, as well as for many other uses. Light traps have been widely used for monitoring the population dynamics of Lepidoptera and Coleoptera (Wolda 1992; Watt and Woiwod, 1999; Kato et al., 2000). When compared with other sampling methods, light-trap sampling was found to be more efficient for lepidopteran population dynamics (Raimondo et al., 2004). However, many factors affect catches of insects in light traps (Bowden, 1982). Trap design, the light source and its energy, and the attraction efficiency under certain conditions all contribute to sampling errors. The effects of weather conditions and moonlight on light-trap catches are well-documented. For example, trap efficiency for Lepidoptera is positively correlated with temperature and the thickness of cloud cover, and negatively correlated with wind speed, precipitation, and the fullness of the moon on the trap night (Bowden, 1982; Dent and Pawar, 1988; Yela and Holyoak, 1997; Butler et al., 1999). The effect of weather factors on the abundance or species richness of Coleoptera captured by light traps has been reported (Rodriguez-DelBosque, 1998). Networks of light traps have been used for year-round monitoring of moth species and the data used to assess the magnitude and reasons for seasonal, annual and long-term faunal changes and their population dynamics in Britain (Lewis, 1980) and India (Anon 2009), and for weekly larval forecasts on cereal crops in Africa (Odiyo, 1979). Light-trap captures have been used to predict the emergence date of adult beetles from overwintering using a degree day model (Zou et al., 2004) and for prediction of population sizes based on moth catches (Raimondo et al., 2004). Long-term light-trap data is highly useful in studying the seasonal dynamics of pests. For example, regression analyses have indicated that the spring generation of two species of Helicoverpa in eastern cropping zones in Australia could be related to rainfall in putative inland source areas (Zalucki and Furlong 2005). Light-trap catch data is also useful for validation of simulation model outputs (Reji and Chander, 2008).

Pheromone traps (traps baited with these lures) are not used only for controlling pests but they also aid in determining if a pest is present and whether a population is increasing in numbers, peaking, or decreasing. This information is important in resolving when and how often to time control actions. Different traps should not be handled at the same time since cross-contamination will affect their development. Insect traps are not only used for detection and observation of pest problem but also they provide measures regarding pest population and crowdness in the examination area. If conducted consistently over multiple years, insect traps can indicate critical changes in population dynamics and behavior of key pests. Effective deployment and routine checks can provide information for calculating economic thresholds; thus, money invested in long-term monitoring with multiple traps can potentially save thousands of dollars of insecticides and protect the environment.

Pheromone programs have been regulated by several decades around the globe and till date, there is no reports public health proof to suggest that agricultural use of artificial pheromones is harmful to humans or to any other non-target species. However, continuous research is being conducted. In order to reduce the use of pesticide in agricultural crops and with environment sustainability, certain behavioral chemicals could be harmless.

5.6 COMPARISON WITH CHEMICAL PESTICIDES

Ever-increasing use of chemical pesticides has evoked their resistance in these pest populations. This resistance often leads to an increase in the frequency and dosage rate of subsequent pesticide applications, and this "pesticide treadmill" can ultimately lead to yield reduction instead of yield enhancement or yield stabilization. Another serious concern while using pesticides is that they often kill natural parasites and predators of insect pests (Hassan et al., 1987) and in this way, any potential for natural biological control is reduced. Further, removal of these biological agents leads to outbreaks of secondary pests and increases the need for additional pesticide applications.

Increasing complexity of chemical structure and synthesis and also a stricter regulatory climate is increasing public chemo phobia and eventually leading to decrease in arrival of new chemical pesticides in market. Clearly, there exists a need for the development and adoption of alternatives in pest management. Insect pest management is in a transition. Research is focused on the development of integrated pest management (IPM) systems which address the intelligent selection and use of pest control strategies that will ensure favorable economic, ecological, and sociological consequences. Public chemophobia is forcing a rapid re-evaluation of use of pesticides for pest control strategies. As a powerful solution to this, pheromone-based pest control tactics stands high.

5.6.1   *POTENTIAL APPLICATIONS OF PHEROMONES*

Insect pheromone-related technologies for monitoring endemic pest populations, detecting invasive species, mass trapping for population suppression, and mating disruption have had a relatively recent history of development in IPM compared to biological control and insecticide technologies. Pheromone-based control tactics present many advantages. The use of pheromones for pest control promises as an important component of the on-going challenge to develop alternatives that may help to solve major environmental and human health problems associated with chemical pesticide use in agriculture.

The most widespread use of pheromones has been for monitoring endemic pest species' adult populations, detecting invasive species, mass trapping for population suppression, and mating disruption. Practical applications of pheromones in pest management can fit into the following categories:

- Trapping for detection in monitoring and survey.
- Monitoring levels of insecticide resistance using pheromones as attractants.
- Luring insects to areas treated with insecticides.
- Luring insects to areas treated with pathogens which are then spread by the infected individuals to the rest of the population.
- Mass-trapping for population suppression.
- Disrupting communication within the insect population by permeation of the atmosphere with pheromone. When using sex pheromones, mating disruption should result in population suppression.

5.7   CONCLUSION

Botanical insecticides are less hazardous to environment and human health in comparison to synthetic chemical insecticides. They are good alternative of insecticides and pesticides for insect pest management. Pyrethrins, rotenone, sabadilla, ryania, nicotine, d-limonene, linalool, and neem are used as insecticides and pesticides against insect pests. They are more effective in controlling disease, insect pests, and are less harmful against humans. Pheromones, sex pheromones, and pheromone traps are the most popular and widely used tools for pest detection and population monitoring. Pheromone traps are most important and widespread practical applications of sex pheromones in pest management. Light traps are very useful for monitoring the population dynamics of Lepidoptera and Coleoptera. Thus, botanicles, pheromone traps, and light traps are very useful in controlling insect pest and disease.

KEYWORDS

- **botanicals**
- **insect control**
- **insect pests**
- **light trap**
- **pheromones trap**

REFERENCES

Aaltonen, M. A.; Aflatuni; P. Parikka. Limoneenilla kemppi kuriin. *Puutarha Kauppa* **2000,** *4*, 45.

Akhtar, N.; Mandal, K. A. M. S. H. Effects of Caffeine and Castor Oil on Growth of *Tribolium castaneum* Herbst (Coleoptera: Tenebrionidae). *Bangladesh J. Entomol.* **2008,** *19* (1), 1–8.

Anon. Progress Report, 2008, Crop Protection (Entomology, Plant Pathology). All India Coordinated Rice Improvement Programme (ICAR), Directorate of Rice Research, Hyderabad, India, 2009, 2.

Arn, H.; Brauchli, J.; Koch, U. T.; Pop, L.; Rauscher, S. The Need for Standards in Pheromone Technology. *IOBC/WPRS Bull.* **1997,** *20,* 27–34.

Ayalew, G.; Sciarretta, A..; Baumgartner, J.; Ogol, C.; Lohr, B. Spatial Distribution of Diamondback Moth, *Plutella xylostella* L. (Lepidoptera: Plutellidae), at the Field and the Regional Level in Ethiopia. *Int. J. Pest Manag.* **2008,** *54*, 31–38.

Baker, G. H.; Tann, C. R.; Fitt, G. P. A Tale of Two Trapping Methods: *Helicoverpa* spp. (Lepidoptera, Noctuidae) in Pheromone and Light Traps in Australian Cotton Production Systems. *Bull. Entomol. Res.* **2010,** *27*, 1–15.

Baldwin, I. T.; Karb, M. J.; Callahan P. Foliar and Floral Pyrethrins of *Chrysanthemum cinerariaefolium* are not Induced by Leaf Damage. *J. Chem. Ecol.* **1993,** *19*, 2081–2087.

Bartell, R. J. Behavioral Responses of Lepidoptera to Pheromones. In *Chemical Control of Insect Behavior. Theory and Application*; H. H. Shorey, Ed.; Wiley and Sons: New York, New York, 1977; pp 201–213.

Bowden, J. An Analysis of Factors Affecting Catches of Insects in Light Traps. *Bull. Entomol. Res.* **1982,** *72*, 535–556.

Brewer, J. G. Microhistological Examination of the Secretory Tissue in Pyrethrum Florets. *Pyrethrum Post* **1973,** *12*, 17–22.

Butenandt, A. Wirkstoffe des Insektenveiches. *Naturwissenschaften* **1959,** *46*, 461–471.

Butler, L.; Kondo, C.; Barrows, E. M.; Townsend, E. C. Effects of Weather Conditions and Trap Types on Sampling for Richness and Abundance of Forest Macrolepidoptera. *Environ. Entomol.* **1999**, *28*, 795–811.

Cannon, R. J. C.; Koerper, D.; Ashby, S. Gypsy Moth, *Lymantria dispar*, Outbreak in Northeast London, 1995–2003. *Int. J. Pest Manag.* **2004**, *50*, 259–273.

Casida, J. E. *Pyrethrum, the Natural Insecticide*; Academic: New York, 1973.

Casida, J. E; Quistad, G. B. *Pyrethrum Flowers: Production, Chemistry, Toxicology, and Uses*; Oxford University Press: New York, 1995.

Coats, J. R.; Karr, L.L.; Drewes C. D. Toxicity and Neurotoxic Effects of Monoterpenoids: in Insects and Earthworms. In *Naturally Occurring Pest Bioregulators;* Hedin, P. A., Ed.; ACS Symposium Series, American Chemical Society: Washington, DC, 1991; pp 305–316.

Code of Federal Regulations. Tolerances and Exemptions from Tolerances for Pesticide Chemicals in Food. *Title 40*, 2003, 21.

Cork, A.; Kamal, Q. N.; Alam, S. N.; Choudhury, S. C. J.; Talekar, N. S. Pheromones and their Application to Insect Pest Control-A Review. *Bangladesh J. Entomol.* **2003**, *13*, 1–13.

Cork, A.; Alam, S. N.; Talekar, N. S. In *Development and Commercialization of Mass Trapping for Control of Brinjal Borer, Leucinodes orbonalis in South Asia*. Proceedings of National Symposium on Recent Advances in Integrated Management of Brinjal Shoot and Fruit Borer, 2005, 29–33.

Davies T. G.; Field L. M.; Usherwood P. N. DDT, Pyrethrins, Pyrethroids and Insect Sodium Channels. *Williamson MSIUBMB Life* **2007**, *259* (3), 151–62.

Dent, D. R; Pawar, C. S. The Influence of Moonlight and Weather on Catches of *Helicoverpa armigera* (Hübner) (Lepidoptera, Noctuidae) in Light and Pheromone Traps. *Bull. Entomol. Res.* **1988**, *78*, 365–377.

Eileen, A. B.; Sydney, G. Park-Brown. *Natural Products for Insect Pest Management*; Institute of Food and Agricultural Sciences, University of Florida, 2002.

El-Sayed, A. M.; Suckling, D. M.; Wearing, C. H.; Byers, J. A. Potential of Mass Trapping for Longterm Pest Management and Eradication of Invasive Species. *J. Econ. Entomol.* **2006**, *99*, 1550–1564.

Fadamiro, H. Y. Pest Phenology and Evaluation of Traps and Pheromone Lures for Monitoring Flight Activity of Oblique Banded Leaf Roller (Lepidoptera: Tortricidae) in Minnesota Apple Orchards. *J. Econ. Entomol.* **2004**, *97*, 530–538.

Fang, N.; Casida J. E. Cubé resin Insecticide: Identification and Biological Activity of 29 Rotenoid Constituents. *J. Agric. Food Chem.* **1999**, *47* (5), 2130–6.

Fields, P. G.; Xie, Y. S.; Hou, X. Repellent Effect of Pea (*Pisum sativum*) Fractions Against Stored-product Pests. *J. Stored Prod. Res.* **2001**, *37*, 359–70.

Florida Chemical Company, Inc. Florida Chemical Company, Inc., Winter Haven, FL. http://www. ßoridachemical.com, 2004. (accessed June 17, 2016).

Forget, G.; Goodman, T.; de-Villiers, A. *Impact of Pesticide Use on Health in Developing Countries*; Int. Dev. Res. Centre: Ottawa, 1993; p 335.

Gallardo, A.; Ocete, R.; Lopez, M. A.; Maistrello, L.; Ortega, F.; Semedo, A.; Soria, F. J. Forecasting the Flight Activity of *Lobesia botrana* (Denis and Schiffermüller) (Lepidoptera, Tortricidae) in Southwestern Spain. *J. Appl. Entomol.* **2009**, *133*, 626–632.

Gurrero, A.; Reddy, G. V. P. Optimum Timing of Insecticide Applications Against Diamondback Moth Plutella xylostella in Cole Crops Using Threshold Catches in Sex Pheromone Traps. *Pest Manag. Sci.* **2001**, *57*, 90–94.

Gwadi, K. W.; Dike, M. C.; Amatobi, C. I. Seasonal Trend of flight Activity of the Pearl Millet Stemborer, *Coniesta ignefusalis* (Lepidoptera: Pyralidae) as Indicated by Pheromone Trap

Catches and Its Relationship with Weather Factors at Samara, Nigeria. *Int. J. Trop. Insect Sci.* **2006**, *26*, 41–47.

Hassan, S. A.; Albert, R.; Bigler, F.; Blaisinger, P.; Bogenschutz, H.; Boiler, E.; Brun, J.; Chiverton, P.; Edwards, P.; Englert, W.D.; Huang, P.; Inglesfield, C.; Naton, E.; Oomen, P. A.; Overmeer, W. P. J.; Rieckmann, W., Samsoe-Petersen, L.; Staubli, A.; Tuset, J. J.; Viggiani, G.; Vanwetswinkel, G. Results of the Third Joint Pesticide Testing Program by the IOBC/WPRS-Working Group "Pesticides and Beneficial Organisms." *J. Appl. Entomol.* **1987**, *103*, 92–107.

Hopkinson, R. F.; Soroka, J. J. Air Trajectory Model Applied to an In-Depth Diagnosis of Potential Diamondback Moth Infestations on the Canadian prairies. *Agric. Forest* Meteorol. **2010**, *150*, 1–11.

Ibrahim, M. A.; Kainulainen, P.; Aflatuni, A.; Tiilikkala, K.; Holopainen, J. K. Insecticidal, Repellent, Antimicrobial Activity and Phytotoxicity of Essential Oils: With Special Reference to Limonene and its Suitability for Control of Insect Pests (Review). *Agric. Food Sci. Fin.* **2001**, *10*, 243–259.

Isaacs, R.; van Timmeren, S. Monitoring and Temperature-based Prediction of the Whitemarked Tussock Moth (Lepidoptera: Lymantridae) in Blueberry. *J. Econ. Entomol.* **2009**, *102*, 637–645.

Islam, M. A.; Begum, S. Quantitative Analysis of α–Mangostin in Mangosteen Fruit Rind Extract. *Int. J. Agril. Res. Innov. Tech.* **2011**, *1* (1–2), 55–59.

Islam, M. A. Pheromone Use for Insect Control: Present Status and Prospect in Bangladesh. *Int. J. Agril. Res. Innov. Tech.* **2012**, *2* (1), 47–55.

Islam, M. A. Study on the Lepidopteran Sex Pheromones Including Multiple Double Bonds. Ph.D. Thesis, Graduate School of Bioapplications and System Engineering, Tokyo University of Agriculture and Technology, Tokyo, Japan, 2009, p 147.

Isman, M. B. Botanical Insecticides, Deterrents, and Repellents in Modern Agriculture and an Increasingly Regulated World. *Ann. Rev. Entomol.* **2006**, *51*, 45–66.

Karlson, P.; Luscher, M. Pheromones: A New Term for a Class of Biologically Active Substances. *Nature* **1959**, *183*, 55–56.

Kato, M.; Itioka, T.; Sakai, S.; Momose, K.; Yamane, S.; Hamid, A. A.; Inoue, T. Various Population Fluctuation Patterns of Light-attracted Beetles in a Tropical Lowland Dipterocarp Forest in Sarawak. *Pop. Ecol.* **2000**, *42*, 97–104.

Knutson, A. E.; Muegge, M. A. A Degree-day Model Initiated by Pheromone Trap Captures for Managing Pecan Nut Casebearer (Lepidoptera: Pyralidae) in Pecans. *J. Econ. Entomol.* **2010**, *103*, 735–743.

Lewis, T. Britain's Pest Monitoring Network for Aphids and Moths. *EPPO Bull.* **1980**, *10*, 39–46.

Liebhold, A. M.; Tobin, P. C. Population Ecology of Insect Invasions and Their Management. *Ann. Rev. Entomol.* **2008**, *53*, 387–408.

Mamun, M. S. A.; Shahajahan, M.; Ahmad, M. Laboratory Evaluation of Some Indigenous Plant Extracts as Repellent Against Red Flour Beetle, *Tribolium castaneum* Herbst. *Bangladesh J. Entomol.* **2008**, *18* (1), 91–99.

Marco, G. J.; Hollingworth, R. M.; Durham, W. *Silent Spring Revisited*; American Chemical Society: Washington, DC, 1987; p 214.

Monobrullah, M.; Bharti, P.; Shankar, U.; Gupta, R. K.; Srivastava, K.; Ahmad, H. Trap Catches and Seasonal Incidence of *Spodoptera litura* on Cauliflower and Tomato. *Ann. Plant Prot. Sci.* **2007**, *15*, 73–76.

Mordue, A. J.; Nisbet, A. J. Azadirachtin from the Neem Tree *Azadirachta indica*: Its Action Against Insects. *Ann. Soc. Entomol. Bras.* **2000**, *29* (4), 615–635

National Research Council. *The Future Role of Pesticides in US Agriculture*; National Academy Press: Washington, DC, 2000; p 301.

Odiyo, P. O. Forecasting Infestations of a Migrant Pest: The African Armyworm *Spodoptera exempta (Walk.)*. Philosophical Transactions of the Royal Society. *Biol. Sci.* **1979**, *287*, 403–413.

Orange Guard, Inc. Orange Guard, Home Pest Control Water Based Formula. http://www.orangeguard.com., 2005. (accessed Feb 20, 2016).

Perry, A. S.; Yamamoto, I.; Ishaaya, I.; Perry, R. Y. *Insecticides in Agriculture and Environment: Retrospects and Prospects*; Springer-Verlag: Berlin, 1998; p 261.

Prasad, N. V. V. S. D.; Mahalakshmi, M. S.; Rao, N. H. P. Monitoring of Cotton Bollworms Through Pheromone Traps and Impact of Abiotic Factors on Trap Catch. *J. Entomol. Res.* **2008**, *32*, 187–192.

Raimondo, S.; Strazanac, J. S.; Butler, L. Comparison of Sampling Techniques used in Studying Lepidoptera Population Dynamics. *Environ. Entomol.* **2004**, *33*, 418–425.

Reardon, B. J.; Sumerford, D. V.; Sappington, T. W. Impact of Trap Design, Windbreaks, and Weather on Captures of European Corn Borer (Lepidoptera: Crambidae) in Pheromone-baited Traps. *J. Econ. Entomol.* **2006**, *99*, 2002–2009.

Regnault-Roger, C.; Philogene, B. J. R.; Vincent, C., Eds. *Biopesticides of Plant Origin*; Lavoisier: Paris, France, 2005.

Reji, G.; Chander, S. A Degree-day Simulation Model for the Population Dynamics of the Rice Bug, *Leptocorisa acuta (Thunb.)*. *J. Appl. Entomol.* **2008**, *132*, 646–653.

Rodriguez-Del-Bosque, L. A. A Sixteen-year Study on the Bivoltinism of *Anomala favipennis* (Coleoptera, Chrysomelidae) in Northern Mexico. *Environ. Entomol.* **1998**, *23*, 1409–1415.

Schoenig, G. P. Mammalian Toxicology of Pyrethrum Extract. In *Pyrethrum Flowers Production, Chemistry, Toxicology, and Uses*; Casida, J. E., Quistad, G. B., Eds.; Oxford University Press: New York, 1995; pp 249–257.

Spear-O'Mara, J.; Allen, D. C. Monitoring Populations of Saddled Prominent (Lepidoptera: Notodontidae) with Pheromone-baited Traps. *J. Econ. Entomol.* **2007**, *100*, 335–342.

Thacker, J. M. R. *An Introduction to Arthropod Pest Control*; Cambridge University Press: Cambridge, UK, 2002; p 343.

Tiberi, R.; A. Niccoli, M.; Curini, F.; Epifano, M. C.; Marcotullio; O. Rosati. The Role of the Monoterpene Composition in Pinusspp. Needles, in Host Selection by the Pine Processionary Caterpillar, Thaumetopea pityocampa. *Phytoparasitica* **1999**, *27*, 263–272.

U.S. Patent and Trademark Office. Patent No. 5,653,991. United States Patent and Trademark Office. 1997. http://www.uspto.gov/patft/. (accessed June 13, 2016).

Vettorazzi, G. *International Regulatory Aspects for Pesticide Chemicals;* CRC Press: USA, 1979; pp 89–90.

Wall, C.; Garthwaite, D. G.; Blood Smyth, J. A.; Sherwood, A. The Efficacy of Sex-attractant Monitoring for the Pea Moth, *Cydia nigricana*, in England, 1980–1985. *Ann. Appl. Biol.* **1987**, 110, 223–229.

Ware, G. W. *Pesticides. Theory and Application*; Freeman: San Francisco, 1883; p 308.

Watt, A. D.; Woiwod, I. P. The Effects of Phenological Asynchrony on Population Dynamics: Analysis of Fluctuations of British Macrolepidopters. *Oikos* **1999**, *87*, 411–416.

Witzgall, P.; Kirsch, P.; Cork, A. Sex Pheromones and Their Impact on Pest Management. *J. Chem. Ecol.* **2010**, *36*, 80–100.

Wolda, H. Trends in Abundance of Tropical Forest Insects. *Oecologia* **1992,** *89*, 47–52.

Yela, J. L.; Holyoak, M. Effects of Moonlight and Meteorological Factors on Light and Bait Trap Catches of Noctuid Moths (Lepidoptera: Noctuidae). *Environ. Entomol.* **1997,** *26*, 1283–1290.

Zalucki, M. P.; Furlong, M. J. Forecasting Helicoverpa Populations in Australia: A Comparison of Regression Based Models and a Bioclimatic Based Modeling Approach. *Insect Sci.* **2005,** *12*, 45–46.

Zou, L.; Stout, M. J.; Ring, D. R. Degree-day Models for Emergence and Development of the Rice Water Weevil (Coleoptera: Curculionidae) in Southwestern Louisiana. *Environ. Entomol.* **2004,** *33*, 1541–1548.

CHAPTER 6

Plant Growth Promoting Rhizobacteria for the Control of Soil-Borne Diseases

POOJA VERMA[1,*], NEELAKANTH S. HIREMANI[1], RACHNA PANDE[1], and MD. SHAMIM[2]

[1]Indian Council of Agricultural Research-Central Institute for Cotton Research, Shankar Nagar, Nagpur 440010, Maharashtra, India

[2]Department of Molecular Biology and Genetic Engineering, Dr. Kalam Agricultural College, Kishanganj, Bihar Agricultural University, Sabour, Bhagalpur 813210, Bihar, India

*Corresponding author. E-mail: poojaverma1906@gmail.com

ABSTRACT

In nature, soil represents the physical covering of the earth's surface, comprises three material states namely, solid (including geological and dead biological materials), liquid (including water), and gas (with air in soil pores). It acts as an excellent niche for the growth of several microbial faunae such as protozoa, fungi, viruses, and bacteria. Rhizosphere, region surrounding the plant roots, has several microorganisms that promote the growth of plants in a mutualistic way. Plants allow the microorganisms to colonize the root by secreting several chemicals into the rhizosphere, which facilitate the colonization. In several literatures, the interaction between plants and microorganisms has been classified as pathogenic, saprophytic, and beneficial. In this chapter, we have discussed about plant growth promoting rhizobacteria, mechanism of action, and their contribution toward control of several soil-borne diseases.

6.1 INTRODUCTION

Plant growth promoting rhizobacteria (PGPR) are the soil bacteria inhabiting around/on the root surface, contributing in promotion of plant growth and development directly or indirectly via production and secretion of various regulatory chemicals in the environs of the rhizosphere. PGPR can be grouped into two categories depending on their relationship with the plants: symbiotic bacteria and free-living rhizobacteria (Khan, 2005). As reviewed by Glick (1995, 2001), Hallman et al. (1997), Hall (2002), Sturzet et al. (2000), Lucy et al. (2004), Welbaum et al. (2004), and Compant et al. (2005), a lot of work has been carried out on the mechanisms and principles of the PGPR-plant relationship, and widely accepted as rhizosphere effect (Hiltner, 1904; Kennedy, 2005). Free-living rhizobacteria occupying the rhizosphere and rhizoplane of plants can be placed in a number of different groupings. Some consider PGPR and biocontrol agents (BCAs) to be completely separate groups while others consider BCAs to be a subgroup of PGPR. Bashan and Holguin (1998), for example, proposed PGPR to be fallen into two groups namely "biocontrol-plant growth promoting bacteria" and "plant growth promoting bacteria."

Soil biology is directly coupled to agricultural sustainability since it is the most important driving force behind decomposition processes; need to break down complex organic compounds, molecules, and substances to the simple forms and making them available to plants (Friedel et al., 2001). Most of the soil-borne pathogens are difficult to control by conventional strategies such as the use of resistant cultivars and synthetic pesticides (Weller et al., 2002). On the other hand, soil application of fungicides is much expensive and lethal to nontarget microflora. Hence, biological control, nowadays, has become a critical factor for plant disease management and has proven to be a practical and safe approach in various crops (Patel and Anahosur, 2001).

6.2 PROCESS OF ROOT COLONIZATION AND FACTORS AFFECTING IT

To establish themselves on the plant roots and moving along with the growing roots is the foremost requirements for PGPR for their field application in order to control soil-borne diseases. Sufficient populations of the PGPR on the rhizoplane ensure the desired effects on growth promotion and disease control. Abundance and diversity of PGPR in the rhizosphere are very much dependent on the plant species due to differences in plant-root exudates and rhizodeposition (Marschner et al., 2004).

6.2.1   ROOT COLONIZATION: DISTRIBUTION, LOCALIZATION, AND MECHANISM INVOLVED

To ensure the effective establishment of PGPR for their beneficial effects, attempts have been made during past years to measure the external or internal amount of bacteria that colonize root and the studies revealed that, in this process, the whole root system is being utilized. One most used technique to measure colonization rate is performed using dilution plating; especially in the case of fluorescent pseudomonades (Ongena et al., 2000; Gamalero et al. 2004). Root colonization can be easily visualized by several techniques such as immunofluorescence colony (IFC) staining technique (Schober and van Vuurde 1997; Raaijmakers et al., 1995), Immunofluorescence microscopy (Troxler et al., 1997; Gamalero et al., 2004), scanning microscopy (Chin-A-Woeng et al., 1997; Tokala et al., 2002), and confocal laser scanning electron microscopy (Bloemberg et al., 2000; Bolwerk et al., 2003; Gamalero et al., 2004, 2005). Among all, Colony staining approach (IFC) is more instructive and informative because it combines quantification along with visualization in planta (Paulitz, 2000). Further, genetically engineered soil bacteria can be quantified using bioluminescence genes method (*lux* gene; de Weger et al., 1997).

6.2.2   FACTORS AFFECTING ROOT COMPETENCY OF PGPR

6.2.2.1   BIOTIC AND ABIOTIC FACTORS

Chemical nature of root exudates may play a dual role, deterring one organism while attracting another, whereas two very different microorganisms may be attracted to the same plant with differing consequences. Rhizosphere acidic environment as a result of exudation of different organic acids from roots also decides the surrounding population of PGPR (Dakora and Phillips, 1996). pH is also known to be a decisive factor for the coaggregation among different PGPR strains (Joe et al., 2009). Similarly, an increase in soil temperature was observed to decrease the density of population of *Pseudomonas* strains on seeds and roots at different soil depth levels (Beauchamp et al., 1993). Apart from that, PGPR also need to protect themselves against predators. Postma et al. (1990, 1991) studied the detailed relationship between the adsorption of *Rhizobium* to soil particles and its survival in three different soils; natural, sterile, and reinoculated. It was observed that smaller pores, which were physically protected, had more number of particle-associated bacteria and the degree of protection got improved on the addition of clay (Kloepper et al., 1989; Lynch 1990).

6.2.2.2   BACTERIAL FACTORS

Bacterial flagella and pili also control the root colonization ability of PGPR. Motility, because of flagella and pili, is considered as an important factor for the establishment of PGPR in the host root zone. It has been proved that active bacterial motility toward the root hair zone is very much vital for the initiation of root colonization by *Azospirillum brasilense* (Van de et al., 1998). Budzik et al. (2008) described that the chemical bonds required for assembling BcpA pilin subunits on the surface of *Bacillus cereus* are also essential for root colonization.

6.3   FUNCTIONS OF PGPR IN PLANT HEALTH MANAGEMENT

PGPR have been proven worthy enough to increase the growth, yield, and productivity of many agricultural crops in response to their inoculation. Their performance has been well documented in many cereal crops such as rice (Ashrafuzzaman et al., 2009), wheat (Khalid et al., 2004, Cakmakci et al., 2007); horticulture crops such as black pepper (Dastager et al., 2010), and banana (Mia et al., 2010), and other commercial crops including cotton (Anjum et al., 2007). PGPR choose different mechanisms to serve the purpose and accordingly they have been termed as "bioprotectants" (suppression of plant disease), "biostimulants" (for phytohormone production), and "biofertilizers" (for improving nutrient acquisition; Sureshbabu et al., 2016). The functions performed by PGPR in control of soil-borne disease (Fig. 6.1) can be categorized into two mechanisms as follows:

6.3.1   DIRECT MECHANISM

A major contribution by PGPR to support plant growth and disease management involves direct and forward mechanism, which includes phytohormones production, solubilization of nutrients nitrogen fixation, and increasing iron availability by producing siderophores. These mechanisms influence the plant growth activity in different ways depending upon plant species as well as bacterial PGPR strains.

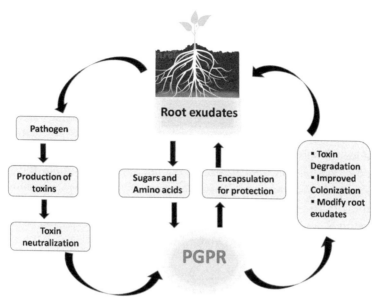

FIGURE 6.1   (See color insert.) Role of PGPR in plant growth promotion and disease protection.

6.3.1.1   PRODUCTION/ALTERATION OF PHYTOHORMONE CONCENTRATION

The endogenous pool of plant indole acetic acid (IAA) gets alter after the acquisition of IAA, being secreted by soil rhizobacteria or PGPR (Glick, 2012; Spaepen et al., 2007). This altered level of IAA interferes with the many plant developmental processes because PGPR possess different routes for the synthesis of IAA, that is, either via L-tryptophan-dependent and independent pathways, and such three L-tryptophan dependent pathways are known. Majority of the PGPR utilize L-tryptophan as a precursor for producing IAA. Rare examples of IAA produced by L-tyrptophan-indepen-dent pathway are also known, one such organism *Azospirillum brasilense*, produces IAA via this route, where more than 90% of IAA is produced by L-tryptophan independent pathway and left over 10% IAA by L-tryptophan-dependent pathway. Although L-tryptophan independent pathway is in reports, but the exact pathway and the enzymes involved in IAA synthesis by this route are yet to uncover (Jha and Saraf, 2015).

Other phytohormones such as cytokinins or gibberellins (GA) or both have been reported to be produced by many soil bacteria in general, and

PGPR in particular (Nieto et al., 1989). Cytokinins are known to promote cell division in plant roots and shoots including cell growth and differentiation as their main function, whereas Gibberellins act as the master controller of many plant developmental processes like germination dormancy, flowering, stem elongation, sex expression, and senescence. Till date, 28 GAs from seven different fungal species, and four GAs (GA1, GA3, GA4, and GA20) from seven bacterial species have been identified (MacMillan, 2001). Two Bacillus strains, namely *B. pumilus* and *B. licheniformis*, are well known to produce gibberellins (Gutierrez-Manero et al., 2001). Another *Bacillus* strain; *Bacillus megaterium* has been documented to promote the growth of *Arabidopsis thaliana* and *Phaseolus vulgaris* seedlings via cytokinins production (Ortíz-Castro et al., 2008). *Proteus, Klebsiella, Escherichia, Pseudomonas,* and *Xanthomonas* are some other PGPR reported to have the capacity to produce cytokinins (Maheshwari et al., 2015).

6.3.1.2   SOLUBILIZATION AND FIXATION OF NUTRIENTS

PGPR are well known for solubilization of some crucial nutrients especially phosphorus (P), the second most important nutrient after nitrogen for plant growth which otherwise remained available only in the limited amount due to the rapid conversion of phosphatic fertilizers into insoluble complexes in the soil (Mckenzie and Roberts, 1990). Microorganisms owing the phosphate solubilizing activity, termed as phosphate solubilizing microorganisms (PSM) are the great options, providing the solubilized forms of P to the plants and thus present a best possible substitute to chemical phosphatic fertilizers (Khan et al., 2006). PSMs lower the pH of the surrounding medium by secretion of organic acids such as acetic, gluconic, lactic, malic, succinic, tartaric, 2-ketogluconic, oxalic, and citric acids to aid in solubilization of fixed phosphorous (Patel et al., 2015). Among the ectorhizospheric soil PGPR, which resides on roots and in rhizospheric soil, strains of *Pseudomonas* and *Bacilli,* and out of endosymbiotic PGPR (residing within the roots/nodules), *Rhizobia* have been found as effective phosphate solubilizers (Goswami et al., 2015). Some of the other organic acids, such as oxalic acid, glycolic acid, malonic acid, citric acid, succinic acid, and propionic acid, have also been identified as one of the phosphate solubilizers. Along with, PSMs have proven themselves capable of augmenting the plant growth by stimulating the efficiency of biological nitrogen fixation, as well as improving the availability of other trace elements as they are capable of synthesizing important plant growth promoting substances (Suman et al., 2001; Zaidi et al., 2009).

A range of bacterial species belonging to different genera comprising symbiotic, such as *Rhizobium*, free-living nitrogen fixers, such as *Azotobacter* and *Azospirillumas*, as well as nonsymbiotic, such as *Azoarcus, Gluconacetobacter, diazotrophicus, Herbaspirillium*, and *Azotobacter*can, provide nitrogen directly to the plant. These nitrogen-fixing strains are commercialized as biofertilizers and considered important for agriculture (Goswami et al., 2015).

6.3.1.3   PRODUCTION OF SIDEROPHORES

PGPR also put forth their antagonistic activity by means of the secretion of the low-molecular-mass iron chelators, siderophores. About 500 siderophores are known till date, of which 270 have been characterized with their chemical structures (Hider and Kong, 2010). Under aerobic conditions, iron exists as Fe^{3+} and is much likely to convert into insoluble forms such as hydroxides or oxyhydroxides, thus making them unavailable to both plants as well as microorganisms (Rajkumar et al., 2010). Siderophore-producing PGPR plays role in improving plant health at various levels, by enhancing iron nutrition, preventing other microorganisms to grow by releasing their antibiotic molecule, and hindering the growth of pathogens by restraining the iron availability for them (Shen et al., 2013).

6.3.2   INDIRECT MECHANISM

PGPR also opt for the indirect mechanism by acting as biocontrol agents (Glick, 2012), which is an environment-friendly approach too. The chief modes of indirect mechanism involve the production of antifungal metabolites like hydrogen cyanide (HCN), phenazines, viscosinamide, and tensin (Bhattacharyya and Jha, 2012), antibiotics production, synthesis of cell wall lytic enzymes, and by induced systemic resistance (ISR).

6.3.2.1   PROTECTION BY ANTIBIOTICS AND ANTIFUNGAL METABOLITES

PGPR respond to the proliferation of several pathogens by producing different antibiotics, specific to pathogens. PGPR have the ability to produce either one or more antibiotics that are critical to major enzymes and metabolism of other microorganisms, thus reducing the deleterious effects of plant pathogens on the plant growth (Glick et al., 2007). Experimental studies of biocontrol PGPR support this activity as these antibiotics have been reported to suppress the soilborne infections of cereal crops including rice, maize, wheat, chickpea, and barley (Raaijmakers et al., 2002). DeSouza et al. (2003) reported the mechanism of these antibiotics by causing membrane damage to pathogens such as *Pythium* spp. and inhibiting the formation of zoospores. Other antifungal compounds have also been reported to be produced by many rhizobacteria like HCN, 2,4-diacetylphloroglucinol, phenazines, pyrrolnitrin, pyoluteorin, tensin, etc. (Bhattacharyya and Jha, 2012). HCN or cyanide has been most studied among them and bacterial strains producing cyanide poses no deleterious effect on host plants when being inoculated, but acts as good biological weed control agents (Zeller et al., 2007).

6.3.2.2   PROTECTION BY CELL WALL DEGRADING ENZYMES

PGPR serve the same purpose of improving the plant health by secreting cell wall-degrading enzymes such as chitinase, cellulase, β-1,3-glucanase, and protease secreted by biocontrol strains of PGPR, posing an inhibitory outcome on the hyphal intensification of fungal pathogens by degrading their cell wall. Strains of *Paenibacillus* and *Streptomyces* spp. synthesize β-1, 3-glucanase to lyse fungal cell walls of pathogenic *F. oxysporum*. Similarly, *Bacillus cepacia* synthesizes β-1,3-glucanase for destroying the cell walls of the different pathogens such as *P. ultimum*, *R. solani*, and *S. rolfsi* (Compant et al. 2005).

6.3.2.3   INDUCTION OF SYSTEMIC RESISTANCE

An alternate strategy provided by PGPR for plant protection from pathogens and diseases is via induced ISR, the term given by van Loon et al. (1998), where treating plants by PGPR elicits host defense as indicated by a drop in severity of disease in form of disease symptoms. Many bacterial components have been involved in inducing ISR, such as cyclic lipopeptides, flagella, lipopolysaccharides, homoserine lactones, siderophores, and volatiles such as, acetoin and 2,3-butanediol (Lugtenberg and Kamilova, 2009).

6.4   ROLE OF PGPR IN SOIL-BORNE PLANT DISEASE MANAGEMENT

PGPR have the potential to control soil-borne as well as leaf spot pathogens in a wide range of agri-horticultural crops (Wei et al., 1996; Ramamoorthy et al., 2002; Vivekananthan et al., 2004). Most of the previously studied PGPR are *Pseudomonas spp.*, (Ramamoorthy et al., 2002) or *Bacillus spp.*,

(Schisler et al., 2004). Moreover, Kumar et al. (2001) suggested the potential use in inducing plant growth and disease suppression under sustainable agriculture production systems. It is well documented that the microbiota, for example, *Rhizobacteria, Trichoderma,* and *Pseudomonas spp.,* frequently occurring in decomposing organic matter enhance the growth and yield of many crops (Sylvia, 2004) by producing growth hormones along with chemical compounds, such as siderophores, tannins, phenols, which are antagonistic to many soil-borne pathogens (Antonio et al., 2008; Table 6.1). It is known that inoculation of PGPR enhances the plant properties like total biomass, leaf area, chlorophyll content, etc. (Baset Mia et al., 2010). On the other hand, Uppal et al. (2008) reported that *P. fluorescens* Biotype F isolate DF37 significantly reduced the disease incidence and severity of *Verticillium* wilt. Besides, it is evident that *P. fluorescens* strain Pf1 induces resistance against many pathogens in various crops, namely, *Rhizoctonia solani* (Nandakumar, 1998), *Colletotrichum falcatum* in sugarcane (Viswanathan, 1999).

6.4.1   ROLE OF PGPR IN CEREALS AND PULSES

The fact that PGPR shows effective management against many soil-borne diseases is well established. Plant growth promotion as well as biological control of plant pathogens using different organisms, such as fungal, actinomycete, and bacterial antagonists, has been studied in the past. For instance, strains of *Streptomyces, Bacillus, Pseudomonas* and *Trichoderma* were found effective in controlling plant pathogens apart from aiding plants mobilize and acquire nutrients (Perner et al., 2006; Borriss et al., 2011). PGP traits of actinomycetes have been reported on different crops like pea, bean, wheat, and rice (Tokala et al., 2002; Nassar et al., 2003; Sadeghi et al., 2012; Gopalakrishnan et al., 2012). In another study, five strains of *Streptomyces* spp. (CAI-24, CAI-121, CAI-127, KAI-32, and KAI-90) were proved to be helpful in controlling *Fusarium* wilt in chickpea (Gopalakrishnan et al., 2011) and also found to be producing siderophore, IAA (except KAI-90), hydrocyanic acid, cellulase (only KAI-32 and KAI-90), and protease (only for CAI-24 and CAI-127).

Use of PGPR in managing sheath blight disease in rice has gained popularity as an alternative to the chemical fungicides. But for this, PGPR must have good colonization potential in the rhizosphere and or phyllosphere so as to control further spread of ShB under field conditions. When 70 PGPR were tested, only 31 strains significantly reduced the ShB lesions as compared to

TABLE 6.1 Crop-specific PGPR Having Potential Biocontrol Properties.

Sr. no.	Crop	Disease	PGPR
1.	Tomato	Damping off tomato/tomato root rot	*Pseudomonas fluorescens* F113, Pf-5, Q2-87, CHA0, etc.
		Wilt	*Bacillus amyloliquefaciens* FZB42
2.	Cotton	Damping off	*Pseudomonas fluorescens* F113, Pf-5, Q2-87, CHA0, etc./*Bacillus subtilis* QST713/*P. fluorescens* HV37aR2/*Pseudomonas fluorescens* Pf-5/ *P. cepacia*
			Bacillus subtilis
3.	Apple	Root knot	*Pseudomonas fluorescens* strain PfA506
4.	Dicot plants	Fire blight	*Agrobacterium radiobacter*
5.	Groundnut	Crown gall	*Bacillus subtilis* AU195
		Contamination	*Bacillus subtilis* 168
		Collar rot	
6.	Grape	Damping off	*Bacillus subtilis* QST713
7.	Papaya	Damping off	*Bacillus subtilis* BBG100
8.	Wheat	Take-all	*Pseudomonas fluorescens* 2–79, 30–84
9.	Rice	Damping off and rice blast	*P. cepacia*
		Sheath blight	*Pseudomonas fluorescens*, *Burkholderia* sp./*Pseudomonas fluorescens* strain Pf1 and Fp7
		Blue mold	*S. marcescens* 90-1, *Bacillus pumilus* SE34
10.	Alfalfa	Damping off	*Bacillus cereus* UW85
11.	Beans	Halo blight	*Pseudomonas fluorescens* strain 97
		Stem rot	*Pseudomonas cepacea*
12.	Barley	Powdery mildew	*Bacillus subtilis*
13.	Carnation	Fusarium wilt	*Pseudomonas* spp. (WCS 417r)
14.	Cucumber	Anthracnose	*Pseudomonas putida* (89B-27)
		Damping off	*Pseudomonas cepacea*
15.	Green gram	Wilt	*Pseudomonas* spp.
16.	Mung bean	Root knot	*Pseudomonas aeruginosa*, *Bacillus subtilis*

PGPR, plant growth promoting rhizobacteria.

the control. The disease severity in these significant PGPR strains ranged from 2.9% to 93.3% whereas, maximum inhibition of lesion development was obtained with *B. subtilis* MBI 600 with 2.9% of disease severity (Vijay Krishna et al., 2011).

In a study, 29 rice-associated bacteria (RAB) isolated from rice plants were tested against *Rhizoctonia solani* (sheath blight) and *Burkholderia glumae* (bacterial panicle blight); out of them, 26 were confirmed to have antimicrobial activities. According to the 16S rDNA sequence identity, 12 antagonistic RAB were closest to *Bacillus amyloliquefaciens*, while seven were *B. methylotrophicus* and remaining seven *B. subtilis*. The five RAB showing the highest antimicrobial activities (RAB6, RAB9, RAB16, RAB17S, and RAB18) selected were not only found to inhibit the sclerotial germination of *R. solani* and the lesion development in vitro but also significantly suppressed the sheath blight and bacterial panicle blight development in field condition (Shrestha et al., 2016).

In another study, three PGPR strains WPR-51, WPR-42 and WM-30 were selected to test *in planta* antagonistic activity on two wheat varieties infected with *R. solani*. Strain WPR-51 and a mixture of all three strains showed maximum inhibition of *R. solani* growth and it was observed that these strains not just increased the germination, biomass, and root–shoot length but also inhibited *R. solani* growth when tested in pot experiments (Fatima et al., 2009). Collar rot caused by *Sclerotium rolfsii* in chickpea is one of the devastating soil-borne diseases causing 10–30% annual yield losses. PGPR have shown high efficacy in managing this disease in vitro as well as in the field. When two PGPR (*Pseudomonas fluorescens* strain 4 and *P. aeruginosa*) were used as a foliar spray with the fresh and heat inactivated microorganisms. *P. fluorescens* strain 4 (10^8 cfu/mL) was more effective in reducing plant mortality as compared to the control. Foliar application of fresh and heat inactivated (121°C for 10 min) *P. fluorescens* strain 4 reduced the plant mortality by 15–25% as compared to *P. aeruginosa*, which showed disease control of 10–15% only. Moreover, it was also observed that fresh and heat-inactivated *P. fluorescens* strain 4 showed maximum efficacy with respect to plant growth promotion followed by fresh and heat inactivated *P. aeruginosa* as compared to the control (Maurya et al., 2008). The effects of *Pseudomonas putida*, *Pseudomonas alcaligenes*, and a *Pseudomonas* isolate (Ps28) against *Macrophomina phaseolina* and the production of siderophores, HCN, and IAA were estimated. It was seen that *P. putida* colonized roots more effectively and also produced the greatest amounts of siderophores, IAA and HCN, as compared to *P. alcaligenes* and Ps28. Plant inoculations of these bacterial isolates increased plant growth and the number of seed pods in diseased plants while reducing root-rot disease index (Akhtar and Siddiqui, 2009).

Some bioactive metabolites (BM1-BM4) produced by *Pseudomonas aeruginosa* strain RRLJ 04 and *Bacillus cereus* strain BS 03 were screened against wilt of pigeon pea (*Fusarium udum*) under gnotobiotic and nursery condition. Seeds treated with BM 1 (50 µL seed^{-1}), BM 2 (30 µL seed^{-1}), and BM 3 (70 µL seed^{-1}) were grown in pathogen-infested soil and it was observed that wilt disease was suppressed in addition to enhanced growth of the plants. After 90 days of growth, the disease control was highest in BM 2 (90%) followed by BM 1 and BM 3 (87% and 83%, respectively) and it was evident that BM 2 treated plants were more resistant to the pathogen (Dutta et al., 2014).

6.4.2   ROLE OF PGPR IN FRUITS AND VEGETABLE CROPS

Vegetables containing several important nutrients, including vitamins and antioxidants, act as an important part of human dietary systems. Injudicious use of agrochemicals into vegetable production practices adversely affects soil fertility, plant health, and making it unfit for human consumption. Some of the pathogens like *P. capsici* infects various susceptible hosts as around 50 plant species like cucurbits, solanaceous, and legume crops It may cause yield losses up to 100% as infecting the plant at any growth stage throughout the growing season (Lee et al., 2001; Hausbeck and Lamour, 2004). Fungicides can kill the causal organism, but as the side effects, they also hamper the survival and growth of beneficial rhizosphere (Carson et al., 1962; Hussain et al., 2009; Heckel, 2012). To avoid these losses, an alternate and inexpensive approach has been introduced into the vegetable production system is to use of PGPR, which has been found to protect the vegetables from various diseases and improve qualitative and quantitative yield (Kloepper, 1992; Rizvi et al., 2017).

Application of PGPR, in tomato, showed enhanced resistance against the speck and spot causal bacteria (Kavitha and Umesha, 2007), simultaneously affecting the quality of fruit size and texture positively (Hortencia et al., 2007). High-level induction of defense enzymes in response to *Pythium aphanidermatum* attack was recorded in tomato plants treated with *Pseudomonas fluorescens* Pf1 (Ramamoorthy et al., 2002). Jinnah et al. (2002) observed on the tomato crop that during the management of bacterial wilt with applying *Pseudomonas fluorescens* affects the plant growth characters like height, number of branches, and yield positively. Kalita (2002) reported that the combined methods of application (biocontrol agents *P. fluorescens* and *T. viride*) managed bacterial wilt of tomato to a greater degree than the single method. Tomato root, when inoculated with *P. alcaligenes*, reduced the wilt caused by *F. oxysporum f.* spp. *vasinfectum* (Gamliel and Katan, 1993). Almaghrabi et al. (2013) reported that in tomato white fly-transmitted diseases, tomato mottle virus (ToMoV), was best managed by *Serratia marcescens* (Murphy et al., 2000; Zehnder et al., 2001) and against cucumber mosaic virus (CMV) (Zehnder et al., 2001). Six strains of PGPR including *Pseudomonas putida, Pseudomonas fluorescens, Serratia marcescens, Bacillus amyloliquefaciens, Bacillus subtilis*, and *Bacillus cereus* significantly reduced the infection of root-knot nematode in tomato by reducing gall formation and egg masses (Martinez-Ochoa, 2000; Zehnder et al., 2001; Lucy et al., 2004; Kloepper and Ryu, 2006) and resulted increase in yield (Siddiqui et al., 2001; Kokalis-Burelle and Dickson, 2003). Damping-off of tomato caused by *Rhizoctonia solani* was controlled by *B. subtilis* strains (Kumar 1999). Seeds treatment of tomato with powder formulation of PGPR (*Bacillus subtilis, B. pumilus*) reduced the severity of tomato mosaic virus and increased the fruit yield (Murphy et al., 2000). In pea pretreatment with PGPR had shown induction of phenolics against *F. oxysporum f.* spp. *pisi* (Benhamou et al., 1996, 2000).

In brinjal seedlings, there was a significant reduction in disease incidence when treated with consortia of biocontrol agents and PGPR (Mohan, 2006). In brinjal integration of seed treatment, root dip and soil application of *P. fluorescens* caused higher reduction of wilt incidence (Gohain, 2001; Chakravarty and Kalita, 2012; Peixto et al., 1995).

Cucumber (*Cucumis sativus*) is very much prone to crown rot and blight (Kim et al., 2008; Maleki et al., 2011). Some fluorescent pseudomonads (antifungal) controlled crown and root rot of cucumber (Shirzad et al., 2012). In cucumber pretreatment with PGPR had shown induction of phenolics against *Pythium ultimum* (Benhamou et al., 1996, 2000). Seed treatment with PGPR strain significantly reduces the anthracnose disease; CMV, and bacterial wilt disease in cucumber (Wei et al., 1991, 1996; Raupach et al., 1996; Kloepper et al., 1993). Specifically, *P. putida* strain 89B-27, *Flavomonas oryzihabitans* strain INR-5, *S. marcescens* strain 90-166, and *Bacillus pumilus* strain INR-7 protect the cucumber against angular leaf spot caused by *Pseudomonas syringae* pv. *lachrymans* systemically after seed treatment (Liu et al., 1995; Wei et al., 1996). In the management of *Fusarium* wilt of cucumber caused by *F. oxysporum* f. spp., a *cucumerinum* combination of *Paenibacillus* spp. 300 and *Streptomyces* spp. 385 in the different ratio (1:1 or 4:1) were more effective than individually (Singh et al., 1999). PGPR isolated from cucumber (*Pseudomonas stutzeri, Bacillus subtilis, Stenotrophomonas maltophilia,* and *Bacillus amyloliquefaciens*) significantly suppressed Phytophthora crown rot caused by *Phytophthora capsica* and increased the germination rate and nitrogen content in root and shoot tissue (Islam et al., 2016). Seed treatment with PGPR resulted in significantly lower numbers of cucumber beetles, *Diabrotica undecimpunctata howardi* and *Acalymma vittatuni* (vector of the bacterial wilt), and the lower incidence of bacterial wilt (Zehnder et al., 1997). *S. cepivorum* causal organism of onion white rot was inhibited by isolates of *Bacillus subtilis* with an increase in onion emergence and yield (Utkhede and Rahe, 1980).

Against the fungal root pathogen *F. oxysporum* f. spp. *raphani,* in radish, seed treatment with *P. fluorescens* strain WCS 417 induces the systemic resistance (Hoffl et al., 1996). PGPR *Pseudomonas* were effective for many crops like in mulberry and apricot (Esitken et al., 2003) potato (*Solanum tuberosum* L.; Schippers et al., 1987), radish (*Raphanus sativus* L.; Kloepper and Schroth, 1978), sugar beet (*Beta vulgaris* L.; Suslow and Schroth, 1982), and lettuce (Chabot et al., 1993). Complete reduction of *Rhizoctonia solani,* as well as a significant reduction of *Macrophomina phaseolina, Fusarium oxysporum,* and *Fusarium solani,* was observed in okra when *P. aeruginosa* and *P. lilacinus* with cotton cake applied in the field as soil application (Shafique et al., 2015). A mixture of PGPR namely, *Azotobacter chroococcum, Bacillus megaterium, Pseudomonas fluorescens, Bacillus subtilis,* and *Trichoderma harzianum* showed enhanced, total biomass, seedling vigor, least disease incidence, and more biocontrol efficiency in cabbage (Sudharani et al., 2014). Crop grown under green house condition like squash showed that PGPR strains (separately or in combinations) had the potential to suppress Phytophthora blight (Zhang et al., 2010). In carrot against *R. solani* PGPR, *Streptomyces griseus* isolate with biocontrol abilities was reported by Merriman et al. (1974).

PGPR improves nutritional quality as well as the shelf life of fruits (Loganathan et al., 2014). The greater amount of the enzymes against *Colletotrichum gloeosporioides* infection was recorded in mango trees sprayed with *P. fluorescens* FP7 (Vivekanathan et al., 2004). A mixture of *Pseudomonas* (NFP6), *P. fluorescens* (Pf3a), and *B. subtilis* (BS1) induced defense reaction against crown rot caused by *Lasiodiplodia theobromae* and *Colletotrichum musae* under in vivo conditions in banana (Sangeetha et al., 2010).

6.4.3 ROLE OF PGPR IN COMMERCIAL AND OILSEED CROPS

Pseudomonas fluorescens produced an antibiotic which controlled the damping-off of cotton seedlings caused by *R. solani* (Howell, 1979). Under the field condition, *Pseudomonas* spp. suppressed the *Rhizoctonia solani* and *P. ultimum* causal organism of seedlings disease in cotton by doing furrow application at the rate of 14 mL/m (Hagedorn et al., 1993). In cotton isolated endophytic bacterial strains reduced galling in roots of cotton caused by the root-knot nematode, *Meloidogyne incognita*, which also make and increase the entry sites for endophytic bacteria (Hallmann et al., 1997).

For ground nut cultivation, some biofertilizers are plant growth-promoting and yield-enhancing PGPR (PGPR1, PGPR2, and PGPR4; Dey et al., 2004); highly competitive groundnut rhizobia (NRCG4 and NRCG9); and beneficial bacterial consortia (consortia 1 and consortia 2) were identified. Charcoal rot in soybean was reduced by 80% in the presence of *P. polymyxa* HKA-15 (Senthilkumar et al., 2009). Accumulation of phenolics and PR proteins was induced by the foliar application of *P. fluorescens* strain Pf1 in groundnut (Meena et al., 2000). In another study under field condition, Meena et al. (2001) reported that seed treatment with the powder formulation of *P. fluorescens* significantly suppressed the root rot incidence of groundnut. Seed treatment of groundnut with *P. fluorescens* supports the antagonist colonization in groundnut rhizosphere (Meena et al., 2006).

Bacterized seeds with a mixture of PGPR and rhizobia of various crops and ornamental plants resulted in profuse growth and disease resistance (Zehnder et al., 2001). In case of coriander and fenugreek, PGPR (*Pseudomonas putida* [FK14], and *Microbacterium paraoxidans* [FL18]) application enhanced the growth and yield ranged from 10% to 20% (Anandaraj and Bini, 2011). Ahmed et al. (2003) demonstrated that *P. capsici* suppressed by the bacterial isolates of sweet pepper in vitro. In greenhouse assay, an endophytic bacterium (*B. megaterium* IISRBP17) isolated from black pepper stem and roots, suppressed *P. capsici* on blackpepper (Aravind et al., 2009). PGPR bioformulations enhanced resistance in tea against blister disease (Saravanakumar et al., 2007). PGPR-mediated ISR against red rot disease in sugarcane was reported by Viswanathan and Samiyappan (1999, 2001) against *Colletotrichum falcatum*. *Bacillus amyloliquefaciens* Ba33 worked as an antiviral agent against tobacco mosaic virus (TMV; Shen et al., 2012). *P. fluorescens* strain CHAO showed the ability to suppress tobacco necrosis virus concomitant in tobacco (Maurhofer et al., 1994; Maurhofer et al., 1998) and enhanced root growth and suppressed black root rot caused by *Thielaviopsis basicola* in tobacco. *Bacillus thuringiensis* induced accumulation of PR proteins against *Hemileia vastatrix* (Guzzo and Martins, 1996).

Dipping of tulip bulbs in Pseudomonas suspension suppressed tulip root rot caused by *P. ultimum* (Weststeijn, 1990). Under the greenhouse condition, sunflower wilt disease was suppressed by *P. cepacia* strain N24 when applied to the seedbeds of 500 mL/m² (Hebber et al., 1991). Applications of *P. cepacia* show that at least three bacterial sprays are effective in the control of Rhizoctonia stem rot of Poinsettia (Kelly Cartwright, 1995).

6.4.4   ROLE OF PGPR IN MEDICINAL PLANTS

The critical time to achieve more bioactive secondary metabolites is flowering which can be stimulated by the use of PGPR. PGPR also improve plant growth, increased the rate of seed germination and seedling emergence (Shaukat et al., 2006). The seed of *Salvia officinalis* treated using PGPR namely, *P. fluorescens* (PF-23) and *P. putida* (PP-41, PP-108, and PP-159) affect germination and vigor parameters (Ghorbanpour and Hatami, 2014). In artichoke (*Cynara scolymus*) combination of PP-168, *Azotobacter*, and *Azospirillum* strains increase % germination, a number of normal plants, radicle and shoot weight, shoot length, and vigority (Jahanian et al., 2012). A mixture of four PGPR strains (*Azospirillum lipoferum, Azotobacter chroococcum, P. fluorescens*, and *Bacillus megaterium*) caused a significant increase in germination rate and vigor index of *Catharanthus roseus* (Lenin and Jayanthi, 2012). Five bacterial strains (TR1–TR5) isolated from the root nodules of fenugreek (*Trigonella foenum-graecum*) showed different efficacy as maximum increments in vigor index, nodule number, and root and shoot biomass in TR1 + TR2 (Kumar and Maheshwari, 2011). TR2 isolate denotes the presence of *Rhizobium leguminosarum*, and the other four as *Ensifer meliloti*. Positive effects on the phenotypic traits of germination of *Acacia Senegal* were observed when the seed was treated with *B. licheniformis* or *Sinorhizobium saheli* PGPR individually or in combination (Singh et al., 2011).

In *Catharanthus roseus* root inoculation increased chlorophyll and nutrient (N, P, and K) content with different PGPR strains like *Azospirillum lipoferum, Azotobacter chroococcum, P. fluorescens*, and *B. megaterium* of (Lenin and Jayanthi, 2012). Inoculation (single or co-) of *P. fluorescens* and *Azospirillum brasilense* in Mexican marigold (*Tagetes minuta*) revealed that essential oil increased by 70% (Cappellari et al., 2013).

6.5 CONSTRAINTS TO COMMERCIALIZATION

Commercialized product of PGPR is available in the market as for ready to use to suppress the disease incidence. Examples of PGPR and biocontrol products available in the market are: Nogall® (*Agrobacterium radiobacter* K1026), Sonata® ™ (*Bacillus pumilus* QST 2808), YieldShield® (*B. pumilus* GB34), Kodiak® (*B. subtilis* GBO3), BlightBan C9-1® (*Pantoea agglomerans* C9-1), Bloomtime® (*P. agglomerans* E325), Tx-1-Spot-Less® T (*Pseudomonas aureofaciens*), Bio-save® (*P. syringae* ESC-10 and ESC-11), BlightBan® (*P. fl uorescens* A506), Cedomon® (*P. chlororaphis* MA 342), Mycostop® (*Streptomyces griseoviridis* K61) and Actinovate® (*S. lydicus* WYEC 108) (Figueiredo et al., 2010). But point to be considered that success of any biological agents depends on the quality of the product, easy availability with acceptable shelf life, proper promotion of the product, its acceptability among the users, and demand in the market. A large number of factors are prevailing that can limit the efficacious commercialization of PGPR. Some are as follows:

- Genuineness of the selected choice.
- Delay in registration of the products.
- The dominance of locally developed biocontrol agents and its competitive trait.
- Inconsistent performance in biocontrol efficacy.
- Lack of awareness about the mode of actions.
- Little awareness and curiosity among the farmers about the potential of products.
- Alteration and loss of desirable traits.
- The poor delivery system for biocontrol PGPR.
- Possible ecological consequences.
- Quality and stability of the products.
- Hazards related to the multiplication at large scale.
- Stiff encounters from environment protection agencies and activists.
- Reliability of the selection.
- Skill development related to mass multiplication.

6.6 CONCLUSION

Microbial inoculants, which were found to acquire different functions in plants, lead to promising solutions for an ecofriendly sustainable agriculture.

Recently, biopesticides are gaining global attention for the sustainability of agriculture. Researchers are oriented toward the selection of bacteria able to antagonize most deleterious plant pathogens. However, the effectiveness and/or success of any biocontrol agent mainly depends on various factors like its interaction with other microbes, the plant pathogen to be controlled, the host habitat, and the availability of resources.

Future strategies should be aimed toward advanced studies like cloning the genes involved in the production of siderophores, antibiotics and other metabolites, and to manipulate them into the strains with high colonization potential considering its other beneficial characteristics. Besides, the application of molecular tools will enhance profoundly the ability to understand and manage the rhizosphere and will pave the way to identify new products with much effectiveness. Recent studies are focusing on the increased use of biorational, screening processes to identify microorganisms with potential for biocontrol, testing in semicommercial and commercial production systems. Increased emphasis is being given on combining biocontrol strains with each other and with other control methods, integrating biocontrol into an overall system. Research interests should be diverted on the development of improved formulations, effective adjuvant, and compatible protectants for microbial pesticides. Focused and precise studies on the survival of the antagonists under field conditions based on biochemical and molecular methods have to be strengthened.

KEYWORDS

- **cereals**
- **plant growth promoting rhizobacteria**
- **rhizobacteria**
- **soil borne**
- **vegetables**

REFERENCES

Ahmed, A.; Ezziyyani, M.; Egea-Gilabert, C.; and Candela, M. E. Selecting Bacterial Strains for Use in the Biocontrol of Diseases Caused by *Phytophthora capsici* and *Alternaria alternate* in Sweet Pepper Plants. *Biol. Plant.* **2003**, *47*, 569–574.

Akthar, M. S.; Siddiqui, Z. A. Use of Plant Growth-Promoting Rhizobacteria for the Biocontrol of Root-Rot Disease Complex of Chickpea. *Australasian Plant Pathol.* **2009**, *38*, 44–50.

Almaghrabi, O. A.; Massoud, S. I.; Abdelmoneim, T. S. Influence of Inoculation with Plant Growth Promoting Rhizobacteria (PGPR) on Tomato Plant Growth and Nematode Reproduction under Greenhouse Conditions. *Saudi J. Biol. Sci.* **2013**, *20*, 57–61.

Anandaraj, M.; Bini, Y. K. In *Potential of PGPR Application for Seed Spices with Special Reference to Coriander and Fenugreek in India.* Proceedings of the 2nd Asian PGPR Conference, Aug 21–24, 2011, Beijing, P. R. China.

Anjum, M. A.; Sajjad, M. R.; Akhtar, N.; Qureshi, M.A.; Iqbal, A.; Jami, A. R.; Hasan, M. Response of Cotton to Plant Growth Promoting Rhizobacteria (PGPR) Inoculation Under Different Levels of Nitrogen. *J. Agric. Res.* **2007**, *45*, 135–143.

Antonio G. F.; Carlos, C. R.; Reiner, R. R.; Miguel, A. A.; Angela, O. L. M.; Cruz, M. J. G.; Dendooven, L. Formulation of a Liquid Fertilizer for sorghum (*Sorghum bicolour* (L.) Moench) Using Vermicompost Leachate. *Bioresour. Technol.* **2008**, *99*, 6174–6180.

Aravind, R.; Kumar, A.; Eapen,S. J.; Ramana, K. V. Endophytic Bacterial Flora in Root and Stem Tissues of Black Pepper (*Piper nigrum* L.) Genotype: Isolation, Identification, and Evaluation Against *Phytophthora capsici. Lett. Appl. Microbiol.* **2009**, *48*, 58–64. https://doi.org/10.1111/j.1472-765X.2008.02486.x.

Ashrafuzzaman, M.; Hossen, F. A.; Razi-Ismail, M.; Hoque, M. A.; Zahurul-Islam, M.; Shahidullah S. M; Meon S. Efficiency of Plant Growth-Promoting Rhizobacteria for the Enhancement of Rice Growth. *Afr. J. Biotechnol.* **2009**, *8*, 1247–1252.

Balakrishnan, S.; Humayun, P.; Srinivas, V.; Vijayabharathi, R.; Ratnakumari, B.; Rupela, O. Plant Growth-Promoting Traits of *Streptomyces* with Biocontrol Potential Isolated from Herbal Vermicompost. *Biocontrol Sci. Technol.* **2012**, *22*, 1110–1199.

Baset Mia, M. A.; Shamsuddin, Z. H.; Wahab, Z.; Marziah, M. Effect of Plant Growth Promoting Rhizobacterial (PGPR) Inoculation on Growth and Nitrogen Incorporation of Tissue-Cultured *Musa* Plantlets Under Nitrogen-Free Hydroponics Condition. *Aust. J. Crop Sci.* **2010**, *4*, 85–90.

Bashan, Y.; Holguin, G. Proposal for the Division of Plant Growth Promoting Rhizobacteria into two Classifications: Biocontrol-PGPB (Plant Growth Promoting Bacteria) and PGPB. *Soil Biol. Biochem.* **1998**, *30* (9), 1225–1228.

Beauchamp, C. J.; Kloepper, J. W.; Antoun, H. Detection of Genetically Engineered Bioluminescent Pseudomonads in Potato Rhizosphere at different temperatures. *Microbiol. Release.* **1993**, *1*, 203–207.

Benhamou, N.; Belanger, R. R.; Paulitz, T. C. Induction of Differential Host Responses by *Pseudomonas fluorescens* in Ri T-DNA Transformed Pea Roots after Challenge with *Fusarium oxysporum* f. sp. *pisi* and *Pythium ultimum. Phytopathology* **1996**, *86*, 114–118.

Benhamou, N.; Gagné, S.; Quere, D. L.; Dehbi, L. Bacterial Mediated Induced Resistance in Cucumber: Beneficial Effect of the Endophytic Bacterium *Serratia plymuthica* on the Protection Against Infection by *Pythium ultimum. Phytopathology* **2000**, *90*, 45–56.

Bloemberg, G. V.; Wijfjes, A. H. M.; Lamers, G. E. M.; Stuurman, N.; Lugtenberg, B. J. J. Simultaneous Imaging of *Pseudomonas fluorescens* WCS365 Populations Expressing Three Different Auto Fluorescent Proteins in the Rhizosphere: New Perspectives for Studying Microbial Communities. *Mol. Plant Microbe Interact.* **2000**, *13*, 1170–1176.

Bolwerk, A.; Lagopodi, A. L.; Wijfjes, A. H. M.; Lamers, G. E. M.; Chin-A-Woeng, T. F. C.; Lugtenberg, B. J. J.; Bloemberg, G. V. Interactions in the Tomato Rhizosphere of Two Pseudomonas Biocontrol Strains with the Phytopathogenic Fungus *Fusarium oxysporum* f. sp. *radicis-lycopersici. Mol. Plant Microbe Interact.* **2003**, *11*, 983–993.

Borriss, R.; Chen, X. H.; Rueckert, C.; Blom, J.; Becker, A.; Baumgarth, B.; Fan, B.; Pukall, R.; Schumann, P.; Sproer, C.; Junge, H.; Vater, J.; Puhler, A.; Klenk, H. P. Relationship of *Bacillus amyloliquefaciens* Clades Associated with Strains DSM7T and FZB42T: A Proposal for *Bacillus amyloliquefaciens* subsp. *amyloliquefaciens* subsp. nov. and *Bacillus amyloliquefaciens* subsp. *plantarum* subsp. nov. Based on Complete Genome Sequence Comparisons. *Int. J. Sys. Evol. Microbiol.* **2011**, *61*, 1786–1801.

Bouizgarne, B. Bacteria for Plant Growth Promotion and Disease Management. In *Bacteria in Agrobiology: Disease Management*; Maheshwari, D. K., Ed.; Springer: Heidelberg, 2013. https://doi.org/10.1007/978-3-642-33639-3-2.

Budzik, J. M.; Marraffini, L. A.; Souda, P. et al. *Proc. Natl. Acad. Sci. USA*, **2008**, *105*, 10215–10220.

Cakmakci, R.; Kantar, F.; Algur, O. F. Sugar Beet and Barley Yields in Relation to *Bacillus polymyxa* and *Bacillus megaterium* var. Phosphaticum Inoculation. *J. Plant Nutr. Soil Sci.* **1999**, *162*, 437–442.

Cappellari, L. D. R.; Santoro, M. V.; Nievas, F.; Giordano, W.; Banchio, E. Increase of Secondary Metabolite Content in Marigold by Inoculation with Plant Growth-Promoting Rhizobacteria. *Appl. Soil Ecol.* **2013**, *70*, 16–22.

Carson, R.; Darling, L.; Darling, L. *Silent Spring*. Riverside Press: Cambridge, MA, 1962.

Cartwright, K. D. Comparison of Pseudomonas Species and Application Techniques for Biocontrol of Rhizoctonia Stem Rot of Poinsettia. *Plant Dis.* **1995**, *79*, 309–313.

Chabot, R.; Antoun, H.; Cescas, M. Stimulation de la Croissance du maï¨s et de la Laitue Romaine par des Microorganismes Dissolvant le Phosphore Inorganique. *Can. J. Microbiol.* **1993**, *39*, 941–947.

Chakravarty, G.; Kalita, M. C. Biocontrol Potential of *Pseudomonas fluorescens* against Bacterial Wilt of Brinjal and Its Possible Plant Growth Promoting Effects. *Ann. Biol. Res.* **2012**, *3* (11), 5083–5094

Chin-A-Woeng, T. F. C.; de Priester, W.; van der Bij, A. J.; Lugtenberg, B. J. J. Description of the Colonization of a Gnotobiotic Tomato Rhizosphere by *Pseudomonas fluorescens* Biocontrol Strain WCS365, Using Scanning Electron Microscopy. *Mol. Plant Microbe Interact.* **1997**, *10*, 79–86.

Guzzo, S.D.; Martins, E. M. F. Local and Systemic Induction of b-1, 3-Glucanase and Chitinase in Coffee Leaves Protected Against *Hemileia vastatrix* by *Bacillus thuringiensis*. *J. Phytopathol.* **1996**, *144*, 449–454.

Compant, S.; Duffy, B.; Nowak, J.; Clement, C. Use of Plant Growth-Promoting Bacteria for Biocontrol of Plant Diseases: Principles, Mechanisms of Action, and Future Prospects. *Appl. Environ. Microbiol.* **2005**, *71*, 4951–4959.

Dakora, F. D.; Phillips, D. A. Diverse Functions of Isoflavonoids in Legumes Transcend Anti-Microbial Definitions of Phytoalexins. *Physiol. Mol. Plant Pathol.* **1996**, *49*, 1–20.

Dastager, S. G.; Deepa, C. K.; Pandey, A. Potential Plant Growth Promoting Activity of *Serratia nematophila* NII-0.928 on Black Papper (*Piper nigrum* L). *World J. Microbiol. Biotechnol.* **2010**, *27*, 259–265.

De Souza, J. T.; Weller, D. M.; Raaijmakers, J. M. Frequency, Diversity and Activity of 2,4-Diacetylphloroglucinol Producing Fluorescent *Pseudomonas spp.* in Dutch Take-All Decline Soils. *Phytopathology* **2003**, *93*, 54–63.

de Weger, L. A.; Kuipe, I.; van der, Bij A. J.; Lugtenberg, B. J. J. Use of a Lux-Based Procedure to Rapidly Visualize Root Colonization by *Pseudomonas fluorescens* in the Wheat Rhizosphere. *Antonie Leeuwenhoek* **1997**, *72*, 365–372.

Dey, R.; Pal, K. K.; Bhatt D. M.; Chauhan, S. M. Growth Promotion and Yield Enhancement of Peanut (*Arachis hypogaea* L.) by Application of Plant Growth Promoting Rhizobacteria. *Microbiol. Res.* **2004,** *159* (4), 371–394.

Dutta, S.; Morang, P.; Nishanth Kumar, S.; Dileep Kumar, B. S. Fusarial Wilt Control and Growth Promotion of Pigeon Pea Through Bioactive Metabolites Produced by Two Plant Growth Promoting Rhizobacteria. *World J. Microbiol. Biotechnol.* **2014,** *30,* 1111–1121.

Esitken, A.; Karlidag, H.; Ercisli, S.; Turan, M.; Sahin, F. The Effect of Spraying a Growth Promoting Bacterium on the Yield, Growth, and Nutrient Element Composition of Leaves of Apricot (*Prunus armeniaca* L. cv. *Hacihaliloglu*). *Aust. J. Agric. Res.* **2003,** *54,* 377–380.

Fatima, Z.; Saleemi, M.; Zia, M.; Sultan, T.; Aslam, M.; Rehman, R.; Chaudhary, M. F. Antifungal Activity of Plant Growth-Promoting Rhizobacteria Isolates against *Rhizoctonia solani* in Wheat. *Afr. J. Biotechnol.* **2009,** *8,* 219–225.

Friedel, J. K.; Gabel, D.; Stahr, K. Nitrogen Pools and Turnover in Arable Soils Under Different Durations of Organic Farming. II: Source- and -Sink Function of the Soil Microbial Biomass or Competition with Growing Plants? *J. Plant Nutr. Soil Sci.* **2001,** *164* (4), 421–429.

Gamalero, E.; Lingua,G; Tombolini, R.; Avidano, L.; Pivato, B.; Berta, G. Colonization of Tomato Root Seedling by *Pseudomonas fluorescens* 92 rkG5: Spatio-Temporal Dynamics, Localization, Organization, Viability, and Culturability. *Microbiol. Ecol.* **2005,** *50,* 289–297.

Gamliel, A.; Katan, J. Suppression of Major and Minor Pathogens by Fuorescent Pseudomonads in Solarized and Non-Solarized Soils. *Phytopathology* **1993,** *83,* 68–75.

Ghorbanpour, M.; Hatami, M. Biopriming of *Salvia officinalis* L. Seed with Plant Growth Promoting Rhizobacteria (PGPR) Changes the Invigoration and Primary Growth Indices. *J. Biol. Environ. Sci.* **2014,** *8,* 29–36.

Glick, B. R. Phytoremediation: Synergistic Use of Plants and Bacteria to Clean Up the Environment. *Biotechnol. Adv.* **2001,** *21,* 383–93.

Glick, B. R. The Enhancement of Plant Growth by Free-Living Bacteria. *Can. J. Microbiol.* **1995,** *41,* 109–17.

Glick, B. R. Plant Growth-Promoting Bacteria: Mechanismsand Applications. *Scientifica* **2012,** *2012,* 15.

Glick, B. R.; Karaturovi, D. M.; Newell, P. C. A Novel Procedure for Rapid Isolation of Plant Growth-Promoting Pseudomonads. *Can. J. Microbiol.* **1995,** *41,* 533–536.

Glick, B. R.; Cheng, Z.; Czarny, J; Duan, J. Promotion of Plant Growth by ACC Deaminase Producing Soil Bacteria. *Eur. J. Plant Pathol.* **2007,** *119,* 329–339.

Gohain, R. Application of Microbial Antagonists for Management of Bacterial Wilt of Brinjal. M.Sc. (Agri) Thesis, Assam Agriculture University: Jorhat, Assam, India; 2001.

Gopalakrishnan, S.; Pande, S.; Sharma M.; Humayun P.; Kiran B. K.; Sandeep D.; Vidya M. S.; Deepthi K.; Rupela, O. Evaluation of Actinomycete Isolates Obtained from Herbal Vermicompost for Biological Control of *Fusarium* wilt of Chickpea. *Crop Prot.* **2011,** *30,* 1070–1078.

Goswami, D.; Patel, K.; Parmar, S.; Vaghela, H.; Muley, N.; Dhandhukia, P.; Thakker, J. N. Elucidating Multifaceted Urease Producing Marine *Pseudomonas aeruginosa* BG as a Cogent PGPR and Biocontrol Agent. *Plant Growth Regulat.* **2015,** *75,* 253–263.

Guzzo, S. D.; MartinsE, Local and Systemic Induction of β-1,3-Glucanase and Chitinase in Coffee Leaves Protected Against *Hemileia vastatrix* by *Bacillus thuringiensis*. *J. Phytopathol.* **1996,** *144* (9–10), 449–454.

Hagedorn, C.; Gould, W. D.; Bardinelli, T. R. Field Evaluation of Bacterial Inoculants to Control Seedling Disease Pathogens on Cotton. *Plant Dis.* **1993,** *77,* 278–282.

Hall, J. L. Cellular Mechanisms for Heavy Metal Detoxification and Tolerance. *J. Exp. Bot.* **2002**, *53*, 1–11.

Hallman, J.; Quadt-Hallman, A.; Mahafee, W. F.; Kloepper, J. W. Bacterial Endophytes in Agricultural Crops. *Can. J. Microbiol.* **1997**, *43*, 895–914.

Hallmann, J.; Rodríguez-Kábana, R.; Kloepper, J. W. Nematode Interactions with Endophytic Bacteria. In *Plant Growth-Promoting Rhizobacteria Present Status and Future Prospects;* Ogoshi, A., Kobayashi, K., Homma, Y., Kodama, F., Kondo, N., Akino, S., Eds.; Nakanishi Printing: Sapporo, 1997; pp 243–245.

Harish Kumar, R. C. D.; Maheshwari, D. K. Effect of Plant Growth Promoting Rhizobia on Seed Germination, Growth Promotion and Suppression of Fusarium Wilt of Fenugreek (*Trigonella foenum*-graecum L.). *Crop Prot.* **2011**, *30*, 1396–1403.

Hausbeck, M. K.; Lamour, K. H. *Phytophthora capsica* on Vegetable Crops: Research Progress and Management Challenges. *Plant Dis.* **2004**, *88*, 1292–1303. https://doi.org/10.1094/PDIS.2004.88.12.1292.

Hebber, P.; Berge, O.; Heulin, T.; Singh, S. P. Bacterial Antagonists of Sunflower (*Helianthus annuus* L.) Fungal Pathogens. *Plant Soil.* **1991**, *133*, 131–140.

Heckel, D. G. Insecticide Resistance After Silent Spring. *Science* **2012**, *337*, 1612–1614. https://doi.org/10.1126/science.1226994.

Hider, R. C.; Kong, X. Chemistry and Biology of Siderophores. *Nat. Prod. Rep.* **2010**, *27*, 637–657.

Hiltner, L. Uber neue Erfahrungen und Probleme auf dem Gebiet der Bodenbakteriolgie und unter Besonderes Berucksichtigung der Grundugungen und Brauche. *Arb. Dtsch. Landwirt. Ges. Berl.* **1904**, *98*, 59–78.

Hoffland, E.; Hakulinem, J.; van Pelt, J. A. Comparison of Systemic Resistance Induced by Avirulent and Nonpathogenic *Pseudomonas* species. *Phytopathology* **1996**, *86*, 757–762.

Hortencia, G. M.; Olalde, V.; Violante, P. Alteration of Tomato Fruit Quality by Root Inoculation with Plant Growth-Promoting Rhizobacteria (PGPR): *Bacillus subtilis* BEB-13bs. *Sci. Hortic.* **2007**, *113*, 103–106.

Howell, C. R.; Stipanovic, R. D. *Phytopathology* **1979**, *69*, 480–482.

Hussain, S.; Siddique, T.; Saleem, M.; Arshad, M.; Khalid, A. Impact of Pesticides on Soil Microbial Diversity, Enzymes, and Biochemical Reactions. In *Advances in Agronomy*; Sparks, D. L., duPont, S. H., Eds.; Elseveir: Amsterdam, 2009; Vol. 102, pp 159–200.

Islam, S.; Akanda, A. M.; Prova, A.; Islam, M. T.; Hossain, M. M. Isolation and Identification of Plant Growth Promoting Rhizobacteria from Cucumber Rhizosphere and Their Effect on Plant Growth Promotion and Disease Suppression. *Front. Microbiol.* **2016**, *6*, 1360. https://doi.org/10.3389/fmicb.2015.01360.

Jahanian, A.; Chaichi, M. R.; Rezaei, K.; Rezayazdi, K.; Khavazi, K. The Effect of Plant Growth Promoting Rhizobacteria (PGPR) on Germination and Primary Growth of Artichoke (*Cynara scolymus*). *Int. J. Agric. Crop Sci.* **2012**, *4*, 923–929.

Jha, C. K.; Saraf, M. Plant Growth Promoting Rhizobacteria (PGPR): A Review. E3 *J. Agric. Res. Dev.* **2015**, *5*, 108–119.

Joe, M. M.; Jaleel, C. A.; Sivakumar, P. K.; et al. *J. Taiwan Inst. Chem. Eng.* **2009**, *40*, 491–499.

Kalita, B. C. Management of Bacterial Wilt of Tomato through Bioagents and Host Resistance. Ph.D. Thesis. Assam Agriculture University: Jorhat, Assam, India; 2002.

Kavitha, R.; Umesha, S. Prevalence of Bacterial Spot in Tomato Fields of Karnataka and Effect of Biological Seed Treatment on Disease Incidence. *Crop Prot.* **2007**, *26*, 991–997.

Kennedy, A. C. Rhizosphere. In *Principles and Applications of Soil Microbiology*, 2nd edn.; Sylvia, D. M., Fuhrmann, J. J., Hartel, P. G., Zuberer, D. A., Eds.; Pearson: Prentice Hall, Upper Saddle River, NJ, 2005; pp 242–262.

Khalid, A.; Arshad, M.; Zahir, Z. A. Screening Plant Growth Promoting Rhizobacteria for Improving Growth and Yield of Wheat. *J. Appl. Microbiol.* **2004,** *96*, 473–480.

Khan, A. G. Role of Soil Microbes in the Rhizospheres of Plants Growing on Trace Metal Contaminated Soils in Phytoremediation. *J. Trace Elem. Med. Biol.* **2005,** *18*, 355–364.

Khan, M. S.; Zaidi, A.; Wani, P. A.; Role of Phosphate Solubilizing Microorganisms in Sustainable Agriculture: A Review. *Agron. Sustain. Dev.* **2006,** *27*, 29–43.

Kim, Y. C.; Jung, H.; Kim, K.Y.; Park, S. K. An Effective Biocontrol Bioformulation Against Phytophthora Blight of Pepper Using Growth Mixtures of Combined Chitinolytic Bacteria Under Different Field Conditions. *Eur. J. Plant Pathol.* **2008**, *120*, 373–382 https://doi.org/10.1007/s10658-007-9227-4.

Kloepper, J. W. *Soil Microbiology Ecology: Applications in Agricultural and Environmental Management*; Marcel Dekker Inc.: New York, USA, 1993; Vol. 19, pp 255–274.

Kloepper, J. W.; Schroth, M. N. In *Plant Growth-Promoting Rhizobacteria on Radishes*, Plant Pathogenic Bacteria, Proceedings of the 4th International Conference, Vol 2. Station de Pathologie Végétale et de Phytobactériologie, INRA, Angers: France, 1978; pp 879–882.

Kloepper, J. W. Plant Growth-Promoting Rhizobacteria as Biological Agents. In *Soil Microbial Ecology. Applications in Agricultural and Environmental Management*; Meeting, F. B. Jr., Ed.; Marcel Dekker Inc.: NY, USA, 1992; pp 255-274.

Kloepper, J. W.; Lifshitz, R.; Zablotowicz, R. M. Free-Living Bacterial Inocula for Enhancing Crop Productity. *Trends Biotechnol.* **1989,** *7*, 39–43.

Kloepper, J. W.; Tuzun, S.; Liu, L.; Wei, G. Plant Growth-Promoting Rhizobacteria as Inducers of Systemic Disease Resistance. In *Pest Management: Biologically Based Technologies*; Lumsden, R. D., Waughn, J., Eds.; American Chemical Society Books: Washington, DC, 1993; pp 156–165.

Kloepper, J. W.; Ryu, C. M. Bacterial Endophytes as Elicitors of Induced Systemic Resistance. In *Microbial Root Endophytes*; Schulz, B., Boyle, C., Siebern, T., Eds.; Springer-Verlag: Heildelberg, 2006; pp 33–51.

Kokalis-Burelle, N.; Dickson, D. W. In *Effects of Soil Fumigants and Bioyield tm on Root-Knot Nematode Incidence and Yield of Tomato*, Methyl Bromide Alternatives and Emissions Reductions, Proceedings of the International Research Conference., 2003, *50*, 1–50.

Kumar, B. S. D. Fusarial Wilt Suppression and Crop Improvement through Two Rhizobacterial Strains in Chick Pea Growing in Soils Infested with *Fusarium oxysporum* f. sp. *ciceris. Biol. Fertil. Soil.* **1999,** *29*, 87–91

Kumar, B. S. D.; Berggren, I.; Martensson, A. M. *Plant Soil* **2001,** *229*, 25–34.

Lee, B. K.; Kim, B. S.; Chang, S. W.; Hwang, B. K. Aggressiveness to Pumpkin Cultivars of Isolates of *Phytophthora capsica* from Pumpkin and Pepper. *Plant Dis.* **2001,** *85*, 497–500. DOI:10.1094/PDIS.2001.85. 5.497.

Lenin, G.; Jayanthi, M. Efficiency of Plant Growth Promoting Rhizobacteria (PGPR) on Enhancement of Growth, Yield, and Nutrient Content of *Catharanthus roseus. Int. J. Res. Pure Appl. Microbiol.* **2012,** *2*, 37–42.

Liu, L.; Kloepper, J. W.; Tuzun, S. Induction of Systemic Resistance in Cucumber Against Bacterial Leaf Spot by Plant Growth Promoting Rhizobacteria. *Phytopathology* **1995,** *85,* 843–847.

Loganathan, M.; Garg, R.; Venkataravanappa, V.; Saha, S.; Rai, A. B. Plant Growth Promoting Rhizobacteria (PGPR) Induces Resistance Against Fusarium wilt and Improves Lycopene Content and Texture in Tomato. *Afr. J. Microbiol. Res.* **2014**, *8* (11), 1105–1111.

Lucy, M.; Reed, E.; Glick, B. R. Applications of Free Living Plant Growth-Promoting Rhizobacteria. *Antonie van Leeuwenhoek.* **2004**, *86*, 1–25.

Lugtenberg, B.; Kamilova, F. Plant Growth Promoting Rhizobacteria. *Annu. Rev. Microbiol.* **2009**, *63*, 541–556.

Maheshwari, D. K.; Dheeman, S.; Agarwal, M. Phytohormone-Producing PGPR for Sustainable Agriculture. In *Bacterial Metabolites in Sustainable Agroecosystem*; Maheshwari, D. K., Ed.; Springer International: Germany, 2015; pp 159–182. DOI:10.1007/978-3-319-24654-3_7.

Maleki, M.; Mokhtarnejad, L.; Mostafaee, S. Screening of Rhizobacteria for Biological Control of Cucumber Root and Crown Rot Caused by *Phytophthora drechsleri*. *Plant Pathol. J.* **2011**, *27*, 78–84. DOI:10.5423/PPJ.20 11.27.1.078.

Marschner, P.; Crowley, D.; Yang, C. H. Development of Specific Rhizosphere Bacterial Communities in Relation to Plant Species, Nutrition, and Soil Type. *Plant Soil.* **2004**, *261*, 199–208.

Martinez-Ochoa, N. Biological Control of the Root-Knot Nematode with Rhizobacteria and Organic Amendments. PhD. Dissertation, Auburn University: Alabama, 2000, 120.

Maurhofer, M.; Hase, C.; Meuwly, P.; Metraux, J. P.; Defago, G. Induction of Systemic Resistance of Tobacco to Tobacco Necrosis Virus by the Root-Colonizing *Pseudomonas fluorescens* Strain CHAO: Influence of the gacA Gene and of Pyoverdine Production. *Phytopathology* **1994**, *84*, 139–146.

Maurhofer, M.; Reimmann, C.; Sacherer, S. P.; Heeb, S.; Haas, D.; Defago, G. Salicylic acid Biosynthetic Genes Expressed in *Pseudomonas fluorescens* Strain P3 Improve the Induction of Systemic Resistance in Tobacco Against Tobacco Necrosis Virus. *Phytopathology* **1998**, *88*, 678–684.

Maurya, S.; Singh, R.; Singh, D. P.; Singh, H. B.; Singh, U. P.; Srivastava, J. S. Management of Collar Rot of Chickpea (*Cicer arietinum*) by *Trichoderma harzianum* and Plant Growth Promoting Rhizobacteria. *J. Plant Prot. Res.* **2008**, *48* (3), 348–354.

McKenzie, R. H.; Roberts, T. L. Soil and fertilizers Phosphorusupdate. In *Proceedings of Alberta Soil Science Workshop*, Feb 20–22, 1990; Alberta: Edmonton, pp 84–104.

Meena, B.; Marimuthu, T.; Vidhyasekaran, P.; Velazhahan, R. Biological Control of Root Rot of Groundnut with Antagonistic *Pseudomonas fluorescens* Strains. *J. Plant Dis. Prot.* **2001**, *108*, 369–381.

Meena, B.; Radhajeyalakshmi, R.; Marimuthu, T.; Vidhyasekaran, P.; Doraiswamy, S.; Velazhahan, R. Induction of Pathogenesis-Related Proteins, Phenolics and Phenylalanine Ammonia-Lyase in Groundnut by *Pseudomonas fluorescens*. *J. Plant Dis. Prot.* **2000**, *107*, 514–527.

Meena, B.; Marimuthu, T.; Velazhahan, R. Role of fluorescent Pseudomonads in Plant Growth Promotion and Biological Control of Late Leaf Spot of Groundnut. *Acta Phytopathol. Entomol. Hung.* **2006**, *4*, 203–212.

Merriman, P. R.; Price, R. D.; Kollmorgen, J. F.; Piggott, T.; Ridge, E. H. Effect of Seed Inoculation with *Bacillus subtilis* and *Streptomyces griseus* on the Growth of Cereals and Carrots. *Aust. J. Agric. Res.* **1974**, *25*, 219–226.

Mia, M.A.B.; Shamsuddin, Z. H.; Wahab, Z.; Marziah, M. Effect of Plant Growth Promoting Rhizobacterial (PGPR) Inoculation on Growth and Nitrogen Incorporation of Tissue Cultured Musa Plantlets under Nitrogen-Free Hydroponics Condition. *Austr. J. Crop Sci.* **2010**, *4*, 85–90.

Mohan, S. M.; Biological Control of Damping Off Disease of Solanaceous Crops in Commercial Nurseries. M. Sc. Thesis, University of Agriculture Science: Bangalore, 2006.

Murphy, J. F.; Zehnder, G. W.; Schuster, D. J.; Sikora, E. J.; Polstan, J. E.; Kloepper, J. W. Plant Growth Promoting Rhizobacteria-Mediated Protection in Tomato Against Tomato Mottle Virus. *Plant Dis.* **2000,** *84,* 779–784.

Nandakumar, R. Induction of Systemic Resistance in Rice with Flourescent Pseudomonads for the Management of Sheath Blight Disease. M.Sc. Dissertation. Tamil Nadu Agricultural University: Coimbatore, Tamil Nadu, India, 1998; 105.

Nassar, A. H.; El-Tarabily, K. A.; Sivasithamparam, K. Growth Promotion of Bean (*Phaseolus vulgaris* L.) by a Polyamine Producing Isolate of *Streptomyces griseoluteus*. *Plant Growth Reg.* **2003,** *40,* 97–106. https://doi.org/10.1023/A:102423330352.

Ngoma, L.; Babalola, O. O.; Faheem, A. Ecophysiology of Plant Growth Promoting Bacteria. *Sci. Res. Essay.* **2012,** *7,* 4003–4013.

Nieto, K. F.; Frankenberger, W. T. Jr. Biosynthesis of Cytokinins by *Azotobacter chroococcum*. *Soil Biol. Biochem.* **1989,** *21,* 967–972.

Ortíz-Castro, R.; Valencia-Cantero, E.; López-Bucio, J. Plant Growth Promotion by *Bacillus megaterium* Involves Cytokinin Signaling. *Plant Signal. Behav.* **2008,** *3,* 263–265. DOI: 10.1094/MPMI-20-2-0207

Pal, K. K.; Gardener, B. M. Biological Control of Plant Pathogens. *Plant Health Instruct.* **2006,** *2,* 1117–1142.

Patel, S. T.; Anahosur, K. H. Potential Antagonism of *Trichoderma harzianum* against *Fusarium* spp. and *Macrophomina phaseolina* and *Sclerotium rolfsii*. *J. Mycol. Plant Pathol.* **2001,** *31,* 365–366.

Patel, K.; Goswami, D.; Dhandhukia, P.; Thakker, J. Techniques to Study Microbial Phytohormones. In *Bacterial Metabolites in Sustainable Agroecosystem*; Maheshwari, D. K., Ed.; Springer International: Germany, 2015; pp 1–27. DOI:10.1007/978-3-319-24654-3_1

Paulitz, T. C. Population Dynamics of Biocontrol Agent and Pathogens in Soils and Rhizospheres. *Eur. J. Plant Pathol.* **2000,** *106,* 401–413.

Peixto, A. R.; Mariana, R. L. R.; Michereff, S. J.; Oliveira, S. M. A.; De' Oliveira, S. M. A. *Summa—Phytopathologia* **1995,** *21,* 219–224.

Perner, H.; Schwarz, D.; George, E. Effect of Mycorrhizal Inoculation and Compost Supply on Growth and Nutrient Uptake of Young Leek Plants Grown on Peat-Based Substrates. *Hortic. Sci.* **2006,** *41,* 628–632.

Raaijmakers, J. M.; Leeman, M.; van Oorschot, M. M. P.; van der Sluis, I.; Schippers, B.; Bakker P. A. H. M. Dose–Response Relationships in Biological Control of Fusarium Wilt of Radish by *Pseudomonas* spp. *Phytopathology* **1995,** *85,* 1075–1081.

Raaijmakers, J. M.; Vlami M.; de Souza J. T. Antibiotic Production by Bacterial Biocontrol Agent. *Anton van Leeuwenhoek* **2002,** *81,* 537–547.

Rajkumar, M.; Ae, N.; Prasad, M. N. V.; Freitas, H. Potential of Siderophore-Producing Bacteria for Improving Heavy Metal Phytoextraction. *Trends Biotechnol.* **2010,** *28,* 142–149.

Ramamoorthy, V.; Raguchander, T.; Samiyappan, R. Enhancing Resistance of Tomato and Hot Pepper to Pythium Diseases by Seed Treatment with Fluorescent Pseudomonads. *Eur. J. Plant Pathol.* **2002,** *108,* 429–441.

Raupach, G. S.; Liu, L.; Murphy, J. F.; Tuzun, S.; Kloepper, J. W. Induced Systemic Resistance in Cucumber and Tomato Against Cucumber Mosaic Cucumo Virus Using Plant Growth Promoting Rhizobacteria (PGPR). *Plant Dis.* **1996,** *80,* 891–894.

Rizvi, A.; Zaidi, A.; Khan, M. S.; Saif, S.; Ahmed, B.; Shahid, M. Growth Improvement and Management of Vegetable Diseases by Plant Growth-Promoting Rhizobacteria. *Microb. Strat. Veg. Prod.* **2017,** 99–123.

Sadeghi, A.; Karimi, E.; Dahazi, P. A.; Javid, M. G.; Dalvand, Y.; Askari, H. Plant Growth Promoting Activity of an Auxin and Siderophore Producing Isolate of *Streptomyces* under Saline Soil Condition. *World J. Microbiol. Biotechnol.* **2012,** *28,* 1503–1509. https://doi.org/10.1007/s11274-011-0952-7.

Sangeetha, G.; Thangavelu, R.; Usha Rani, S.; Muthukumar, A.; Udayakumar, R. Induction of Systemic Resistance by Mixtures of Antagonist Bacteria for the Management of Crown Rot Complex on Banana. *Acta Physiol. Plant* **2010,** *32* (6), 1177–1187.

Saravanakumar, D.; Vijayakumar, C.; Kumar, N.; Samiyappan, R. PGPR-Induced Defense Responses in the Tea Plant Against Blister Blight Disease. *Crop Prot.* **2007,** *26,* 556–565.

Schippers, B.; Bakker, A. W.; Bakker, P. A. H. M. Interactions of Deleterious and Beneficial Rhizosphere Microorganisms and the Effect of Cropping Practices. *Annu. Rev. Phytopathol.* **1987,** *25,* 339–358.

Schisler, D. A.; Slininger, P. J.; Behle, R. W.; Jackson, M. A. Formulation of *Bacillus* spp. for Biological Control of Plant Diseases. *Phytopathology* **2004,** *94,* 1267–1271.

Schobe, R. B. M.; van Vuurde, J. W. L. Detection and Enumeration of *Erwinia carotovora* subsp. *atroseptica* Using Spiral Plating and Immunofluorescence Colony Staining. *Can. J. Microbiol.* **1997,** *43,* 847–853.

Senthilkumar, M.; Swarnalakhmi, K. V.; Govindasamy, Y. K. L.; Annapurna, K. Biocontrol Potential of Soybean Bacterial Endophytes against Charcoal rot Fungus *Rhizoctonia bataticola. Curr. Microbiol.* **2009,** *58,* 288–293.

Shafique H. A.; Sultana, V.; Ara, J.; Haque, S. E.; Athar, M. Role of Antagonistic Microorganisms and Organic Amendment in Stimulating the Defense System of Okra Against Root Rotting Fungi Polish. *J. Microbiol.* **2015,** *64* (2), 157–162.

Shaukat, K.; Affrasayab, S.; Hasnain, S. Growth Responses of *Helianthus annus* to Plant Growth Promoting Rhizobacteria Used as a Biofertilizer. *J. Agric. Res.* **2006,** *1,* 573–581.

Shen, L.; Wang, F.; Liu, Y.; Qian, Y.; Yang, J.; Sun, H. Suppression of Tobacco Mosaic Virus by *Bacillus amyloliquefaciens* Strain Ba33. *J. Phytopathol.* **2012,** *161,* 293–294.

Shen, X.; Hu, H.; Peng, H.; Wang, W.; Zhang, X. Comparative Genomic Analysis of Four Representative Plant Growth-Promoting Rhizobacteria in Pseudomonas. *BMC Genomics* **2013,** *14,* 271. https://doi.org/10.1186/1471-2164-14-271

Shirzad, A.; Fallahzadeh-Mamaghani, V.; Pazhouhandeh, M. Antagonistic Potential of Fluorescent Pseudomonads and Control of Crown and Root Rot of Cucumber Caused by *Phytophthora drechsleri. Plant Pathol. J.* **2012,** *28,* 1–9. https://doi.org/10.5423/PPJ.OA.05.2011.0100.

Shrestha, B. K.; Karki, H. S.; Groth, D. E.; Jungkhun, N.; Ham, J. H. Biological Control Activities of Rice-Associated *Bacillus* sp. Strains Against Sheath Blight and Bacterial Panicle Blight of Rice. *PLoS One* **2016,** *11,* 1–18.

Siddiqui, I.A.; Ehetshamul-Haque, S.; Shaukat, S.S., Use of Rhizobacteria in the Control of Root Rot-Root Knot Disease Complex of Mungbean. *J. Phytopathol.* **2001,** *149,* 337–346.

Singh, P. P.; Shin, Y. C.; Park, C. S.; Chung, Y. R. Biological Control of *Fusarium* Wilt of Cucumber by Chitinolytic Bacteria. *Phytopathology* **1999,** *89,* 92–99.

Singh, S. K.; Pancholy, A.; Jindal, S. K.; Pathak, R. Effect of Plant Growth Promoting Rhizobia on Seed Germination and Seedling Traits in Acacia Senegal. *Ann. For. Res.* **2011,** *54,* 161–169.

Spaepen, S.; Vanderleyden, J. Auxin and Plant-Microbe Interactions. *Cold Spring Harb. Perspect. Biol.* **2011,** *3* (4). https://doi.org/10.1101/cshperspect.a001438.

Sturz, A. V.; Christie, B. R.; Nowak, J. Bacterial Endophytes. Potential Role in Developing Sustainable Systems of Crop Production. *Crit. Rev. Plant Sci.* **2000,** *19,* 1–30.

Sudharani, M.; Shivaprakash M. K.; Prabhavathi, M. K. Role of Consortia of Biocontrol Agents and PGPRs in the Production of Cabbage Under Nursery Condition. *Int. J. Curr. Microbiol. Appl. Sci.* **2014,** *3* (6), 1055–1064.

Sureshbabu, K.; Amaresan, N.; Kumar K. Amazing Multiple Function Properties of Plant Growth Promoting Rhizobacteria in the Rhizosphere Soil. *Int. J. Curr.Microbiol. App. Sci.* **2016,** *5* (2), 661–683.

Suslow, T. V.; Schroth, M. N. Rhizobacteria of Sugar Beets: Effects of Seed Application and Root Colonization on Yield. *Phytopathology* **1982,** *72,* 199–206.

Sylvia, E.W. The Effect of Compost Extract on the Yield of Strawberries and Severity of *Botrytis cinerea. J. Sustain. Agric.* **2004,** *25,* 57–68.

Tokala, R. K.; Strap, J. L.; Jung, C. M.; Crawford, D. L.; Salove, M. H.; Deobald, L. A.; Bailey, J. F.; Morra, M. J. Novel Plant-Microbe Rhizosphere Interaction Involving *Streptomyces lydicus* WYEC108 and the Pea Plant (*Pisum sativum*). *Appl. Environ. Microbiol.* **2002,** *68,* 2161–2171. https://doi.org/10.1128/AEM.68.5.2161-2171.2002.

Troxler, J.; Berling, C. H.; Moe¨nne-Loccoz, Y.; Keel, C.; Défago, G. Interactions between the Biocontrol Agent *Pseudomonas fluorescens* CHA0 and *Thielaviopsis basicola* in Tobacco Roots Observed by immunofluorescence Microscopy. *Plant Pathol.* **1997,** *46,* 62–71.

Uppal, A. K.; El Hadrami, A.; Adam, R.; Tenuta, M.; Daayf, F. Biological Control of Potato Verticillium Wilt under Controlled and field Conditions Using Selected Bacterial Antagonists and Plant Extracts. *Biol. Control* **2008,** *44,* 90–100.

Utkhede, R. S.; Rahe, J. E. Biological Control of Onion White Rot. *Soil Biol. Biochem.* **1980,** *12,* 101–104.

Van de, A. B.; Lambrecht, A.; Vanderleyden, M. *J. Microbiol.* **1998,** *144,* 2599–2606.

van Loon, L. C.; Bakker, P. A. H. M.; Pieterse, C. M. J. Systemic Resistance Induced by Rhizosphere Bacteria. *Annu. Rev. Phytopathol.* **1998,** *36,* 453–483. https://doi.org/10.1146/annurev.phyto.36.1.453.

Vijay Krishna Kumar, K.; Reddy, M. S.; Yellareddygari, S. K. R.; Kloepper, J. W.; Lawrence, K. S.; Zhou, X. G.; Sudini, H.; Miller, M. E. Evaluation and Selection of Elite Plant Growth-Promoting Rhizobacteria for Suppression of Sheath Blight of Rice Caused by *Rhizoctonia solani* in a Detached Leaf Bio-Assay. *Int. J. Appl. Biol. Pharm. Technol.* **2011,** *2,* 488–495.

Viswanathan, R. Induction of Systemic Resistance Against Red Rot Disease in Sugarcane by Plant Growth Promoting Rhizobacteria. Ph.D. Dissertation. Tamil Nadu Agricultural University: Coimbatore, Tamil Nadu, India, 1999; p 175.

Viswanathan, R.; Samiyappan, R. Induction of Systemic Resistance by Plant Growth Promoting Rhizobacteria against Red Rot Disease Caused by *Colletotrichum falcatum* Went in Sugarcane. *Proc. Sugar Technol. Assoc. India* **1999,** *61,* 24–39.

Viswanathan, R.; Samiyappan, R. Antifungal Activity of Chitinase Produced by Some Fluorescent Pseudomonads against *Colletotrichum falcatum* Went Causing Red Rot Disease in Sugarcane. *Microbiol. Res.* **2001,** *155,* 309–314.

Vivekananthan, R.; Ravi, M.; Ramanathan, A.; Samiyappan, R. Lytic Enzymes Induced by *Pseudomonas fluorescens* and Other Biocontrol Organisms Mediate Defense Against the Anthracnose Pathogen in Mango. *World J. Microb. Biotechnol.* **2004,** *20,* 235–244.

Wei, G.; Kloepper, J. W.; Tuzun, S. Induction of Systemic Resistance of Cucumber to *Colletotrichum orbiculare* by Select Strains of Plant Growth Promoting Rhizobacteria. *Phytopathology* **1991**, *81*, 1508–1512.

Wei, G.; Kloepper, J. W.; Tuzun, S. Induced Systemic Resistance to Cucumber Diseases and Increased Plant Growth by Plant Growth Promoting Rhizobacteria Under Field Conditions. *Phytopathology* **1996**, *86*, 221–224.

Weller, D. M.; Raaijimakers, J. M.; Gardener, B. B. M.; Thomashow, L. S. Microbial Populations Responsible for Specific Soil Suppressiveness to Plant Pathogens. *Ann. Rev. Phytopathol.* **2002**, *40*, 309–348.

Weststeijn, W. A. Fluorescent Pseudomonads Isolate E11-2 as Biological Agent for Pythium Root Rot in Tulips. *Neth. J. Plant Pathol.* **1990**, *96*, 262–272.

Zaidi, A.; Khan, M. S.; Ahemad, M.; Oves, M. Plant Growth Promotion by Phosphate Solubilizing Bacteria. *Acta Microbiol. Immunol. Hung.* **2009**, *56*, 263–284.

Zehnder, G.; Kloepper, J.; Tuzun, S.; Yao, C.; Wei, G.; Chambliss, O.; Shelby, R. Insect Feeding on Cucumber Mediated by Rhizobacteria-Induced Plant Resistance. *Entomologia Experimentalis et Applicata* **1997**, *83*, 81–85.

Zehnder, G. W.; Murphy, J. F.; Sikora, E. J.; Kloepper, J. W. Application of Rhizobacteria for Induced Resistance. *Eur. J. Plant Pathol.* **2001**, *107*, 39–50.

Zeller, S. L.; Brand, H.; Schmid, B. Host-Plant Selectivity of Rhizobacteria in a Crop/Weed Model System. *Plos One* **2007**, *2* (9), e846.

Zhang, S.; White,T. L.; Martinez, M. C.; McInroy, J. A.; Kloepper, J. W.; Klassen, W. Evaluation of Plant Growth-Promoting Rhizobacteria for Control of Phytophthora Blight on Squash under Green House Conditions. *Biol. Control* **2010**, *53*, 129–135. https://doi.org/10.1016/j.biocontrol.20 09.10.015.

CHAPTER 7

Botanicals and Their Application in Control of Insect Pests

SALMAN AHMAD[1,*], NADEEM KHAN[1], RABIA BASRI[2], SABA SIDDIQUI[1], MD. ABU NAYYER[1], and DEEPTI SRIVASTAVA[1]

[1]*Integral Institute of Agricultural Science and Technology, Integral University, Lucknow 226021, Uttar Pradesh, India*

[2]*Department of Plant Protection, Faculty of Agricultural Science, Aligarh Muslim University, Aligarh 202001 Uttar Pradesh, India*

Corresponding author. E-mail: salmanamd@iul.ac.in

ABSTRACT

Use of synthetic insecticides for controlling insect pest creates several problems such as their persistent toxicity in food grains, the subsequent development of resistance in insect populations, effects the non-target organisms and other adverse environmental impacts. Many plant species produce substances that protect them by killing or repelling the insects that feed on them. Botanicals are naturally occurring chemicals (insect toxins) extracted or derived from plants or minerals used as natural insecticides. Thus, sound management of insect pest by other natural insecticides has more interest and is eco-friendly to nature with less reduced negative effects on the environment. However, in very early time of insect control, tobacco extract was used as a plant spray in parts of Europe and tobacco product nicotine was discovered in France to kill aphids. Botanical insecticides have different chemical structures and modes of action. Based on the physiological activity, there are six groups of botanical insecticides, namely; repellents, feeding deterrents/antifeedants, toxicants, natural grain protectants, chemosterilants/reproduction inhibitors and insect growth, and development inhibitors. These natural pesticides have many advantages over synthetic ones and may be more cost-effective as a whole, considering the environmental cost of chemical alternatives.

7.1 INTRODUCTION

Today and since the beginning of agriculture, pests are the major constraints in crop production. Pests include any organism which causes economic damage to the crop. Due to over rising population of the world (about 7.6 billion people) and more than half of the population is living below poverty line, this situation becomes more critical in the present time. The productivity of agricultural crops in India is 3 t/ha in comparison to the global average of 4 t/ha; out of which insect pests account for 26% of the total loss (Anonymous, 2015a) and stored food grain accounts for 20–25% damage by storage pests (Rajashekar et al., 2010). The discovery of dichlorodiphenyltrichloroethane (DDT) in 1939 (which was first synthesized in 1874) brought a revolution in pest control. Since then, a number of pesticides including carbamates, organocholines, arsenicals, diamide, flourines, inorganic insecticides, juvenile hormones (JH), JH mimics moulting hormones (MH), MH inhibitors, isoxazolines, neonicotinoids, heterocyclic organothiophosphates, synthetic pyrethoids, phosphonates, pyraoles, aliphatic nitrogen fungicides, amides, aromatic fungicides, arsenical fungicides, benzimidazole fungicides, and so on came into existence. In the beginning, DDT and the other pesticides were used to fulfill the purpose of pest control. The chemicals were applied without difficulty, were fast acting and killed mostly target pests. In the due passage of time, the indiscriminate use of these pesticides lead to harmful effects on mankind as well as beneficial organisms and cause ecological imbalances.

Above and beyond, drawbacks in the use of the pesticides include unfavorable effects on environment, development of resistance against a number of insecticides in insects, resurgence of minor pest as major pests, ecological inequity, residue of pesticides in food, and pollution of environment. For example, due to indiscriminate use of insecticides, *Helicoverpa armigera* has developed resistance to almost all groups of insecticides (Armes et al., 1996; Bues et al., 2005; Ugurlu and Gurkan, 2007). The alternative methods of pest management include botanicals (or biopesticides) which came into existence and proved to be much effective to regulate the pest population in safer way. Biopesticides are derived from natural products which may be plant, animal, or microbial origin. natural toxins (or botanical pesticides) are source of new chemical classes of pesticides, which are environmentally and toxicologically safer molecules than many of the currently used pesticide (Duke et al., 2010). They are usually less toxic than conventional pesticides and generally affect only the target pests and closely related organisms and so far no resistance is developed in insects against them. They are very effective in small quantities, are environmental friendly, and often decompose easily resulting in minimum or no pollution of the environment. Most botanicals have low to moderate mammalian toxicity. Insecticidal properties are present in many plants either in their roots, seeds, kernals, or leaves which are utilized against pests and diseases for a long time. Some of the plants metabolites are repellents, antifeedants in action like azadirachtin, rape seed extract, some are toxic such as pyrethrum, nicotine, rotenone and so on, some like *Acorus calamus* act as sterilants (Ignatowicz and Wesolowska, 2015). Asian countries including India have plentiful of these plants which are still traditionally utilized by the rural people for preparations against insect control (Talukder and Howse, 1993). These plant products are utilized either singly or as a part of an integrated pest management program (IPM program; Ahmad et al., 2015; Lal et al., 2017). They can be used against agricultural pests (Isman, 2006, Ahmad et al., 2013, 2015), storage pests (Lal et al., 2017), production of compost through solid state fermentation (Ballardo et al., 2017), malarial vectors (Blanford et al., 2011), household insects (Pawar, 2013) and are harmless to beneficial and non-target organisms (Kima et al., 2017). Botanicals decompose rapidly in moisture, sunlight, and air by detoxification through enzymes. Rapid breakdown indicates less persistence and minimum threat to non-target organisms. However, accurate timing and/or more recurrent applications may be required. Most botanicals are non-phytotoxic. However, nicotine sulfate may be toxic to some vegetables and ornamentals (Duke et al., 2010).

7.2 PESTICIDES: MERITS AND DEMERITS

Pesticides or chemicals are supposed to control destructive pests such as insects, diseases, nematodes, rodents, weeds and so on. Agrochemicals have facilitated to more than double the food production during the last century, and the existing need to maximize the food production to nourish the rapidly growing population sustains pressure on the rigorous use of pesticides and fertilizers (Carvalho, 2017). There are different categories of pesticides that each are meant to be effective against specific pests. Some examples are included in Table 7.1.

Crop protection market in India is taken over by insecticides, which account to almost 70% of domestic crop protection chemicals market. The standard application rates of pesticides per hectare of cultivated land as worked out by FAO and the highest average values occurred in Asia and in some countries of South America (Carvalho, 2017). The major reason for increasing use of pesticides is that these substances are the final contribution for any crop cultivation. Farmers spend on land, soil, seeds, labor, and fertilizers and in the last on pesticides. Pesticides are the concluding input in the agricultural process, which saves all other inputs when substantial investment is already done. Hence they provide full probability to save the crops from pests and diseases. Pesticides have been an essential part of the pest management program because they reduce the direct losses from the insect pests, weeds, diseases, rodents that can markedly reduce the amount of harvestable produce. The general benefits of pesticides include control of pests, improving productivity, protection of crop losses/yield reduction, vector disease control, increasing quality of food, improving storage losses, and so on (Aktar et al., 2009).

TABLE 7.1   List of Pesticides and Their Effectiveness Against Specific Pests.

Category	Effectiveness
Insecticides	Control insects
Fungicides	Control fungal pathogens
Acaricides	Control tics and mites
Nematicides	Control nematodes
Herbicides	Kill or inhibit the growth of unnecessary plants
Antimicrobials	Control germs and microbes like bacteria
Insect Growth Regulators	Disrupt growth and reproduction of insects
Rodenticides	Kill rodents like mice, rats, and household pests
Disinfectants	Control germs and microbes such as bacteria and viruses
Algaecides	Kill and/or reducing the growth and development of algae
Molluscicides	Control molluscs

The pattern of pesticide consumption in India is different from that of the world which can be seen in Figure 7.1. Pesticide use in India differs with the cropping prototype, strength of insect pests, and diseases and agro-ecological regions. The state-wise and crop wise consumption also vary significantly.

In addition to benefits, chemical pesticides also account for enormous hazardous effect on environment, mankind, and agricultural crops. Synthetic pesticides along with the target organism also kill beneficial natural enemies and other non-target organisms; thereby enhance pest problems as they are important in pest control (Lal et al., 2017). They also pose adverse effects on environment, development of resistance in insects, pest resurgence, ecological imbalance, residue in food, and environmental contamination. Most of the pesticides have been connected with issues related to health and environment (WHO, 2017; Hayes et al., 2006; Alewu et al., 2011; Nicolopoulou-Stamati et al. 2016), and the agricultural use of certain pesticides has been discarded (Hayes et al., 2006). Human exposure to pesticides can be through contact with the skin, mouth (ingestion), or nose (inhalation). The duration and route of contact, nature of the pesticide, and the person's health status (e.g., nutritional deficiencies and healthy/damaged skin) are influential factors in the probable health outcome (Alewu et al., 2011; Nicolopoulou-Stamati et al., 2016). Within a human or animal body, pesticides may be stored, metabolized, excreted out, or bioaccumulated in fat bodies (WHO, 2017; Alewu et al., 2011; Pirsaheb et al., 2015).

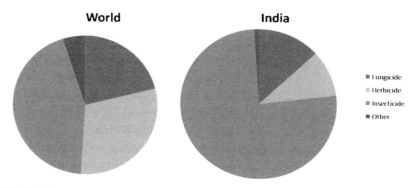

FIGURE 7.1 Consumption patterns of pesticides.

Source: Adapted with permission from Aktar et al. (2009). https://creativecommons.org/licenses/by/4.0

Resistance to pesticides in the pests is yet another problem faced by the injudicious and indiscriminate use of pesticides. Pesticide resistance is described as the declined susceptibility of a pest population to a pesticide that was earlier efficient in controlling that pest. Control failures with the use of a number of insecticides have been reported frequently for populations of *H. armigera* (Armes et al., 1996; Bues et al., 2005; Ugurlu and Gurkan, 2007). Srinivas et al. (2004) reported that highest seasonal average percentage survival was recorded by fenvalerate (65.0%) followed by cypermethrin (62.4%). Saleem et al. (2008) reported several fold resistance in *Spodoptera litura*. For chlorocyclodiene and pyrethroids tested, the resistance ratios compared with Lab-PK were in the range of 10–92-fold for endosulfan, 5–111-fold for cypermethrin, 2–98-fold for deltamethrin, and 7–86-fold for beta-cyfluthrin. For organophosphates and carbamates, resistance ratios were in the range of 3–169-fold for profenofos, 18–421-fold for chlorpyrifos, 3–160-fold for quinalphos, 6–126-fold for phoxim, 7–463-fold for triazophos, and 10–389-fold for methomyl and 16–200-fold for thiodicarb.

7.3 BOTANICALS AS PESTICIDES

The scientific society is very much concerned about the inherent toxic effect of insecticide used by the farmers on human beings, environment, and in various crop ecosystem for decades. Integrated management of pests is facing ecological and economic disputes throughout the globe due to human and environmental hazards caused by majority of the synthetic pesticides. The use of biopesticides is an alternate and noble chance for minimizing the deleterious effect on environment and human health as well as managing insect pests by avoiding resistance and resurgence developed among insect against chemical pesticides. These compounds offer a lot of environmental benefits (Wakeil, 2013). The plant kingdom synthesizes numerous products that are used to defend plants against different pests and is recognized as the most efficient producer of chemical compounds, (Isman, 2006; Isman and Aktar, 2007). Plants have evolved for over million years and to secure themselves from pests they have developed protection mechanisms such as repellents and even insecticidal effects. The insecticidal use of plant extracts and powdered plant products goes as far as the Roman Empire. For example, there are reports that in 400 B.C. during Persian king Xerxes' reign, the delousing procedure for children was with a powder obtained from the dry flowers of pyrethrum (*Tanacetum cinerariaefolium*).

Presently, there are number plants which have insecticidal properties and are being marketed like neem, pyrethrum, grapefruit seeds, nicotine, lantana, aloevera, garlic, and so on. Some major plants having insecticidal properties are listed in Table 7.2. The benefits of botanicals as pesticides recline in their quick degradation and lack of persistence and bioaccumulation in the environment, which is one of the major problems in synthetic pesticides. The variety and redundancy of botanicals extracts are also useful. Redundancy is the presence of number analogs of one compound which are recognized to boost the overall effectiveness of extracts through analog synergism, reduces the metabolic rate of compounds, and prevents resistance against the particular pesticide when selection occurs after several generations.

TABLE 7.2 List of Some Tested Botanicals for Their Insecticidal Activity.

Name of plant	Pests	References
Neem (*Azadirachta indica*)	Armyworms, stemborers, bollworms, leaf miners, diamond backmoth, caterpillars, storage pests (moth), aphids, whiteflies, leaf hoppers, psyllids, scales, maize tassel, beetle, thrips, weevils, and flour beetle	Isman, (2006); Xu et al. (2010); Ahmad et al. (2015)
Chrysanthemum cinerariifolium	Lepidopterous pests	Schleier and Peterson (2011)
Tobacco (*Nicotiana tabacum*)	Lepidopteran pests	Isman (2006); Vandenborre et al. (2010)
Derris spp. (Fabaceae) and *Lonchocarpus* spp (Rotenone)	Lepidopterous pests	Draper et al. (1999)
Garlic (*Allium sativum*)	Caterpillars, cabbage worms, and aphids	Yang et al. (2010)
Lantana (*Lantana camara*)	*Callosobruchus maculates. Sitophilus* sp.	Rajashekhar et al. (2014)
Aloevera (*Aloe* sp.)	*Tribolium castaneum*, mosquito	Omotoso et al. (2005)
Marigold (*Tagetes minuta*)	*Aedes aegypt, Anopheles stephensi*	Amer and Mehlhorn (2006a,c)
Ocimum basilicum	Cattle tick, *Anopheles stephensi, Aedes aegypti*	Prajapati et al. (2005); Martinez Velazquez et al. (2011)
Datura stramonium	*Anopheles stephensi; Sitophilus oryzae*	Senthilkumar et al. (2009); Jawalkar et al. (2016)
Eucalyptus camaldulensis	*Aedes aegypti; Anopheles stephensi*	Senthilkumar et al. (2009); Maciel et al. (2010)
Ipomea carnea	*Aedes aegypti; Anopheles stephensi*	Senthilkumar et al. (2009)
Ricinus communis	*Aedes aegypti; Anopheles stephensi*	Senthilkumar et al. (2009)

7.3.1   NEEM (AZADIRACHTA INDICA; AZADIRACHTIN)

Neem, *Azadirachta indica* A. Juss (Meliaceae), is a key ingredient in botanical-based management providing a natural choice to synthetic pesticides.

Azadirachtin (tetranortriterpenoids; Fig. 7.2) is the major active component extracted from neem leaves and seeds and is one of the most promising plant products for integrated pest management (Schmutterer, 1990; Ahmad et al., 2013). Azadirachtin has been reported to control >400 insect species, including pests of economic importance such as *Pieries brassicae, Plutella xylostella* L., *Spodoptera* spp., *H. armigera*, leafminers, aphids, and whiteflies (Schmutterer, 1990; Isman, 2006; Pineda et al., 2009; Ahmad et al., 2015). In fact, the neem seeds contain a number of insecticidal analogs, which are likely to have a small involvement toward overall effectiveness (Isman, 2006).

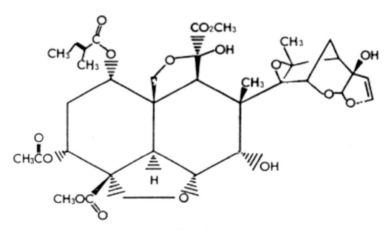

FIGURE 7.2   Azadirachtin (active ingredient of neem).

The major action of azadirachtin is to inhibit synthesis and release of molting hormone (MH) from the prothoracic gland, leading to defective ecdysis in immature stages of insects which prevents adult formation. In addition, it acts as a strong antifeedant, oviposition deterrent, repellent, and growth inhibitor against most insects (Schmutterer, 1990; Mordue and Blackwell, 1993; Isman, 2006; Hasan and Ansari, 2011; Ahmad et al., 2013; 2015). Azadirachtin was first used on the desert locust based on its exceptional antifeedant activity, and this substance remains the most potent antifeedant against locust discovered till date. Unlike pyrethrins, azadirachtin has defied total synthesis to this point. Promoted in the United States by Robert Larson (with assistance from the United States Department of Agriculture), neem quickly became the contemporary concept for development of botanical insecticides (Isman, 2006). Azadirachtin being chemically complicated has not been synthesized. It is structurally similar to the natural insect hormone ecdysone. It is safer for human beings, fishes, pollinators and biocontrol agents (Isman, 2006). Presently, a number of commercial formulations of neem are commercially available like neemix, neemban, neemazal, neemcure, neem gold, econeem, neemark, and azatin.

7.3.2 *Chrysanthemum cinerariifolium* (PYRETHRIN)

Pyrethrins are the active ingredient present in *C. cinerariifolium* which represent the most important group of commercial botanical insecticides. Pyrethrins are often extracted from the flowers of *C. cinerariaefolium,* and are a set of esters with similar structures. Pyrethrins of type I are the most plentiful compound and having maximum insecticidal property in comparison to others (Schleier and Peterson, 2011). Pyrethrins produce "knock-down" effect causing hyperactivity of insects followed by spasms. These symptoms result from a neurotoxic method that opens Na^+ channels in the nerve membranes of the insects, causing hyperactivity, and excitement (Davies et al., 2007). The insecticidal action of the pyrethrins is distinguished by a quick knockdown effect, particularly in lepidopteran insects, and hyperactivity and convulsions in most of the insects (Wakeil, 2013).

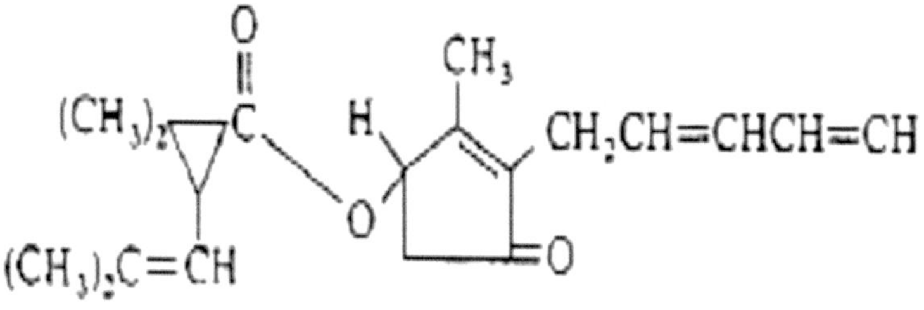

FIGURE 7.3 Pyrethrin I (active ingredient of *Chrysanthemum cinerariifolium*).

7.3.3 *Derris* spp. AND *Lonchocarpus* spp. (ROTENONE)

The active ingredient of *Derris* spp. and *Lonchocarpus* spp. which is used as botanical pesticide is rotenone. Rotenone is a flavonoid extracted from the roots of *Derris* spp. and *Lonchocarpus* spp. *Derris* spp. may have up to 13% of rotenone while *Lonchocarpus* spp. may have about 5%. Rotenone is an ingestion compound and contact in action with a good repellent effect on insects. Its mechanism of action involves the inhibition of the electron transport at the mitochondrial level, and thus blocking phosphorylation of ADP in to ATP and by this means inhibits insect metabolism (Draper et al. 1999; Li et al., 2003). Rotenone is usually sold as a dust formulation containing 1–5% active ingredients for home and garden use, but liquid formulations used in organic agriculture may have as much as 8% rotenone and 15% rotenoids. Pure rotenone is equivalent to DDT and other synthetic insecticides in terms of its acute toxicity to mammals (rat oral LD_{50} = 132 mg kg^{-1}; Isman, 2006; Wakeil, 2013).

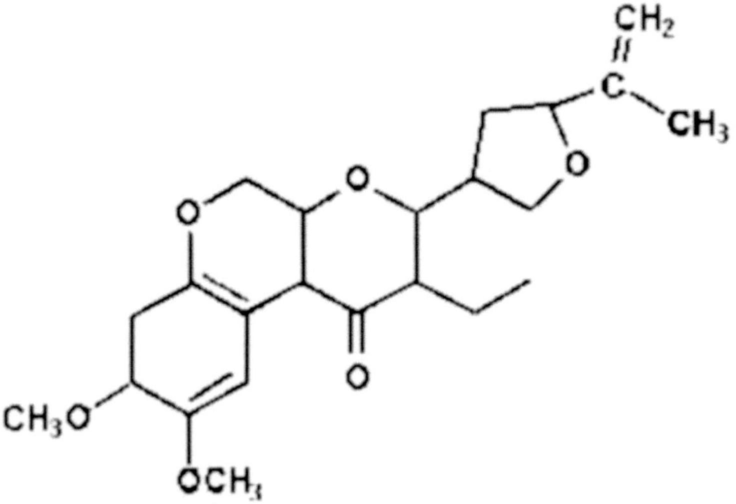

FIGURE 7.4   Molecular structure of rotenone.

7.3.4   *TOBACCO (NICOTIANA TABACUM; NICOTINE)*

The active ingredient of tobacco plant that has pesticidal properties is an alkaloid Nicotine. The insecticidal properties of nicotine were distinguished in the first half of the 16th century. An excellent review on nicotine as an insecticide was presented by Schmeltz (1971). Nicotine is essentially a non-persistent contact insecticide. The mode of action includes mimicking acetylcholine after binding with its receptor in the post-synaptic membrane of the muscular combination. The acetylcholinic receptor is a site of action of the postsynaptic membrane which reacts with acetylcholine and alters the membrane porosity (Narahashi et al. 2000; Wakeil, 2013). Gunasena et al. (1990) found that when nicotine was included into artificial diet it significantly reduced the 7-d larval and pupal weight and prolonged the pupation time of larval tobacco bud worms, *Heliothis virescens* (F.) but parasitoid larval development after regression was not affected. There are reports that due to the excessive toxicity of pure nicotine to mammals (rat oral LD_{50} is 50 mg/kg) and its rapid absorption through skin in human beings, nicotine has seen declining use, primarily as a fumigant in greenhouses against soft-bodied pests (Wakeil, 2013).

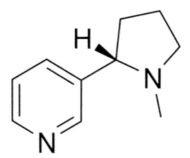

FIGURE 7.5   Molecular structure of nicotine.

7.3.5   MARIGOLD (Tagetes minuta)

Tagetes is a herbaceous plant that includes some species of the family composite. This genus is native to Central and South America and was established in other tropical and subtropical regions. *Tagetes* spp. is commonly known as marigolds (Andreotti et al., 2014). A number of insecticidal compounds have been isolated from *Tagetes* spp. Compounds extracted from the leaves and flowers of *T. minuta* are toxic to *Aedes aegypti* (L.) larvae (Amer and Mehlhorn, 2006a, c) and tick infestation (Andreotti et al., 2014). The roots of marigold contain the alpha terthienyl which is a light-sensitive compound and suppresses nematode populations and improve the growth of plants like tobacco (Miller and Ahrens, 1969) and tomatoes (Ijani and Mmbaga, 1988). The chemicals extracted from marigold belong to essential oils, carotenoids, thiophenes, phenolic compounds, and flavonoids. A preliminary phytochemical screening of the crude successive extracts of the roots of *T. erecta* revealed the presence of glycosides, sterols, gums, and mucilages (Gupta et al., 2009).

7.3.6   Eucalyptus camaldulensi

Eucalyptus belongs to family Myrtaceae, native of Australia, represented by about 700 species and is a genus of tall, evergreen, and wonderful trees cultivated worldwide for its oil, pulp, medicine, timber, gum, and aesthetic value (Batish et al., 2008). The essential oil also possesses a broad spectrum biological activity comprising insecticidal/insect repellent, antimicrobial, fungicidal, herbicidal, acaricidal, and nematicidal (Batish et al., 2008). The major ingredients of the essential oil with pesticidal activity extracted from *Eucalyptus* species include γ-terpinene, 1,8-cineole, eucamalol; citronellol, citronellyl acetate (Watanabe et al., 1993; Su et al., 2008; Liu et al., 2008). Eucalyptus oil has also been used as an antifeedant, predominantly against insects' bite. Trigg (1996a, b) found that eucalyptus-based products can be applied on human skin as insect repellent can protect from biting of insects up to 8 h depending upon the concentration of the essential oil.

7.3.7   Datura stramonium

Datura or *Datura stramonium* is a popular herb and is well known for its toxic effect, medicinal purposes, cosmetic applications, and in bio fertilizers and pesticides. The leaf extract can be taken orally for the treatment of asthma, sinus infections, and stripped bark is applied exteriorly to cure swelling, burns, and ulcers. The growing plants repel insects, which also protects neighboring plants from insects (Das et al., 2012; Jawalkar et al., 2016). The active ingredient of the plants having insecticidal properties includes alkaloids, phenols, tannins, flavonoids, steroids, saponins, and glycosides (Shagal et al., 2012). All the plant parts are lethal to insects, but the ripe seeds contain the highest amount of alkaloids (Shagal et al., 2012; Jawalkar et al., 2016). The insecticidal property of seed extracts is also well established against the rice weevil, *Sitophilus oryzae* (Jawalkar et al., 2016).

7.4   CONCLUSION

Synthetic pesticides, while significant for their convenience and effectiveness also pose certain problems, such as phytotoxicity and effect on non-target organisms, environmental degradation, and health hazards. Botanical pesticides have long been touted as attractive substitute to synthetic chemicals for pest management because they pose a little or no risk to the environment and human health. There are undoubtedly, a number of plants having pesticidal properties which are being utilized commercially and yet to be identified and characterized. The use of botanicals as pesticide is a wide topic for research to find solutions to a number of problems caused by synthetic pesticides as well as economic problems. The actual advantages of botanicals as insecticides can be best realized in developing countries like India where farmers may not be able to afford synthetic insecticides and the traditional utilization of plants and their derivatives for safeguard of stored products is long established. There is massive potential in plants and their extracts to act as a substitute for synthetic pesticides but more studies are to be carried out in this field. Research efforts should not only focus on their effectiveness, but also on mode of action, stability of the compound, toxicity to mammalian, environment, and non-target organisms.

KEYWORDS

- **biopesticide**
- **botanicals**
- **synthetic pesticide**
- **toxicity**
- **insecticidal properties**

REFERENCES

Ahmad, S.; Ansari, M. S.; Moraiet, M. A. Demographic Changes in *Helicoverpa armigera* After Exposure to Neemazal (1% EC Azadirachtin). *Crop Protect.* **2013,** *50*, 30–36.

Ahmad S.; Ansari, M. S.; Muslim, M. Toxic Effects of Neem Based Insecticides on the Fitness of *Helicoverpa armigera* (Hübner). *Crop Protect.* **2015,** *68*, 72–78.

Aktar, M. W.; Sengupta, D.; Chowdhury, A. Impact of Pesticides Use in Agriculture: Their Benefits and Hazards. *Interdisc. Toxicol.* **2009,** *2* (1), 1–12.

Amer, A.; Mehlhorn, H. Larvicidal Effects of Various Essential Oils Against Aedes, Anopheles, and Culex larvae (Diptera, Culicidae). *Parasit. Res.* **2006,** *99*, 466–472.

Amer, A.; Mehlhorn, H. Persistency of Larvicidal Effects of Plant Oil Extracts Under Different Storage Conditions. *Parasit. Res.* **2006,** *99*, 473–477.

Andreotti, R.; Garcia, M. V.; Matias, J.; Barros, J. C.; Cunha, R. C. *Tagetes minuta* Linnaeus (Asteraceae) as a Potential New Alternative for the Mitigation of Tick Infestation. *Med. Arom. Plants.* **2014,** *3* (4).

Anonymous. Ushering in the 2nd Green Revolution: Role of Crop Protection Chemicals. *A Report on Agrochemical Industry, FICCI.* **2015,** p 51.

Armes, N. J.; Jadhav, D. R.; DeSouza, R. A Survey of Insecticide Resistance in *Helicoverpa armigera* in the Indian Subcontinent. *Bull. Entomol. Res.* **1996,** *5,* 86, 499–514.

Ballardo, C.; Barrena, Y.; Artola, A.; Sánchez, A. A Novel Strategy for Producing Compost with Enhanced Biopesticide Properties Through Solid-state Fermentation of Biowaste and Inoculation with *Bacillus thuringiensis. Waste Manage.* **2017,** *70,* 53–58.

Blanford, S.; Shi, W.; Christian, R.; Marden, J. H.; Koekemoer, L. L.; Brooke, B. D.; Coetzee, M.; Read, A. F.; Thomas, M. B. Lethal and Pre-Lethal Effects of a Fungal Biopesticide Contribute to Substantial and Rapid Control of Malaria Vectors. *Plos One* **2011,** *6,* 8.

Bues, R.; Bouvier, J. C.; Boudinhon, L. Insecticide Resistance and Mechanisms of Resistance to Selected Strains of *Helicoverpa armigera* (Lepidoptera: Noctuidae) in the South of France. *Crop Protect.* **2005,** *24,* 814–820.

Carvalho, F. P. Pesticides, Environment, and Food Safety. *Food Ener. Sec.* **2017,** *6* (2), 48–60.

Copping, L. G.; Menn, J. J. Biopesticides: A Review of Their Action, Applications and Efficacy. *Pest Manag. Sci.* **2000,** *56,* 651–676.

Das, S.; Kumar, P.; Basu, S. P. Phytoconstituents and Therapeutic Potentials of *Datura stramonium* Linn. *J. Drug Deliv. Ther.* **2012,** *2* (3), 4–7.

Davies, T. G.; Field, L. M.; Usherwood, P. N.; Williamson, M. S. DDT, Pyrethrins, Pyrethroids and Insect Sodium Channels. *IUBMB Life.* **2007,** *59,* 151–162.

Draper, W. M.; Dhoot, J. S.; Perera, S. K. Determination of Rotenoids and Piperonyl Butoxide in Water, Sediments and Pesticide Formulations. *J. Environ. Monit.* **1999,** *1,* 519–524.

Duke, S. O.; Cantrell, C. L.; Meepagala, K. M.; Wedge, D. E.; Tabanca, N.; Schrader, K. K. Natural Toxins for Use in Pest Management. *Toxins (Basel)* **2010,** *2* (8), 1943–1962.

Fournier, J. *Chimie des Pesticides, Cultures et Techniques*; Nantes et Agence de Cooperation Culturelle et Tech-´nique: Paris, 1988; p 351.

Gunasena, G. H.; Vinson, S. B.; Williams, H. J. Effects of Nicotine on Growth, Development, and Survival of the Tobacco Budworm (Lepidoptera: Noctuidae) and the Parasitoid *Campoletis sonorensis* (Hymenoptera: Ichneumonidae). *J. Econ. Entomol.* **1990,** *83* (5), 1777–1782.

Gupta, P.; Vasudeva, N.; Sharma, S. K. Pharmacognostical Study and Preliminary Phytochemical Screening of the Roots of *Tagetes erecta* Linn. *Hamdard Med.* **2009,** *52,* pp 153–160.

Hasan, F.; Ansari, M. S. Toxic Effects of Neem-based Insecticides on *Pieris brassicae* (Linn.). *Crop Prot.* **2011,** *30,* 502–507.

Hayes, T. B.; Case, P.; Chui, S.; Chung, D.; Haeffele, C.; Haston, K. et al. Pesticide Mixtures, Endocrine Disruption, and Amphibian Declines: Are We Underestimating the Impact? *Environ. Health Perspect.* **2006,** *114,* 40–50. DOI: 10.1289/ehp.8051.

Ignatowicz, S.; Wesolowska, B. In *Potential of Common Herbs as Grain Protectants: Repellent Effect of Herb Extracts on the Granary Weevil, Sitophilus granarius L.* Proceedings of 6th International Working Conference on Stored-products Protection, 2015, 2, 790–794.

Ijani, M. S. A.; Mmbaga, M. T. Studies on the Control of Root-knot Nematodes (*Meloidogyne* species) on Tomato in Tanzania using Marigold Plants (Tagetes Species), Ethylene Dibromide and Aldicarb. *Trop. Pest Manag.* **1988,** *34,* 147–149.

Isman, M. B. Botanical Insecticides, Deterrents, and Repellents in Modern Agriculture and an Increasingly Regulated World. *Ann. Rev. Entomol.* **2006,** *51,* 45–66.

Isman, M. B.; Aktar, Y. Plant Natural Products as Source for Developing Environmentally Acceptable Insecticides. In *Insecticides Design Using Advanced Technologies*; Shaaya, I.; Nauen, R.; Horowitz, A. R., Eds.; Springer: Berlin, Heidenberg, 2007; pp 235–248.

Jawalkar, N.; Zambare, S.; Zanke, S. Insecticidal Property of *Datura stramonium* L. Seed Extracts Against *Sitophilus oryzae* L. (Coleoptera: Curculionidae) in Stored Wheat Grains. *J. Entomol. Zool. Stud.* **2016**, *4* (6), 92–96.

Kima, S. Y.; Ahnb, H. G.; Hab, P. J.; Limc, U. T.; Lee, H. J. Toxicities of 26 Pesticides Against 10 Biological Control Species. *J. Asia Pac. Entomol.* **2017**, *10*, 15.

Lal, M.; Ram, B.; Tiwari, P. Botanicals to Cope Stored Grain Insect Pests: A Review. *Int. J. Curr. Microbiol. App. Sci.* **2017**, *6* (6), 1583–1594.

Li, N.; Ragheb, K.; Lawler, G.; Sturgis, J.; Rajwa, B.; Melendez, J. A.; Robinson, J. P. Title? *J. Biol. Chem.* **2003**, *278* (10), 8516–8525.

Liu, X.; Chen, Q.; Wang, Z.; Xie, L.; Xu, Z. Allelopathic Effects of Essential Oil From *Eucalyptus grandis E. urophylla* on Pathogenic Fungi and Pest Insects. *Front. Forest. China* **2008**, *3*, 232–236.

Maciel, M. V.; Morais, S. M.; Bevilaqua, C. M. L.; Silva, R. A.; Barros, R. S.; Sousa, R. N.; Sousa, L. C.; Brito, E. S.; Souza-Neto, M. A. Chemical Composition of *Eucalyptus* spp. Essential Oils and Their Insecticidal Effects on Lutzomyia longipalpis. *Veter. Paras.* **2010**, *167*, 1–7.

Martinez-Velazquez, M.; Castillo-Herrera, G. A.; Rosario-Cruz, R.; Flores-Fernandez, J. M.; Lopez-Ramirez, J.; Hernandez-Gutierrez, R.; Lugo-Cervantes, E.; Del, C. Acaricidal Effect and Chemical Composition of Essential Oils Extracted from *Cuminum cyminum*, *Pimenta dioica* and *Ocimum basilicum* Against the Cattle Tick *Rhipicephalus* (Boophilus) *microplus* (Acari: Ixodidae). *Parasit. Res.* **2011**, *108*, 481–487.

Miller, M. P.; Ahrens, J. F. Influence of Growing Marigolds, Weeds, Two Covercrops and Fumigation on Subsequent Populations of Parasitic Nematodes and Plant Growth. *Plant Dis. Rep.* **1969**, *53*, 642–646.

Mordue, A. J.; Blackwell, A. Azadirachtin, An Update. *Insect Physiol.* **1993**, *39*, 903–924.

Narahashi, T.; Fenster, C. P.; Quick, M. W.; Lester, R. A.; Marszalec, W.; Aistrup, G. L.; Sattelle, D. B.; Martin, B. R.; Levin, E. D. Symposium Overview: Mechanism of Action of Nicotine on Neuronal Acetylcholine Receptors, From Molecule to Behavior. *Toxicol. Sci.* **2000**, *57* (2), 193–202.

Nicolopoulou-Stamati, P.; Maipas, S.; Kotampasi, C.; Stamatis, P.; Hens, L. Chemical Pesticides and Human Health: The Urgent Need for a New Concept in Agriculture. *Front Pub. Health.* **2016**, *4*, 148. doi: 10.3389/fpubh.2016.00148.

Omotoso, O. T. Insecticidal and Insect Productivity Reduction Capacities of *Aloe vera* and *Bryophyllum pinnatum* on *Tribolium castaneum* (herbst). *Afr. J. Appl. Zool. Environ. Biol.* **2005**, *7*.

Pawar, R. Effect of Curcuma longa (Turmeric) on Biochemical Aspects of House Fly, *Musca domestica* (Diptera: Muscidae). *Int. J. Sci. Res. Public* **2013**, *3* (5).

Pineda, S.; Martínez, A. M.; Figueroa, J. I.; Schneider, M. I.; Estal, P. D.; Viñuela, E.; Gómez, B.; Smagghe, G.; Budia, F. Influence of Azadirachtin and Methoxyfenozide on Life Parameters of *Spodoptera littoralis* (Lepidoptera: Noctuidae). *J. Econ. Entom.* **2009**, *102*, 1490–1496.

Pirsaheb, M.; Limoee, M.; Namdari, F.; Khamutian, R. Organochlorine Pesticides Residue in Breast Milk: A Systematic Review. *Med. J. Islam. Repub. Iran.* **2015**, *29*, 228.

Prajapati, V.; Tripathi, A. K.; Aggarwal, K. K.; Khanuja, S. P. S. Insecticidal, Repellent and Oviposition-deterrent Activity of Selected Essential Oils Against *Anopheles stephensi*, *Aedes aegypti* and *Culex quinquefasciatus*. *Biores. Techn.* **2005**, *96*, 1749–1757.

Rajashekar, Y.; Gunasekaran, N.; Shivanandappa, T. Insecticidal Activity of the Root Extract of *Decalepis hamiltonii* Against Stored Product Insect Pests and Its Application in Grain Protection. *J. Food Sci. Tech.* **2010**, *47* (3), 310–314.

Rajashekar, Y.; Ravindra, K. V.; Bakthavatsalam, N. Leaves of *Lantana camara* Linn. (Verbenaceae) as a Potential Insecticide for the Management of Three Species of Stored Grain Insect Pests. *J. Food Sci. Tech.* **2014**, *51*(11), 3494–3499.

Regnault-Roger, C.; Philogène, B. J. R. Past and Current Prospects for the Use of Botanicals and Plant Allelochemicals in Integrated Pest Management. *Pharmac. Biol.* **2008**, 46, *1–2*, 41–52.

Saleem, M. A.; Ahmad, M.; Ahmad, M.; Aslam, M.; Sayyed, A. H.. Resistance to Selected Organochlorin, Organophosphate, Carbamate and Pyrethroid, in Spodoptera litura (Lepidoptera: Noctuidae) from Pakistan. *J. Econ. Entomol.* **2008**, *101* (5), 1667–1675.

Schleier, J. J. Peterson, R. K. Pyrethrins and Pyrethroid Insecticides. *Green Trends Insect Contr.* **2011**, 94–131.

Schmeltz, I. Nicotine and Other Tobacco Alkaloids. In *Naturally Occurring Insecticides*, Jacobson, M.; Crosby, D. G., Eds.; Marcel Dekker: New York, 1971, pp 99–136.

Schmutterer, H. Properties and Potential of Natural Pesticides from the Neem Tree. *Ann. Rev. Entomol.* **1990**, *35*, 27–J298.

Senthilkumar, N.; Varma, P.; Gurusubramanian, G. Larvicidal and Adulticidal Activities of Some Medicinal Plants Against the Malarial Vector, *Anopheles stephensi* (Liston). *Parasitol. Res.* **2009**, *104*, 237–244.

Shagal, M. H.; Modibbo, U. U.; Liman, A. B. Pharmacological Justification for the Ethnomedical use of *Datura stramonium* Stem-bark Extract in Treatment of Diseases Caused by Some Pathogenic Bacteria. *Intern. Res. Pharm. Pharmacol.* **2012**, *2* (1), 016–9.

Srinivasa, R.; Udikeri, S. S.; Jayalakshmi, S. K.; Sreeramulu, K. Identification of Factors Responsible for Insecticide Resistance in Helicoverpa armigera. *Compar. Biochem. Physiol. Part C Toxic Pharmac.* **2004**, *137* (3), 261–269.

Su, Y. C.; Ho, C. L.; Wang, I. C.; Chang, S. T. Antifungal Activities and Chemical Compositions of Essential Oils from Leaves of Four Eucalypts. *Taiwan J. Forens. Sci.* **2006**, *21*, 49–61.

Talukder, F. A.; Howse, O. E. Deterrent and Insecticidal Effects of Extracts of Pithraj, *Aphanamixis polystachya* (Meliaceae) Against *Tribolium castaneum* in Storage. *J. Chem. Eco.* **1993**, *19* (11), 2463–2471.

Trigg, J. K. Evaluation of Eucalyptus-based Repellent Against *Anopheles* spp. in Tanzania. *J. Amer. Mosquito Cont. Assoc.* **1996a**, *12*, 243–246.

Trigg, J. K. Evaluation of Eucalyptus-based Repellent Against *Culicoides impunctatus* (Diptera: Ceratopogonidae) in Scotland. *J. Amer. Mosquito Cont. Assoc.* **1996b**, *12*, 329–330.

Ugurlu, S.; Gurkan, M. O. Insecticide Resistance in *Helicoverpa armigera* from Cotton-growing Areas in Turkey. *Phytoparasitica* **2007**, *35*, 376–379.

Vandenborre, G.; Groten, K.; Smagghe, G.; Lannoo, N.; Baldwin, I. T.; Van Damme, E. J. M. *Nicotiana tabacum* Agglutinin is Active Against Lepidopteran Pest Insects. *J. Exper. Botany* **2010**, *61* (4), 1003–1014.

Wakeil, N. E. Botanical Pesticides and Their Mode of Action. *Gesunde Pflanzen.* **2013**, *65*, 125–149.

Watanabe, K.; Shono, Y.; Kakimizu, A.; Okada, A.; Matsuo, N.; Satoh, A.; Nishimura, H. New Mosquito Repellent from *Eucalyptus camaldulensis*. *J. Agric. Food Chem.* **1993**, *41*, 2164–2166.

WHO Global Nutrition Report: From Promise to Impact 2017, Downloaded from https://data. unicef.org/wp-content/uploads/2016/06/130565-1.pdf.

WHO Pesticide Residues in Food Fact Sheet, 2017, Downloaded from: http://www.who.int/ mediacentre/factsheets/pesticide-residues-food/en/.

Xu, J.; Fan, Q. J.; Yin, Z. Q.; Li, X. T.; Du, Y. H.; Jia, R. Y.; Wang, K. Y.; Lv, C.; Ye, G.; Geng, Y.; Su, G.; Zhao, L.; Hu, T. X.; Shi, F.; Zhang, L.; Wu, C. L.; Tao, C.; Zhang, Y. X.; Shi, D. X. The Preparation of Neem Oil Microemulsion (*Azadirachta indica*) and the Comparison of Acaricidal Time Between Neem Oil Microemulsion and Other Formulations in vitro. *Veterin. Parasit.* 2010.

Yang, F. L.; Liang, G. W.; Xu, Y. J.; Lu, Y. Y.; Zeng, L. Diatomaceous Earth Enhances the Toxicity of Garlic, *Allium sativum*, Essential Oil Against Stored-product Pests. *J. Stored Prod. Res.* **2010,** *46*, 118–123.

CHAPTER 8

Symbiotic Nitrogen Fixation and Pulses Yield

RAMBALAK NIRALA[1], MAYANK KUMAR[2], RAVI RANJAN KUMAR[3],
BISHUN DEO PRASAD[3], AWADHESH KUMAR PAL[4], VINOD KUMAR[3],
VIJAY KUMAR JHA[5], and TUSHAR RANJAN[3,*]

[1]*Department of Plant Breeding and Genetics,
Bihar Agricultural University, Sabour, Bhagalpur 813210, Bihar, India*

[2]*Amity Institute of Biotechnology, Amity University, Mumbai 410206,
India*

[3]*Department of Molecular Biology and Genetic Engineering,
Bihar Agricultural University, Sabour, Bhagalpur 813210, Bihar, India*

[4]*Department of Biochemistry and Crop Physiology, Bihar Agricultural
University, Sabour, Bhagalpur 813210, Bihar, India*

[5]*Department of Botany, Patna University, Patna 800005, Bihar, India*

*Corresponding author. E-mail: mail2tusharranjan@gmail.com

ABSTRACT

Legumes fix nitrogen by absorbing N_2 from the soil atmosphere into tiny nodules of their roots and the bacteria (rhizobia) in the nodules transform the atmospheric N_2 into ammonia (NH_3). The ammonia is then translated into organic compounds by the plant and used for their growth and development. The nitrogen-rich residues and exudates from the legumes add to the N of the soil to be used by other nonleguminous crops, such as cereals. The amount of N_2 fixed by legumes is strongly linked to the productivity of the pulses. Pulses must be well nodulated for maximum N_2 fixation and rotational benefits. The use of legumes grown in rotations or intercropping is nowadays regarded as an alternative and sustainable way of introducing N into low-input cropping systems. Currently, at the global scale, symbiotic N fixation by legumes only accounts for 13% of the total fertilization on arable land. The potential of symbiotic N fixation is largely underexploited.

8.1   INTRODUCTION

Nitrogen is an important nutrient for all plants. Approximately 80% of the atmosphere is nitrogen gas (N_2). Unfortunately, N_2 is unusable by most living organisms. Plants, animals, and microorganisms can die of nitrogen deficiency, surrounded by N_2 they cannot use. All organisms use the ammonia (NH_3) form of nitrogen to manufacture amino acids, proteins, nucleic acids, and other nitrogen-containing components necessary for life. Biological nitrogen fixation (BNF) is the process that changes inert N_2 to biologically useful NH_3. While there is an abundance of nitrogen in the atmosphere, plants are unable to convert N_2 into a usable form. Fixation of nitrogen gas into ammonia is an ability restricted to nitrogen-fixing bacteria, which contribute most of the inorganic nitrogen to the Earth's nitrogen cycle. This process is mediated in nature only by bacteria. Other plants benefit from nitrogen-fixing bacteria when the bacteria die and release nitrogen to the environment, or when the bacteria live in close association with the plant. The symbiosis of legumes with nitrogen-fixing soil bacteria called rhizobia has become a model for our understanding of plant–microbe interactions. In legumes and a few other plants, the bacteria live in small growths on the roots called nodules. Within these nodules, nitrogen fixation is done by the bacteria, and the NH_3 produced is absorbed by the plant. Nitrogen fixation by legumes is a partnership between a bacterium and a plant. BNF can take many forms in nature including blue-green algae (a bacterium), lichens, and free-living soil bacteria. These types of nitrogen fixation contribute significant quantities of NH_3 to natural ecosystems, but not to most cropping systems, with the exception of paddy rice. Their contributions are less than 5 lbs of nitrogen per acre per year. However, nitrogen fixation by legumes can be in the range of 25–75 lb of nitrogen per acre per year in a natural ecosystem, and several hundred pounds in a cropping system. Overall atmospheric N_2 cycle is described in Figure 8.1.

Legumes have been used in crop rotations since the time of the Romans. However, it was not until detailed N balance studies became possible, that they were shown to accumulate N from sources other than soil and fertilizer. In 1886, Hellriegel and Wilfarth demonstrated that the ability of legumes to convert N_2 from the atmosphere into compounds, which could be used by the plant, was due to the presence of swellings or nodules on the legume root, and to the presence of particular bacteria within these nodules.

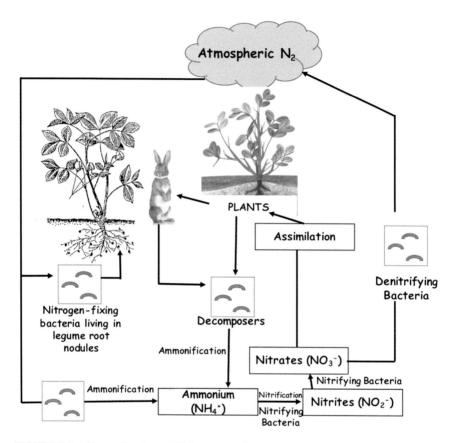

FIGURE 8.1 **(See color insert.)** Nitrogen cycle.

Source: Adapted with permission from https://commons.wikimedia.org/wiki/File:Nitrogen_Cycle.svg.

8.2 MECHANISM OF NITROGEN FIXATION

Nitrogen fixation is a process by which nitrogen in the Earth's atmosphere is converted into ammonia (NH_3) or other molecules available to living organisms (Postgate, 1998). Atmospheric nitrogen or molecular dinitrogen (N_2) is relatively inert: It does not easily react with other chemicals to form new compounds. The fixation process frees nitrogen atoms from their triply bonded diatomic form, $N\equiv N$, to be used in other ways. Nitrogen fixation is essential for some forms of life because inorganic nitrogen compounds are required for the biosynthesis of the basic building blocks of plants, animals, and other life forms, for example, nucleotides for DNA and RNA, the coenzyme nicotinamide adenine dinucleotide for its role in metabolism (transferring electrons between molecules), and amino acids for proteins. Therefore, as part of the nitrogen cycle, it is essential for agriculture and the manufacture of fertilizer. It is also, indirectly, relevant to the manufacture of all chemical compounds that contain nitrogen, which includes explosives, most pharmaceuticals, dyes, etc. Nitrogen fixation is carried out naturally in the soil by nitrogen fixing bacteria such as *Azotobacter*. Some nitrogen-fixing bacteria have symbiotic relationships with some plant groups, especially legumes. Looser relationships between nitrogen-fixing bacteria and plants are often referred to as associative or nonsymbiotic, as seen in nitrogen fixation occurring on rice roots. It also occurs naturally in the air by means of NOx production by lightning. All BNF is carried out by way of metalloenzymes called nitrogenases. These enzymes contain iron, often with a second metal, usually molybdenum but sometimes vanadium. Microorganisms that can fix nitrogen are prokaryotes (both bacteria and archaea, distributed throughout their respective domains) called diazotrophs. Some higher plants, and some animals (termites) have formed associations (symbiosis) with diazotrophs (Slosson, 1919; Hill et al., 1979). BNF was discovered by the German agronomist Hermann Hellriegel and Dutch microbiologist Martinus Beijerinck. BNF occurs when atmospheric nitrogen is converted to ammonia by an enzyme called a nitrogenase (Beijerinck, 1901). The overall reaction for BNF is:

$$N_2 + 8H^+ + 8e^- + 16ATP \rightarrow 2\ NH_3 + H_2 + 16ADP + 16Pi.$$

The process is coupled to the hydrolysis of 16 equivalents of adenosine triphosphate (ATP) and is accompanied by the coformation of one molecule of H_2. The conversion of N_2 into ammonia occurs at a cluster called FeMoco, an abbreviation for the iron-molybdenum cofactor. The mechanism proceeds via a series of protonation and reduction steps wherein the FeMoco active site hydrogenates the N_2 substrate (Lee et al., 2014; Hoffman et al., 2013). In free-living diazotrophs, the nitrogenase-generated ammonium is assimilated into glutamate through the glutamine synthetase/glutamate synthase pathway. The microbial genes required for nitrogen fixation are widely distributed in diverse environments (Gaby and Buckley, 2011; Hoppe et al., 2014). Enzymes responsible for nitrogenase action are very susceptible to destruction by oxygen. For this reason, many bacteria cease production of the enzyme in the presence of oxygen. Many nitrogen-fixing organisms exist only in anaerobic conditions, respiring to draw down oxygen levels, or binding the oxygen with a protein such as leghemoglobin.

This reaction is performed exclusively by prokaryotes (the bacteria and related organisms), using an enzyme complex termed Nitrogenase. The enzyme nitrogenase is an enzyme complex which consists of two metalloproteins.

i) Fe-protein or iron-protein component (previously called as azo ferredoxin) and

ii) Fe Mo-protein or iron-molybdenum protein component (previously called as molybdoferredoxin). None of these two components alone can catalyze the reduction of N_2 to NH_3.

The Fe-protein component of nitrogenase is smaller than its other component and is a Fe–S protein which is extremely sensitive to O_2 and is irreversibly inactivated by it. This Fe-S protein is a dimer of two similar peptide chains each with a molecular mass of 30–72 kDa (depending upon the microorganism). This dimer contains four Fe atoms and four S atoms (which are labile and 12 titrable thiol groups). The MoFe-protein component of nitrogenase is larger of the two components and consists of two different peptide chains that are associated as a mixed $(\alpha_2\beta_2)$ tetramer with a total molecular mass of 180–235 kDa (depending upon the microorganism). This tetramer contains two Mo atoms, about 24 Fe atoms, about 24 labile S atoms, and 30 titrable thiol groups probably in the form of three 24 Fe_4–S_4 clusters. This component is also sensitive to O_2. Because the nitrogenase enzyme complex is sensitive to O_2, BNF requires anaerobic conditions. If the nitrogen-fixing organism is anaerobic then there is no such problem. But, even when the organism is aerobic, nitrogen fixation occurs only when conditions are made to maintain a very low level of O_2 or almost anaerobic conditions prevail inside them around the enzyme nitrogenease. Apart from N_2, the enzyme nitrogenase can reduce a number of other substrates such as N_2O (nitrous oxide), $N3^-$ (azide), C_2H_2 (acetylene), protons $(2H^+)$ and catalyse the hydrolysis of ATP. Direct measurement of nitrogen fixation is performed by mass spectroscopy. However, for comparative studies, reduction of acetylene can be measured rather easily by the gas chromatography method. The electrons are transferred from reduced ferredoxin or flavodoxin or other effective reducing agents to Fe-protein component which gets reduced. From reduced Fe-protein, the electrons are given to MoFe-protein component, which, in turn, gets reduced and is accompanied by hydrolysis of ATP into adenosine diphosphate (ADP) and inorganic phosphate (Pi).

Two Mg^{++} and two ATP molecules are required per electron transferred during this process.

Binding of two ATPs to reduced Fe-protein and subsequent hydrolysis of two ATPs to two ADP + 2 Pi is believed to cause a conformational change of Fe-protein, which facilitates redox (reduction-oxidation) reactions. From reduced MoFe-protein, the electrons are finally transferred to molecular nitrogen (N_2) and eight protons, so that two ammonia and one hydrogen molecules are produced (Fig. 8.2).

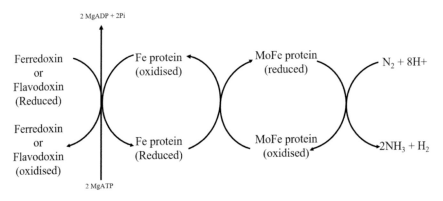

FIGURE 8.2 Mechanism of biological nitrogen fixation.

At first glance, it might be expected that six electrons and six protons would be required for reduction of one N_2 molecule to two molecules of ammonia. But, the reduction of N_2 is obligatorily linked to the reduction of two protons to form one H_2 molecule also. It is believed that this is necessary for the binding of nitrogen at the active site. The electrons for the regeneration of reduced electron donors (ferredoxin, flavodoxin, etc.) are provided by the cell metabolism, for example, pyruvate oxidation. A substantial amount of energy is lost by the microorganisms in the formation of H_2 molecule during nitrogen fixation. However, in some rhizobia, the hydrogenase enzyme is found which splits H_2 to electrons and protons ($H_2 \rightarrow 2H^+ + 2e^-$). These electrons may then be used again in the reduction of nitrogen, thereby increasing the efficiency of nitrogen fixation. Although scientists have tried to explain the mechanism of BNF, but the precise pathway of electron transfer, substrate entry and product release, and source of protons during BNF have not yet been completely elucidated (Lee et al., 2014; Hoffman et al., 2013).

Symbiotic nitrogen fixation occurs in plants that harbor nitrogen-fixing bacteria within their tissues. The best-studied example is the association between legumes and bacteria in the genus Rhizobium. A symbiotic relationship in which both partner benefits is called mutualism. A mutualistic symbiosis is an association between two organisms from which each derives benefit. It is usually a long-term relationship, and with symbiotic nitrogen (N_2) fixation, often involves a special structure to house the microbial partner. Each N_2-fixing symbiotic association involves an N_2-fixing prokaryotic organism, the microsymbiont (e.g., Rhizobium, Klebsiella, Nostoc or Frankia) and a eukaryotic, usually photosynthetic, host (e.g., leguminous or nonleguminous plant, water fern or liverwort; Gaby and Buckley, 2011; Hoppe et al., 2014).

8.3 NITROGEN-FIXING MICROORGANISMS

Diazotrophs are a diverse group of prokaryotes that includes cyanobacteria (e.g., the highly significant Trichodesmium and Cyanothece), as well as green-sulfur bacteria, Azotobacteraceae, rhizobia, and Frankia. Cyanobacteria inhabit nearly all illuminated environments on Earth and play key roles in the carbon and nitrogen cycle of the biosphere. In general, cyanobacteria can use various inorganic and organic sources of combined nitrogen, such as nitrate, nitrite, ammonium, urea, or some amino acids. Several cyanobacterial strains are also capable of diazotrophic growth, an ability that may have been present in their last common ancestor in the Archean eon. Nitrogen fixation by cyanobacteria in coral reefs can fix twice as much nitrogen as on land—around 1.8 kg of nitrogen is fixed per hectare per day (~660 kg/ha/year). The colonial marine cyanobacterium Trichodesmium is thought to fix nitrogen on such a scale that it accounts for almost half of the nitrogen fixation in marine systems globally (Bergman et al., 2012; Latysheva et al., 2012).

8.4 OXYGEN PROTECTION OF BACTERIAL NITROGEN FIXATION

Nitrogenase is highly oxygen sensitive because one of its components, the MoFe-cofactor, is irreversibly denatured by oxygen (Shaw and Brill, 1977). On the other hand, a large amount of energy required for this reaction has to be generated by oxidative processes; thus, there is a high demand for oxygen in nodules. Different strategies are used in different symbiotic interactions to cope with this paradox. In legume nodules, low oxygen tension in the central part of the nodule is achieved by a combination of high metabolic activity of the microsymbiont and an oxygen diffusion barrier in the periphery of the nodule, that is, in the nodule parenchyma (Figure 8.3; Witty et al., 1986). Because oxygen diffuses 104 times faster through the air than through water, it is generally assumed that oxygen diffusion in nodules occurs via the intercellular spaces. The nodule parenchyma contains very few and small intercellular spaces and this morphology is thought to be responsible for the block in oxygen diffusion (Witty et al., 1986). In the nodule parenchyma, nodulin genes such as ENODP are expressed whose protein products might contribute to the construction of the oxygen barrier (Van de Wiel et al., 1990). In the infected cells of the central part of the nodule, high levels of the oxygen carrier protein leghemoglobin facilitate oxygen diffusion. In this way, the microsymbiont is provided with sufficient oxygen to generate energy within a low overall oxygen concentration (Figure 8.3). In the infected cells of the central part of the nodule, high levels of the oxygen carrier protein leghemoglobin facilitate oxygen diffusion. In this way, the microsymbiont is provided with sufficient oxygen to generate energy within a low overall oxygen concentration. In contrast to Rhizobium, Frankia bacteria can form specialized vesicles in which nitrogenase is protected from oxygen.

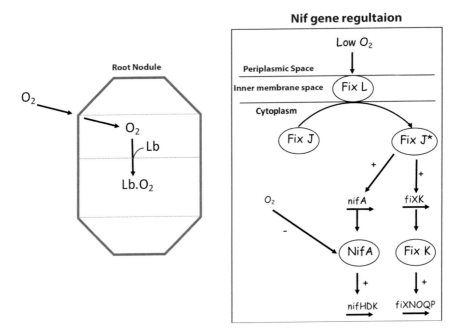

FIGURE 8.3　(See color insert.) Regulation of oxygen in legume nodules.

Source: Adapted with permission from Mylona et al. (1995).

To obtain nitrogen-fixing root nodules, several genes of both symbionts are specifically induced or repressed during nodule development. The use of reporter genes as well as in situ hybridization studies has provided detailed insights into the spatial and temporal regulation of such genes in indeterminate nodules. In such nodules, major, sudden developmental changes occur at the transition of the prefixation zone to the interzone: Starch is deposited in the plastids of the infected cells, and the bacteroid morphology alters. These events are accompanied by changes in bacterial gene expression: transcription of bacterial *nif* genes, which encode enzymes involved in the nitrogen fixation process, is induced, whereas expression of the bacterial outer membrane protein gene ropA is dramatically reduced (Yang et al., 1991). All of these events, together with dramatic changes in plant gene expression, take place within a single cell layer.

8.5 LEGUME NODULES

Establishing a symbiotic N_2-fixing association includes two main stages: root hair infection and nodule organogenesis. During the root hair infection process, the plant root sends signals by exuding various substances. Rhizobia sense the signal and travel relatively close to contact with the root hair. Once bound to the root hair, rhizobia respond by producing their own signals called Nod factors, which deform the root hair to trap rhizobia (Walley, 2013). Rhizobia then invade the root hair through the formation of an infection thread. The infection thread is "an intercellular tube that penetrates the cells of the plant, and the bacteria then enter the root cells through the deformed root hair" (Wang et al., 2012). The infection thread continues to develop from cell-to-cell, extending into the inner cortex of the root. The bacteria multiply within the expanding network of tubes and continue to produce Nod factors which stimulate the root cells to proliferate, eventually forming a root nodule. The bacteria continue to divide and induce the establishment of the N_2-fixing enzyme system including the synthesis of nitrogenase in bacteroids, which are root nodules colonized by thousands of living rhizobia (Downie, 2010). Biological N_2 fixation occurs in symbiosomes, which may contain several or just one bacteroid (Udvardi and Day, 1997). Thus, the symbiotic relationship is established between rhizobia and a host legume plant. Rhizobia can provide N resources to the plant and in exchange, the plant provides carbohydrates to rhizobia.

Plants that contribute to nitrogen fixation include those of the legume family, Fabaceae, with taxa such as kudzu, clovers, soybeans, alfalfa, lupines, peanuts, and rooibos. They contain symbiotic bacteria called rhizobia within nodules in their root systems, producing nitrogen compounds that help the plant to grow and compete with other plants. When the plant dies, the fixed nitrogen is released, making it available to other plants; this helps to fertilize the soil. The great majority of legumes have this association, but a few genera (e.g., *Styphnolobium*) do not. In many traditional and organic farming practices, fields are rotated through various types of crops, which usually include one consisting mainly or entirely of clover or buckwheat (nonlegume family *Polygonaceae),* often referred to as "green manure." The efficiency of nitrogen fixation in the soil is dependent on many factors, including the legume as well as air and soil conditions. For example, nitrogen fixation by red clover can range from 50 to 200 lb/acre depending on these variables (Bergman et al., 2012).

Legume nitrogen fixation starts with the formation of a nodule. A common soil bacterium, Rhizobium, invades the root and multiplies within the cortex cells (Fig. 8.2). The plant supplies all the necessary nutrients and energy for the bacteria. Within a week after infection, small nodules are visible with the naked eye. In the field, small nodules can be seen 2–3 weeks after planting, depending on legume species and germination conditions. When nodules are young and not yet fixing nitrogen, they are usually white or gray inside. As nodules grow in size they gradually turn pink or reddish in color, indicating nitrogen fixation has started. The pink or red color is caused by leghemoglobin (similar to hemoglobin in blood) that controls oxygen flow to the bacteria. Nodules on many perennial legumes such as alfalfa and clover are finger-like in shape. Mature nodules may actually resemble a hand with a center mass (palm) and protruding portions (fingers), although the entire nodule is generally less than 1/2 in. in diameter. Nodules on perennials are long-lived and will fix nitrogen through the entire growing season, as long as conditions are favorable. Most of the nodules (10–50 per large alfalfa plant) will be centered around the taproot. Nodules on annual legumes such as beans, peanuts, and soybeans are round and can reach the size of a large pea. Nodules on annuals are short-lived and will be replaced constantly during the growing season. At the time of pod fill, nodules on annual legumes generally lose their ability to fix nitrogen because the plant feeds the developing seed rather than the nodule. Beans will generally have less than 100 nodules per plant, soybeans will have several hundred per plant, and peanuts may have 1000 or more nodules on a well-developed plant.

Legume nodules that are no longer fixing nitrogen usually turn green, and may actually be discarded by the plant. Pink or red nodules should predominate on a legume in the middle of the growing season. If white, gray, or green nodules predominate, little nitrogen fixation is occurring as a result of an inefficient Rhizobium strain, poor plant nutrition, pod filling, or other plant stress. The nitrogen fixed is not free. The plant must contribute a significant amount of energy in the form of photosynthate (photosynthesis derived sugars) and other nutritional factors for the bacteria. A soybean plant may divert 230% of its photosynthate to the nodule instead of to other plant functions when the nodule is actively fixing nitrogen. Any stress that reduces plant activity will reduce nitrogen fixation. Factors like temperature and water may not be under the control of the farmer. But nutrition stress (especially phosphorus, potassium, zinc, iron, molybdenum, and cobalt) can be corrected with fertilizers. When nutritional stress is corrected, the legume responds directly to the nutrient, and indirectly to the increased nitrogen nutrition resulting from enhanced nitrogen fixation. Poor nitrogen fixation in the field can be easily corrected by inoculation, fertilization, irrigation, or other management practices.

8.6 OPTIMISING N_2 FIXATION TO INCREASE PULSE YIELD THROUGH AGRONOMIC MANAGEMENT

The amount of N_2 fixed by crops varies, for example, pulse crops can be ranked according to their estimated ability to fix N_2: faba bean > pea > chickpea > lentil > dry bean (Walley et al., 2007). Within the constraints of the climate and season, good legume management to maximize productivity will benefit N_2 fixation. Examples of legume management include optimizing nutrient inputs (e.g., P), reducing acidity with lime, managing weeds, disease, and insects. Optimizing the basic agronomy is critical in terms of legume productivity and N_2 fixation. This means maintaining a good cover of stubble on the soil surface in the pre-crop fallow, sowing on time and establishing the appropriate plant density. A management option for cropping that has gained popularity in recent years is no-tillage. The effect was a result of increased soil water and decreased soil nitrate accumulation during the summer (precrop) fallow. Farmers have no control over seasonal weather but they have some control over the efficiency with which water is infiltrated into and stored in the soil (fallowing efficiency), coupled with the efficiency with which the water is used by crops (water-use efficiency). In the rotation experiments, the no-tilled plots had an average of 35 mm additional soil water at sowing.

Pulses must be well nodulated for maximum N_2 fixation and rotational benefits. For the majority of situations, farmers will need to inoculate the seed or soil with the appropriate strain of rhizobia at sowing in order to ensure good levels of nodulation. In other situations, however, there will be adequate numbers of effective rhizobia already in the soil and inoculation will have no effect on either nodulation or crop growth. The NSW Department of Primary Industries and other state Departments of Agriculture take the conservative approach and recommend that all legumes are inoculated at sowing. There are far less problems with unnecessary inoculation than not using inoculants when they are needed. Unnecessary inoculation represents a small cost of production; N-deficient crops can mean substantial reductions in yield and income. Until recently, the commonly-used method for inoculation was to apply a peat-based inoculant, produced and marketed by just one or two manufacturers, as a slurry to the seed just before sowing. Now, there are five manufacturers selling a more diverse range of inoculant products with different modes of application. A farmer's choice of inoculant will depend to a large extent on personal experience and product availability, relative cost, and perceived efficacy. The commercial inoculant such as Peat inoculants (NODULAID™, Nodule N™, N-Prove™), applied to the legume seed as a slurry, remains the most widely used of the formulations and the benchmark for efficacy. Under the right conditions, the freeze-dried formulation (EasyRhiz™) is highly efficacious. The clay and peat granular inoculants applied directly to the soil, are appealing to farmers because of ease-of-use and convenience and in the future may well supplant peat as the inoculant formulation of choice. New coinoculant products, such as TagTeam® and BioStacked®, are exciting new products that promise yield increases and improved gross margins under certain conditions. Although the potential benefits of all formulations and products may be real and appealing, farmers should look for evidence of efficacy in their particular environment.

8.7 CONCLUSION

Biological N fixation is a key factor for both economic viability and environmental performance of low-input farming systems. Many farmers fertilize crops after legumes as they would after any other crop and treat the legume-derived N as a bonus that supports higher yield and protein content than what is normally achieved. Field measurement of biological N_2 fixation is complex and costly and no methods are available for routine on-farm use. A very sturdy demand for a reliable tool to estimate BNF for different legumes in various environmental conditions is needed.

KEYWORDS

- **nitrogen fixation**
- **nodulation**
- **pulses**
- **Rhizobium**
- **nitrogenase**

REFERENCES

Beijerinck, M. W. Über Oligonitrophile Mikroben (On Oligonitrophilic Microbes). *Centralblatt für Bakteriologie, Parasitenkunde, Infektionskrankheiten und Hygiene.* **1901,** *7* (2), 561–582.

Bergman, B.; Sandh, G.; Lin, S.; Larsson, H.; Carpenter, E. J. Trichodesmium: a Widespread Marine Cyanobacterium with Unusual Nitrogen Fixation Properties. *FEMS Microbiol. Rev.* **2012,** *37* (3), 1–17.

Gaby, J. C.; Buckley, D. H. A Global Census of Nitrogenase Diversity. *Environ. Microbiol.* **2011,** *13* (7), 1790–1799.

Hill, R. D.; Rinker, R. G.; Wilson, H. D. Atmospheric Nitrogen Fixation by Lightning. *J. Atmos. Sci.* **1979,** *37* (1), 179–192.

Hoffman, B. M.; Lukoyanov, D.; Dean, D. R.; Seefeldt, L. C. Nitrogenase: A Draft Mechanism. *Acc. Chem. Res.* **2013,** *46,* 587–595.

Hoppe, B.; Kahl, T.; Karasch, P.; Wubet, T.; Bauhus, J.; Buscot, F.; Krüger, D. Network Analysis Reveals Ecological Links between N-Fixing Bacteria and Wood-Decaying Fungi. *PLoS One* **2014,** *9* (2), e88141.

Latysheva, N.; Junker, V. L.; Palmer, W. J.; Codd, G. A.; Barker, D. The Evolution of Nitrogen Fixation in Cyanobacteria. *Bioinformatics* **2012,** *28* (5), 603–606.

Lee, C. C.; Ribbe, M. W.; Hu, Y. Chapter 7. Cleaving the N, N Triple Bond: The Transformation of Dinitrogen to Ammonia by Nitrogenases. In *The Metal-Driven Biogeochemistry of Gaseous Compounds in the Environment. Metal Ions in Life Sciences;* Kroneck, P., Torres, M., Eds; Springer: Dordrecht, 2014, *14*, pp 147–174.

Mylona, P.; Pawlowaski, K.; Bisseling, T. Symbiotic Nitrogen Fixation. *Plant Cell* **1995,** *7,* 869–885.

Postgate, J. *Nitrogen Fixation,* 3rd ed; Cambridge University Press: Cambridge, UK, 1998.

Shaw, V. K.; Brill, W. J. Isolation of an Iron-Molybdenum Cofactor from Nitrogenase. *Proc. Natl. Acad. Sci. USA* **1977,** *74,* 3249–3253.

Slosson, E. E. *Creative Chemistry*; The Century Co.: New York, NY, 1919; pp 19–37.

van de Wiel, C.; Scheres, B.; Franssen, H.; van Llerop, M. J.; van Lammeren, A.; van Kammen, A.; Bisseling, T. The Early Nodulin Transcript ENOD2 is Located in the Nodule Parenchyma (Inner Cortex) of Pea and Soybean Root Nodules. *EMBO J.* **1990,** *9,* 1–7.

Witty, J. F.; Minchin, F. R.; Skbt, L.; Sheely, J. E. Nitrogen Fixation and Oxygen in Legume Root Nodules. Oxford Survey. *Plant Cell Biol.* **1986,** *3,* 275–315.

Yang, W.-C; Horvath, B.; Hontelez, J.; van Kammen, A.; Bissellng, T. In Situ Localization of Rhizobium mRNAs in Pea Root Nodules: nifA and nifH Localization. *MOI Plant-Microbe Interact.* **1991,** *4,* 464–468.

CHAPTER 9

Free Living Nitrogen Fixation and Their Response to Agricultural Crops

PARVEEN FATIMA[1,*], ANURAG MISHRA[1], HARI OM[2],
BHOLANATH SAHA[3], and PUSHPENDRA KUMAR[1]

[1]*Department of Agricultural Biotechnology, Sardar Vallabhbhai Patel University of Agriculture and Technology, Modipuram, Meerut 250110, Uttar Pradesh, India*

[2]*Department of Agronomy, Dr. Kalam Agricultural College, Kishanganj, Bihar Agricultural University, Sabour, Bhagalpur 813210, Bihar, India*

[3]*Department of Soil Science Agricultural Chemistry, Dr. Kalam Agricultural College, Kishanganj, Bihar Agricultural University, Sabour, Bhagalpur 813210, Bihar, India*

*Corresponding author. E-mail: parveenfrizvi@gmail.com

ABSTRACT

The environmental concerns due to the increasing amount of the reactive forms of nitrogen in atmosphere, originating from the manufacture and use of chemical fertilizers have resulted in a re-focus on the importance of biological nitrogen fixation (BNF). The economic costs of the very high use of chemical nitrogenous fertilizers in agriculture system are also a concern in the modern world for the poor farming community. BNF (symbiotic and non-symbiotic), a microbiological process which converts atmospheric nitrogen into a plant-usable form, offers this alternative. Non-symbiotic N_2 fixation (by free-living bacteria in soils or associated with the rhizosphere) has the potential to meet several needs, especially in the lower input cropping systems worldwide. There has been considerable research on nonsymbiotic N_2 fixation, but still, there is much argument about the amount of N that can potentially be fixed by this process. Largely due to shortcomings of indirect measurements; however, isotope-based direct methods indicate agronomically significant amounts of N_2 fixation both in annual crop and perennial grass systems. New molecular technologies offer opportunities to increase our understanding of N_2-fixing microbial communities (many of them non-culturable) and the molecular mechanisms of non-symbiotic N_2 fixation. This knowledge should assist the development of new plant-diazotrophic combinations for specific environments and more sustainable exploitation of N_2-fixing bacteria as inoculants for agriculture. Although the ultimate goal might be to introduce nitrogenase genes into significant non-leguminous crop plants, it may be more realistic in the shorter-term to better synchronize plant-microbe interactions to enhance N_2 fixation when the N needs of the plant are greatest. The review explores possibilities to maximize potential N inputs from non-symbiotic N_2 fixation through improved management practices, identification of better performing microbial strains and their successful inoculation in the field, and plant-based solutions.

9.1   INTRODUCTION

Sustainable agriculture involves the successful management of agricultural resources to satisfy the changing human needs while maintaining or enhancing the environmental quality and conserving natural resources. Consequently, sustainability considerations demand that alternative sources to the fertilizer nitrogen must be explored. The biological nitrogen fixation (BNF) can offer this alternative in farming practices as it uses the capacity of certain nitrogen-fixing bacteria (NFB) to convert atmospheric nitrogen into the plant usable form. Microorganisms which are capable of transforming atmospheric nitrogen into fixed nitrogen (inorganic compound) in the usable form for plants are called NFB. These bacteria are found both in the soil and in the plants' roots. They are divided into two types, the first type includes the free-living (nonsymbiotic) bacteria that are not found in plants or attached to any plant parts particularly they are found in the soil and in water. Examples of this type include the genera Clostridium, Azotobacter, and some blue–green algae. The second type includes mutualistic (symbiotic) bacteria that lives on the roots of certain legumes, like *Rhizobium* genus, which lives in the root nodules of the plants. Non-symbiotic (NS) N_2 fixation includes N_2 fixation by free-living soil bacteria (autotrophic and heterotrophic) that are not in a direct symbiosis with plants, and associative N_2-fixation (e.g., associated with the rhizosphere of grasses and cereals). Free-living N_2 fixation can also be associated with decomposing plant residues, aggregates with decomposable particulate organic matter, and in termite habitats globally, the demand for N fertilizers is expected to exceed 112 million tonnes in 2015 (FAO, 2015) and much of this is produced by the Haber–Bosch process (Jenkinson, 2001), a process which uses large amounts of fossil fuel (Jensen, 2003). This, together with the increasing demand for organically grown agricultural and horticultural products and the need to address economic and environmental concerns, has rekindled interest in promoting biological N_2 fixation in non-leguminous crops. Nitrogen is a critical element for sustainable agriculture but inappropriate use of fertilizer results in lower efficiency and has the potential to contribute to (1) greenhouse gas loads such as N_2O thereby contributing to climate change and (2) leaching of N from agricultural lands as NO^{-3} causing eutrophication of rivers, lakes and oceans, and reducing the quality of water supplies (Good, 2011).

Nitrogen is the only major essential nutrient available through the process called BNF. This process of BNF is aided by the members of the family Rhizobiaceae, Bradirhizobiaceae, and Phyllobacteriaceae with the leguminous plants through the formation of N-fixing specialized structure called nodules (Schultze and Kondorosi, 1998). Frankia, a versatile N fixing actinobacteria fixes N in non-legumes under both symbiotic and free-living conditions. It infects the root cells of actinorhizal plants through either intracellular root-hair infection or intercellular root invasion (Benson and Silvester, 1993). Cyanobacteria can fix nitrogen in both aerobic and anaerobic conditions. The aerobic N_2-fixing cyanobacteria have to cope not only with external oxygen but also with that generated intracellularly by the operation of photosystem II (PS II). The importance of nitrogen from the standpoint of the fertility of the soil has long been recognized and our knowledge concerning the nature, distribution, and transformations of nitrogen compounds in soil is extensive. The nitrogen cycle includes several biological and non-biological processes. The biological processes are: ammonification/mineralization, nitrification, denitrification, nitrogen fixation, nitrogen assimilatory reduction, and microbial synthesis of ammonium and organic nitrogen into microbial cells, plant uptake, and conversion of ammonium and nitrate nitrogen into plant proteins. The non-biological processes are ammonia volatilization, leaching of nitrite and nitrate nitrogen to groundwater, ammonium fixation into soil clay minerals, precipitation of nitrate and ammonium nitrogen. Nitrification is the oxidation of ammonium nitrogen to nitrites and nitrates. It is the result of metabolism by chemoautotrophic (or chemolithotrophic) organisms. The two groups of organisms that are considered to be the primary nitrifying bacteria are *Nitrosomonas* sp. and

Nitrobacter sp. *Nitrosomonas* carry out the oxidation of ammonium to nitrite to obtain energy (E) and *Nitrobacter* oxidizes nitrite to nitrate for the same purpose. The general oxidative processes involved can be represented by the following equations (Anthonisen, 1976; Ward, 2011):

$$NH^{4+} + 1^{1/2} O_2 \xrightarrow{\text{Nitrogenase}} NO^{-2} + H_2O + 2H^+ + \text{Energy} \quad (9.1)$$

$$NO^{2+} + 1/2\ O_2 \xrightarrow{\text{Nitrogenase}} NO^{-3} + \text{Energy.} \quad (9.2)$$

Both genera use CO_2 as their sole carbon source for growth as follows, as they are obligate autotrophs and strict aerobes (Liu, 2012).

$$NH^{+4} + 5CO_2 + 2H_2ONO^{-3} \xrightarrow{\text{Nitrogenase}} C_5H_7O_2N^+ H^+ + 5O_2 + \text{Energy} \quad (9.3)$$

9.1.1 NITROGEN FIXATION

Although atmospheric nitrogen (N_2) is abundant in the atmosphere accounting for more than three-fourths of the volume, it is not in a readily available form to be used by most organisms including crop plants as N atoms are strongly triple bonded and is hard to break. In Haber–Bosch process, in which chemically synthesized nitrogen is formed, through a catalyst (red hot magnesium) is used at elevated pressures and temperatures to make N_2 reactive (Postgate, 1998). However, in the BNF process, nitrogen is fixed by the microorganisms that are capable of breaking the bond at ambient temperatures and pressures and is called diazotrophic (Postgate, 1998; Kennedy, 2004). The diazotrophs are available in soil both as free-living and in symbiotic association with plants. Diazotrophic microorganisms use the enzyme nitrogenase to carry out the fixation process. There are about 100 species of enzymes but they are very similar in their activities (Postgate, 1998). Nitrogenase reduces dinitrogen to ammonium (NH_4^+). The enzymatic reaction can be described as follows:

$$N_2 + 4H_2O \xrightarrow{\text{Nitrogenase}} 2NH_4^+ + 2O_2. \quad (9.4)$$

Soil can gain small amounts of nitrogen from the rain which falls on them. However, the most important natural process for increasing the nitrogen content of soils is nitrogen fixation by microorganisms living in the soil on and around the roots. A functional classification of the range of NFB in soil is shown in Table 9.1 (Kahindi, 1997). The estimates of nitrogen fixed by different microorganisms are shown in Table 9.2 (Bohlool, 1992). The process of nitrogen fixation is inhibited in the presence of high level of available nitrogen (NH_4^+ or NO_3^-).

The process is also controlled by N:P ratio as phosphorous activate the gene for synthesis (Bohlool, 1992). Nitrogen can be fixed in the soil through symbiotic fixation and non-symbiotic fixation and the rate of nitrogen fixation is related to the rate of photosynthesis (plant growth).

TABLE 9.1 Classification of Nitrogen Fixing Bacteria Contributing to Agriculture.

Types of microorganisms	Nature of nitrogen-fixing bacteria	Symbiont
Free-living	Heterotrophs	
Anaerobic	*Clostridium*	
Microaerophilic	*Frankia, Azospirullum*	
Aerobic	*Bradyrhizobium, Azotobacter, Derxia*	
Root-associated		
Microaerophilic	*Azospirullum, Herbaspirullum*	
Endophytic	*Acetobacter*	Sugar cane (*Saccharum sp.*)
Symbiotic	*Frankia*	*Casuarina sp.* *Alnus sp.*
	Rhizobium, Bradyrhizobium	Many legumes
	Azorhizobium	*Sesbania rostrata*
Free-living	Autotrophs	
Anaerobic	*Chromatium, Chlorobium*	
Microaerophilic	*Rhodospirillum, Bradyrhizobium*	
Aerobic	*Cyanobacteria*	
Symbiotic	*Cyanobacteria*	*Fungi* (lichens), Cycads
	Anabaena azollae	*Azolla sp.*
	Bradyrhizobium	*Aeschynomene sp.*

Source: Adapted with permission from Kahindi (1997).

9.1.1.1   AEROBIC DIAZOTROPHS

Aerobic diazotrophs require oxygen for growth and fix nitrogen in the presence of oxygen (low concentration; *Azotobacter,* methane producing bacteria).

9.1.1.2   FREE-LIVING DIAZOTROPHS

Free-living diazotrophs fix nitrogen both in aerobic and anaerobic conditions (*Bacillus* and *Klebsiella*).

TABLE 9.2   Estimation of Dinitrogen Fixed by Different Nitrogen Fixing Systems.

Nitrogen fixation process	Approximate percentage of nitrogen fixation
Free-living	
Rice-blue green algae	10–80
Rice-bacterial association	10–30
Sugarcane bacterial association	variable
Symbiotic	
Rice-Azolla	20–100
Legume-*Rhizobium*	
Leucaena leucocephala	100–300
Glycine max	0–237
Trifolium repens	13–280
Sesbania rostrata	320–360
Nonlegume-Frankia Casuarina sp.	40–60

Source: Adapted with permission from Bohlool (1992).

9.1.1.3   SYMBIOTIC DIAZOTROPHS

Symbiotic diazotrophs fix nitrogen only by the formation of nodules (*Rhizobium, Bradyrhizobium*, and *Sinorhizobium*). Cyanobacteria such as *Anabaena, Nostoc, Lyngbya, Oscillatoria*, and so on fix N_2 both asymbiotically and symbiotically and it can also be categorized in two ways: sheath form (*Gloeocapsa* sp., *Gloeothece* sp. etc.) and sheathless form (*Synechococcus* sp. and *Cyanothece* sp.; Gallon et al., 1988; Waterbury and Rippka 1989; Huang and Grobbelaar 1995; Colon-Lopez et al., 1997). BNF can be explained in two ways that is, symbiotic and nonsymbiotic N_2 fixation.

9.1.2   FREE-LIVING BACTERIA

Certain bacteria, yeasts, fungi, and actinomycetes living in the soil system could fix nitrogen. Free-living diazotrophic bacteria are those that do not associate with plants and are found in soils that are free from the direct influence of plant roots. Besides carbon energy sources, microorganisms need low levels of available soil N, adequate supply of other mineral nutrients, near neutral pH and suitable moisture for proper growth. After their death, the organic N present in their bodies mineralizes and becomes available to plants. These microorganisms are ubiquitous in terrestrial and aquatic environments and are physiologically very diverse (Reed et al., 2011). Since most soils are C and N limited, the amount of N_2 they fix in soil is restricted by access to energy sources, that is, substrates to generate adenosine triphosphate (ATP) and micronutrients required for the synthesis and functioning of nitrogenase (Reed et al., 2011). BNF by free-living diazotrophs is also limited by the severe oxygen sensitivity of nitrogenase (Postgate, 1998), which is a problem that has been at least partially overcome in different ways by diazotrophs participating in nitrogen-fixing symbioses. Antagonistic microbial interactions such as parasitism and competition for nutrients (Cacciari et al., 1986; Bashan and Holguin, 1997; Bashan et al., 2004) further reduce the amount of N_2 they fix. While the general belief is that free-living diazotrophs do not contribute large quantities of fixed N to most terrestrial ecosystems, perhaps 3–5 kg/ha/year (Postgate, 1998; Newton, 2007), their cumulative N contributions are thought to be important in some tropical and temperate forest ecosystems (Cleveland et al., 1999; Gehring et al., 2005; Reed et al., 2007; 2008).

9.1.3 NON-SYMBIOTIC N_2-FIXATION

Some of the bacteria and most of the cyanobacteria comprise this class of microorganisms. They are also called free-living diazotrophs. Among cyanobacteria unicellular, filamentous nonheterocystous, and filamentous heterocystous fix nitrogen independently. Both aerobic and anaerobic bacteria are free-living diazotrophs. Water, oxygen, and nutrients are required in optimum amount, so that, the microorganism can grow. Cyanobacteria grow mainly in the crop fields. The site of nitrogen fixation in the cyanobacteria is the heterocyst because the enzyme (nitrogenase) required for nitrogen fixation acts under anaerobic condition. Then the question arises how unicellular and non-heterocystous cyanobacteria fix nitrogen? Some cells in these microorganisms become specialized that is, have oxygen level reduced. Typically, they fix nitrogen in dark and photosynthesize in light.

9.2 N_2-FIXING ENZYME SYSTEM

Dinitrogen is reduced to ammonia by an enzyme complex known as nitrogenase (occurs within the bacteroids), a reaction that is dependent on reduced ferredoxin and requires hydrogen and energy from ATP, resulting in the formation of molecular hydrogen.

$$N_2 + 8Fdred + 8H^+ + 16 ATP \, Æ \, NH_3 + H_2 + 8Fdox + 16ADP + 16Pi \quad (9.5)$$

The enzyme, nitrogenase consists of two components, which are highly conserved in sequence and structure, are: dinitrogenase (large, containing molybdenum, iron, and inorganic sulfur) and dinitrogenase reductase (small, containing iron and inorganic sulfur). Nitrogenase is oxygen sensitive. Host plant supplies energy for N_2 fixation. Such energy supplied as photosynthate acts as regulator of N_2 fixation. Nitrogenase reduces one molecule of N_2 to two molecules of NH_3 by the utilization of 16 molecules of ATP and releases one molecule of H_2 as a by-product. The activity and synthesis of nitrogenase are regulated at the level of both enzyme activity and gene expression. Enzyme activity may be regulated by post-translation modification of the Fe protein (Reich et al., 1986; Smith et al., 1987).

9.3 GENES INVOLVED IN NITROGEN FIXATION

The genes involved in nitrogen fixation are called *nif* gene. The structure and organization of nitrogen-fixing genes is studied mainly in *Klebsiella pneumoniae* (Arnold et al., 1988), *Azotobacter vinelandii* (Jacobson et al., 1989), *Bradyrhizobium japonicum* (Fischer et al., 1988), *Rhodobacter capsulatus* (Klipp, 1990), *Clostridium pasteurianum* (Wang et al., 1989), *Anabaena* PCC 7120 (Haselkorn and Buikema, 1992), and cyanobacteria (Kallas et al., 1983; 1985). The genes coding for structural subunit of cyanobacterial nitrogenase (*nif HDK*) has been cloned.

9.3.1 STEPS COMMON TO BOTH SYMBIOTIC AND NON-SYMBIOTIC N_2-FIXATION

1. The reduced ferredoxin donates its electron to Fe-protein component of nitrogenase which is ultimately reduced.
2. The reduced Fe-protein combines with ATP in presence of Mg^{2+} to become activated and to reduce second subunit (Mo-Fe-protein) of enzyme nitrogenase.
3. The reduced Mo-Fe-protein donates its electron to N_2 to reduce N_2 to NH_3 (Fig. 9.1) in the following way:

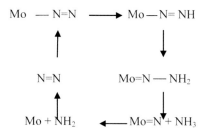

FIGURE 9.1 Reduction of N_2 to NH_3^-.

9.4 FREE LIVING NFB IN AGRICULTURE

The ecosystems, where free-living nitrogen fixing bacteria are found, are vast and varied, and Zechmeister-Boltenstern and Kinzel (1990) stated that the rates of fixation in forests and arable fields are relatively low compared to saline and peat soils with studies documenting remarkable N_2 fixation rates under the latter conditions. The contribution of the FLNFB is dependent on physiochemical properties where only a proportion of the microbes capable of fixing N will be active under given conditions (Wakelin et al., 2010). In addition, overall microbial activity plays a role, as shown in Patra et al. (2007), where contribution of FLNFB was increased when soil microbial activity decreased. Thus, it can be deduced that FLNFB is less competitive in the rhizosphere. Still, studies have been undertaken to quantify the amount of N fixed by these bacteria in agricultural systems. Herridge et al. (2008) cited studies where it was estimated that FLNFB fixed 25 kg N ha^{-1} year^{-1} in sugar cane crops and <5 kg N ha^{-1} year^{-1} in croplands other than used for legumes and rice; their conclusions on both studies was that large variations were evident in the data or that sufficient data was not available for decisive values. While models exist to calculate potential N-fixation rates, measuring and predicting actual rates is extremely difficult. These values are underwhelming compared to the ~100 kg N ha^{-1} year^{-1} of N fertilizer applied to many modern agriculture systems (Reed et al. 2011). Using this rate as a reference point, we are beginning to quantify FLNFB's impact and contribution to an agricultural production system. Herridge et al. (2008) then show in stark contrast the N-fixation rates that occur in leguminous species with the symbiotic relationship of the rhizobium bacteria. Not only are the models used to predict fixation rates based on sound historical crop yield data, but actual fixation rates are easier to measure in both cropping and pasture/fodder systems (cropping more so). Herridge et al. (2008) offer the following table summarizing the differences in fixation rates (Table 9.3):

TABLE 9.3 Summary of Estimates of N Fixed Annually in Agricultural Systems Comparing the Agent by Which it Occurs.

Agent/association	Agricultural system	Approximate rate of N_2 fixation (kg N ha^{-1} year^{-1})	Crop N fixed (Tg year^{-1})
Free-living bacteria	Croplands (other than legumes and rice)	<5	<4
Free-living bacteria	Extensive savannas used for grazing	<10	<14
Legume rhizobia	Crop (pulse and oilseed legumes)	115	21
Legume rhizobia	Pasture and fodder legumes	110–227	12–25

Source: Adapted with permission from Herridge et al. (2008).

It is worthwhile noting, however, that symbiotic associations cost the plant significant photosynthates (De Mita, 2012) and reduce the total yields of legumes. Despite the lesser amount of N that FLNFB actually fixes when compared to common fertilizer rates and symbiotic associations, products exist to help boost the numbers of these microorganisms to play their part (Webb, 2015). While inoculating the soil with mass numbers of these FLNFB may increase the amount of N they fix, it would seem unlikely that it would ever match the amount of N required of other inputs for agricultural systems. Still, increasing numbers and managing the soils to promote their activity can have beneficial effects (Wakelin et al., 2010). While acknowledging that the majority of N-fixation occurs in symbiosis with leguminous plants, Wakelin et al. state that FLNFB plays an important role between crop cycles, contributing the N needed to decompose crop residues which generally have wide C:N ratios. So while their contribution to the N needed in agricultural production systems may not be their primary role, they clearly contribute to overall soil fertility and nutrient cycling and thus should not be ignored.

9.5 NITROGEN FIXATION IN RICE

9.5.1 AUTOCHTHONOUS BNF SYSTEMS

The indigenous or autochthonous free-living and associative BNF systems offer an alternative to the input-based exogenous BNF technologies (Roger and Watanabe, 1986).

9.5.2 FREE-LIVING PHOTOTROPHS AND HETEROTROPHS

Different groups of free-living N_2-fixing microorganisms (aerobes, faculta-tive anaerobes, heterotrophs, and phototrophs) grow in submerged and/or wetland rice fields and contribute to plant available N. These BNF systems include cyanobacteria and photosynthetic bacteria inhabiting floodwater and the soil surface, and heterotrophic bacteria in the root zone and in the soil. Semi-quantitative data suggest that free-living diazotrophs have a low-to-moderate potential to supply N to rice (Table 9.1). Heterotrophic rhizosphere bacteria appear to contribute less BNF than the phototrophs (Table 9.1).

9.5.3 ASSOCIATIVE HETEROTROPHS

These heterotrophs form an association intracellular as well as intercellular with plant roots, particularly in rice, sugarcane, and maize. Even though several types of diazotrophs inhabit the rhizosphere of rice (James, 2000), some bacteria, such as *Azospirillum* spp., *Azorhizobium caulinodans*, *Kleb-siella* spp., *Pseudomonas diazotrophicus*, and *Enterobacter* spp. form loose associations with the plant surface (James, 2000; Ladha, 1986). Nitrogen fixa-tion by free-living diazotrophs occurs in the immediate vicinity of roots that is, in the soils system, and as such, it is not immediately available to the plant and subjected to losses after its fixation by the bacteria. Unlike the N fixa-tion by free-living diazotrophs, some of the N fixed by associative bacteria may still be immediately available to the rice plant. Soil and plant N balance studies (the sum of all gains and losses) in the greenhouse revealed significant differences in the ability of rice cultivars to support a positive N balance (App, 1986). The possible mechanisms by which rice genotypic diversity associated with 15 N dilution may influence N_2^- fixing bacteria in the rhizosphere are not yet elucidated. It has been hypothesized that diazotrophs associate with roots in response to a specific chemical compound/attractant excreted by the roots. Malic acid in the root exudates has been identified as the most important attractant to diazotrophs associated with C_4 crops (Neyra, 1978). In rice, a C_3 crop grown under flooded conditions, the main organic acids in root exudates are citric and oxalic acids (Vancura, 1964; Lin and You, 1989), although the nature of the main attractant in the chemotaxis of diazotrophs to rice is not yet identified. Furthermore, although it has long since been reported that most (culturable) bacteria (80%) found in rice roots are N_2^- fixing (Watanabe, 1979), the physiological and biochemical mechanisms that operate in this association have not yet been deciphered (Ladha, 1986).

9.6 NON-SYMBIOTIC FIXATION OF NITROGEN

Several experiments have been conducted during the last more than five decades in an effort to determine which kinds and species of soil micro-organisms possess the ability to fix atmospheric nitrogen, their distribution, and the amount of nitrogen fixed under a variety of environmental conditions (Jensen, 1965). There exist several controversies in the process of N-fixing ability of non-symbiotic microorganisms (bacteria as well as blue–green algae). This can be attributed primarily to the limited nitrogen-fixing ability of many microorganisms and to the inadequacy of the methods used for detection of the ability. Recently, isotopic studies with "N-labelled" nitrogen on genes involved in the process has settled most of the arguments and shown that nitrogen-fixing ability is indeed widespread, even if limited in amount, among microorganisms. Reference should be made to the discussions by Jensen (1965) and Alexander (1961) for details regarding the nitrogen-fixing ability of specific groups and species of microorganisms. The discussion here will deal primarily with the more important of these organisms and the ones that have received the most study. These include bacteria that live heterotrophically and blue–green algae that obtain their energy by photosynthesis. Claims have been made from time to time that certain fungi, yeasts, and actinomycetes are able to use atmospheric nitrogen but there is no general agreement as to which species if any possess this faculty. If any of these organisms are nitrogen fixers the amount fixed is too small to be of much economic importance.

9.6.1   HETEROTROPHIC MICROORGANISMS (THE ORGANISMS INVOLVED)

9.6.1.1   AEROBIC BACTERIA

The most important of the free-living heterotrophic aerobic nitrogen fixers is *Azotobacter* and significantly enhance soil fertility through BNF under diverse climatic conditions. This genus has long been studied partly because of its large cells and unusual morphological characteristics, but also because of its comparatively high nitrogen-fixing ability in the presence of a suitable source of energy. In laboratory culture media, a fixation of 10–20 mg nitrogen per gram of sugar oxidized is not uncommon, and of course, a fixation of this magnitude is readily demonstrated by ordinary Kjeldahl analysis, contrary to experience with many other microorganisms.

Azotobacter is, aside from its unusual morphological features, a typical soil inhabitant although rarely present in large numbers. The most commonly found species are *A. chroococcum* and *A. beijerinckia,* although *A. vinelandii* is not uncommon. In the laboratory, *Azotobacter* is usually grown on simple sugars or mannitol in the presence of the usual mineral elements, but it is capable of utilizing a number of other low-molecular-weight compounds. The presence of a few parts per millions of molybdenum to serve as a catalyst is required for vigorous fixation, although vanadium acts as a less efficient substitute. There is sometimes a considerable response to traces of other elements, especially Fe, Ca, and possibly B, but there is some uncertainty as to the essentiality of these elements. Apparently various species respond differently to these nutrients. *Azotobacter* has the highest rate of respiration of any known living organism with QO, values ranging up to 4000. Apparently much of the energy consumed constitutes wasteful consumption if considered from the standpoint of nitrogen fixation. The organism fixes more nitrogen per unit of sugar oxidized if the partial pressure of oxygen is held at near 1% instead of the usual 20%.

The optimum temperature for *Azotobacter* is approximately 30°C or slightly higher for some strains. In this respect it behaves like most soil bacteria. The optimum reaction is usually given as around neutrality with a lower growth limitation at about pH 6.0 or slightly lower. Closely allied to *Azotobacter* is the organism known as *Beijerinckziz.* This organism is usually considered as a separate genus but others have merely considered it as a slight variation of *Azotobacter.* Its chief distinguishing feature is that it is acid-tolerant to as low as pH 3–4. It resembles *Azotobacter* very closely in nitrogen-fixing ability and in growth. It is found most commonly in tropical soils, doubtless because it can tolerate their low pH, but has been reported also in cooler climates. Red lateritic soils of the Tropics are low in calcium and high in iron and aluminum, which conditions are unfavorable for *Azotobacter.* One would expect these conditions to limit the growth of *Beijerinckia* also, but such seems not to be the case even though the explanation is not known.

9.6.1.2 ANAEROBIC BACTERIA

The chief anaerobic nitrogen fixers found in soils belong to the genus *Clostridium.* They are widespread and present in much larger numbers than are *Azotobacter.* Since they are anaerobic, they are commonly present largely as spores in the better-aerated soils. Greenwood (1963) reported that even so-called well-aerated soils are not strictly aerobic inside of aggregates or perhaps in some cases between them. If further research supports this idea, we may need to reassess our views as to the importance of these anaerobes with respect to nitrogen fixation. The numbers of clostridia in arable, unaerated and unamended soils is commonly at least as high as 10,000 per gram and maybe 10–100 times this number. In the deeper layers of soils, or when they have an excess of water, the numbers may be much higher. Jensen (1940) showed that under conditions of high moisture content clostridia may dominate the soil flora within 2 or 3 days and the vegetative cells may reach numbers of 100–600 million per gram provided suitable energy material is present. Clostridia can utilize a fairly wide variety of energy sources including both hexose and pentose sugars, pectic substances, polysaccharides, and so on. These are not oxidized completely to carbon dioxide and water as in the case of *Azotobacter* but only partially to organic acids, hydrogen, alcohol, and acetone. Thus, most of the energy is left unused in these end products. In the older literature it was frequently stated that clostridia will fix only 2 or 3 mg nitrogen per gram of sugar used but more recent work has shown that some species are capable of fixing as much as 6–8 mg nitrogen per gram of sugar. Fixations of this magnitude are considerably lower per gram of sugar than those obtained with *Azotobacter,* but on the basis of energy utilized are higher. Jensen states that clostridia can co-exist with aerobic organisms that reduce the redox potential of the medium to a level favorable for the initiation of growth by the anaerobes. If the aerobes happen to be *Azotobacter* they can utilize the end products of anaerobic growth for nitrogen fixation.

9.6.2 AUTOTROPHIC MICROORGANISMS

The autotrophic nitrogen-fixing microorganisms that are of primary interest as soil and water inhabitants are the blue–green algae. These organisms can grow normally, even if at a comparatively slow rate, on a liquid medium containing the usual mineral elements without fixed nitrogen or carbohydrate if sunlight or other suitable light source is provided. They obtain their energy supply via photosynthesis and hence require a supply of carbon dioxide. If available fixed nitrogen is added to the medium they will utilize it, as do other nitrogen-fixing microorganisms, but growth is usually little accelerated. Since their growth requirements are so simple, these algae are likely to be found growing in nature where nutrient conditions are unsuitable for other organisms provided that moisture is not limiting. Prior to that it had been shown that mixed cultures of algae and bacteria sometimes fixed nitrogen but most workers were inclined to attribute the fixation to the bacteria that were supposedly living cooperatively with green or blue–green algae. Most blue–green algae are surrounded by a gelatinous secretion in which bacteria thrive, making it very difficult to obtain bacteria-free cultures. They are confined almost wholly to the soil surface where light is available, but at this location moisture is so limited that growth is almost impossible. Also cultivation disturbs any growth that they may be able to make. There are indications, however, that in grassland or in dark moist places they may add appreciable amounts of nitrogen to the soil. Their ability to grow under low light intensities would favor such fixation. In this connection it is interesting to note that Cameron and Fuller (1960) demonstrated that algal and lichen crusts taken from the surface of semi-arid Arizona soils contain nitrogen-fixing blue–green algae. When these crusts were incubated in the laboratory in illuminated moist chambers for 4 weeks, nitrogen gains of 0.024–0.04% (240–400 ppm) were found. Under natural semi-arid conditions presumably the algae would be present largely or wholly as spores, and since a few days are required after a rain for vegetative cells or filaments to develop, the quantity of nitrogen fixed per year under natural conditions where rains are infrequent would probably be negligible. It is, however, amazing that the algae should even be present in such an environment. It is in rice paddies, or other frequently submerged lands, that nitrogen-fixing algae are believed to be a very important factor in agriculture. Much work has been done in this area in recent years but we are still uncertain as to the amount of nitrogen fixed under field conditions (MacRae and Castro, 1967). De and Mandal (1956) estimated a fixation of up to 44 Ib. of nitrogen per acre per year. Under many conditions it seems that the nitrogen fixed by algae is the main source of this element for the rice crop, but it is well not to overstress the importance of these organisms. Some studies have shown that for unknown reasons, nitrogen-fixing species are not always present in rice paddies.

9.6.3 *AZOTOBACTER CHROOCOCCUM AND ITS POTENTIAL FOR CROP PRODUCTION*

Biofertilizer plays key role in productivity and sustainability of soil health management in terms of solubilizing plant nutrients; both micro and macronutrients, and stimulating plant growth. Use of biofertilizer in agriculture protect the environment as they are eco-friendly and economical for the farmers (Khosro and Yousef, 2012). Increasing and extending the application of biofertilizers can reduce the negative impact of chemical fertilizers and decrease the adverse environmental effects of agrochemicals. Biofertilizers or bio-inoculants are minute organisms which are beneficial to the growth of plants and responsible for enhanced crop yield. The various naturally available free-living bacteria which are very beneficial for the overall plant growth are collectively known as plant growth promoting rhizobacteria (PGPR; Kloepper, 1994). These PGPRs are mainly involved in metabolic process related to nitrogen fixation, phosphate solubilization and overall plant growth promotion. Since the advent of green revolution and high input agriculture practices, chemical fertilizer have become the major source of nitrogen for crop plants (Peoples et al., 1995). BNF plays an important role in maintaining of nitrogen status of the soil. Biofertilizer are able to fix atmospheric nitrogen as single microbial inoculants or as consortia. Microbes that are able to fix atmospheric nitrogen are grouped into three broad categories as: (1) non-symbiotic bacteria, (2) blue–green algae, and (3) symbionts (Rajendra et al., 1998).

The genus *Azotobacter*, belonging to family *Azotobacteriaceae*, represents the main group of heterotrophic non-symbiotic NFB principally inhabiting neutral or alkaline soils (Sartaj et al., 2013). The non-symbiotic free living *Azotobacter* is largely associated with nitrogen fixation in plant rhizosphere (Lakshminarayana, 2000). It is also involved in various other metabolic pathways of plant growth promotion (Gonzalez-Lopez et al., 1986), antagonistic activity against plant pathogens in soil (Verma et al., 2001) and activation of potential rhizospheric bio-inoculants for enhancement of plant yield (Lakshminarayana et al., 2000). *Azotobacter* sp. and *Azospirillum* sp. are the most commonly used biofertilizers in different agricultural crops either as combined or as single inoculation (Vessey, 2003). The estimated contribution of these free-living nitrogen fixing bacteria to the nitrogen input of soil ranges from 0–60 kg/ha per year. Studies have shown that inoculation of maize crop with *Azotobacter* significantly increased its plant height, grain weight and grain yield. Inoculation of *Azotobacter* increased grain yield in maize up to 35% over the non-inoculated treatment (Bandhu and Parbati, 2013). Therefore, in the development of sustainable agricultural use of *Azotobacter* as biofertilizer has great importance in improved of nutrient profile of plant and soil and increased crop yield accompanied by protection of environmental pollution and soil contamination (Saini et al., 2004; Namvar et al., 2012; Rana et al., 2012).

9.6.4 CHARACTERISTIC FEATURES OF AZOTOBACTER CHROOCOCCUM

The genus *Azotobacter,* belonging to the family *Azotobacteriaceae* is an aerobic, heterotrophic, and nonsymbiotic biological nitrogen-fixing microbe. It has been found that some *Azotobacter* species exist in association with some crops especially cereals (Martyniuk et al., 2003). *Azotobacter* sp. is generally present in natural and alkaline soil with its most commonly occurring species found in arable soils. The genus *Azotobacter* comprises different species: *A. chroococcum, A. vinelandii, A. beijerinckii, A. paspali, A. armeniacus, A. nigricans*, and *A. salinestri*. The free-living, Gram-negative, motile, and Mesophilic *Azotobacter* spp. are capable of fixing on an average 20 kg N/ha/per year (Rawia et al., 2009). *Azotobacter* spp. was commonly found as small, medium, or large rod-shaped cells. The colonies developed on Jensen's agar medium were raised, spherical flat, and with irregular margins. The colony size varied from 2 to 5 mm in 7 days. The colony characters such as colony-margin, size, color, and consistency also differed among the different species of *Azotobacter* spp. (Tejera, et al., 2005; Ahmad, et al., 2008). The agronomic importance of *Azotobacter* spp. is due to its the capability of BNF, synthesis of antibiotics, plant growth hormones (Pandey et al., 1998), vitamins, exopolysaccharides, and pigment (Jimenez et al., 2011) and also its antifungal activity (Sudhir et al., 1983). *A. chrococcum* is a common nitrogen-fixing microbe found in the rhizosphere of agricultural crops. The first representative of the genus *A. chroococcum*, was discovered and described in 1901 by the Dutch microbiologist and botanist Martinus Beijerinck (Martyniuk and Martyniuk, 2003). Nitrogen fixation process is highly sensitive to O_2, but *Azotobacter* spp. have special mechanism of oxidases and catalases to reduce the concentration of O_2 in the cells (Shank et al., 2005). *Azotobacter* species have two types of nitrogenase which is, molybdenum—iron nitrogenase, vanadium—iron nitrogenase (Neeru et al., 2000).

9.7 EFFECT OF NITROGENOUS BIOFERTILIZERS ON THE PLANT GROWTH AND YIELD

Applications of nitrogenous biofertilizers are of well-known practice by farmers in almost all agro-climatic condition, particularly in tropical climate. Nitrogen is a macro plant nutrient that plays an important role in the nutrition and yield of agricultural crops as well as for sustainability of the environment. Besides nitrogen, phosphorus and potassium are required by crops to develop a natural ability to tolerate different biotic and abiotic stress (Tsai et al., 2007). Biofertilizers also includes organic fertilizers, which are rendered in an available form with interaction of microorganisms and plants through plant-microbe interaction (Sujanya and Chandra, 2011). Mehrotra and Lehri (1971) observed the effect of *Azotobacter* inoculation on growth and yield of vegetable crops. They found that inoculation of seeds or roots with *Azotobacter* cultures seems to influence the fertility of the soil and distinctly increased the yield of the plant. *Azotobacter* inoculation on tomato seeds resulted increase in fruit yield vigorous growth in terms of plant spread, number of leaves, as compared to control. An increase in total plant protein content and chlorophyll content were absorbed in tomato plants/seeds inoculated with *Azotobacter* culture as compared to control. This was because of the vigorous growth in terms of plant spread and number of leaves in the plants treated with *Azotobacter* (Ramakrishnan and Selvakumar, 2012). Singh (2001) observed the same response of significantly increased tuber yield in *Azotobacter* treated potato grown under north-eastern hill conditions. Combined application of microorganisms along with 150 kg N/ha resulted in highest yield of tuber productions which was significantly higher than the control. In a study on cabbage with various treatments including application of poultry manure and biofertilizers, reported that application of *A. chroococcum* saved 25% nitrogen in cabbage cultivation (Devi et al., 2003). Kumar et al. (2008) studied the response of knolkhol (*Brassica oleracea* L. var.) to various integrated nutrient management practices and found that the application of *Azotobacter* as an earliness seedling treatment was best for getting higher yield and earliness is a prerequisite for high-intensity cropping system.

9.8   NEED OF BIO-FERTILIZERS

In recent year's intensive farming practices have led to high crop yield which has been achieved by extensive use of chemical fertilizers and these chemical fertilizers are environmentally and economically inhibitory. Continuous use of chemical fertilizers has led to the environmental pollution and contamination of the soil, pollution of water basins, and disruption of flora–fauna of ecosystem and ultimately reduced soil fertility (Mishra et al., 2013). The field of eco-friendly biofertilizers is creating advancement in growing level of concern toward environmental safety sustainable agricultural practice (Debojyoti et al., 2014).

9.8.1   SIGNIFICANCE OF BIOFERTILIZER IN AGRICULTURE

Biofertilizer could be used as a source of nutrient for agricultural crops for improved crop yield and quality and low availability at a production costs (Caalcante et al., 2012). Nitrogen-fixing microorganisms plays an important role in increasing yield by making atmospheric nitrogen available to the plant in soluble ammonia from either symbiotic or non-symbiotic microbes. The symbiotic bacteria *Rhizobia* are associated with legumes and nitrogen fixation occurs within the root where the bacterium resides (Saikia and Jain, 2007). The non-symbiotic bacteria as *Azotobacter* and *Azospirillum* are most important N-fixing bacteria in non-leguminous crops. Under appropriate conditions, *Azotobacter* and *Azospirillum* can enhance plant growth development and yield of several agricultural important crops in different agro climatic soils (Okon and Labendera-Gonzalez, 1994). *Azotobacter* along with other biofertilizers has important function in soil as they can fix 15–20 kg nitrogen/ha/crop and give about 10–15% increase in yield as compared to singly treated *Azotobacter* (Singh and Singh, 2007).

9.8.2   ENHANCING NON-SYMBIOTIC N_2 FIXATION IN AGRICULTURE

Biotechnological approaches used to enhance nitrogen fixation and crop productivity had been of limited usefulness under field conditions. Non-symbiotic (NS) N_2 fixation includes N_2 fixation by free-living soil bacteria (autotrophic and heterotrophic) that are not in a direct symbiosis with plants, and associative N_2 fixation (e.g. associated with the rhizosphere of grasses and cereals). Free-living N_2 fixation can also be associated with decomposing plant residues, aggregates with decomposable particulate organic matter and in termite habitats. Globally, the demand for N fertilizers is expected to exceed 112 million tonnes in 2015 (FAO, 2015) and much of this is produced by the Haber-Bosch process (Jenkinson, 2001), a process which uses large amounts of fossil fuel (Jensen, 2003). This, together with the increasing demand for organically grown agricultural and horticultural products, and the need to address economic and environmental concerns, has rekindled interest in promoting biological N_2 fixation in non-leguminous crops. Nitrogen is a critical element for sustainable agriculture but inappropriate use of fertilizer results in lower efficiency and has the potential to contribute to (1) greenhouse gas loads such as N_2O thereby contributing to climate change and (2) leaching of N from agricultural lands as NO^{-3} causing eutrophication of rivers, lakes and oceans, and reducing the quality of water supplies (Good, 2011). The need to maximize the nutrient (N) inputs from natural processes such as biological N_2 fixation is greater today than ever before (Beatty and Good 2011). Recent increases in fossil fuel costs have resulted in significant increases in N fertilizer costs and the increased variability in rainfall patterns escalates the risk associated with higher input costs in the rainfed farming systems. Cleveland et al. (1999) estimated that the potential global biological N_2 fixation (symbiotic and NS) in natural ecosystems is between 100 and 290 million tonnes N year^{-1}. In soils under agricultural production, estimates of biologically fixed N range from approximately 33 million tonnes N year^{-1} (Smil, 1999) to 50–70 million tonnes N year^{-1} (Herridge, 2008). Kennedy and Islam (Kennedy, 2001) expressed an optimism that up to half the N requirements of some cereal crops might be met from NS N_2 fixation in the future through the use of genetic tools and inoculant biofertilizers. In addition, Beatty and Good (2011) proposed two other strategies: (1) developing root nodule symbioses in important cereal crops such as wheat, rice and maize and (2) introducing nitrogenase genes into a plant organelle.

There have been many studies on inoculation with N_2-fixing bacteria of non-legumes (predominantly cereals and grasses), with reported above- and below-ground increases in total plant growth and N content (Bashan, 1990). The most successful inoculation responses have been in pot trials under controlled conditions (Alam, 2001; Aly,1999; Negi, 1987; Yanni, 1999), but inoculation experiments in the field have been less consistent (Baldani, 1987). Andrews et al. (2003) concluded that currently no NS N_2-fixing bacterial inoculant is available that can match the consistency of N fertilizers for reducing soil N deficiencies. One of the difficulties of inoculating soils with bacteria is that the inoculants generally decline rapidly due to competition with the native microflora (Rao, 1987; Schank, 1984). Inoculants compete with other microflora for available nutrients or become food for indigenous micro- and macro-fauna (Gupta, 2002). Hence the ultimate test for even the most effective beneficial organism is the ability to survive and colonize plant roots in the presence of much larger populations of indigenous microorganisms (Bashan, 1990). Inoculum formulation and application technology, that is, along with organic matter (compost or peat) or micro-granulated inoculum, are likely to be crucial for inoculant survival and success (Fages, 1992).

Endophytes are more likely to be successful inoculants because they can escape competition from indigenous microflora and can directly access the required energy source from the plant. Increased success with endophytic N_2-fixing inoculants may be possible through genetic manipulation. For example, An et al. (2007) suggested that manipulation of the promoter of the *nif* a gene in a N$_2$-fixing bacterium that has a high colonization competence may achieve stable associative N$_2$ fixation in cereals. A similar approach has been put forward (Bloemberg, 2007). Other advances using molecular strategies may be possible, for example, the creation of ammonium excreting mutant diazotrophs, in which the mechanisms by which ammonium inhibits N$_2$ fixation, are disarmed (Colnaghi, 1997) or the increased production of nitrogenase reductase such as an *Azospirillum brasilense* mutant (de Campos, 2006). However, the survival of such mutants in the field is uncertain (James, 2000).

9.9 CONCLUSION

Management of nitrogen inputs into agricultural systems is of increasing interest due to the intensive use of synthetic fertilizers in modern cash crop systems. Nitrogen-fixing plants offer an economically attractive and ecologically sound means of reducing external inputs and improving the quality and quantity of internal resources. Environmental and management factors play an enormous role in the contribution of N from beneficial microbial function. New technologies using molecular approaches, particularly when combined with isotope methods, are broadening our understanding of NS N$_2$ fixation, and the molecular mechanisms of plant-diazotrophs interactions. Generally, there is a good understanding of the environmental factors controlling N$_2$ fixation and this can be helpful in designing farming systems that promote N inputs from fixation, but most estimates of N$_2$ fixation, particularly in the field were determined more than 20 years ago. Since then, farming practices have evolved toward intensive cropping (particularly with cereals), no-tillage and stubble retention, and further evaluation in terms of quantity of N fixed and identity of significant members of the N$_2$-fixing community is needed. New research using molecular techniques will reveal the true diversity of diazotrophic bacteria in agricultural and natural ecosystems and their potential to be used as inoculants in agricultural systems. Additionally, co-occurrence network analysis using *nif*H sequence data indicated the presence of complex co-occurrence patterns in the free-living diazotrophs than that known in symbiotic diazotrophs. Such novel insights into the ecology of diazotrophs may lead to development of inoculant mixtures that promote overall N$_2$ fixation. Re-introduction into modern varieties of traits that promote the colonization of highly efficient diazotrophic populations should further contribute biologically fixed N to agricultural systems, particularly in non-leguminous crops. Finally, N_2 fixation provides an attractive option as an environmentally responsible alternative fertilizer source for sustainable food production, especially in lower organic matter and low fertility soils worldwide. There are still many unknowns in the scientific understanding of N_2 fixation. Research into the basic mechanisms of the process is an important goal for improving N_2 fixation in the future, but few, if any, of these unknowns are constraining implementation of the existing BNF technologies. So much of the existing knowledge is not being used, especially in developing countries, that it is essential to devote major efforts toward adoption of what is already known.

KEYWORDS

- blue–green algae
- diazotrophs
- free living bacteria
- nitrogen fixation
- cyanobacteria

REFERENCES

Ahmad, F.; Ahmad, I.; Khan, M. S. Screening of Free-living Rhizospheric Bacteria for their Multiple Plant Growth Promoting Activities. *Microbiol. Res.* **2008**, *163*, 173–181.

Alam, M. S.; Cui, Z.; Yamagishi, T.; Ishii. R. Grain Yield and Related Physiological Characteristics of Rice Plants (*Oryza sativa L.*) Inoculated with Free-living Rhizobacteria. *Plant Prod Sci.* **2001**, *4*, 125–30.

Alexander, M. *Introduction to Soil Microbiology*; Wiley: New York, NY, 1961; p 472.

Aly, S. S.; Soliman, S.; El-Akel, E. A.; Ali, M. E. Significance of Free N_2-fixing Bacteria and Nitrification Inhibitors on Saving the Applied Nitrogen to Wheat Plants. *Bull. Facul. Agric. Cairo Univ.* **1999**, *50*, 347–65.

An, Q.; Dong, Y.; Wang, W.; Li, Y; Li, J. Constitutive Expression of the *nifA* Gene Activates Associative Nitrogen Fixation of *Enterobacter gergoviae* 57-7, an Opportunistic Endophytic Diazotroph. *J. Appl Microbiol.* **2007**, *103* (3), 613–20.

Andrews, M.; James, E.; Cummings, S. Use of Nitrogen Fixing Bacteria Inoculants as a Substitute for Nitrogen Fertilizer for Dry Land Graminaceous Crops: Progress Made, Mechanisms of Action and Future Potential. *Symbiosis* **2003**, *35* (1), 209–29.

Anthonisen, A. C.; Loehr, R. C.; Prakasam, T. B.; Srinath, E. G. Inhibition of Nitrification by Ammonia and Nitrous Acid. *J. Water Pollut. Control Fed.* **1976**, *48*, 835–852.

App, A.; Watanabe, I.; Ventura, T. S.; Bravo, M.; Jurey, C. D. The Effect of Cultivated, Wild Rice Varieties on the Nitrogen Balance of Flooded Soil. *Soil Sci.* **1986,** *141,* 448–452.

Argandoña, M.; Fernández-Carazo, R.; Llamas, I. The Moderately Halophilic Bacterium *Halomonas maura* Is a Free-living Diazotroph. *FEMS Microbiol. Lett.* **2005,** *244* (1), 69–74.

Arnold, W.; Rump, A.; Klipp, W.; Priefer, U. B.; Puhler, A. Nucleotide Sequence of a 24206-base-pair DNA Fragment Carrying the Entire Nitrogen Fixation Gene Cluster of *Klebsiella pneumoniae. J. Mol. Biol.* **1988,** *203,* 715–738.

Baldani, V.; Baldani, J.; Döbereiner, J. Inoculation of Field-grown Wheat (*Triticum aestivum*) with *Azospirillum* spp. in Brazil. *Biol. Fertil Soils* **1987,** *4* (1–2), 37–40.

Bandhu, R. B.; Parbati, A. Effect of Azotobacter on Growth and Yield of Maize. *SAARC J. Agri.* **2013,** *11* (2), 141–147.

Bashan, Y.; Levanony, H. Current Status of Azospirillum Inoculation Technology: Azospirillum as a Challenge for Agriculture. *Can. J. Microbiol.* **1990,** *36* (9), 591–608.

Bashan, Y.; Holguin, G. *Azospirillum*-plant Relationships: Environmental and Physiological Advances (1990–1996). *Can. J. Microbiol.* **1997a,** *43,* 103–121.

Bashan, Y.; Holguin, G.; de-Bashan, L. E. *Azospirillum*-plant Relationships: Physiological, Molecular, Agricultural, and Environmental Advances (1997–2003). *Can. J. Microbiol.* **2004,** *50,* 521–577.

Beatty, P. H.; Good, A. G. Future Prospects for Cereals that Fix Nitrogen. *Science* **2011,** *333* (6041), 416–417.

Benson, D. R.; Silvester, W. B. Biology of Frankia Strains, Actinomycetes Symbionts of Actinorhizal Plants. *Microbiol. Mol. Biol. Rev.* **1993,** *57,* 293–319.

Bloemberg, G. V. Microscopic Analysis of Plant-bacterium Interactions Using Auto Fluorescent Proteins. *Eur. J. Plant Pathol.* **2007,** *119,* 301–309.

Boddey, R. M. Methods for Quantification of Nitrogen Fixation Associated with Gramineae. *Crit. Rev. Plant Sci.* **1987,** *6* (3), 209–266.

Bohlool, B. B.; Ladha, J. K.; Garrity, D. P.; George, T. Biological Nitrogen Fixation for Sustainable Agriculture: A Perspective. *Plant and Soil* **1992,** *141,* 1–11.

Bürgmann, H.; Widmer, F.; Von, Sigler W.; Zeyer, J. New Molecular Screening Tools for Analysis of Free-living Diazotrophs in Soil. *Appl. Environ. Microbiol.* **2004,** *70* (1), 240–7.

Cacciari, I.; Del Gallo, M.; Ippoliti, S.; Lippi, D.; Pietrosanti, T.; Pietrosanti, W. Growth and Survival of *Azospirillum brasilense* and *Arthrobacter giacomelloi* in Binary Continuous Culture. *Plant Soil.* **1986,** *90,* 107–116.

Cameron, R. E.; Fuller, W. H. Nitrogen Fixation by Some Algae by Arizona Soils. *Soil Sci Soc. Am. Proc.* **1960,** *24,* 353–356.

Chalk, P. The Contribution of Associative and Symbiotic Nitrogen Fixation to the Nitrogen Nutrition of Non-legumes. *Plant Soil.* **1991,** *132* (1), 29–39.

Cleveland, C. C.; Townsend, A. R.; Schimel, D. S.; Fisher, Howarth, R. W. H. Global Patterns of Terrestrial Biological Nitrogen (N_2) Fixation in Natural Ecosystems. *Global Biogeochem. Cycles* **1999,** *13,* 623–45.

Colnaghi, R.; Green, A.; He, L.; Rudnick, P.; Kennedy, C. Strategies for Increased Ammonium Production in Free-living or Plant Associated Nitrogen Fixing Bacteria. *Plant Soil.* **1997,** *194* (1–2), 145–54.

Colon-Lopez, M. S.; Sherman, D. M.; Sherman, L. A. Transcriptional and Translational Regulation of Nitrogenase in Light-dark- and Continuous-light-grown Cultures of the Unicellular Cyanobacterium *Cyanothece* sp. Strain ATCC 51142. *J. Bacteriol.* **1997,** *179,* 4319–4327.

Dart, P. Nitrogen Fixation Associated with Non-legumes in Agriculture. *Plant Soil* **1986,** *90,* 303–34.

De, P. K.; Mandal, L. N. Fixation of Nitrogen by Algae in Rice Soil. *Soil Sci.* **1956,** *81,* 453–458.

De Mita, S. For Better or for Worse: Cooperation and Competition in the Legume-rhizobium Symbiosis. *New Phytol.* **2012,** *194,* 885–887.

Debojyoti, R.; Manibrata, P.; Sudip, K. B. A Review on the Effects of Biofertilizers and Biopesticides on Rice and Tea Cultivation and Productivity. *Int. J. Sci. Eng. Technol.* **2014,** *2* (8), 96–106.

Devi, H. J.; Maity, T. K.; Paria, N. C. Effect of Different Sources of Nitrogen on Yield and Different Biofertilizers. *Ann. Agric. Sci.* **2003,** *57* (1), 53–62.

Fages, J. An Industrial View of *Azospirillum* Inoculants: Formulation and Application Technology. *Symbiosis* **1992,** *13*, 15–26.

FAO. FAO Current World Fertilizer Trends and Outlook to 2015; Food and Agriculture organization of the United Nations: Rome **2015**, 1–41.

Fischer, H. M.; Bruderer, T.; Hennecke, H. Essential and Non-essential Domains in the *Bradyrhizobium japonicum* NifA Protein: Identification of Indispensable Cysteine Residues Potentially Involved in Redox Activity and Metal Binding. *Nucl. Acids Res.* **1988,** *16*, 2207–2224.

Gallon, J. R.; Perry, S. M.; Rajab, T. M. A.; Flayeh, K. A. M.; Yunes, J. S.; Chaplin, A. E. Metabolic Changes Associated with the Diurnal Pattern of Nitrogen Fixation in *Gloeothece*. *J. Gen. Microbiol.* **1988,** *134*, 3079–3087.

Gehring, C.; Vlek, P. L. G.; de Souza, L. A. G.; Denich, M. Biological Nitrogen Fixation in Secondary Regrowth and Mature Rainforest of Central Amazonia. *Agric. Ecosyst. Environ.* **2005,** *111*, 237–252.

Giller, K. E.; Merckx, R. Exploring the Boundaries of N_2-fixation in Cereals and Grasses: An Hypothetical and Experimental Framework. *Symbiosis* **2003,** *35* (1–3), 3–17.

Gonzalez-Lopez, J.; Salmeron, V.; Martinez-Toledo, M. V.; Ballesteros, F.; Ramos-Cormenzana, A. Production of Auxins, Gibberellins and Cytokinins by *Azotobacter vinelandii* ATCC 12837 in Chemically-defined Media and Dialysed Soil Media. *Soil Biol. Biochem.* **1986,** *18*, 119–120.

Greenwood, D. J. Nitrogen Transformations and the Distribution of Oxygen in a Soil. *Chem. Ind. Lond.* **1963,** 799–803.

Gupta, V. V. S. R.; Roper, M. M. In *Protozoan Diversity in Soil and its Influence on Microbial Functions that Determine Plant Growth*, 17th World Congress of Soil Science Abstracts, 2002; Vol. 1, p 275.

Gupta, V. V. S. R.; Roper, M. M. Protection of Free-living Nitrogen-fixing Bacteria Within the Soil Matrix. *Soil Tillage Res.* **2010,** *109* (1), 50–54.

Haselkorn, R.; Buikema, W. J. Nitrogen Fixation in Cyanobacteria. In *Biological Nitrogen Fixation*, Stacey, G., Burris, R. H., Evans, H. J., Eds.; Chapman and Hall: New York, 1992; pp 166–190.

Herridge, D. F.; Peoples, M. B.; Boddey, R. M. Global Inputs of Biological Nitrogen Fixation in Agricultural Systems. *Plant Soil* **2008,** *311* (1–2), 1–18.

Huang, T.-C.; Grobbelaar, N. The Circadian Clock in the Prokaryote *Synechococcus* RF-1. *Microbiology* **1995,** *141*, 535–540.

Jacobson, M. R.; Brigle, K. E.; Bennett, L. T.; Setterquist, R. A.; Wilson, M. S.; Cash, V. L.; Beynon, J.; Newton, W. E.; Dean, D. R. Physical and Genetic Map of the Major *nif* Gene Cluster from *Azotobacter vinelandii*. *J. Bacteriol.* **1989,** *171*, 1017–1027.

James, E. Nitrogen Fixation in Endophytic and Associative Symbiosis. *Field Crops Res.* **2000,** *65* (2), 197–209.

James, E. K.; Gyaneshwar, P.; Barraquio, W. L.; Mathan, N.; Ladha, J. K. In The Quest for Nitrogen Fixation in Rice; Ladha, J. K., Reddy, P. M., Eds.; International Rice Research Institute: Makati City, Philippines, 2000; pp 119–140.

Jenkinson, D. The Impact of Humans on the Nitrogen Cycle, with Focus on Temperate Arable Agriculture. *Plant Soil* **2001,** *228* (1), 3–15.

Jensen, H. L. Contributions to the Nitrogen Economy of Australian Wheat Soils, with Particular Reference to New South Wales. *Proc. Linn. Soc. New South Wales* **1940,** *65,* 1–122.

Jensen, H. L. Non-symbiotic Nitrogen Fixation. In *Soil Nitrogen Agronomy*; Bartholomew, W. V., Clark, F. E., Eds.; 1965, *10*, 437–480.

Jensen, H. L.; Swaby, R. J. Nitrogen Fixation and Cellulose Decomposition by Soil Microorganisms. II. The Association Between *Azotobacter* and Facultative Anaerobic Cellulose Decomposers. *Proc. Linn. Soc. N.S.W.* **1941,** *66*, 89–106.

Jensen, E. S.; Hauggaard-Nielsen, H. How Can Increased Use of Biological N2 Fixation in Agriculture Benefit the Environment? *Plant Soil* **2003,** *252* (1), 177–86.

Jimenez, D. J.; Jose, S. M.; Maria, M. M. Characterization of Free Nitrogen Fixing Bacteria of the Genus Azotobacter in Organic Vegetable-grown Colombian Soils. *Braz. J. Microbiol.* **2011,** *42*, 846–858.

Kahindi, J. H. P.; Woomer, P.; George, T.; de, Souza.; Moreira, F. M.; Karanja, N. K.; Giller, K. E. Agricultural Intensification, Soil Biodiversity and Ecosystem Function in the Tropics: The Role of Nitrogen-fixing Bacteria. *Appl. Soil Ecol.* **1997,** *6*, 55–76.

Kallas, T.; Rebiere, M. C.; Rippka, R.; Marsac, N. T. The Structural *nif* Genes of the Cyanobacteria *Gloeothece* sp. and *Calothrix* sp. Share Homology with those of *Anabaena* sp., but the *Gloeothece* Genes have a Different Arrangement. *J. Bacteriol.* **1983,** *155*, 427–431.

Kallas, T.; Coursin, T.; Rippka, R. Different Organization of *nif* Genes in Nonheterocystous and Heterocystous Cyanobacteria. *Plant Mol. Biol.* **1985,** *5*, 321–329.

Kennedy, I. R.; Choudhury, A. T. M. A.; Kecskes, M. L. Non Symbiotic Bacterial Diazotrophs in Crop Farming Systems: Can their Potential Plant Growth Promotion be Better Exploited? *Soil Biol. Biochem.* **2004,** *36*, 1229–1244.

Khosro, M.; Yousef, S. Bacterial Biofertilizers for Sustainable Crop Production: A Review. *ARPN J. Agric. Biol. Sci.* **2012,** *7*, 307–316.

Klipp, W. Organization and Regulation of Nitrogen Fixation Genes in *Rhodobacter capsulatus*. In *Nitrogen Fixation: Achievements and Objectives*; Gresshoff, P. G. M., Roth, L. E., Stacey, G., Newton, W. E., Eds.; Chapman and Hall: New York, 1990; pp 467–474.

Kloepper, J. W. Plant Growth Promoting Bacteria (Other Systems). In *Azospirillum/Plant Association*; J. Okon, Ed.; CRC Press: Boca Raton, FL, 1994; pp 137–154.

Kumar, S.; Sharma, J. P.; Kumar, S. Response of Knolkhol (*Brassica oleracea* L. var. GONGYLODES) cv. G-40 to Various Integrated Nutrient Management Practices. *J. Res. SKUAST-J* **2008,** *7* (2), 257–261.

Ladha, J. K. Studies on N2 Fixation by Free Living and Rice Plant Associated Bacteria in Wetland Rice Field. *Bionature* **1986,** *6*, 47–58.

Lakshminarayana, K. R.; Shukla, B.; Sindhu, S. S.; Kumari, P.; Narula, N.; Sheoran, R. K. Analogue Resistant Mutants of *Azotobacter chroococcum* Depressed for Nitrogenase Activity and Early Ammonia Excretion Having Potential as Inoculants for Cereal Crops. *Ind. J. Exp. Biol.* **2000,** *38*, 373–378.

Latty, S. E.; Von Fischer, J.; Elseroad, A.; Wasson, M. Global Patterns of Terrestrial Biological Nitrogen (N_2) Fixation in Natural Ecosystems. *Global Biogeochem. Cycles.* **1999**, *13*, 623–645.

Li, R.; MacRae, I. Specific Association of Diazotrophic Acetobacters with Sugarcane. *Soil Biol. Biochem.* **1991**, *23*(10), 999–1002.

Liu, G.; Wang, J. Probing the Stoichiometry of the Nitrification Process Using the Respirometric Approach. *Water Res.* **2012**, *46*, 5954–5962.

Martyniuk, S.; Martyniuk, M. Occurrence of *Azotobacter* Spp. In Some Polish Soils. *J. Environ. Stud.* **2003**, *12* (3), 371–374.

Mehrotra, C. L.; Lehri, L. K. Effect of Azotobacter Inoculation on Crop Yields. *J. Ind. Soc. Soil Sci.* **1971**, *19*, 243–248.

Mishra, D. J.; Singh, R.; Mishra, U. K.; Shahi, S. K. Role of Bio-Fertilizer in Organic Agriculture: A Review. *Res. J. Recent Sci.* **2013**, *2*, 39–41.

Namvar, A.; Khandan, T.; Shojaei, M. Effects of Bio and Chemical Nitrogen Fertilizer on Grain and Oil Yield of Sunflower (*Helianthus annuus* L.) Under Different Rates of Plant Density. *Ann. Biol. Res.* **2012**, *3* (2), 1125–1131.

Neeru, N.; Kumar, V.; Behl, R. K.; Deubel, A.; Granse, A.; Merbach, W. Effect of P-Solubilizing *A. chroococcum* on NPK Uptake in P Responsive Wheat Genotypes Grown Under Green House Conditions. *J. Plant Nutr. Soil Sci.* **2000**, *163*, 93–398.

Negi, M.; Sadasivam, K.; Tilak, K. A Note on the Effect of Non-symbiotic Nitrogen Fixers and Organic Wastes on Yield and Nitrogen Uptake of Barley. *Biol. Wastes* **1987**, *22* (3), 179–85.

Newton, W. E. Physiology, Biochemistry and Molecular Biology of Nitrogen Fixation. In *Biology of the Nitrogen Cycle*; Bothe, H., Ferguson, S. J., Newton, W. E., Eds.; Elsevier: Amsterdam, The Netherland, 2007; pp 109–130.

Neyra, C. A.; Hagerman, R. H. Relationship Between CO_2, Malate and Nitrate Accumulation and Reduction in Corn (*Zea maize* L.) Seedlings. *Plant Physiol.* **1978**, *58*, 726–730.

Okon, Y.; Labandera-González, C. Agronomic Applications of Azospirillum: An Evaluation of 20 years Worldwide Field Inoculation. *Soil Biol. Biochem.* **1994**, *26*, 1591–1601.

Pandey, A.; Sharma, E.; Palni, L. Influence of Bacterial Inoculation on Maize in Upland Farming Systems of the Sikkim Himalaya. *Soil Biol. Biochem.* **1998**, *30*, 379–384.

Patra, A.; Le Roux, X.; Abbadie, L.; Clays-Josserand, A.; Poly, F.; Loiseau, P.; Louault, F. Effect of Microbial Activity and Nitrogen Mineralization on Free-living Nitrogen Fixation in Permanent Grassland Soils. *J. Agron. Crop Sci.* **2007**, *193*, 153–156.

Peoples, M. B.; Herridge, D. F.; Ladha, J. K. Biological Nitrogen Fixation: An Efficient Source of Nitrogen for Sustainable Agricultural Production. *Plant Soil* **1995**, *174*, 3–28.

Postgate, J. *Nitrogen Fixation*, 3rd ed.; Cambridge University Press: Cambridge, UK, 1998.

Rajendra, P.; Singh, S.; Sharma, S. N. Interrelationship of Fertilizers Use and Other Agricultural Inputs for Higher Crop Yields. *Fert. News* **1998**, *43*, 35–40.

Rana, A.; Joshi, M.; Prasanna, R.; Shivay, Y. S.; Nain, L. Bio Fortification of Wheat Through Inoculation of Plant Growth Promoting Rhizobacteria and Cyanobacteria. *Eur. J. Soil Biol.* **2012**, *50*, 118–126.

Rao, V. R.; Jena, P.; Adhya, T. Inoculation of Rice with Nitrogen-fixing Bacteria-Problems and Perspectives. *Biol. Fertil Soils* **1987**, *4* (1–2), 21–6.

Rawia, E. A.; Nemat, M. A.; Hamouda, H. A. Evaluate Effectiveness of Bio and Mineral Fertilization on the Growth Parameters and Marketable Cut Flowers of *Matthiola incana* L. Am. *Eurasian J. Agric. Environ. Sci.* **2009**, *5*, 509–518.

Reed, S. C.; Cleveland, C. C.; Townsend, A. R. Functional Ecology of Free-living Nitrogen Fixation: A Contemporary Perspective. *Annu. Rev. Ecol. Evol. Syst.* **2011,** *42,* 489–512.

Reed, S. C.; Cleveland, C. C.; Townsend, A. R. Tree Species Control Rates of Free Living Nitrogen Fixation in Tropical Rain Forest. *Ecology* **2008,** *89* (10), 2924–34.

Reed, S. C.; Seastedt, T. R.; Mann, C. M.; Suding, K. N.; Townsend, A. R.; Cherwin, K. L. Phosphorus Fertilization Stimulates Nitrogen Fixation and Increases Inorganic Nitrogen Concentrations in a Restored Prairie. *Appl. Soil Ecol.* **2007,** *36* (2), 238–42.

Reich, S.; Almon, H.; Böger, P. Short-term Effect of Ammonia on Nitrogenase Activity of *Anabaena variabilis* (ATCC29413). *FEMS Microbiol. Lett.* **1986,** *34,* 53–56.

Roper, M. M.; Ladha, J. Biological N2 Fixation by Heterotrophic and Phototrophic Bacteria in Association with Straw. *Plant Soil.* **1995,** *174,* 211–24.

Saikia, S. P.; Jain, V. Biological Nitrogen Fixation with Non-legumes: An Achievable Target or a Dogma? *Curr. Sci.* **2007,** *92,* 317–322.

Saini, V. K.; Bhandari, S. C.; Tarafdar, J. C. Comparison of Crop Yield, Soil Microbial C, N and P, N-fixation, Nodulation and Mycorrhizal Infection in Inoculated and Non-inoculated Sorghum and Chickpea Crops. *Field Crops Res.* **2004,** *89,* 39–47.

Sartaj, A. W.; Subhash, C.; Tahir, A. Potential Use of *Azotobacter chroococcum* in Crop Production: An Overview. *Curr. Agric. Res. J.* **2013,** *1* (1), 35–38.

Schank, S.; Smith, R. In *Status and Evaluation of Associative Grass-Bacteria N-fixing Systems in Florida.* Proceedings-Soil and Crop Science Society of Florida; USA, 1984; Vol. 43, pp 120–23.

Schultze, M.; Kondorosi, A. Regulation of Symbiotic Root Nodule Development. *Ann. Rev. Genet.* **1998,** *32,* 33–57.

Shank, Yu.; Demin, O.; Bogachev, A. V. Respiratory Protection Nitrogenase Complex in *Azotobacter vinelandii. Suc. Biol. Chem.* **2005,** *45,* 205–234.

Singh, K. Response of Potato (*Solanum tuberosum*) to Bio-fertilizer and Nitrogen Under North-eastern Hill Conditions. *Ind. J. Agron.* **2001,** *46* (2), 375–379.

Singh, D.; Singh, A. *Role of Biofertilizers in Vegetable Production. Intensive Agriculture*; Krishi Vigyan Kendra Kumher, Bhartapur), Rajasthan, 2007, 24–26.

Smil, V. Nitrogen in Crop Production: An Account of Global Flows. *Glob. Biogeochem. Cyc.* **1999,** *13* (2), 647–62.

Smith, R. L.; Van Baalen, C.; Tabita, F. R. Alteration of the Fe Protein of Nitrogenase by Oxygen in the Cyanobacterium *Anabaena* sp. Strain CA. *J. Bacteriol.* **1987,** *169,* 2537–2542.

Sujanya, S.; Chandra, S. Effect of Part Replacement of Chemical Fertilizers with Organic and Bio-organic Agents in Ground Nut, *Arachis hypogeal. J. Algal Biomass Util.* **1983,** *2* (4), 38– 41.

Tan, Z.; Hurek, T.; Reinhold-Hurek, B. Effect of N-fertilization, Plant Genotype and Environmental Conditions on *nif*H Gene Pools in Roots of Rice. *Environ. Microbiol.* **2003,** *5* (10), 1009–15.

Tejera, N.; Lluch, C.; Martínez, M.; González, J. Isolation and Characterization of Azotobacter and Azospirillum Strains from the Sugarcane Rhizosphere. *Plant Soil* **2005,** *270,* 223–232.

Tsai, S. H.; Liu, C. P.; Yang, S. S. Microbial Conversion of Food Wastes for Biofertilizer to *Rhizoctonia solani. Eur. J. Plant Pathol.* **2007,** *89,* 91–197.

Tu, Q.; Zhou, X.; He, Z.; Xue, K.; Wu, L.; Reich, P.; Hobbie, S.; Zhou, J. The Diversity and Co-occurrence Patterns of N2-fixing Communities in a CO_2-enriched Grassland Ecosystem. *Microb. Ecol.* **2016,** *71* (3), 604–15.

Vancura, V. Root Exudates of Plants. I. Analysis of Root Exudates of Barley and Wheat in their Initial Phases of Growth. *Plant Soil* **1964**, *21*, 231–248.

Verma, S.; Kumar, V.; Narula, N.; Merbach, W. Studies on In Vivo Production of Antimicrobial Substances. *J. Plant Dis. Protect.* **2001**, *108*, 152–165.

Wakelin, S.; Gupta, V.; Forrester, S. Regional and Local Factors Affecting Diversity, Abundance and Activity of Free-living, N2 Fixing Bacteria in Australian Agricultural Soils. *Pedobiologia (Jena)* **2010**, *53*, 391–399.

Wang, S.-Z.; Chen, J.-S.; Johnson, J. L. Nucleotide and Deduced Amino Acid Sequence of *nif E* from *Clostridium pasteurianum*. *Nucl. Acids Res.* **1989**, *17*, 3299.

Ward, B. B. Measurement and Distribution of Nitrification Rates in the Oceans. *Methods Enzymol.* **2011**, *486*, 307–323.

Watanabe, I.; Barraquio, W. L.; De Guzman, M. R.; Cabrera, D. A. Nitrogen Fixing (Acetylene Reduction) Activity and Population of Aerobic Heterotrophic Nitrogen Fixing Bacteria Associated with Wetland Rice. *Appl. Environ. Microbiol.* **1979**, *37*, 813–819.

Waterbury, J. B.; Rippka, R. Sub-section I. Order *Chroococcales* Wettstein 1924, emend. Rippka et al., 1979. In *Bergey's Manual of Systematic Bacteriology*; Staley, J. T., Bryant, M. P., Pfenning, N., Holt, J. G., Eds.; Williams and Wilkins: Baltimore, 1989; Vol. 3, pp 1728–1746.

Widmer, F.; Shaffer, B. T.; Porteous, L. A.; Seidler, R. J. Analysis of *nif* H Gene Pool Complexity in Soil and Litter at a Douglas Fir Forest Site in the Oregon Cascade Mountain Range. *Appl. Environ. Microbiol.* **1999**, *65* (2), 374–80.

Yanni, Y.; El-Fattah, F. Toward Integrated Bio Fertilization Management with Free Living and Associative Dinitrogen Fixers for Enhancing Rice Performance in the Nile Delta. *Symbiosis* **1999**, *27* (3–4), 319–31.

Zechmeister-Boltenstern, S.; Kinzel, H. Non-symbiotic Nitrogen Fixation Associated with Temperate Soils in Relation to Soil Properties and Vegetation. *Soil Biol. Biochem.* **1990**, *22*, 1075–1084.

CHAPTER 10

Associative Nitrogen Fixation

MANURAJ PANDEY[1], AKANKSHA NIGAM[2], SHIVAM PRIYA[3], and
MAYANK KAASHYAP[4,*]

[1]*Molecular Mechanism and Biomarkers Group, International Agency for
Research on Cancer, 150 Cours Albert Thomas, Lyon 69008, France*

[2]*Department of Microbiology and Molecular Genetics, IMRIC,
The Hebrew University Hadassah Medical School,
Jerusalem, 911120, Israel*

[3]*Laboratory of Developmental Biology, Institute of Dental Science,
Hebrew University, Hadessah Ein Kerem, P.O. Box 12272,
Jerusalem 91120, Israel*

[4]*The Pangenomics Group, School of Science, RMIT University,
Melbourne, VIC 3083, Australia*

[*]*Corresponding author. E-mail: mayankkaashyap@gmail.com*

ABSTRACT

To meet the demand of expanding world, food production has to be increased.
But, due to the increased use of chemical inputs, fertility of soil is decreasing
day by day. To overcome the losses, due to chemical fertilizers, there is an urgent
need to explore the potential role of free-living nitrogen-fixing heterotrophic
organisms in increasing soil fertility. Interactions between plants and associative nitrogen-fixing bacteria, which are measured a detachment of plant growth-
promoting rhizobacteria, are the simplest form of nitrogen-fixing symbiosis.

The number of N_2-fixing plant-associated bacteria identified is still
growing, but we are far from having a complete view of the ecological
impact of these associations. Understanding and optimizing N_2-fixing plant
bacteria associations have promising prospective for sustainable agriculture.
Isolation and characterization of effective and competitive strains tolerant
to high temperature, drought, nitrate, acidity, and other abiotic stresses is

of high priority. There is also a need to build up populations by addition of organic materials as well as repeated inoculation of the desired strains.

10.1 INTRODUCTION

In the plant growth and development, nitrogen (N) plays an important role. Eighty percent of nitrogen is present in atmosphere, but in plant dry matter it constitutes only 2% due to the inability of plants to directly use the atmospheric nitrogen. Nitrogen is the essential element for life as it is an important component of amino acids, protein, nucleic acid, etc. Plants take nitrogen in the form of ammonium and nitrates through their roots (Santi et al., 2013). Through biological nitrogen fixation (BNF) inert N_2 is converted into useful NH_3. Few prokaryotes including bacteria and certain species of actinomycetes help in this process (Lam et al., 1996; Franche et al., 2009; Santi et al., 2013). BNF was first discovered by Beijerinck in 1901. Nitrogenase enzyme helps in the conversion of atmospheric nitrogen to ammonia. These organisms utilize the nitrogenase enzyme to catalyze the conversion of atmospheric nitrogen (N_2) to ammonia (NH_3). Symbiotic association between rhizobia and legumes is the major source of fixed nitrogen. In the symbiotic association both partners get benefits, such as plant supply carbon source in the form of dicaboxylic acid and in return receives ammonium. Nitrogen fixation occurs inside the cells of de novo formed organs; nodules that usually develop on roots but occasionally on stems. Nitrogen fixed through the N_2-fixation is the major source of biosphere nitrogen has an important ecological and agronomical role and widely used in agriculture. There were various reports that suggested that legume symbioses play an important role in environment-friendly agriculture (Sant et al., 2013; Rao, 2014). On the other hand, non-symbiotic N_2-fixation includes fixation of nitrogen by free-living soil bacteria (autotrophic and heterotrophic) that are not in a direct symbiosis with plants, and associative N_2-fixation (e.g., associated with the rhizospheres of grasses and cereals). Free-living N_2-fixation is also associated with decomposition of plant residues, aggregates of decomposable organic matter etc. (Roper and Gupta, 2016). Associative nitrogen fixing microorganisms are those diazotrophs that live in the rhizosphere of plants and obtain energy from plants. Several reports have shown that the existence of association between tropical grasses and nitrogen fixing bacteria contributes significantly in the nitrogen economy of these plants (Rinaudo et al., 1971; Dobereiner et al., 1972; Dobereiner and Day, 1976; von Bulow and Dobereiner, 1975; Table 10.1). Thus, utilization of associative BNF technology in grass and cereal crop has an immense importance in the development of profitable agriculture technologies. Although associative N_2-fixation has an immense importance in agriculture and forestry but quantification of their potential has not been established. When nitrogen from the soil depleted then associative nitrogen fixers function vigorously such as *Azotobacter* spp. and *Azospirillum* spp., but they are measured as of minor agricultural significance. Two other free-living diazotrophs, namely, *Acetobacter diazotrophicus* and *Herbaspirillum* spp., live endophytically in the vascular tissue of sugarcane so that they can take abundant sucrose as a possible source of energy for N_2-fixation (Döbereiner et al., 1993) and reduce the energy required for production of ethanol from sugarcane (Reinhold-Hurek and Hurek, 1998). Diazotrophs that colonize intercellularly in the inner tissues of plants are referred as edtophytic (e.g., *Herbaspirillum* spp., *Gluconacetobacter* spp., *Azoarcus* spp., *Burkholderia* spp.). *Azospirillum* spp. have a more "associative" or "facultative endophytic" lifestyle. They sometimes enter the root tissues, but are also found in large numbers on the root surface. Epiphytic diazotrophic bacteria like *Beijerinckia fluminensis* and *Azorhizophilus paspali* (*Azotobacter paspali*) are mainly isolated from rhizoplane (Reinhold-Hurek and Hurek, 1998; Baldani and Baldani, 2005). In this chapter, we have discussed the different types of associative nitrogen fixing bacteria and their role of associative N_2-fixation.

TABLE 10.1 Early Important Landmarks in N_2-fixation Research with Special Reference to Free-living, Associative, and Endophytic N_2-fixing Bacteria.

S. no.	Major events in soil microbiology and associative N_2-fixation	References
1.	Isolation of *Nostoc* and *Anabaena* by Drewes	Drewes (1928)
2.	Discovery of production of phytohormones by bacteria	Boysen Jensen (193)
3.	Isolation of *Beijerinckia* spp. by Starkey and De	Döbereiner and Pedrosa (1987)
4.	Clark proposed the term "rhizoplane" for the microbiology of root surface	Starkey (1950) and Rovira (1991)
5.	Nitrogenase activity is obtained in cell to free extract of *C. pasteurianum* by Carnahan, Mortenson, Mower, and Castle	Wilson (1969)
6.	Association *Azotobacter paspali-Paspalum notatum*	Döbereiner (1974)
7.	Acetylene reduction technique to assay nitrogenase activity by Schollhorn and Burris and by Dilworth	Hardy et al. (1968)
8.	Clarification of the taxonomic status of *Azospirillum* spp.	Tarrand et al. (1978)
9.	Nitrogen-fixation in *Pseudomonas stutzeri*	Krotzky and Werner (1987)
10.	Isolation of (*Glucon-*)*Acetobacter diazotrophicus*	Döbereiner (1992)
11.	Development of the N_2-fixing endophytes concept	Döbereiner (1992)

10.2 MAJOR N$_2$-FIXING BACTERIA ASSOCIATED WITH PLANTS AND THEIR ROLE

10.2.1 AZOSPIRILLUM

Azospirillum is rod-shaped, slightly curved, Gram-negative bacteria that were isolated from soil in 1925 at the Netherlands (Beijerinck, 1925). *Azospirilla* are mainly found in tropical soils, but also in several temperate zones, in tundra and semi-desert sites of the Canadian High Arctic (Nosto et al., 1994). Among the nitrogen fixing bacteria, *Azospirilla* group is discovered earliest and best characterized genus of plant-growth promoting rhizobacteria. Other free-living diazotrophs are also identified in association with roots, namely, *Acetobacter diazotrophicus, Herbaspirillum seropedicae, Azoarcus* spp., and *Azotobacter. Azospirillum* derived its name from two words Azo= N$_2$-fixing capability and spirillum= spiral movement of cell (Tarrand et al., 1978). Natural habitat, plant–root interaction, N$_2$-fixation, and biosynthesis of plant growth hormones are the four characteristic features of *Azospirillum* plant–root interaction. The rhizosphere of Digitaria and *Zea mays* also reported with the associative N$_2$-fixation bacteria, which helps in the nitrogen fixation (Steenhoudt and Vanderleyden, 2000). *Azospirilla* colonize both annual and perennial crops as well as plants and weeds. *Azopirillum* has also been isolated from leaves of marine mangroves, lake or pond water etc. (Hartmann and Baldani, 2003). Eight species of *Azospirillum* that are: *A. brasilense, A. lipoferum, A. halopraeferens, A. irakense, A. largimobile, A. doebereinerae, A. oryzae,* and *A. amazonense* (Xie and Yokota, 2005) have been identified. Interaction of *Azospirillum* resulted in the better plant growth, crop yield, and increase in N content of plant (Bashan and Holguin, 1997; Okon, 1985; Eckert et al., 2001). Although there is no significant report that fixed N$_2$ contributes to the N content of cereals. For significant associative N$_2$-fixation, available C source, N concentration, and O$_2$ tension at the root surface are the limiting factors. However, NH$_4$ produced during N$_2$-fixation available to plants when bacteria died. A high affinity NH$_4$ uptake system and assimilation of NH$_4$ by the glutamine synthetase facilitated the bacteria to use fixed N for their own growth. *Azospirillum* generally affects the plant growth due to its ability to produce plant growth regulating substances such as auxin that increases the lateral roots as well as enlarge root hairs. It is reported that concentration auxin and other plant hormones in roots promote the higher nutrient uptake and improve water status of the plant under suboptimal conditions for normal plant root development. Yield enhancing capacity of *A. lipoferum* and *A. brasilense* strains is widely used in commercial inocula. (Steenhoudt and Vanderleyden, 2000).

10.2.2 *HERBASPIRILLUM*

Another genus of associative nitrogen fixative bacteria is *Herbaspirillum,* which is a N_2-fixer and colonizes in various plants endophytically. Diazotrophic *H. seropedicae* were first isolated from diverse species of the *Gramineae* family like maize (*Zea mays*), sorghum (*Sorghum bicolor*), sugarcane (*Saccharum officinarum*), and wild rice (*Oryza sativa*; Elbeltagy et al., 2001). Later, they were also found in association with dicotyledonous plants, root nodules of legumes, and in roots of monocots outside the grass family as well as stems of different cultivars of banana (*Musa* spp.) and pineapple [*Ananas comosus* (L.) Merril; Schmid et al., 2005]. *Herbaspirillum* spp. colonizes the interior of the root and establishing themselves in the cortex and vascular tissues of roots along with systemically in the whole plant. Histochemical analysis of seedlings of maize, sorghum, wheat, and rice grown in vermiculite showed the same. Using axenic systems of different plants, a significant stimulation of root development due to inoculation by *H. seropedicae* and *H. frisingense* was established. Root exudation sites such as axils of secondary roots and intercellular spaces of the root cortex are firstly colonized, and then vascular tissue and expression of *nif* genes occurs in roots, stems, and leaves. *Nif* gene is also expressed in the bacterial cells colonizing the external mucilaginous root material (Roncato-Maccari et al., 2003). Nitrogenase activity of *H. seropedicae* was only present when C source is added in rice seedlings (James et al., 2002). Presently, the *H. seropedicae* genome is being sequenced by the Brazilian GENOPAR program (http://www.genopar.org/).

10.2.3 *Azotobacter vinelandii* AND *Pseudomonas* spp.

Azotobacter vinelandii is a Gram-negative, aerobic, and free-living and widely distributed soil bacterium. *Azotobacter vinelandii* has capacity to nurture carbohydrates, alcohols, organic acids, alginate production, and N_2-fixation (Rosenblueth and Martínez-Romero, 2006). In contrast to most diazotrophs, *A. vinelandii* has the capability to fix N_2 in the presence of atmospheric O_2. In order to prevent nirtogenase from O_2 inactivation is a complex process in which high respiration rates, formation of a complex with Shethna protein, autoprotection by reduction of O_2 to H_2O_2, morphological protection, and the formation of alginate plays an important role (Becking, 1999; Sabra et al., 2000). Earlier, N_2-fixing ability of *A. vinelandii* was a major physiological characteristic to differentiate it from *Pseudomonas* spp.; *Pseudomonas* is mostly found in the rhizosphere of cereal crops, weeds sugarcane (Ladha et al., 1983; Seldin et al., 1984; Young, 1992). Many new nitrogen fixing species were isolated from roots of various grasses, sugarcane, and maize discovered by Johanna Döbereiner. Nitrogen fixation in *Pseudomonas* spp. has been discussed but nowadays various strains of *Pseudomonas* are identified (Vermeiren et al., 1998; Rediers, et al., 2004). *P. stutzeri* strain A15 (formerly *Alcaligenes faecalis*) is a N_2-fixing *pseudomonas* frequently isolated as a predominant diazotrophic strain in the rhizosphere of Chinese paddy field rice and observed to express *nifH* in the roots of rice. *P. stutzeri* strains have been isolated from sorghum (*Sorghum bicolor*) and caper (*Capparis spinosa*) (Vermeiren et al., 1999). Also, other plants have been reported to have *pseudomonas* strains in their rhizosphere (Lalucat et al., 2006).

10.2.4 Gluconacetobacter (FORMERLY ACETOBACTER) diazotrophicus

Gluconacetobacter diazotrophicus (formerly *Acetobacter diazotrophicus*) is a Gram-negative, rod-shaped, aerobic, obligate endophytic, and diazotrophic nitrogen-fixing bacterium identified in monocotyledon and sugarcane plants (Eskin et al., 2014) discovered by Johanna Döbereiner (1924–2000). It has numerous physiological properties for nitrogen-fixation, such as tolerance to low pH and to high sugar concentrations. *Gluconacetobacter diazotrophicus* has no assimilatory nitrate reductase, so N_2-fixation continues in the presence of NO_3^-, and its nitrogenase is only partially inhibited by addition of NH_4^- via an unidentified process that does not involve covalent modification (ADP-ribosylation) of its Fe⁻ protein component. These properties have extensive agricultural significance as biological N_2-fixation might be supplemented by N fertilizer (Fisher and Newton, 2005). This bacterium hardly survives in soil and most preferably grow in the plant tissue (Shanti et al., 2003). Earlier, *G. diazotrophicus* was identified in sugarcane and other sucrose accumulating plants, such as sweet potato, Cameroon grass etc., but later it was also identified in endophytic association with various other plants, such as *Coffea arabica, Eleusine coracana,* and *Ananas comosus* (Estrada-De et al., 2001). *G. diazotrophicus* identified in roots, stems, and aerial part of Australian and Brazilian sugarcane cultivars. Low PO_2 in the xylem sap and intercellular spaces of sugarcane favors the expression of nitrogenase due to which this bacterium is found in this location. (Tejera et al., 2004). In addition to the capability to fix N_2 in association with sugarcane plants, G. diazotrophicus also have the other interesting physiological capacities such as phytohormone production (Sevilla et al., 2001).

10.2.5   KLEBSIELLA

Klebisella is another diazotrophs among these *Klebsiella pneumoniae* used as a model system for the study of organization and regulation of nitrogen-fixation (*nif*) genes in diazotrophs (Dixon et al., 1986; (Chen et al., 2015). *Klebsiella* isolated from the root surface of various plants. *K. pneumoniae, K. oxytoca,* and *K. planticola* are all capable of fixing N_2 and are classified as associative N_2-fixers. *Klebsiella* isolated from leaves of rice, grassland soil, sweet potato etc. (Grimont et al., 1999; Asis and Adachi, 2004). Some strains of *K. pneumonia* have the ability to colonize the rhizosphere as well as inside the seedling of legumes like alfalfa (*Medicago sativa*), *Medicago truncatula,* colonize *Arabidopsis thaliana*, wheat (*Triticum aestivum*), and rice (*Oryza sativa*; Dong et al., 2003).

   K. pneumonia fixes N_2 and raises total N concentration in wheat (*Triticum aestivum* L.); however in maize, this bacterium is not able to relieve nitrogen deficiency but increases the yield significantly (Riggs et al., 2001). With the help of immunolocalization, it was observed that *K. pneumonia* is found in the intercortical layers of stem but nitrogenase activity is found only in roots when bacteria were supplied with exogenous C source (Chelius and Triplett, 2000).

10.2.6   AZOARCUS

The genus *Azoarcus* (type species, *Azoarcus indigens*), proposed by Rein-hold-Hurek et al. (1993), belongs to the family Rhodocyclaceae of the order Rhodocyclales in the class Betaproteobacteria (Reinhold-Hurek and Hurek, 2006). Some species fix nitrogen and then require microaerobic conditions for growth on N_2 (Reinhold-Hurek and Hurek, 2006; Chen et al., 2013). Associative N_2-fixing species of the genus *Azoarcus* were first found in association with Kallar grass [*Leptochloa fusca* (L.) Kunth]. Afterward, they were also isolated from field-grown rice cultivated in Nepal and from resting stages of plant associated fungi found in rice field soil from Pakistan. *A. indigens*, *A. communis,* and *A.* sp. BH72 occur inside roots or on the root surface of Gramineae and have never been isolated from root-free soil so far, apart from the members of *A. communis* originating from a petroleum refinery oily sludge in France or from a compost biofilter in Canada. This is in disparity to the soil-borne strains of the genus, such as *A. tolulyticus, A. toluvorans, A. toluclasticus, A. evansii, A. buckelii,* or *A. anaerobius*, which do not instigate from living plants but frequently from soil and sediments. While currently plant-associated *Azoarcus* spp. have been isolated simply from a limited number of samples, they may be more broadly distributed than assumed. In molecular-ecological studies on root material or fungal spores, *Azoarcus* 16S rDNA genes or *nifH* genes have been retrieved which did not correspond to genes of cultivated strains or species (Reinhold-Hurek and Hurek, 2004). In situ hybridization studies demonstrated the expression of *Azoarcus* nitrogenase in the root cortex of Kallar grass and inside rice roots (Reinhold-Hurek and Hurek 1998). Nitrogen-fixation was studied in more detail in *Azoarcus* sp. BH72. Compared to *Azospirillum* spp.; *Azoarcus* is more tolerant to O_2. Similarly to *Azospirillum*, the expression of nitrogenase genes is transcriptionally regulated in response to O_2. A particular trait of N_2-fixation in this strain is the formation of so-called "diazosomes."

10.2.7 *BURKHOLDERIA*

The genus *Burkholderia* I is another diaazotroph, but it also includes soil bacteria, plant-growth-promoting rhizobacteria, and human and plant pathogens. (Caballero-Mellado et al., 2004). Strains of *Burkholderia cepacia* can endure within vacuoles of free-living amoebae (Marolda et al., 1999).

   B. vietnamiensis was the first species that fixes N_2, and found in association with roots of rice plants but also observed inside maize roots, in the rhizosphere as well as rhizoplane of maize and coffee plants (Estrada-De Los Santos et al., 2001). *B. brasilensis* and *B. tropicalis* were isolated as N_2-fixing bacteria from banana (*Musa* spp.) and pineapple [*Ananas comosus* (L.) Merril; Magalhaes Cruz et al., 2001]; however other N_2-fixing bacteria such as *B. kururiensis, B. tropica, B. unamae, B. silvatlantica* (Caballero-Mellado et al., 2004), and *B. silvatlantica* were also identified. Initially, it was thought that only members of proteobacteria were capable for the nodulation of leguminous plants but later *B. tuberum, B. phymatum,* and *B. caribensis* strains were found to be N_2-fixing legume symbionts (Chen et al., 2003) as well as they possess nodulation genes (Moulin et al., 2001).

10.3 OTHER ASSOCIATIVE N_2-FIXING BACTERIA

N_2-fixing bacteria, *Serratia marcescens* strains have been isolated from surface-sterilized roots and stems of different rice varieties and in vitro on inoculation with this bacterium. Nitrogen-fixation was measured only when external carbon (e.g., malate, succinate, or sucrose) was added to the rooting medium (Gyaneshwar et al., 2001). Similarly, *Pantoea* species are basically plant pathogens, which can grow a wide range of temperature, pH, and salt concenteration (Loiret et al., 2004) and also isolated endophytically and epiphytically from a wide variety of crops, such as corn, cotton (*P. agglomerans* (*Enterobacter agglomerans, Erwinia herbicola*)), wheat (Asis and Adachi, 2004). β-Proteobacteria of the orders *Enterobacteriales* or *Pseudomonadales* such as *Pantoea agglomerans, Enterobacter kobei, Enterobacter cloacae, Leclercia adecarboxylata, Escherichia vulneris* (Benhizia et al., 2004), and *pseudomonas* sp. were found in the nodules of wild *Hedysarum* legume species. Besides finding nodulating strains outside of the *Rhizobiales*, rhizobia may also occurs as endophytes in the roots of cereals, such as rice, wheat, and maize, without nodule formation. They are able to promote the growth of these non-legumes, but this growth promotion is related to mechanisms independent of biological N_2-fixation, such as phosphate-solubilizing ability, and neither root nodules nor N_2-fixation are observed during these interactions (Somers et al., 2004). Nitrogen-fixing *Paenibacillus* has been also established in the rhizosphere of various plants and important crop species. They include: *P. polymyxa, P. macerans, P. durus* (synonyms: *Paenibacillus azotofixans, Bacillus azotofixans, Clostridium durum, respectively*), *P. peoriae, P. borealis, P. brasilensis, P. graminis,* and *P. odorifer*. Strains belonging to these species are considered to be important for agriculture. They can influence plant growth and health directly by the production of phytohormones, by providing nutrients, by fixing N_2, and/or by the suppression of deleterious microorganisms through antagonistic functions. Moreover, their input as N_2-fixers, numerous of these strains are also of industrial significance for the production of chitinases, amylases, proteases, and antibiotics (Coelho et al., 2003). Anaerobic clostridia have been found to fix N_2 in a consortium with diverse non-diazotrophic bacteria in various gramineous plants. A major feature of these consortia is that N_2-fixation by the anaerobic clostridia is supported by the elimination of O_2 by the accompanying bacteria in the culture. These consortia are widespread in wild rice species and pioneer plants, which are able to grow in unfavorable locations (Minamisawa et al., 2004). With the use of cultivation-independent methods, based on the detection of the *nifH* gene, a remarkable diversity of up to now uncultured diazotrophs was detected in association with roots of grasses (Tan et al., 2003).

10.4 ASSOCIATIVE N_2-FIXING BACTERIA

Interaction between *Azospirillum* and other rhizospheric bacteria and their host plants is known as "associative symbiosis," "rhizocoenoses," and "associative nitrogen fixation." The degree of interaction between plant and bacteria as well as process may vary from species to species. Association between soil nitrogen fixing bacteria and roots of grasses termed as associative nitrogen fixing bacteria, and nitrogen fixing PGPR due to the growth promoting effects. However, on roots no differentiated structures are identified that were induced by bacteria, and rhizoplane is not always well-defined due to which benefit of the association is questioned.

10.5   ROLE OF ASSOCIATIVE N₂-FIXING BACTERIA

In the family poaceae (rice, wheat, corn, oats, barley, etc.), most of the associative type of microorganism resides in the close association with the rhizosphere region and fix high amount of nitrogen (20–25% of total nitrogen in rice and maize, Montanez et al., 2012). Among the most N_2-fixation species, *Azospirillum* (Saikia and Jain, 2007) is most important. Besides these several other species are also reported which we have already discussed. *Nostoc* strains are also capable of fixing nitrogen and isolated from *Gunnera* and *Anthoceros* (Nilsson et al., 2005) and as well as from rice roots, where it showed increased N_2-fixation rate (Nilsson et al., 2002). Role of nitrogen-fixing bacteria with graminaceous plants was reported around 50 years ago (Döbereiner and Day, 1976; Okon and Labandera-Gonzales, 1994). Increase in yield is also reported from bacteriazation experiments in *Azotobacter* (Brown, 1974), *Azosiprillum* (Boddey and Döbereiner, 1982), cynobacteria (Roger and Kulasooriya, endophytes (Döbereiner et al., 1993). Nitrogen-fixation was also reported in clove and pea by Jean-Baptiste Boussingault in 1838. At the same time, Georges Ville positive effects of N_2-fixation in legumes, wheat, rye, and watercress, and to avoid any controversy he also repeated the experiments under the observation of committee mandated by the French Academy (Dumas et al., 1855). During earlier period, *Azotobacter* and *Clostridium* were thought to be the only genus that were capable of fixing nitrogen in free-living state and *Nostoc* was only known blue-green alage (Stewart, 1969; Wilson, 1969). But, after that several other soil bacteria were also reported to produce plant growth promoters and increasing the yield. Beneficial effect of non-leguminous crops by bacteria is given term "bacteriazation" (Brown, 1974; Macura, 1966). Earlier, *Azotobacter* was observed as aerobic non-symbiotic bacterium that has ability to fix nitrogen (Winogradsky, 1949). Winogradsky also developed various methods for the isolation of *Azotobacter* and measurement for the estimation of density of bacteria in the soil by the use of silica gel plates that were devoid of nitrogen or shifted to carbon source (Pochon and Tchan, 1948). *Beijerinckia* was the other important soil bacteria that was discovered in tropical acid soil (Döbereiner and Pedrosa, 1987). Later, anaerobic N_2-fixation was also observed by *C. pasteurianum* (soil bacteria) other than *Azotobacter* (Chang and Knowles, 1965). In general, free-living heterotrophs contributes insignificant quantities of nitrogen and amount of fix nitrogen increases with the availability of organic substances, such as grass cuttings, straw, or other plant residues. Increased rate of N_2-fixation was observed with the addition of soluble organic substrate (Delwiche and Wijler, 1956). These observations lead to the development of inoculants industry and later these inouculants used in non-leguminous crops, such as cereals and vegetables. There were also reports where bacillus is used for phosphate mineralization (phosphobacterin; Rovira, 1991).

10.6 NITROGEN FIXATION IN NON-LEGUME CROPPING SYSTEMS

Non-symbiotic N_2-fixation was studied by Berkum and Bohlool in 1980 where he suggested the effect of salt marshes, fallow fields pasture field on N_2-fixation. Rice and sugarcane from different parts of the World have shown to be benefited from N_2-fixation by unknown means while they have not received fertilizers over the several centuries (Boddey and Döbereiner, 1982). To estimate the N_2-fixation in natural ecosystem was made possible with the help of reduction of acetylene by nitrogenase (Hardy et al., 1973; Bergersen, 1980). Although determination of acetylene reduction on excised roots or on plant soil cores was not accepted widely because that did not reflect the actual rate of N_2-fixation in whole plant in their natural conditions (van Berkum and Bohlool, 1980), and in most cases there was time lag in the detection of nitrogenase activity due to which huge variation was observed. In situ acetylene reduction for grass (*Panicum maximun*), rice, and peanut was determined and confirmed the existence of non-symbiotic N_2-fixation along with plant-specific diurnal variations in N_2-fixation rates (Balandreau et al., 1974). In the associative N_2-fixation discovery of *Azotobacter paspali* plays an important role which is specific for *Paspalum notatum* cv. Batatasis, approximately 15–90 kg N/ha/year was fixed by this species (Döbereiner et al., 1972b). In the other study, it is reported that 10% of total nitrogen accumulated in *Paspalum notatum* is produced by N_2-fixation (Boddey et al., 1983).

Beijerinck in 1923 isolated sprilillum like bacteria on which little attention was paid and after that bacteria were rediscovered by Becking (1963, 1982); and finally Dahey (1976) explained the association of these bacteria with grasses and many cereal crops and later named as *Azospirillum* (Tarrand et al., 1978) *Azotobacter paspali* isolated from the grass (Döbereiner, 1974) followed by *Herbaspirillum* in 1986 and *Gluconacetobacter* in 1988. As we have earlier reported that acetylene reduction is the major determination of nitrogenase activity, but in further studies it is reported that it is not always associated with nitrogenase activity, as *A. paspali* improved the growth of *P. notatum* by producing phytohormones rather than by N_2-fixation. In rice about 20–30% of total N in rice plants originated from biological N_2-fixation (Boddey and Döbereiner, 1982; Watanabe and Roger, 1984) while in wetland rice, to estimate the contribution of heterotrophic bacterial N_2-fixation is difficult because of complexity of microflora (Table 10.2). In the other study, it is measured that N_2-fixation is cultivar-dependent which was measured through the acetylene reduction rates (Sano et al., 1981).

TABLE 10.2   Classification of Associative N_2-fixing Bacteria in Rice Root System.

S. no.	Bacteria character	Examples
1.	**Autotroph**	*Rhodobacter (rhodopseudomonas)*
	–Photosynthetic N_2-fixing bacteria	*Rhodospirillum*
2.	**Heterotroph**	*Azotobacter, Azotomonas*
	–Autofixing Bacteria/aerobic	*Derxia, Methylomonas*
	–Autofixing Bacteria/slightly aerobic	*Bacillus*
	–Autofixing Bacteria/anaerobic-aerobic	*Clostridium,*
	–Autofixing Bacteria/anaerobic	*Desulfotomaculum, Desulfovibrio*
	–Associative nitrogen fixing bacteria/aerobic	*Beijerinckia*
	–Associative nitrogen fixing bacteria/slightly	Anaerobic
		Alcaligenes, Arthrobater, Azospirillum, Flavobacterium, Pseudomonas
	–Associative N_2-fixing bacteria/ oxidative–reductive	*Enterobacter, Klebsiella*

Azospirillum genus is widely distributed and associated with various diverse plants (van Berkum and Bohlool, 1980), among them seven species were widely known and their physiology and genetics has been studied including the genetics of *nif* gene, colonization of root, phytohormone production, and response of plants to inoculation (Elmerich et al., 1992, 1997; Okon, 1994, 1985; Costacurta and Vanderleyden, 1995; Steenhoudt and Vanderleyden, 2000). These bacteria are also known to produce bacteriocions which might serve as biocontrol agents. New species of *Alcaligenes, Azoarcus, Burkholderia, Campylobacter, Gluconacetobacter, Herbaspirillum,* and *Paenibacillus* were discovered in rice, coffee, sugarcane, maize, pineapple, Kallar grass, and sorghum. Proliferation of fermentive bacteria (*Clostridium* spp.) such as methanogenic archaea, and sulfate-reducing bacteria (*Desulfovibrio*) is favored by anoxic compartment of bulk soil. Members of all of these groups are known to fix nitrogen (Postgate, 1981; Young, 1992). Anoxic soil is that soil where methane is produced due to the complex interaction between bacteria in rice field (Watanabe and Roger, 1984; Liesack et al., 2000).

10.7 PLANT-GROWTH PROMOTION UNDER ASSOCIATIVE N$_2$-FIXATION

For the efficient N$_2$-fixation, sufficient supply of substrate is needed from the host plant under suitable environmental conditions as well as transfer of fixed nitrogen to the host plant (Klucas, 1991). High rates of N$_2$-fixation were observed by the association of *Azospirillum* with maize (Boddey and Döbereiner, 1982; Klucas, 1991). Various reports suggested that inoculation with *Azospirillum* increases the yield significantly (5–30%) even under low chemical fertilizer rate. Increase in yield also resulted from the better development of root system due to which it facilitates the increase in uptake of water and mineral. *Azospirillum* also possesses various pathways for the indole-3-acetic acid (IAA) synthesis and sometimes gibberellins although nitrogen fixing capacity of these bacteria is thought to contribute little to plant growth.

10.8 CONCLUSION

Frequent use of chemical fertilizers leads to harmful effect on environment as well as causes major dependency for nitrogen source. In order to reduce the dependency on chemical fertilizers for the nitrogen source and eco-friendly environment, there is a need of alternate source for nitrogen. Nitrogen fixation in plants provides nitrogen to the plants. Earlier reports on N$_2$-fixation suggested that it frequently occurs in leguminous crops but later in non-leguminous crops N_2-fixation also reported. But for good yield, non-leguminous crops such as cereals etc. require high dose of nitrogen. On enetic manipulation of these crops, could be effectively used as symbiotic or associative nitrogen fixing systems as well as also remove the major dependency on nitrogen fertilizer. Although it is not clear that which species has contributed in more N_2-fixation and how the N_2-fixation activity is influenced from the association with plants and microbial communities. Transformation of both plants and microorganisms provides new opportunities, new symbiotic, and associative bacterial N_2-fixation systems. Nitrogen-fixing bacteria also secrete plant-growth-promoting substances that help in the better root development, enhanced resistance to pathogens, heath of plant, phosphate solubilization which facilitate enhanced mineral uptake and also benefit human health. Although there will always the challenge persists to integrate all these factors (genetic, biotic, and abiotic) that effect the associative N_2-fixation and finding the cost-effective methods.

KEYWORDS

- **associative nitrogen-fixers**
- **colonization**
- **endophytes**
- **Gram-negative bacteria**
- **nitrogen-fixation**
- **rhizosphere**

REFERNCES

Asis, C. A.; Adachi, K. Isolation of Endophytic Diazotroph *Pantoea agglomerans* and Nondiazotroph *Enterobacter asburiae* from Sweet Potato Stem in Japan. *Lett. Appl. Microbiol.* **2004,** *38,* 19–23.

Balandreau, J.; Millier, C. R.; Dommergues, J. Diurnal Variations of Nitrogenase Activity in the Field. *Appl. Microbiol.* **1974,** *27,* 662–665.

Baldani, J. I.; Baldani, V. L. D. History on the Biological Nitrogen Fixation Research in Graminaceous Plants: Special Emphasis on the Brazilian Experience. *An. Acad. Bras. Cienc.* **2005,** *77,* 549–579.

Bashan, Y.; Levanony, H. Current Status of *Azospirillum* Inoculation Technology: *Azospirillum* as a Challenge for Agriculture. *Can. J. Microbiol.* **1990,** *36,* 591–607.

Bashan, Y.; Holguin, G. Azospirillum: Plant Relationships: Environmental and Physiological Advances. *Can. J. Microbiol.* **1997,** *43* (1990–1996), 103–121.

Becking, J. H. Fixation of Molecular Nitrogen by an Aerobic *Vibrio* or *Spirillum. Antonie. Van Leeuwenhoek.* **1963,** *29,* 326.

Becking, J. H. *Azospirillum lipoferum,* a reappraisal. In *Azospirillum, Genetics, Physiology, Ecology;* Klingmüller, W., Ed.; Birkhäuser Verlag: Basel, Switzerland, 1982; Experientia Suppl. 42, pp 130–149.

Becking, J. H. The Family *Azotobacteraceae.* In *The Prokaryotes: An Evolving Electronic Resource for the Microbiological Community;* Dworkin, M., et al., Eds.; Springer-Verlag: New York, 1999. http://link.springer-ny.com/link/service/books/10125/. (accessed Apr 2, 2016).

Beijerinck, M.W. über ein *Spirillum,* Welches Freien Stickstoff Binden Kann? *Centralbl. Bakt. II* 1925, *63,* 353–357.

Benhizia, Y.; Benhizia, H.; Benguedouar, A.; Muresu, R.; Giacomini, A.; Squartini, A. Gamma Proteobacteria can Nodulate Legumes of the Genus *Hedysarum. Syst. Appl. Microbiol.* **2004,** *27*: 462–468.

Bergersen, F. *Methods for Evaluating Biological Nitrogen Fixation;* John Wiley and Sons: Chichester, UK, 1980.

Boddey, R. M.; Chalk, P. M.; Victoria, R. L.; Matsui, E.; Döbereiner, J. The Use of the 15N Isotope Dilution technique to Estimate the Contribution of Associated Nitrogen Fixation Nutrition of *Paspalum notatum* cv. Batatais. *Can. J. Microbiol.* **1983,** *29,* 1036–1045.

Boddey, R. M.; Döbereiner, J. In *Association of Azospirillum and Other Diazotrophs with Tropical Gramineae,* Transactions of the 12th International Congress of Soil Science, Non Symbiotic Nitrogen Fixation and Organic Matter in the Tropics, Indian Society of Soil Science, New Delhi, India, 1982; pp 28–47.

Boddey, R. M.; Urquiaga, S.; Alves, B. J. R.; Reis, V. Endophytic Nitrogen Fixation in Sugarcane: Present Knowledge and Future Applications. *Plant Soil.* **2003,** *252,* 139–149.

Boysen Jensen, P. Über Wachstumsregulatoren Bei Bakterium. *Biochem. Zeit.* **1931,** *236,* 205–210.

Brown, M. E. Seed and Root Bacterization. *Annu. Rev. Phytopath.* **1974,** *12,* 181–197.

Brown, M. E. Role of *Azotobacter paspali* in Association with *Paspalum notatum. J. Appl. Bacteriol.* **1976,** *40,* 341–348.

Caballero-Mellado, J.; Martínez-Aguilar, L.; Paredes-Valdez, G.; Santos, P. E. Burkholderia unamae sp. nov., an N_2-fixing Rhizospheric and Endophytic Species. *Int J Syst Evol Microbiol.* **2004,** *54* (4), 1165–1172.

Chang, P. C.; Knowles, R. Non-symbiotic Nitrogen Fixation in Some Quebec Soils. *Can. J. Microbiol.* **1965,** *11,* 29–38.

Chelius, M. K.; Triplett, E. W. Immunolocalization of Dinitrogenase Reductase Produced by *Klebsiella pneumoniae* in Association with *Zea mays* L. *Appl. Environ. Microbiol.* **2000,** 66, 783–787.

Chen, W. M.; Moulin, L.; Bontemps, C.; Vandamme, P.; Bena, G.; Boivin-Masson, C. Legume Symbiotic Nitrogen Fixation by Beta-proteobacteria Is Widespread in Nature. *J. Bacteriol.* **2003,** *185,* 7266–7272.

Chen, M.-H.; Sheu, S.-Y.; James, E. K.; Young, C.-C.; Chen, W.-M. Azoarcus Olearius sp. nov., a Nitrogen-fixing Bacterium Isolated from Oil-contaminated Soil. *Int. J. Syst. Evol. Microbiol.* **2013,** *63,* 3755–3761.

Chen, M.; Li, Y.; Li, S.; Tang, L.; Zheng, J.; An, Q. Genomic Identification of Nitrogen-fixing *Klebsiella variicola,* *K. pneumoniae* and *K. quasipneumoniae. J. Basic Microbiol.* **2016,** *56,* 78–84.

Coelho, M. R.; von der Weid, I.; Zahner, V.; Seldin, L. Characterization of Nitrogen-fixing *Paenibacillus* Species by Polymerase Chain Reaction- Restriction Fragment Length Polymorphism Analysis of Part of Genes Encoding 16S rRNA and 23S rRNA and by Multilocus Enzyme Electrophoresis. *FEMS Microbiol. Lett.* **2003**, *222*, 243–250.

Costacurta, A.; Vanderleyden, J. Synthesis of Phytohormones by Plant-associated Bacteria. *Crit. Rev. Microbiol.* **1995**, *21*, 1–18.

Delwiche, C. C.; Wijler, J. Non-symbiotic Nitrogen Fixation in Soil. *Plant Soil,* **1956**, *7*, 113–129.

Dixon, R.A.; Buck, M.; Drummond, M.; Hawkes, T.; Khan, H.; MacFarlane, S.; Merrick, M.; Postgate, J. R. Regulation of the Nitrogen Fixation Genes in Klebsiella Pneumonia: Implications for Genetic Manipulation. *Plant and Soil* **1986**, *90*, 225–233.

Döbereiner, J. Nitrogen-fixing Bacteria of the Genus *Beijerinckia* Derx in the Rhizosphere of Sugarcane. *Plant Soil.* **1961**, *15*, 211–216.

Döbereiner, J. Nitrogen-fixing Bacteria in the Rhizosphere. In *The Biology of Nitrogen Fixation;* Quispel, A. Ed.; North Holland Publishing Company: Amsterdam, The Netherlands, 1974; pp 86–120.

Döbereiner, J. History and New Perspectives of Diazotrophs in Association with Non-leguminous Plants. *Symbiosis,* **1992**, *13*, 1–13.

Döbereiner, J.; Day, J. M. In *Associative Symbioses in Tropical Grasses: Characterization of Microorganisms and Dinitrogen Fixing Sites*, Proceedings of the 1st international symposium on nitrogen fixation; Newton W. E., Nyman, C. J., Eds.; Washington State University Press: Pullman, WA, 1976; pp 518–538.

Döbereiner, J.; Pedrosa, F.O. *Nitrogen-fixing Bacteria in Non-leguminous Crop Plants;* Springer Verlag: Berlin; Science Tech. Publishers: Madison, WI, 1987.

Döbereiner, J.; Day, J. M.; Dart, P. J. Nitrogenase Activity and Oxygen Sensitivity of the *Paspalum notatum-Azotobacter Paspali* Association. *J. Gen. Microbiol.* **1972b**, *71*, 103–116.

Döbereiner, J.; Reis, V. M.; Paula, M. A.; Olivares, F. Endophytic Diazotrophs in Sugarcane, Cereal and Tuber Plants. In *New horizons in nitrogen fixation;* Palacios, R., Mora, J., Newton, W. E. Eds.; Kluwer Academic Publishers: Dordrecht, The Netherlands, 1993; pp 671–676.

Dommelen, A. V.; Vanderleyden, J. Associative Nitrogen Fixation. *Biology of the Nitrogen Cycle*; Bothe, H., Ferguson, S. J., Newton, W. E., Elsevier; pp 180–192.

Dong Y.; Iniguez, A. L.; Triplett, E. W. Quantitative Assessments of the Host Range and Strain Specificity of Endophytic Colonization by *Klebsiella pneumoniae* 342. *Plant Soil* **2003**, *257*: 49–59.

Drewes, K. Über die Assimilation des Luftstickstoffe durch Blaualgen. *Zentralbl. Bakteriol.* **1928**, *76*, 88–101.

Dumas, J. B.; Regnault, H. V.; Payen, A.; Decaine, J.; Peligot, E. M. Rapport Sur un travail de M. Georges Ville dont l'objet est de prouver que le gaz de l'azote de l'air s'assimile aux végétaux. *C R. Acad. Sci. Paris* **1855**, *41*, 757–775.

Eckert, B.; Weber, O. B.; Kirchhof, G.; Halbritter, A.; Stoffels, M.; Hartmann, A. *Azospirillum doebereinerae* sp. *nov.*, a Nitrogen-fixing Bacterium Associated with the C4-grass Miscanthus. Int. *J. Syst. Evol. Microbiol.* **2001**, *51*, 17–26.

Elbeltagy, A.; Nishioka, K.; Sato, T.; Suzuki, H.; Ye, B.; Hamada, T.; Tsuyoshi, I.; Mitsui, H.; Minamisawa, K. Endophytic Colonization and in Planta Nitrogen Fixation by a *Herbaspirillum* sp. Isolated from Wild Rice Species. *Appl. Environ. Microbiol.* **2001**, *67* (11), 5285–5293.

Elmerich, C.; Zimmer, W.; Vieille, C. Associative Nitrogen Fixing Bacteria. In *Biological nitrogen fixation;* Stacey, G., Burris, R. H., Evans, H. J., Eds.; Chapman and Hall*:* New York, NY, 1992; pp 212–258.

Eskew, D. L.; Focht, D. D.; Ting, I. P. Nitrogen Fixation, Denitrification, and Pleomorphic Growth in a Highly Pigmented *Spirillum lipoferum*. *Appl. Environ. Microbiol.* **1977,** *34,* 582–585.

Eskin, N.; Vessey, K.; Tian, L. Research Progress and Perspectives of Nitrogen Fixing Bacterium*, Gluconacetobacter diazotrophicus*, in Monocot Plants. *Int. J. Agron.* **2014,** *2014,* 208–383

Estrada-De Los Santos, P.; Bustillos-Cristales, R.; Caballero-Mellado, J. *Burkholderia*, A Genus Rich in Plant-associated Nitrogen Fixers with Wide Environmental and Geographic Distribution. *Appl. Environ. Microbiol.* **2001,** *67*, 2790–2798.

Fisher K.; Newton, W. E. Nitrogenase Proteins from *Gluconacetobacter diazotrophicus*, A Sugarcane-colonizing Bacterium. *Biochim. Biophys. Acta* **2005,** *1750*, 154–165.

Grimont, F.; Grimont, P. A. D.; Richard, C. The Genus *Klebsiella*. In *The Prokaryotes: An Evolving Electronic Resource for the Microbiological Community;* Dworkin, M. et al., Eds.; Springer-Verlag: New York, 1999. http://link.springer-ny.com/link/service/books/10125/. (accessed Feb 21, 2016).

Gyaneshwar, P.; James, E. K.; Mathan, N.; Reddy, P. M.; Reinhold-Hurek, B.; Ladha, J. K. Endophytic Colonization of Rice by a Diazotrophic Strain of *Serratia marcescens*. *J. Bacteriol.* **2001,** *183* (8), 2634–2645.

Hardy, R. W. F.; Holsten, R. D.; Jackson, E. K.; Burns, R. C. The Acetylene-ethylene Assay for N$_2$ fixation: Laboratory and Field Evaluation. *Plant Physiol.* **1968,** *43,* 1185–1207.

Hardy, R. W. F.; Burns, R. C.; Holsten, R. D. Application of the Acetylene-ethylene Reduction Assay for Measurement of Nitrogen Fixation*. Soil. Biol. Biochem.* **1973,** *5,* 47–81.

Hartmann, A.; Baldani, J. I. The genus *Azospirillum*. In *The Prokaryotes: An Evolving Electronic Resource for the Microbiological Community*; Dworkin, M. et al., Eds.; Springer-Verlag: New York, 2003. http://link.springer-ny.com/link/service/books/10125/. (accessed July 22, 2016).

James, E. K.; Gyaneshwar, P.; Mathan, N.; Barraquio, W. L.; Reddy, P. M.; Iannetta, P. P.; Olivares, F. L.; Ladha, J. K. Infection and Colonization of Rice Seedlings by the Plant Growth-promoting Bacterium *Herbaspirillum seropedicae* Z67. *Mol. Plant Microbe Interact.* **2002,** *15*, 894–906.

Minamisawa, K.; Nishioka, K.; Miyaki, T.; Ye, B.; Miyamoto, T.; You, M.; Saito, A.; Saito, M.; Barraquio, W.; Teaumroong, N.; Sein, T.; Tadashi, T.; Anaerobic Nitrogen-fixing Consortia Consisting of Clostridia Isolated from Gramineous Plants. *Appl. Environ. Microbiol.* **2004,** *70*, 3096–3102.

Klucas, R. V. Associative Nitrogen Fixation in Plants. In *Biology and Biochemistry of Nitrogen Fixation;* Dilworth, M. J.; Glenn, A. R., Eds.; Elsevier: Amsterdam, The Netherlands, 1991; pp 187–198.

Krotzky, A.; Werner, D. Nitrogen Fixation in *Pseudomonas stutzeri*. *Arch. Microbiol.* **1987,** *147,* 48–57.

Ladha, J. K.; Barraquio, W. L.; Watanabe, I. Isolation and Identification of Nitrogen Fixing *Enterobacter cloacae* and *Klebsiella planticola* Associated in Plant Rice. *Can. J. Microbiol.* **1983,** *29*, 1301–1308.

Lalucat, J.; Bennasar, A.; Bosch, R.; Garcia-Valdes, E.; Palleroni, N. J. Biology of *Pseudomonas stutzeri*. *Microbiol. Mol. Biol. Rev.* **2006,** *70*, 510–547.

Liesack, W.; Schnell, N.; Revsbech, N. P. Microbiology of the Flooded Rice Paddies. *FEMS Microbiol. Rev.* **2000**, *24*, 625–645.

Loiret, F. G.; Ortega, E.; Kleiner, D.; Ortega-Rodes, P.; Rodes, R.; Dong, Z. A Putative New Endophytic Nitrogen-fixing Bacterium *Pantoea* sp. from Sugarcane. *J. Appl. Microbiol.* **2004**, *97*, 504–511.

Macura, J. Rapport général. *Ann. Inst. Pasteur (Paris)* **1966**, *111*(Suppl. 3), 9–38.

Magalhaes Cruz, L.; de Souza, E. M.; Weber, O. B.; Baldani, J. I.; Dobereiner, J.; Pedrosa, F.O. 16S Ribosomal DNA Characterization of Nitrogen-fixing Bacteria Isolated from Banana (*Musa* spp.) and Pineapple (*Ananas comosus* (L.) Merril. *Appl. Environ. Microbiol.* **2001**, *67*, 2375–2379.

Marolda, C. L. B.; Hauroder, B.; John, M. A.; Michel, R.; Valvano, M. A. Intracellular Survival and Saprophytic Growth of Isolates from the *Burkholderia cepacia* Complex in Free-living Amoebae. *Microbiology* **1999**, *145*, 1509–1517.

Moulin, L.; Munive, A.; Dreyfus, B.; Boivin-Masson, C. Nodulation of Legumes by Members of the β-Subclass of Proteobacteria. *Nature* **2001**, *411*, 948–950.

Nosto, P.; Bliss, L. C.; Cook, F. D. The Association of Free-living Nitrogenfixing Bacteria with the Roots of High Arctic Graminoids. *Arct. Alp. Res.* **1994**, *26*, 180–186.

Okon, Y. Azospirillum as a Potential Inoculant for Agriculture. *Trends Biotechnol.* **1985**, 3, 223–228.

Okon, Y. *Azospirillum/Plant Associations*; CRC Press Inc.: Boca-Raton, FL, 1994.

Okon, Y.; Labandera-Gonzales, C. A. Agronomic Applications of *Azospirillum*: An Evaluation of 20 years World-wide Field Inoculation. *Soil Biol. Biochem.* **1994**, *26*, 1591–1601.

Pochon, J.; Tchan, Y.-T. *Précis de Microbiologie du sol.* Paris; Masson et Cie Editeurs: France, 1948.

Postgate, J. Microbiology of the Free-living Nitrogen-fixing Bacteria, Excluding Cyanobacteria. In *Current Perspectives in Nitrogen Fixation;* Gibson, A. H.; Newton, W. E., Eds.; Elsevier/North-Holland Biomedical Press: Amsterdam, The Netherlands, 1981; pp 217–228.

Rao, D. L. N. Recent Advances in Biological Nitrogen Fixation in Agricultural Systems. *Proc. Indian Natn. Sci. Acad.* **2014**, *80* (2), 359–378.

Rediers, H.; Vanderleyden, J.; De Mot, R. *Azotobacter vinelandii*: A *Pseudomonas* in Disguise? *Microbiology* **2004**, *150*, 1117–1119.

Reinhold-Hurek, B.; Hurek T.; Life in Grasses: Diazotrophic Endophytes, *Trends Microbiol.* **1998**, *6*, 139–144.

Reinhold-Hurek, B.; Hurek, T.; Gillis, M.; Hoste, B.; Vancanneyt, M.; Kersters, K.; De Ley, J. Azoarcus gen. nov.; Nitrogen-fixing Proteobacteria Associated with Roots of Kallar Grass (Leptochloa fusca (L.) Kunth), and Description of Two Species, *Azoarcus indigens* sp. nov. and *Azoarcus communis* sp. nov. *Int. J. Syst. Bacteriol.* **1993**, *43*, 574–584.

Reinhold-Hurek, B.; Hurek, T. The Genera *Azoarcus, Azovibrio, Azospira,* and *Azonexus*. In *The Prokaryotes: An Evolving Electronic Resource for the Microbiological Community;* Dworkin, M. et al. Eds.; Springer-Verlag: New York, 2004. http://link.springer-ny.com/link/service/books/10125/. (accessed May 22, 2016).

Reinhold-Hurek, B.; Hurek, T. The Genera Azoarcus, Azovibrio, Azospira and Azonexus. In *The Prokaryotes*, 3rd ed.; Dworkin, M.; Falkow, S.; Rosenberg, E.; Schleifer, K. H.; Stackebrandt, E., Eds.; Springer: New York, 2006; Vol. 5; pp 873–891.

Riggs, P. J.; Chelius, M. K.; Iniguez, A. L.; Kaeppler, S. M.; Triplett, E. W. Enhanced Maize Productivity by Inoculation with Diazotrophic Bacteria. *Aust. J. Plant Physiol.* **2001**, *28*: 829–836.

Roger, P. A.; Kulasooriya, S. A. *Blue Green Algae and Rice*; IRRI Publications: Los Baños, The Philippines, 1980.

Roncato-Maccari, L. D. B.; Ramos, H. J. O.; Pedrosa, F. O.; Alquini, Y.; Chubatsu, L. S.; Yates, M. G.; Rigo, L. U.; Steffens, M. B. R.; Souza, E. M. Endophytic *Herbaspirillum seropedicae* Expresses *nif* Genes in Gramineous Plants. FEMS *Microbiol. Ecol.* **2003,** *45,* 39–47.

Roper M. M.; Gupta V. V. S. R. Enhancing Non-symbiotic N₂ Fixation in Agriculture. *Open Agric. J.* **2016,** *10,* 7–27.

Rosenblueth, M.; Martínez-Romero, E. Bacterial Endophytes and Their Interactions with Hosts. *Mol. Plant-Microbe Interac.* **2006,** *19,* 827–837.

Rovira, A. D. Rhizosphere Research, 85 Years of Progress and Frustration. In *The Rhizosphere and plant growth*; Keister, D. L., Cregan, P. B., Eds.; Kluwer Academic Publishers: Dordrecht, The Netherlands, 1991; pp 3–13.

Sabra W.; Zeng, A. P.; Lnsdorf, H.; Deckwer, W. D. Effect of Oxygen on Formation and Structure of *Azotobacter vinelandii* Alginate and its Role in Protecting Nitrogenase. *Appl. Environ. Microb.* **2000,** *66,* 4037–4044.

Sano, Y.; Fujii, T.; Iyama, S.; Hirota, Y.; Komataga, K. Nitrogen Fixation in the Rhizosphere of Cultivated and Wild Rice Strains. *Crop Sci.* **1981,** *21,* 758–761.

Schmid, M.; Baldani, J. I., Hartmann, A. The genus *Herbaspirillum*. In *The Prokaryotes: An Evolving Electronic Resource for the Microbiological Community;* Dworkin, M. et al., Eds.; Springer-Verlag, New York, 2005. http://link.springer-ny.com/link/service/books/10125/. (accessed Mar 31, 2016).

Seldin, L.; van Elsas, J. D.; Penido, E. G. *Bacillus azotofixans* sp. nov., A Nitrogen Fixing Species from Brazilian Soils and Grass Roots. *Int. J. Syst. Bacteriol.* **1984,** *34,* 451–456.

Sevilla M.; Burris R. H.; Gunapala, N.; Kennedy, C. Comparison of Benefit to Sugarcane Plant Growth and 15N₂ Incorporation Following Inoculation of Sterile Plants with *Acetobacter diazotrophicus* Wild-type and Nif- Mutants Strains. *Mol. Plant Microbe Interact.* **2001,** *14,* 358–366.

Somers E.; Vanderleyden J.; Srinivasan M. Rhizosphere Bacterial Signalling: A Love Parade Beneath Our Feet. *Crit. Rev. Microbiol.* **2004,** *30,* 205–240.

Starkey, R. L. Interrelations Between Micro-organisms and Plant Roots in the Rhizosphere. *Bacteriol. Rev.* **1958,** *22,* 154–172.

Steenhoudt, O.; Vanderleyden, J. Azospirillum, a Free-living Nitrogen-fixing Bacterium Closely Associated with Grasses: Genetic, Biochemical and Ecological Aspects. *FEMS Microbiol. Rev.* **2000,** *24* (4), 487–506.

Stewart, W. D. P. Biological and Ecological Aspects of Nitrogen Fixation by Free-living Microorganisms. *Proc. Roy. Soc. B.* **1969,** *172,* 367–388.

Subba Rao, N. S. *Biofertilizers in Agriculture*; Oxford and IBH Publishing Co.: New Delhi, India, 1982.

Tan Z.; Hurek T.; Reinhold-Hurek B. Effect of N-fertilization, Plant Genotype and Environmental Conditions on *nifH* Gene Pools in Roots of Rice. *Environ. Microbiol.* **2003,** *5,* 1009–1015.

Tarrand, J. J.; Krieg, N. R.; Döbereiner, J. A Taxonomic Study of the *Spirillum lipoferum* Group with Description a New Genus, *Azospirillum* gen. nov., and Two Species, *Azospirillum lipoferum* (Beijerinck) comb. nov. and *Azospirillum brasilense* sp. nov. *Can. J. Microbiol.* **1978,** *24,* 967-980.

Tejera, N. A.; Ortega, E.; Rodes, R.; Lluch, C. Influence of Carbon and Nitrogen Sources on Growth, Nitrogenase Activity, and Carbon Metabolism of *Gluconacetobacter diazotrophicus*. *Can. J. Microbiol.* **2004,** 50, 745–750.

van Berkum, P.; and Bohlool, B. B. Evaluation of Nitrogen Fixation by Bacteria in Association with Roots of Tropical Grasses. *Microbiol. Rev.* **1980,** *44,* 491–517.

Vermeiren, H.; Willems, A.; Schoofs, G.; de Mot, R.; Keijers, V.; Hai, W., *et al.* The Rice Inoculant Strain A15 Is a Nitrogen-fixing *Pseudomonas stutzeri* Strain. *System. Appl. Microbiol.* **1999,** *22,* 215–224.

Wagner, S. C. Biological Nitrogen Fixation. *Nat. Educ. Knowl.* **2011,** *3* (10), 15.

Watanabe, A.; Nishigaki, S.; Konishi, C. Effect of Nitrogen-fixing Blue-green Algae on the Growth of Rice Plants. *Nature* **1951,** *168,* 748–749.

Watanbe, I.; Roger, P. A. Nitrogen Fixation in Wetland Rice Field. In *Current Developments in Biological Nitrogen Fixation;* Subba Rao, N. S., Ed.; Oxford and IBH Publishing Co.: New Delhi, India, 1984; pp 237–276.

Wilson, P. W. On the Sources of Nitrogen of Vegetation etc. *Bacteriol. Rev.* **1957,** *21,* 215–226.

Wilson, P. W. First Steps in Biological Nitrogen Fixation. *Proc. Roy. Soc. B (London),* **1969,** *172,* 319–325.

Winogradsky, S. *Microbiologie du sol, problèmes et méthodes, cinquante ans de recherches*; Masson and Cie. Editeurs: Paris, France, 1949.

Xie C.H.; Yokota A. *Azospirillum oryzae* sp. nov., A Nitrogen-fixing Bacterium Isolated from the Roots of the Rice Plant *Oryza sativa*. *Int. J. Syst. Evol. Microbiol.* **2005,** *55*, 1435–1438.

Young, J. P. W. Phylogenetic Classification of Nitrogen-Fixing Organisms. In *Biological nitrogen fixation;* Stacey, G., Burris, R. H., Evans, H. J., Eds.; Chapman and Hall: New York, NY, 1992; pp 43–86.

CHAPTER 11

Cyanobacteria and *Azolla* in Rice Cultivation/Improving Biological N$_2$-Fixation System in Rice

DHARMENDRA KUMAR VERMA[1,*] and SANTOSH KUMAR PATEL[2]

[1]*Department of Soil Science and Agricultural Chemistry, Bihar Agricultural University, Sabour, Bhagalpur 813210, Bihar, India*

[2]*Department of Agronomy, Mata Gujri College, Fatehgarh Sahib 140406, Punjab, India*

[*]*Corresponding author. E-mail: dkvermabhu@gmail.com*

ABSTRACT

In the recent challenges concerning agroecosystem and ecology of the environment, we have required sustainable crop production without any type of pollution. In agricultural field, N$_2$-fixing cyanobacteria and *Azolla* are often the dominant group of microflora and play a major role for enhancing the soil fertility level and sustainable development of the crop. These are algae which are considered very valuable in paddy croplands as biofertilizers. In the presence of water or good moist soil, they develop properly and fix atmospheric nitrogen in plant root. Cyanobacteria have different unique features like high biomass yield, oxygenic photosynthesis, different varieties that were found in water sources (contaminated and polluted water), and production of beneficial byproducts and biofuels for the sustainable agriculture development. *Azolla* is a fern and floats on water. It belongs to the Nostocaceae family. It is found to be outstanding in its symbiotic relation and ability to fix nitrogen in plant at high rate. It is widely found and also cultivated in the Asian regions as a biofertilizer in agricultural crop to fix and enhance the natural nitrogen in soil. Paddy crop is most suitable for the *Azolla* cultivation because water plays an important role for its proper growth and development. *Azolla* is transplanted into the soil before paddy transplanting or after 1 week. It fixed approximate 25–30 kg/ha nitrogen into the soil.

11.1 INTRODUCTION

Out of the atmospheric air content, about 79% N_2 is free in the atmosphere. Biological nitrogen fixation (BNF) is the biochemical process through which atmospheric nitrogen is converted into inorganic forms by several species of bacteria, such as a few actinomycetes and blue–green algae. This process is to reduce elemental nitrogen to ammonia. Plants require nitrogen for the development of proteins and it is a continuous natural process. However, nitrogen present in the atmosphere in the form of N_2, cannot be used by plants. Symbiotic (Rhizobium) and nonsymbiotic (*Azolla*) type of organism have the ability to convert atmospheric nitrogen to a usable form such as ammonia with the association of leguminous and nonleguminous plant.

Cyanobacteria use chlorophyll and sunlight for their growth and development; these are prokaryotes that produce energy. These are generally found in rivers, ponds, springs, wetlands, streams, and lakes and play a major role in the nitrogen, carbon, and oxygen dynamics of many aquatic environments. The present agricultural activity is mostly dependent on the application of chemical fertilizers, pesticides, insecticide, heavy tillage operations, and excess irrigation practices, which have increased food production. The use of excess fertilizer and pesticide results in environmental pollution and health-related problems. It also causes different types of issues like reduces the soil fertility, contaminates the ground water resources, contaminates food and deteriorates its quality, and increases the cost of agricultural production. The blue–green algae *Anabaena* is the nitrogen-fixing partners and its symbiosis relation with each other. *Azolla–Anabaena* is a complex symbiotic in which the endophytic algae. The blue–green algae were found in symbiotic relation with *Azolla* and it is important in paddy cultivation, the *Azolla–Anabaena* symbiosis has always attracted the attention of agriculturist.

In the world, about six species of *Azolla* was identified as *A. filiculoides, A. caroliniana, A. mexicana, A. microphylla, A. nilotica,* and *A. pinnata.* Their taxonomies are based primarily on vegetative and reproductive structures. A unique feature of the *Azolla–Anabaena* symbiosis is the presence of symbiont in the host mega sporocarp during its sexual cycle. This continuous association between *Anabaena azollae* and *Azolla* eliminates the need for a free-living stage of the symbiont.

11.2 CYANOBACTERIA

Rice is the important food crop of the world and more than 40% of the world's population depends on rice as a rich source of calories. In India, about 1990 kg/ha rice is produced against a maximum of 3346 kg/ha in Punjab. The paddy production in Punjab is the highest in the country, whereas it is quite low compared to China (5807 kg/ha) and Japan (6273 kg/ha). Such type of diversity was found in yields due to mismanagement and unavailability of the nitrogenous fertilizer. Nitrogen is present as elemental nitrogen in the atmosphere and constitutes more than 79% of total gases. Nitrogen is a primary nutrient required as a large amount for plant growth. The aforementioned constitutes more than 79% in the atmosphere of the total gases. Still, crop plants cannot utilize nitrogen in elemental form. Cyanobacteria constitute the largest, most diverse, and most widely distributed group of photosynthetic prokaryotes (Stanier et al., 1977). N_2-fixing forms contribute by keeping the soil fertility of natural and cultivated ecosystems. N_2-fixing cyanobacteria also occur in symbiotic relation with the aquatic fern *Azolla*, which has been recycled as green manure for rice since the eleventh century in Vietnam and the 14th century in China (Lumpkin and Plucknett, 1982).

11.3 DISTRIBUTION OF CYANOBACTERIA

An N_2-fixing cyanobacterium is a usual fertility of flooded rice field soils and it maintains the process of BNF by blue–green algae (cyanobacteria). Extents in 396 countries ranged from 10 to 10^7 g^{-1} dry soil (median: 2×10^4; Roger et al., 1987). Cyanobacteria are one of the earliest forms of life on Earth and are known to be the primary colonizers in most uncongenial habitats. Subsequently, the first report of cyanobacteria in rice fields, enormous numbers of forms has been reported from diverse habitats. The paddy field ecosystem provides an environment encouraging for the growth of cyanobacteria with deference to their requirement for light, water, temperature, humidity, and nutrient availability. According to reports, while countrywide cyanobacteria account to about 33% of total algae population, in some of the southern and eastern states this reaches up to 50%.

11.4 FILAMENTOUS, HETEROCYSTOUS CYANOBACTERIA

Through the early 1960s, there was no clear idea as to whether any specific cell/s in filamentous cyanobacteria accomplish N_2 fixation, and the role of heterocysts was not entirely understood (Cox, 1966). The classical research article by Fay et al. (1968) brought a new facet proposing that a specific cell, the heterocyst is the site of N_2 fixation in cyanobacteria. Accordingly, the major adaptation for N_2 fixation under aerobic conditions among the filamentous cyanobacteria performed to be due to structural changes of their cells.

A majority of filamentous cyanobacteria that exhibit aerobic nitrogen fixation possess heterocysts. Compared to vegetative cells heterocysts are generally larger in size, daintier in color, have thicker cell walls have thickenings called polar nodules at the points of supplement to adjacent cells (Fogg et al., 1956). A thick envelope that consists of an outer polysac-charide fibrous layer, a middle homogenous layer, and an inside glycolipid laminated layer. This cell envelope has been shown to be critically important for N_2 fixation and also limits the ingress of O_2 into the heterocysts. The isolated heterocysts contained very little chlorophyll-a and were devoid of phycocyanin and phycoerythrin pigments, which are associated with the O_2 progressing photosystem-II of photosynthesis (Thompson, 2012).

11.5   NONHETEROCYSTOUS CYANOBACTERIA (UNICELLULAR CYANOBACTERIA)

The unicellular cyanobacteria were first report by Wyatt and Silvey in 1969. They reported the function and capability of unicellular cyanobac-teria species creature of nitrogen fixation that of demonstrated nitrogenase activity in *Gloeocapsa* (syn. *Gloeothece, Cyanothece*). In appraisal of 133 species of nonheterocystous strains, Rippkain (1971) found nitrogenase activity in 46 strains and they also proposed the inclusion of all nitrogen-fixing unicellular species into a single genus *Cyanothece*. It has been proposed that N_2 fixation in *Cyanothece* takes place in the shady, supported by ATP and carbon substrates produced during the preceding light period and they are under a quotidian.

11.6   ROLE AND IMPORTANCE OF CYANOBACTERIA IN PADDY CROP

The presently used outmoded agriculture management practices heavily rely on the application of chemical fertilizers and pesticides, and prac-tices like exhaustive tillage and excess irrigation which otherwise lead to ever-increasing cost of cultivation, overexploitation of natural resources like soil and water, and also increase the environmental pollution (Kumar et al., 2012). Cyanobacteria remained N_2 produces in paddy crop through the mineralization after their decease since N excreted during the build-up of bloom is either re-immobilized or lost by NH_3 volatilization. Contingent on the nature of the material (fresh vs. dried), the method of application, and the presence or absence of soil fauna, recovery of cyanobacteria N by rice varies from 13% to 50% (Roger, 1996). Now, there is necessity to adopt such sustainable agricultural practices that are not only eco-friendly but are also cost-effective, and really help us attain long-term food security. Some of the major objectives of sustainable agriculture include the production of safe and healthy foods, conservation of natural resources, economic viability, restoration, and conservation of ecosystem services. It may be suggested that if the four major ecosystems processes, that is, energy flow, water cycle, mineral cycles, and ecosystem dynamics, function together without disturbing the harmony or homeostasis of individual components, can eventually reduce the cost of cultivation. The application of cyanobacteria in management of soil and environment includes the economic benefits (reduced input cost), nutrient cycling, N_2 fixation, bioavailability of phosphorus, water storage and movement, environmental safety, and prevention of pollution and land degradation especially through decreasing the use of agrochemicals, and reprocessing of nutrients and restoration of soil fertility through reclamation.

Following are their possible beneficial effects on rice through the use of cyanobacteria:

- Antagonism with weeds.
- Amplified soil organic matter content.
- Elimination of organic acids that increase P availability to rice.
- Reserve of sulfide injury in sulfate-reduction-prone soils by O_2 production.
- Improved solubilization and mobility of nutrients of limited supply.
- Complexion of heavy metals and xenobiotics to bound their mobility and transport in plants system. Mineralization of simpler organic molecules such as amino acids for direct uptake.
- Safety of plants from pathogenic insects and diseases as biocontrol agents.
- Motivation of the plant growth due to their plant growth promoting attributes.
- Improving the physicochemical situations the soil.

11.7 CYANOBACTERIA AS BIOFERTILIZERS

Cynobacteria play a significant role in agriculture, they reduce the application of the chemical fertilizer. Cyanobacteria fix atmospheric N_2 by the two forms, that is, free-living and symbiotic associations with partners such as water fern *Azolla,* cycads, *Gunnera,* and so forth. In agriculture farmers use as a biofertilizer, soil-based mixed algal cultures having strains of *Tolypothrix, Nostoc, Anabaena, Aulosira, Cylindrospermum,* and so on. They are considered as main nitrogen fixers and hence and can be used as algal biofertilizers for the growth of paddy plant. The application of soil-based algal mixed culture 10–15 kg/ha is generally recommended for paddy.

11.8 CYANOBACTERIA AND ITS SUSTAINABLE MANAGEMENT

Cyanobacteria (blue–green algae) are one of the major constituents of the nitrogen-fixing biomass in paddy fields. N_2 fixation was prevalent among cyanobacteria all species did not fix nitrogen. Cyanobacteria aptitude to fix atmospheric N_2, decay the organic wastes and excesses, detoxify heavy metals, pesticides, and other xenobiotics, catalyze the nutrient cycling, suppress growth of pathogenic microorganisms in soil and water, and also produce some bioactive mixtures such as vitamins, hormones, and enzymes which contribute to plant growth. Many did so both under aerobic and microaerobic conditions while a lesser number fix N_2 only under aerobic conditions. These bioagents can improve the soil quality and plant growth, and reduce the crop production cost by enhancing the good crop management practices such as crop rotation, use of organic manures, least tillage, and the biocontrol of pests and diseases. The use of cyanobacteria in agriculture promises definite beneficial effects on crop output if used properly. Nitrogen-fixing cyanobacteria were rich in paddy fields of India.

TABLE 11.1 Some Important Biofertilizers and Their Nitrogen (N)-Fixing Capacity.

S. no.	Name of biofertilizer	N-fixing capacity (kg/ha)
1.	Azolla	25–30
2.	Cyanobacteria	15–45
3.	Azotobacter	20–30

11.8.1 AZOLLA

Paddy is the most popular and stable food crop in the world. In total 90% of the world's paddy is produced in Asian countries. Paddy plant required shallow and low land, water logging, and irrigated condition for better crop production. The *Azolla* is applied as a biofertilizer in agricultural output due to its capacity to fix nitrogen at high rates. Since of the growing and highly reasonable concern about conservation of the environment and the necessity for arranging renewable, sustainable resources, one of the most relevant of which is agriculture, applying *Azolla* to crops as a natural source of the vital nutrient nitrogen can be very beneficial to the future of our planet. Moreover, the environmental suitability of the practice of *Azolla,* for multitudes of farmers in many parts of the world who cannot afford chemical fertilizers, *Azolla* application can enhance their economic status, increasing yields while diminishing costs. Experiments conducted at Banaras Hindu University, Varanasi showed beneficial effects of *Azolla* in the cultivation of rice (Bhuvaneshwari, 2012). Application of *Azolla* has been found to significantly improve the physical and chemical properties of the soil especially nitrogen, organic matter, and other cations such as magnesium, calcium, and sodium released into the soil (Bhuvaneshwari and Kumar, 2013).

Over the years, scientists have developed some paddy varieties that require less amount of water compared to others but it requires appropriate nutrient management for its growth. Paddy crop can be nourishing yield of development of the crop plants need nutrients for their growth, development, and end of life cycle. Plants uptake nutrients and water through soil. This free-floating freshwater fern, fixes atmospheric nitrogen through the symbiotic association with *A. azollae* that lives inside the dorsal lobes of *Azolla* leaves, potentially supplying a considerable amount of N_2 to the paddy crop. The genus of *Azolla* species is very sensitive to absence of water in aquatic ecosystems such as stagnant waters, ditches, canals, ponds, or paddy fields. These areas may be seasonally shielded by a mat of *Azolla* associated with other free-floating crops species such as *Lemna minor.*

11.8.2 TAXONOMY OF AZOLLA

The *Azolla* floras are delicate, small, and triangular or polygonal in shape (Fig. 11.1). It is free-floating and aquatic but can grow on moist soils as long as the moisture persists in the soil. A leaf consists of a thick dorsal lobe and a thin ventral lobe. The synergetic blue–green alga is constrained to the dorsal lobe (Peters and Mayne, 1974). Epidermis security the surface of the dorsal lobe and the epidermis has vertical rows of single-celled stomata and trachoma's of one or more cells. The ventral lobe which helps in floating due to its convex surface stirring water has a few stomata and trichomes (Eames, 1936).

a   b

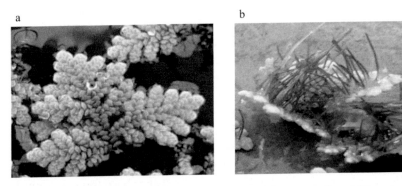

FIGURE 11.1   (See color insert.) *Azolla* (a) dorsal (upper panel) and (b) ventral (lower panel) in water.

Source: Photograph: D.K. Verma.

Azolla macrophyte called a frond varies from 1 to 2.5 cm in length in species such as *A. pinnata* and to 15 cm or more in the largest species, *A. nilotica*. It consists of main rhizome, branching into secondary rhizomes, all of which bear small leaves alternately arranged. Unbranched, adventitious roots hang down into the water from nodes on the ventral surfaces of the rhizomes. The roots absorb nutrients straight from the water, though in very shallow water they may touch the soil, originating nutrients from it. Each leaf consists of two lobes: An aerial dorsal lobe, which is chlorophyllous, and a partially submerged ventral lobe, which is colorless and cup-shaped and provides buoyancy. Each dorsal lobe contains a leaf cavity which houses the symbiotic *A. azollae* (Peters, 1977; Lumpkin and Plucknett, 1980). The inner surface of each leaf cavity is lined with an envelope (Peters, 1976) and covered by a mucilaginous layer of strange composition which is embedded with filaments of *A. azollae* and permeated by multicellular transfer hairs (Shi and Hall, 1988). It has been shown that the mucilage is produced by the symbiont (Robins et al., 1986). The blue–green alga *A. azollae* contains unbranched trichomes containing bead-like, severely pigmented vegetative cells, approximately 6 lam in diameter and 10 lam in length (Van Hove, 1989), and lightly pigmented, intercalary heterocyst, which are marginally larger and have thicker cell walls. According to Hill (1977), heterocyst frequency reaches a maximum of about 30% of the cells in the 15th leaf from the apex. Mature trichomes also contain spores called akinetes. According to Peters (1975), trichomes, on average, consist of 60.9% vegetative cells, 23.1% heterocysts, and 16% akinetes.

11.8.3   MORPHOLOGY OF AZOLLA

Integration of different types of data based on morphology, vegetative characters, and molecular biology to provide a firm footing for the taxonomy of *Azolla* has been suggested (Perreira et al., 2011). Species-specific SCAR primers (sequence characterized amplified region) for the precise identification of different species of *Azolla* is developed recently (Abraham et al., 2013).

11.8.4   GROWTH AND MULTIPLICATION

Azolla found in both temperate and tropical regions and it grows luxuriantly in fresh water ponds, ditches, and paddy fields. The necessity of the soil pH is 7.2 for an ideal growth coupled with a temperature of 32°C. Although studies conducted elsewhere in India have shown that the region-specific selected species for their use as a bio fertilizer, a study was conducted at Centre for Conservation and Utilization of Blue Green Algae, Indian Agricultural Research Institute, New Delhi to compare the biomass production and nitrogen fixation potential of different species of *Azolla* (Arora and Singh, 2003). Based on the experiments, *A. microphylla* was found to perform better and hence selected for mass multiplication. This strain is sustained round the year as it is capable to withstand both high as well as low-temperature environments.

11.8.5   PROPAGATION OF AZOLLA

The dissemination of *Azolla* is generally carried out in soil-based nurseries. However, to maintain the germplasm medium based cultures is sensible. Effects of nutrient position of the medium on the productivity and nitrogen fixation in *Azolla* are reported (Kushari and Taheruzzaman, 1990). *Azolla* production can be carried out in nursery plots, ponds, ditches, canals, concrete tanks, and polythene lined ditches. The field selected for *Azolla* cultivation needs to be thoroughly prepared and leveled equally. Generally, 20 m × 2 m size plots or 20 m × 3 m are made in the field with suitable bunds and irrigation channels with a water depth of at least 10 cm and in each plot water (20 L) is added and inoculated with *Azolla* (8–10 kg). Single super phosphate (100 g) in 2–3 split doses is applied at an interval of 4 days to each plot. Carbofuran (3%) can be applied in the plots (100 g plot^{-1}) with or after a week of inoculation. The 100–150 kg fresh *Azolla* can be harvested from each plot after 15 days from each plot. As per requirement it can be produced in bigger plots. The requirement of inputs applied as per recommendation. Depending on the availability and need of the inoculation, *Azolla* can also be maintained in a nursery in trays or earthen or cemented pots of any dimension. If the production is carried out in a pond or canal fertilizers and insecticides are not applied. Cattle slurry and animal dung for the production and consumption of *Azolla* as biofertilizer for rice has been reported (Singh et al., 1993). The cattle slurry and animal dung are effective as phosphorous fertilizers and application of *Azolla* is found to enhance the C, N, and available P content of the soil.

11.8.6   METHODS OF APPLICATION

The most collective mode of application of *Azolla* in the field is as green manure or as a dual crop along with paddy. In case of application as green manure, *Azolla* collected directly from ponds/ditches is applied in the field. This is grown in nurseries as stated earlier and can also be applied in the field. A thick mat of *Azolla* will be formed after application in about 2–3 weeks' time and can be incorporated in the soil. Paddy can also be transplanted in the field successively. Single super phosphate (25–50 kg ha^{-1}) is applied in split doses. After analyzing the soil P-status the dosage of the same can be reduced. Cattle dung or slurry may also be used as an alternative of single superphosphate. In case of pest infestation or attack, pest control measures have to be undertaken. *Azolla* application by this mode contributes around 20–40 kg N ha^{-1}.

In double cropping, *Azolla* is grown along with rice and each crop of *Azolla* contributes on an average 30 kg N ha^{-1}. After 7–10 days of transplantation, fresh inoculums of *Azolla* are applied in the field at the rate of 0.50–1.0 ton ha^{-1}. Single super phosphate is applied at the rate of 20 kg ha^{-1} in split doses. In about 15–20 days time a thick mat of *Azolla* is formed. *Azolla* thus incorporated decomposes in about 8–10 days time and release the fixed nitrogen. Another crop of *Azolla* can be raised in a similar way during the crop cycle of rice. *Azolla* production technology is simple and not very exclusive and at the same time, it is very efficient in terms of biomass accumulation and nitrogen fixation. The rice growing season is also conducive for the growth of *Azolla* plants. The dual application does not have any negative effect on the rice crop.

11.8.7   *EFFECT OF AZOLLA IN CROP (RICE) PRODUCTIVITY*

The best appropriate crop for the application of *Azolla* is low-lying rice, and then both plants require similar environmental conditions most particularly, a flooded territory and they grow together compatibly. *Azolla* may be applied on rice either as a monocrop or an intercrop. As an intercrop it is usually inoculated into the field just after the transplanting of paddy and, after a period of growth, may be incorporated into the mud or permitted to die naturally by fungal rot or light malnourishment (Lumpkin, 1987a). However, a mixture of applications is usually recommended. In particular, value to rice can be maximized by first rising *Azolla* as a monocrop, integrating it, and then, after transplanting the rice, applying it again as an intercrop with one or more successive incorporations. For many centuries, *Azolla* has been used to successfully raise rice yield in Vietnam and southern China (Fogg et al., 1973; Watanabe and Liu, 1992). *Azolla* grows in low-lying rice fields in Indonesia, India, China, Vietnam, Thailand, Senegal, and other tropical countries. Many trials have validated the effectiveness of *Azolla* as a biofertilizer on rice, though the magnitude of the benefit varies significantly according to the climate, the method of application, the species of *Azolla* used, and many other factors. In California, Peters (1978) reported that the use of *Azolla* improved rice yields by 112% over unfertilized controls when applied as a monocrop during the fallow season, by 23% when applied as an intercrop with paddy, and by 216% when applied both as a monocrop and an intercrop. Tung and Shen (1985) found that *Azolla* grown with rice seemed to suppress the growth of rice in the early stages, perhaps due to the competition. However, at maturity, though rice grown with *Azolla* did not have better height or tiller number, straw and grain yield were higher, mostly grain yield which was 42–55% higher than the controls where no *Azolla* was applied. Sisworo et al. (1990) found that *Azolla* was equally as effective as urea on rice when both were applied at the rate of 30 kg N ha^{-1} at transplanting and at maximum tillering stage. In Tanzania, Wagner (1996) applied *A. niloticain* with various trials as an intercrop, with one incorporation, obtaining an increase of 19–103% in grain yield and of 1–23% in straw yield. Significant increase was also found in rice height and the number of tillers per hill. In contrast with the application of chemical nitrogenous fertilizers, the benefits brought about by green manures such as *Azolla* and *Sesbania* sp. are long-term. Satapathy (1993) reported that increasing soil fertility was first culturing *Azolla* as a monocrop, incorporating it before transplanting, and subsequently culturing it as an intercrop with two incorporations. Organic matter in the form of green manure and biofertilizers has been found useful instead of the inorganic fertilizers (Nayaket al., 2004; Bhuvaneshwari et al., 2012).

Azolla can supply more than half of the required nitrogen to the rice crop and besides providing nitrogen it is beneficial in wetland rice fields for bringing a number of changes which include preventing the rise in pH, reducing water temperature, curbing NH_3 volatilization, suppressing weeds and mosquito proliferation (Pabby et al., 2004). Van Hove (1989) found that *Azolla* improves soil structure when incorporated because of its high productivity, which supplies large quantities of organic matter. Rice grain analysis for Iron, Zinc, Manganese and Copper contents showed a significant increase in these essential ion contents in the treatments having two or more organic amendment added altogether over control. Similar results have been obtained by Bhattacharya and Chakraborty (2005). Use of *Azolla* as basal or dual or basal plus dual influenced the rice crop positively where the use of fern as basal plus dual was superior and served the nitrogen requirement of rice. There was marked increase in plant height, a number of effective tillers, dry mass, and nitrogen content of rice plants with the use of *Azolla* and N-fertilizers alone and other combinations. The use of *Azolla* also increased organic matter and potassium contents of the soil. *Azolla* has long been used as both green manure for rice and as fodder for poultry and livestock in China and Vietnam (Kamalasamana et al., 2005).

Wagner (1996) conducted experiments in which *Azolla nilotica* was applied to rice grown in outdoor cement tanks with soil collected from paddy, closely simulating natural rice paddy ecosystems. Thus, when the fern decomposes, it acts indirectly as a potassium fertilizer (Van Hove, 1989). Many studies have been conducted that compare the effectiveness of *Azolla* with other types of biofertilizers. Singh et al. (1992) found that rice grain yield was highest with the application of *Azolla* + 120 kg N ha^{-1} (5.01 t ha^{-1}), fb blue–green algae + *Azolla* + 60 kg N ha^{-1} (4.62 t ha^{-1}) and, lastly, by 120 kg N ha^{-1} (4.61 t ha^{-1}). Watanabe and Ventura (1992) at Philippines, reported that Azolla gave higher rice grain yield (3.3–3.9 t ha^{-1}) during the dry season than did *Sesbaniaor* urea (both 1.8–2.5 t ha^{-1}). During the wet season, there was no difference among *Azolla*, *Sesbania*, and urea treatments, although each of these yielded 1.7 t ha^{-1} more than the control. Singh and Singh (1990) found that the addition of phosphorus fertilizer increased growth and nitrogen fixation of both blue–green algae and *Azolla* biofertilizers. The effect of *Azolla* on rice was similar to that of 30 kg N ha^{-1}, while the effect of blue–green algae was comparatively less. At Central Rice Research Institute, Cuttack, India field trials demonstrated that the use of *Azolla* enhanced crop yield and crop N uptake significantly as compared to treatments without *Azolla* (Manna and Singh, 1989). Bhuvaneshwari and Kumar (2013) at Varanasi carried out in order to investigate the effect of *Azolla* on the physicochemical properties of the soil in which the soil was treated with *Azolla* at 0, 10, 30, 60, and 90 g/kg. The treated soils were nursed in the dark at 25°C for 60 days in the laboratory. The application of *Azolla* improved the soil pH, organic matter, nitrogen, phosphorus, potassium, calcium, magnesium, and sodium with the level of the *Azolla*. There was a decrease in soil bulk density but increased soil porosity.

Azolla plants in huge numbers sequestered significant quantities of atmospheric CO_2 and converted it directly into the biomass of *Azolla* (Speelman et al., 2009). Rowndel and Sonbeer (2017) reported that *Azolla* is known for efficiently sequestering CO_2 without the requirement of soil-based nitrogen source. This makes *Azolla* an important green manuring crop in rice-based farming systems. Extensive studies have established that elevated CO_2 profoundly affects growth and development, physiology, and productivity in rice plant. This study was showed to conclude the effect of the elevated CO_2 on rice supplemented by *Azolla* as a nitrogen source. Paddy was grown under different CO_2 concentrations and *Azolla* treatments in environment-controlled chambers from June to November 2016. Rice plants responded mostly positively to elevate CO_2.

11.9 CONCLUSION

Azolla considerably boosts the quality of nitrogenous fertilizers existing to rising rice and it has been used for thousands of years as a green biofertilizer for nitrogen to enhance crop production. As nitrogen is one of the most limiting factors in rice cultivation. It is strongly affecting the crop yield. Thus, symbiotic of *Azolla*, that is, Anabaena is used as a sequester atmosphere nitrogen fixating agent. *Azolla* is extensively grown in the puddled rice field to raise rice production more than 50%. Several other important characters *Azolla* also has like they minimize the ammonia volatilization that occurs following the submission of nitrogen fertilizers by 20–50%. This is owed to the circumstance that the *Azolla* cover reduces light penetration into the flood water, thus, hampering the rise of pH which normally inspires ammonia volatilization in an *Azolla*-free paddy field.

KEYWORDS

- **bioavailability**
- **cyanobacteria**
- **nitrogen fertilizer**
- **soil structure**
- **algal cultures**

REFERENCES

Abraham, G; Pandey, N; Mishra, V; Chaudhary, A. A; Ahmad, A; Singh, R; Singh, P. K. Development of SCAR-Based Molecular Markers for the Identification of Different Species of *Azolla*. *Indian J. Biotechnol.* **2013**, *12* (4), 12489–12492.

Arora, A; Singh, P. K. Comparison of Biomass Productivity and Nitrogen Fixing Potential of *Azolla* spp. *Biomass Bioenergy* **2003**, *24*, 175–178.

Bhattacharya, P; Chakraborty, G. Current Status of Organic Farming in India and Other Countries. *Indian J. Fert.* **2005**, *1* (9) 111–123.

Bhuvaneshwari, K. Beneficial Effects of Blue Green Algae and *Azolla* in Rice Culture. *Environ. Conserv.* **2012**, *13* (1&2), 1–5.

Bhuvaneshwari, K; Kumar, A. Agronomic Potential of the Association *Azolla-Anabaena*. *Sci. Res. Rep.* **2013**, *3* (1), 78–82.

Bhuvaneshwari, K; Singh, P. K; Response of Nitrogen-Fixing Water Fern Azolla Biofertilization to Rice Crop. *3 Biotechnol.* **2015**, *5*, 523–529.

Bocchi, S.; Malgioglio, A. Azolla-Anabaena as a Biofertilizer for Rice Paddy Fields in the Po Valley, a Temperate Rice Area in Northern Italy. *Hindawi Publ. Corp. Int. J. Agron.* **2010**, 1–5. https://doi.org/10.1155/2010/152158.

Eames, A. J. *Morphology of Vascular Plants*; McGraw Hill Book Co: New York, 1936.

Fogg, G. E. The Comparative Physiology and Biochemistry of the Blue-Green Algae. *Bacteriol. Rev.* **1956**, *20*, 148–165.

Fogg, G. E; Stewart, W. D. P; Fay, P; Walsby, A. E. *The Blue-Green Algae*; Academic Press: London, 1973.

Hill, D. J. The Pattern of Development of *Anabaena* in the *Azolla-Anabaena* Symbiosis. *Planta* **1975**, *122*, 179–184.

Kamalasamana, P; Premalatha, Rajamony S. *Azolla:* a Sustainable Feed for Live Stock. *LEISA India* **2005,** *21* (3), 26–27.

Kumar, M.; Bauddh, K.; Sainger, M.; Sainger, P. A.; Singh, J. S.; Singh, R. Increase in Growth, Productivity and Nutritional Status of Rice (*Oryza sativa* L. cv. Basmati) and Enrichment in Soil Fertility Applied with an Organic Matrix Entrapped Urea. *J. Crop Sci. Biotechnol.* **2012,** *15,* 137–144. https://doi.org/10.1007/s12892-012-0024-z.

Kushari, D. P; Taheruzzaman, Q. Multiplication of *Azolla pinnata* in Primary Production Units. In *National Symposiumon Cyanobacterial Nitrogen Fixation*; Kaushik, B. D., Ed.; NFBGAC, IARI: New Delhi, India, 1990; pp 353–358.

Lumpkin, T. A. Collection, Maintenance, and Cultivation of *Azolla.* In *Symbiotic Nitrogen Fixation Technology*; Elan, G. H., Ed.; Marcel Dekker: New York, 1987; pp 55–94.

Lumpkin, T. A; Plucknett, D. L. *Azolla:* Botany, Physiology and Use as a Green Manure *Econ. Bot.* **1980,** *34,* 111–153.

Lumpkin, T. A.; Plucknett, D. L. *Azolla as a Green Manure, Use and Management in Crop Production*; Westview Press, Boulder: CO, USA, 1982.

Manna, A. B; Singh, P. K. Growth and Acetylene Reduction Activity of *Azolla caroliniana* Willd. As Influenced by Split Applications of Urea Fertilizer and Their Response on Rice Yield. *Fert. Res.* **1989,** *18,* 189–199.

Nayak, S; Prasanna, R; Pabby, A; Dominic, T. K; Singh, P. K. Effect of Urea, Blue-Green Algae and *Azolla* on Nitrogen Fixing and Chlorophyll Accumulation in Soil Under Rice. *Biol. Fert. Soils* **2004,** *40,* 67–72.

Pabby, A; Prasanna, R; Singh, P. Biological Significance of Azolla and Its Utilization in Agriculture. *Proc. Indian Natl. Sci. Acad.* **2004,** *70,* 299–333.

Perreira, A. L; Martins, M; Oliviera, M. M; Carrapico, F. Morphological and Genetic Diversity of the Family Azollaceae Inferred from Vegetative Characters and RAPD Markers. *Plant Sys. Evol.* **2011,** *297,* 213–226.

Peters, G. A. Blue-Green Algae and Algal Associations. *Bio Sci.* **1978,** *28,* 580–585.

Peters, G. A. In *Studies on the Azolla-Anabaena symbiosis.* Proceedings of the International Symposium on Nitrogen Fixation; Newton, W. E., Nyman, C. J.; State University Press: Washington, 1976; pp 592–610.

Peters, G. A. The *Azolla-Anabaena azollae* Relationship III. Studies on Metabolic Capacities and a Further Characterization of the Symbiont. *Arch. Microbiol.* **1975,** *103,* 113–122.

Peters, G. A. The *Azolla-Anabaena azollae* Symbiosis. In Genetic Engineering for Nitrogen Fixation; Hollaender, A., Ed.; Plenum Press: New York, 1977; pp 231–258.

Peters, G. A; Mayne, B. C. The *Azolla-Anabaena azollae* Relationship I. Initial Characterization of the Association. *Plant Physiol.* **1974,** *53,* 813–819.

Robins, R. J., D. O.; Hall, D. J.; Shi, R. J. T.; Rhodes, M. J. C. Mucilage Acts to Adhere Cyanobacteria and Cultured Plant Ceils to Biological and Inert Surfaces. *FEMS Microbiol. Lett.* **1986,** *34,* 155–160.

Roger, P.A.; Ladha, J. K. Biological N_2-fixation in Wetland Ricefields, Estimation, and Contribution to Nitrogen Balance. *Plant Soil* **1992,** *141,* 41–55.

Rowndel, K; Sonbeer, C. Response of Rice Crop Under Elevated Carbon Dioxide Concentration with Azolla Treatment, *Int. J. Agric. Sci. Res.* **2017,** *7* (2), 541–546.

Satapathy, K. B. Effect of Different Plant Spacing Pattern on the Growth of *Azolla* and Rice. *Indian J. Plant Physiol.* **1993,** *36,* 98–102.

Saunders, R. M. K; Fowler, K. A morphological Taxonomic Revision of *Azolla* Lain. Section *Rhizosperma* (Mey.) Mett. (Azollaceae). *Bet. L. Linn. Soc.* **1992,** *109,* 329–357.

Shi, D. J; Hall, D. O. The *Azolla-Anabaena* Association: Historical Perspective, Symbiosis and Energy Metabolism. *Bet. Rev.* (Lancaster) **1988**, *54*, 353–386.

Singh P. K; Singh, D. P; Satapathy, K. B. Use of Cattle Slurry and Other Organic Manures for *Azolla* Production and Its Utilization as Biofertilizer for Rice. In *Utilization of Biogas Slurry*; Consortium on Rural Technology: New Delhi, 1993; pp 135–142.

Sisworo, E. L; Eskew, D. L; Sisworo, W. H; Rasjid, H; Kadarusman, H; Solahuddin, S; Soepardi, G. Studies on the Availability *of Azolla* Nitrogen and Urea Nitrogen for Rice Growthusing Nitrogen. *Plant Soil* **1990**, *128*, 209–220.

Speelman, E. N.; Van Kempen, M. M.; Barke, J.; Brinkhuis, H.; Reichart, G. J.; Smolders, A. J.; Roelofs, J. G.; Sangiori, F.; de Leeuw, J. W; Lotter, A. F.; Sinninghe Damste, J. S. The Eocene Arctic Azolla Bloom: Environmental Conditions, Productivity and Carbon Drawdown. *Geobiology* **2009**, *7* (2), 155–170.

Stanier, R. Y.; Cohen-Bazire, G. Phototrophic Prokaryotes, the Cyanobacteria. *Ann. Rev. Microbiol.* **1977**, *31*, 225–274.

Tan, B. C. P; Payawal, I; Watanabe, N; Laedan, Ramirez, C. Modern Taxonomy of *Azolla:* A Review. *Philipp. Agric.* **1986**, *69*, 491–512.

Thompson, A. W.; Foster, R. A.; Krupke, A.; Carter, B. J.; Musat, N.; Vaulot, D.; Kuypers, M. M. M.; Zehr, J. P. Unicellular Cyanobacterium Symbiotic with a Single Celled Eukaryotic Alga. *Science* **2012**, *337*, 1546–1550.

Tung, H. F; Shen, R. C. Studies of the *Azollapinnata-Anabaena azollae* Symbiosis: Concurrent Growth *of Azolla* with Flee. *Aquatic Bot.* **1985**, *22*, 145–152.

Van Hove, C. *Azolla and Its Multiple Uses with Emphasis on Africa.* FAO: Rome, 1989.

Vijayan, D.; Ray J. G. Ecology and Diversity of Cyanobacteria in *Kuttanadu* Paddy Wetlands, Kerala, India. *Am. J. Plant Sci.* **2015**, *6*, 2924–2938.

Wagner, G. M. The *Utricularia-Cyanophyta and Azolla-Anabaena* Associations: Their Ecology, Nitrogen Fixation Rates, and Effects as Biofertilizers on Rice. Ph.D. Thesis, University of Dares Salaam, 1996.

Watanabe, I; Liu, C. C. Improving Nitrogen-Fixing Systems and Integrating Them Into Sustainable Rice Farming. *Plant. Soil* **1992**, *141*, 57–67.

CHAPTER 12

Phosphorus Availability to Crops Through Phosphate-Solubilizing Microorganisms

B. D. KAUSHIK[1,*], DEEPAK KUMAR[1], ISRAR AHMAD[2], PRAMILA PANDEY[3], and K. N. SINGH[4]

[1]*Research and Development Unit, Shri Ram Solvent Extractions Pvt. Ltd. Jaspur, Udham Singh Nagar 244712, Uttarakhand, India*

[2]*Division of Crop Improvement and Biotechnology, ICAR-CISH, Rehmankhera, Lucknow 226016, Uttar Pradesh, India*

[3]*Krishi Vigyan Kendra, Haidergarh, Barabanki 227301, Uttar Pradesh, India*

[4]*Department of Plant Molecular Biology and Genetic Engineering, Narendra Dev University of Agriculture and Technology, Kumarganj, Faizabad 224229, Uttar Pradesh, India*

[*]Corresponding author. E-mail: bdkaushik@hotmail.com

ABSTRACT

Phosphorus is an essential macronutrient needed by plants and plays a vital role in maintaining soil fertility and plant nutrition. Phosphorus in the soil is present in the organic as well as inorganic combinations. The plant available forms of P in marginal soil can be increased with plant growth promoting rhizobacteria (PGPR) or application of vesicular arbuscular mycorrhiza (VAM) fungi. Globally, phosphorus is mined from geological sediments and most of the mined P is used to meet the critical need of crop plants for agronomic productivity. However, recovery of P by plants is comparatively low and major amount of applied P is fixed in the soil creating a need for addition of P fertilizer. Microorganisms play a crucial role in mineral mobilization in the rhizosphere soil. Insoluble phosphates are converted into available forms by phosphate-solubilizing microorganisms (PSMs) via the process of acidification, chelation, exchange reactions and production of organic acid. Several bacterial strains (*Pseudomonas, Bacillus, Rhizobium,* and *Enterobacter*) and fungal strains (*Aspergillus* and *Penicillium*) have so far been recognized as powerful phosphate solubilizers. However, plant inoculations with PSMs during field studies had inconsistent effect on plant growth and crop yields due to variations in soil, crop, and environmental factors affecting the survival and colonization of the rhizosphere. Increasing availability of soil P through microbial inoculation will require identification of the most appropriate strains, preparation of effective formulations, and introduction of efficient agronomic managements to ensure delivery and survival of inoculants and associated improvement of P efficiency.

12.1 INTRODUCTION

Phosphorus (P) is the second most essential macronutrient that limits plant growth, accordingly required for enhancing growth of plants as well as crop production. In contrast to nitrogen, there is no atmospheric source for this important macronutrient that can be made biologically available (Ezawa et al., 2002). Phosphorus plays a significant role in plant metabolism and is important for the functioning of key enzymes that regulate the metabolic pathways (Theodorou and Plaxton, 1993). Next to nitrogen (N), it is the second major inorganic nutrient in term of quantitative requirement for both crop plants and microorganisms. It is a major constituent of nucleic acids in all living systems essential in the accumulation and release of energy during cellular metabolism. It is also a constituent of adenosine triphosphate (ATP) which is often termed as "energy currency" of the plant cell. An adequate supply of P encourages root development, stalk and stem strength, flower and seed formation, crop maturity and production, N-fixation in legumes, etc. Therefore, P deficiency is considered to be the most important chemical factor that restricts plant growth because of its vital role in physiological and biochemical functions of the plant.

The utilization efficiency of P fertilizer throughout the world is less than 25% and concentration of bio-available P in soil is very low reaching the level of 1.0 mg kg^{-1} soil (Isherword, 1998; Goldstein, 2000). In India, on an average total P content in soil is 0.05% of which only 0.1% from the total P forms is available to plants, rest of the P forms become insoluble salt (Bhattacharya and Jain, 1996). The use of chemical fertilizers to circumvent the P deficiency in soil also becomes unavailable to the plants due to the rapid immobilization of P soon after application (Yadav and Tarafdar, 2011). This immobilization is due to the high reactivity of phosphate anions with divalent and trivalent cations (Fe^{3+}, Al^{3+}, Ca^{2+}) in soil solution and precipitates out soon from the soil solution (Sample et al., 1980). The reaction products of soluble P fertilizers with soil cations are variscite, taranakite, apatite, Al- and Fe-phosphates. In these forms, P is highly insoluble and unavailable to plants (Rdresh et al., 2004).

Soil microbes play a key role in soil P dynamics, mobility, and subsequent availability of phosphate to agricultural soils and plants (Richardson, 2001). Inorganic forms of P are solubilized by a group of heterotrophic microorganisms, particularly bacteria, and fungi excreting different organic acids which dissolve phosphatic minerals and/or chelate cationic partners of the P ions directly, releasing P into solution (He et al., 2002). Phosphate-solubilizing microorganisms (PSMs) are being used as biofertilizers since 1950s (Krasilinikov, 1957). Release of P by microorganisms from insoluble and fixed/adsorbed forms is an important aspect regarding P availability in soils. Soil bacteria are capable of transforming unavailable form of soil P to a form available to the plant. Microbial biomass protects soluble P and prevents it from adsorption or soil fixation (Khan and Joergensen, 2009). Microbial community influences soil fertility through enhancing the P availability to plants by mineralizing organic P in soil and by solubilizing unavailable phosphates (Chen et al., 2006, Pradhan and Sukla, 2005). These microorganisms in the presence of labile carbon serve as a sink for P by rapidly immobilizing it (Bünemann, 2004). Subsequently, microorganisms become a source of P to plants upon its release from their cells. The microorganisms reduce P fertilizer application by 50% without any significant reduction of crop yield (Jilani et al., 2007; Yazdani et al., 2009). It also infers that phosphorus-solubilizing microorganism inoculants/biofertilizers hold great prospects for sustaining crop production with optimized P fertilization.

12.2 PHOSPHORUS PROBLEM IN SOIL

After N, P is one of the most essential macronutrients presents in diverse forms in the soil. The major form of P in soil can broadly be categorized as insoluble inorganic phosphorus and insoluble organic phosphorus. However, small fraction (1 ppm or 0.1%) of phosphorus is readily available to plants. The plant roots take up several forms of phosphorus, out of which the greatest part is absorbed in the forms of $H_2PO_4^-$ and HPO_4^{2-} depending upon soil pH (Mahidi et al., 2011). As a consequence of continuous application of high doses of phosphatic fertilizers, most of the agricultural soils generally contain large reserves of accumulated phosphorus (Richardson, 2004) which are not readily available to plants. After application, a large portion of soluble inorganic phosphate applied to soil as chemical fertilizer is rapidly immobilized and becomes unavailable to plants. Thus, P fertilizer shows residual nutrient values as these accumulated P release soluble P in due course of time. When the fertilizer or manure phosphate comes in contact with the soil, a series of different reactions begins which make the phosphate less soluble and less available to plants. However, the degree of fixation and precipitation of phosphorus in soil is highly dependent upon the different parameters, such as soil conditions, namely pH, moisture content, temperature, and the minerals already present in the soil. In the case of acidic soils, free oxides and hydroxides of aluminum and iron play a key role in fixing phosphorus, while in alkaline soils it is fixed by calcium (Toro, 2007) and to some extent magnesium.

12.2.1 PHOSPHATE SOLUBILIZATION

Microorganisms with phosphate-solubilizing abilities have proved to be an economically sound alternative to the more expensive superphosphates and possess a greater agronomic utility under diverse soil and agro-climatic conditions (Khan et al., 2007; Xiao et al., 2009). Inorganic forms of phosphorus are solubilized by a group of heterotrophic microorganisms excreting organic acids that dissolve phosphatic materials and/or chelate cationic partners of the phosphorus ions that is, PO_4^{3-} directly, releasing phosphorus into solution (He et al., 2002). Microorganisms that are involved in the solubilization of insoluble phosphorus are bacteria, fungi, actinomycetes, and arbuscular mycorrhizal (AM) fungi, respectively (Khan et al., 2007; Wani et al., 2007a; Xiao et al., 2009).

Besides proving available form of phosphorus to the plants, the phosphate-solubilizing bacteria (PSB) also facilitate the growth of plants by stimulating the efficiency of nitrogen-fixation, accelerating the accessibility of other trace elements (Mittal et al., 2008), including siderophores (Wani et al., 2007) and antibiotics (Lipping et al., 2008), and providing protection to plants against soilborne pathogens (Hamdali et al., 2008). Accordingly, these microbial communities as biofertilizers when used separately (Chen et al., 2008) or in combination with other rhizosphere microbes (Wani et al., 2007) have shown substantial effects on plants.

12.3 PHOSPHATE-SOLUBILIZING MICROORGANISMS

Several species of fungi and bacteria, commonly known as PSMs help the plants in mobilizing insoluble forms of phosphate. PSMs improve the solubilization of phosphates fixed in soil resulting in their uptake by plants and higher crop yields, and are used as biofertilizers. PSMs are ubiquitous whose numbers vary from soil to soil. PSB constitute 1–50% and fungi 0.1–0.5% of the total respective population in soil (Chen et al., 2006).

Several bacteria, fungi including mycorrhizal fungi and actinomycetes are highly capable of converting insoluble phosphate into soluble inorganic phosphate ion. Among the soil bacterial communities, ectorhizospheric strains from *Pseudomonas* and *Bacilli*, and endosymbiotic rhizobia have been described as effective phosphate solubilizers. Strains from bacterial genera *Pseudomonas*, *Bacillus*, *Rhizobium,* and *Enterobacter* along with *Penicillium* and *Aspergillus* fungi are the most powerful P solubilizers (Sashidhar and Podile, 2010) *acillus megaterium, B. circulans, B. subtilis, B. polymyxa, B. sircalmous, Pseudomonas striata,* and *Enterobacter* could be referred as the most important strains (Table 12.1).

TABLE 12.1 Phosphate-Solubilizing Microorganisms.

S. no.	Microorganisms	Genera/species
1.	Bacteria	*Bacillus megaterium, B. circulans, B. subtilis, B. polymyxa, B. sircalmous, Pseudomonas striata, Enterobacter* sp., *Beggiatoa, Thiomargarita, Leifsonia xyli* FeGl 02, *Burkholderia cenocepacia* FeSu 01, *Burkholderia caribensis* FeGl, *Burkholderia ferrariae* FeGl 01. sp.
2.	Actinobacteria	*Actinobispora yunnanensis, Actinomodura citrea, Microtetrospora astidiosa, Micromonospora echinospora, Sacchromonospora viridis, Saccharopolyspora hirsute, Streptomyces albus, Streptoverticillium album, Streptomyces cyaneus, Thermonospora mesophila*
3.	Fungi	Belonging to genera *Aspergillus* (*A. awamori*) and *Penicillium* (*P. bilaii*).
4.	Mycorrhiza	Belonging to genera *Glomus, Funneliformis, Rhizophagus, Sclerocystis, Claroideoglomus, Gigaspora, Scutellospora, Racocetra, Acaulospora, Entrophospora, Pacispora, Diversispora, Otospora, Paraglomus, Geosiphon, Ambispora, Archaeospora* sp.
5.	Endophytes	Bacteria: *Achromobacter, Acinetobacter, Enterobacter cloacae, Pantoea agglomerans, Pseudomonas* sp. Fungi: *Piriformospora indica*, dark septate endophytes belonging to Ascomycota

12.4   MICROBIAL STRATEGY FOR RELEASE OF UNAVAILABLE FORMS OF PHOSPHORUS

PSMs employ different strategies to convert unavailable forms of phosphate into available forms. In case of most bacteria, production of organic acids is produced to the dissolution of mineral phosphates. Goldstein (1996) reported direct oxidation of glucose to gluconic acid (GA) as the foremost mechanism in Gram-negative bacteria for mineral phosphate solubilization (MPS). Organic acids which released by the microorganisms act as good chelators of divalent cations of Ca^{2+} coupled with the release of phosphates from insoluble complexes (Sashidhar and Podile, 2010). These organic acids may also form soluble complexes with metal ions co-complexed with insoluble P, thereby releasing the P moiety. Many of the PSMs induce a reduction in the pH of the medium either by H^+ extrusion or by secretion of various organic acids (Illmer and Schinner, 1995; Sashidhar and Podile, 2010). In oxidative phosphorylation process which occurred in microorganisms, proton transport from the cytoplasm to the outer surfaces of the microbes may take place in exchange for a cation (especially ammonium) or with the help of ATPase (ABC transporter) located in the cell membrane and uses the energy from ATP hydrolysis, respectively (Sashidhar and Podile, 2010).

12.5   BIOCHEMICAL MECHANISM OF PHOSPHORUS RELEASE

Microbial PSB produced 2-keto gluconic acid from direct oxidation of glucose and play an important role in weathering and solubilization of phosphates in soil. *Erwinia herbicola* and *Pseudomonas cepacia* produced the gluconic (pKa ~3.4) and 2-keto gluconic acids (pKa ~2.4) by direct oxidation of glucose in highly efficient solubilization of rock phosphate (Sashidhar and Podile, 2010). Some bacteria undertake the direct oxidation pathway to such elevated levels that externally added glucose is quantitatively converted to gluconic acid at concentration of 1 mol L^{-1} or higher. Gram-negative bacteria are more efficient to release of several organic acids into the extracellular medium at dissolving mineral phosphates when compared to Gram-positive bacteria (Sashidhar and Podile, 2010). In case of thermotolerant acetic acid producing *Acetobacter* and *Gluconobacter* also have the direct oxidation pathway with thermotolerant glucose dehydrogenase (GDH) and solubilize mineral phosphate.

Except gluconic acid, several other organic acids such as lactic, acetic, malic, succinic, tartaric, oxalic, and citric acids are also produced (Table 12.2). Weak organic acids, namely acetate, malate, and succinate are present in

TABLE 12.2 Individual or Gene Clusters from Different P-solubilizing Bacteria Cloned and Expressed in *E. coli*.

Gene/function	Source	Host	Mineral P solubilized	Organic acid	References
pqqED genes	*Rahnella aquatilis*	*E. coli*	HAP	GA	Kim et al. (1998)
pqqABCDEF genes	*Enterobacter intermedium*	*E. coli* DH5α	HAP	GA	Kim et al. (2003)
PQQ biosynthesis	*Serratia marcescens*	*E. coli*	TCP	GA	Krishnaraj and Goldstein (2001)
pqqE	*Erwinia herbicola*	*Azospirillum* sp.	TCP	GA	Vikram et al., (2007)
Ppts-gcd, P gnlA-gcd	*E. coli*	*Azotobacter vinelandii*	TCP	GA	Sashidhar and Podile (2010)
gabY Putative PQQ transporter	*Pseudomonas cepacia*	*E. coli* HB101	GA	–	Babu-Khan et al. (1995)
gltA/citrate synthase	*E. coli* K12	*Pseudomonas fluorescens* ATCC 13525	DCP	Citric acid	Buch et al. (2009)
gad/gluconate dehydrogenase	*P. putida* KT2440	*E. asburiae* PSI3	RP	GA and 2KG	Kumar et al. (2013)
Unknown	*Synechocystis* PCC 6803	*E. coli* DH5 α	RP	Unknown	Gyaneshwar et al. (2002)
Unknown	*Erwinia herbicola*	*E. coli* HB101	TCP	GA	Goldstein and Liu (1996)

the rhizosphere as fermentation products of rhizobacteria. *Pseudomonas* sp. is known to preferentially utilize these weak organic acids over glucose, sucrose, and fructose (Görke and Stülke, 2008). Similar catabolite repression of glucose metabolism is also found in root nodule bacteria (Mandal and Chakrabartty, 1993). In case of many fluorescent pseudomonads are also known to solubilize mineral phosphates by secretion of gluconic acid (Patel et al., 2011). Presence of malate and succinate has been shown to repress MPS phenotype in fluorescent pseudomonads (Selvakumar, 2009). Similarly, MPS phenotype mediated by oxalic acid in *Klebsiella pneumonia* is repressed by the presence of succinate (Rajput et al., 2013) However, the MPS phenotype to be very effective under field conditions would require higher amount of stronger acids. GDH, the key enzyme, is a quinoprotein that converts glucose to gluconic acid, using the redox cofactor 2,7,9-tricarboxyl- 1 H-pyrrolo [2,3-f] quinine-4.5-dione (PQQ; Duine at al., 1979). GDH requires PQQ and has binding sites for Mg^{2+} (in vitro), Ca^{2+} (in vivo), ubiquinone, and the substrate glucose.

There are two types of GDH, GDH A and GDH B, based on their localization within the cell. GDH B is soluble (s-GDH) and is reported only from *Acinetobacter calcoaceticus* whereas GDH A is more widespread and is a membrane-bound enzyme (m-GDH) (Sashidhar and Podile, 2010). PQQ-dependent GDH is present in several bacterial species such as *P. aeruginosa.* This bacteria *P. aeruginosa* produces the cofactor PQQ, others such as *E. coli* are unable to produce PQQ and require external supply of PQQ for GDH activity. The GDH has been characterized from various bacteria are about 88 kDa monomeric proteins having primary structure similar to each other, differing marginally in some of the properties such as substrate specificity (Yamada et al., 2003). The GDH enzyme has an N terminal hydrophobic domain (residues 1–150) consisting of five transmembrane segments ensuring a strong fastening of the protein to the membrane, and a large conserved PQQ-binding C-terminal domain with the catalytic activity.

12.6 GENOMICS AND PROTEOMICS OF PHOSPHORUS-SOLUBILIZING ENZYME

The MPS characteristic is induced or repressed by the different levels of inorganic phosphate available in the environment. Many individual or *pqq* gene clusters from PSB were cloned and expressed in *E. coli* which enabled them to produce GA leading to MPS phenotype (Table 12.2). Expression of bacterial *gabY* (396 bp) gene in *E. coli* JM109 induced MPS ability and GA production.

This *gabY* gene sequence was similar to membrane-bound histidine permease component which may be a PQQ transporter (Babu-Khan et al., 1995). In case of *Serratia marcenses,* genomic DNA fragment induces GA production in *E. coli* but it does not exhibits any homology with either *gdh* or *pqq* genes (Krishnaraj and Goldstein, 2001). The DNA fragment did not validate GA secretion in either *E. coil gdh* mutant or in other *pqq* producing strain. Therefore, it was postulated that the gene product could be an inducer of GA production. Similarly, other reports showed that genes that are not directly involved in *gdh* or *pqq* biosynthesis could induce MPS ability. Genomic DNA fragment of *Enterobacter agglomerans* showed MPS ability in *E. coli* JM109 without any significant change in pH (Kim et al., 1997). *R. aquatilis* MPS genes exhibited higher GA production and hydroxyapatite dissolution in *E. coli* compared to native strain (Kim et al., 1998). The overexpression of *P. putida* KT 2440 gluconate degydrogenase (*gad*) operon in *E. asburiae* PSI3 improved MPS phenotype by secretion of 2-ketogluconic acid along with GA3.

12.6.1 FACTORS INFLUENCING THE EFFICACY OF PHOSPHORUS RELEASE BY MICROBES IN SOILS

The field studies of phosphate-solubilizing microbe inoculations with plant had inconsistent effect on plant growth and crop yields (Gyaneshwar et al., 2002; Richardson and Simpson, 2011) as this varied due to variations in climate, soil, and environmental conditions prevailing over the area where these inoculations are made influencing the survival and colonization of the rhizosphere. Effectiveness of introduced microorganism in the field condition requires maintenance of minimal number in soils.

12.6.1.1 ROOT COLONIZATION ABILITY

Root colonization is a major factor influencing the success of inoculants. Majority of the microbial population found in the soil is associated with the plant roots where their population can reach up to 10^9 to 10^{12} per gram of soil (Metting, 1992) leading to biomass equivalent to 500 kg ha^{-1} (Bhattacharyya et al., 2013). Abundance of microbes in the rhizosphere is due to secretion of high amount of root exudates. Root-associated bacterial diversity and their growth and activity vary in response to the biotic and abiotic environment of the rhizosphere of the particular host plant (Berg and Smalla, 2009; Bais et al., 2006).

12.6.1.2   SOIL PROPERTIES

Soil properties are differing in term of texture, particle, and pore size of soil. The efficacy of the microorganism and different behavioral pattern in bacteria in soil depend upon pore size distribution when released in different texture soil (Bashan, 1995). It is studied that in loamy sand and silt loam soils showed that survival of inoculated *P. fluorescens* was better in finer-texture soil, that is, silt loam than in the sandy soil (Vassilev et al., 2012).

12.6.1.3   ABIOTIC STRESSES

Phosphate-solubilizing microorganisms need to survive under a variety of abiotic stresses in field conditions which depend on the agro-climatic conditions along with seasonal changes. It is generally observed that the PSMs are mostly found in greater numbers in plant rhizosphere than nonrhizosphere soil, as the root exudates provide readily metabolizable carbon and nitrogen compounds for the growth of heterotrophic organisms. The bacteria and fungi form an associative symbiosis with the root system to get substrates from the roots and in return provide mineral nutrients which otherwise normally could not be absorbed by the roots including phosphorus. Many PSMs have been isolated according to their potential ability under different stress conditions (Chang and Yang, 2009). Efficiency of PSMs differs significantly with high or low temperatures. PSMs in tropical countries need to tolerate 35–45°C temperatures, while temperate regions require cold tolerance. Many *Bacillus*, *Streptomyces*, and *Aspergillus* strains showed very good P solubilization ability at 50°C, which could facilitate composting is known as thermophilc microorganisms (Chaiharn and Lumyong, 2008). The bacteria *Acinetobacter* CR 1.8 could grow up to 25% NaCl, between 25°C and 55°C and at pH 5–9, but maximum solubilization of tricalcium phosphate and aluminum phosphate was obtained at neutral pH, and 37°C (Das et al., 2003).

12.6.1.4   SUBSTRATE AVAILABILITY

The amount brought into solution by heterotrophs varies with the carbohydrate oxidized, and the transformation generally proceeds only if the carbonaceous substrate is converted to organic acids. Plant roots secrete complex mixture of different organic compounds, namely organic acid, amino acid, and sugars. Sugars secreted are glucose, fructose, maltose, ribose, sucrose, arabinose, mannose, galactose, and glucuronic acid. These organic carbon sources are used by phosphate-solubilizing microorganisms as substrate during inorganic phosphate solubilization respectively (Lugtenberg & Kamilova, 2009, van Elsas at al., 1986). Organic acids secreted in rhizosphere are malate and citrate, releasing P from soils and among amino acids, histidine, proline, valine, alanine, and glycine are present, respectively.

12.7 EFFECT OF PSMS ON CROP PRODUCTION

The quality of phosphate rocks and its particle size also affect solubilization. Generally, high-grade rock phosphates, with fine particle sizes easily dissolve than low-grade phosphates of coarse sized particles. Phosphate rock minerals are often too insoluble to provide sufficient P for crop uptake. Use of PSMs can increase crop yields up to 70% depending on the soil and environmental conditions. Phosphorus bio-inoculants have often been projected as vital constituents of integrated nutrient management approaches with specific interest in their potential of solubilizing sparingly available P in order to increase its availability for the crops (Sashidhar and Podile, 2010; Peix et al., 2001). Combined inoculation of VAM fungi and PSB give better uptake of both native P from the soil and P coming from the phosphatic rock (Cabello et al., 2005). Higher crop yields result from solubilization of fixed soil P and applied phosphates by PSB (Zaidi et al., 2003; Mohammadi, 2011). Microorganisms with phosphate solubilizing potential increase the availability of soluble phosphate and enhance the plant growth by improving biological nitrogen fixation (Table 12.3). The analysis results from Canada indicated that inoculation with *P. bilaii* increased in P uptake and yield of spring wheat in only 5 of 47 trials, despite 33 of the trials showing responses to P fertilizer (Rodrıguez and Fraga et al., 1999). Inoculation experiments conducted with phosphobacterin and other PSMs in various crops like oat, wheat, potatoes, peas, tomatoes, and tobacco showed an average 10–15% increase in crop yields in about 30% of the experiments conducted. Several experiments conducted in legume and nonlegume crops by coinoculation of PSM with diazotrophs have shown synergistic effects with regard to increase in population of both bacteria and significant increase in crop yields in comparison to single inoculation.

12.8 CONCLUSION

Solubilization of accumulated phosphorus by microbial inoculants plays a pivotal role in nutrition of higher plants, particularly where the availability

TABLE 12.3 Effect of PSMs on Crop Growth and Yield.

Name of crop	PSMs	Crop growth parameter	Increase in yield (%)	Nutrient uptake	References
Wheat	*Aspergillus awamori*	Grain yield	57.25	P, N	Giddens et al. (1982)
Wheat	AM fungi	Shoot dry matter yield	52.0	P	Bais et al. (2006)
Wheat	AM fungi	Seed grain spike number	19.0	P	Bais et al. (2006)
Wheat	AM fungi	Grain yield	26.0	p	Bais et al. (2006)
Ground nut	*Aspergillus niger, Penicillium notatum*	Dry weight of plant	105	P, N	Bashan (1995)
Ground nut	*A. niger, P. notatum*	Protein content	57.5	P, N	Bashan (1995)
Ground nut	*A. niger, P. notatum*	Oil content	29.5	P, N	Bashan (1995)
Maize	*Penicillium bilaii* and *Penicillium* spp.	Maize yield	20–23	P	Postma et al. (1988)
Maize	*Pseudomonas* sp. CDB 35	Grain yield	85	p	Metting (1992)
Rice	*Bacillus coagulans*	Grain yield	7593.7 kg/h	P	Lugtenberg and Kamilova (2009)
Rice	PSB	Grain yield	1–11	P	Chang and Yang (2009)
Sun flower	*Bacillus* M-13	Seed yield	15	P	Das et al. (2003)
Sun flower	*Bacillus* M-13	Oil yield	24.7	P	Das et al. (2003)
Canola	PSB and *Thiobacillus* sp.	Total yield	60	P	Singh et al. (2004)
Canola	PSB and *Thiobacillus* sp.	Oil yield	39	P	Singh et al. (2004)
Chickpea	*Ps. jessenii* and *Mesorhizobium ciceri*	Seed yield	52	P	Zhu et al. (2011)
Soybean	*Bradyrhizobium japonicum* and PSB	Grain yield	33	P, N	Vassilev et al. (2012)
Sugar beet and barley	*Bacillus* sp.	Yield	20.7–25.9	P	Trivedi and Sa (2008)
Sugarcane	*Bacillus megatherium*	Cane yield	12.6	P	Bhattacharyya et al. (2013)
Kalmegh	*Trichoderma harzianum* + AM (*Glomus mosseae*)	Overall plant growth	49.8	P	Berg and Smalla (2009)
Wheat and bean	*Penicillium bilaii* with VAM fungi	Grain yield	18	P	Chaiharn and Lumyong (2008)

of labile P is low to very low in soil solution for their ready uptake. Several microbes are reported to involve in P-solubilization through soil acidification by producing organic acids such as gluconic acids. Some of these microorganisms can benefit plant growth by several different mechanisms, such as enhancing nitrogen fixation, IAA, and ethylene production, etc. Metal ion chelation and ion exchange reactions in growth environment are endorsed to the phosphate solubilization by PSMs. An extensive understanding of the rhizosphere ecology, multi-trophic interactions, and molecular processes associated with the augmentation of P availability in soils will help in the selection and management of inoculants across diverse cropping systems with consistently better performance. The efficiency of P-solubilizing microorganisms also depends on their extracting potentials in the soil environments and capacities to compete, colonize, survive, and proliferate in the rhizosphere. Comparative genomic, transcriptomic, and proteomic characterization of microbial genotypes with or without P-solubilizing capabilities and analyses of gene expression under extracting conditions requiring P solubilization for growth have potential to identify novel isolates, their functioning, and ultimately their use under field condition in an ecologically more sustainable manner. The limited application of PSMs are due to the isolation, identification, and selection of PSMs have not as yet been successfully commercialized, although PSMs are abundant in many of the soils.

KEYWORDS

- **adenosine triphosphate**
- **glucose dehydrogenase**
- **phosphate solubilization**
- **phosphate-solubilizing microorganisms**
- **organic acid**

REFERENCES

Babu-Khan, S.; Yeo, T. C.; Martin, W. L.; Duron, M. R.; Rogers, R. D.; Goldstein, A. H. Cloning of a Mineral Phosphate Solubilizing Gene from *Pseudomonas cepacia*. *Appl. Environ. Microbiol.* **1995**, *61*, 972–978.

Bais, H. P.; Weir, T. L.; Perry, L. G.; Gilroy, S.; Vivanco, J. M. The Role of Root Exudates in Rhizosphere Interactions with Plants and Other Organisms. *Ann. Rev. Plant Biol.* **2006**, *57*, 233–266.

Bashan, Y. et al. Survival of *Azorhizobium brasilense* in the Bulk Soil and Rhizosphere of 23 Soil Types. *Appl. Environ. Microbiol.* **1995**, *61*, 1938–1945.

Berg, G.; Smalla, K. Plant Species Versus Soil Type: Which Factors Influence the Structure and Function of the Microbial Communities in the Rhizosphere. *FEMS Microbiol. Ecol.* **2009**, *68*, 1–13.

Bhattacharya, P.; Jain, R. K. Phosphorus Solubilizing Biofertilizers in the Whirlpool of Rock Phosphate Challenges and Opportunities. *Fert. News.* **2000**, *45*, 45–49.

Bhattacharyya, P.; Das, S.; Adhya, T. K. Root Exudates of Rice Cultivars Affect Rhizospheric Phosphorus Dynamics in Soils with Different Phosphorus Status. *Commun. Soil Sci. Plant Anal.* **2013**, *44*, 1643–1658.

Buch, A. D.; Archana, G.; Naresh Kumar, G. Enhanced Citric Acid Biosynthesis in *Pseudomonas fluorescens* ATCC 13525 by Overexpression of the *Escherichia coli* Citrate Synthase Gene. *Microbiology* **2009**, *155*, 2620–2629.

Bünemann, E. K.; Bossio, D. A.; Smithson, P. C.; Frossard, E.; Oberson, A. Microbial Community Composition and Substrate Use in a Highly Weathered Soil as Affected by Crop Rotation and P Fertilization. *Soil Biol. Biochem.* **2004**, *36*, 889–901.

Cabello, M.; Irrazabal, G.; Bucsinszky, A. M.; Saparrat, M.; Schalamuck, S. Effect of an Arbuscular Mycorrhizal Fungus, G. Mosseae and a Rock-phosphate-solubilizing Fungus, P. Thomii in Mentha Piperita Growth in a Soilless Medium. *J. Basic Microbiol.* **2005**, *45*, 182–189.

Chaiharn, M.; Lumyong, S. Phosphate Solubilization Potential and Stress Tolerance of Rhizobacteria from Rice Soil in Northern Thailand. *World J. Microbiol. Biotechnol.* **2008**, *25*, 305–314.

Chang, C.-H.; Yang, S. S. Thermo-tolerant Phosphate Solubilizing Microbes for Multi-functional Biofertilizer Preparation. *Biores. Technol.* **2009**, *100*, 1648–1658.

Chen, Y. P.; Rekha, P. D.; Arun, A. B.; Shen, F. T.; Lai, W. A.; Young, C. C. Phosphate Solubilizing Bacteria from Subtropical Soil and their Tricalcium Phosphate Solubilizing Abilities. *Appl. Soil Ecol.* **2006**, *34*, 33–41.

Chen, Z.; Ma, S.; Liu, L. L. Studies on Phosphorus Solubilizing Activity of a Strain of Phospho Bacteria Isolated from Chestnut Type Soil in China. *Bioresour. Technol.* **2008**, *99*, 6702–6707.

Chen, Y. P.; Rekha, P. D.; Arunshen, A. B.; Lai, W. A.; Young, C. C. Phosphate Solubilizing Bacteria from Subtropical Soil and Their Tricalcium Phosphate Solubilizing Abilities. *Appl. Soil Ecol.* **2006**, *34*, 33–41.

Das, K.; Katiyar, V.; Goel, R. P Solubilization Potential of Plant Growth Promoting *Pseudomonas* Mutants at Low Temperature. *Microbiol. Res.* **2003**, *158*, 359–362.

Duine, J. A.; Frank, J.; van Zeeland, J. K. Glucose Dehydrogenase from *Acinetobacter calcoaceticus*: A 'Quinoprotein'. *FEBS Lett.* **1979**, *108*, 7321–7326.

Ezawa, T.; Smith, S. E.; Smith, F. A. P Metabolism and Transport in AM Fungi. *Plant Soil* **2002**, *244*, 221–230.

Giddens, J. E.; Dunigan, E. P.; Weaver, R. W. Legume Inoculation in the Southeastern USA. *South Coop. Ser. Bull.* **1982**, *283*, 1–38.

Goldstein, A. H. Involvement of the Quinoprotein Glucose Dehydrogenase in the Solubilization of Exogenous Phosphates by Gram negative Bacteria. In *Phosphate in*

Microorganisms: Cellular and Molecular Biology; Torriani-Gorini, A.; Yagil, E.; Silver, S., Eds.; ASM Press: Washington, DC, 1996, pp 197–203.

Goldstein, A. H. Bioprocessing of Rock Phosphate Ore: Essential Technical Considerations for the Development of a Successful Commercial Technology, Proc. 4th Int. Fert. Assoc. Tech. Conf. IFA, Paris, 2000; p 220.

Görke, B.; Stülke, J.; Carbon Catabolite Repression in Bacteria: Many Ways to Make the Most Out of Nutrients. *Nat. Rev. Microbiol.* **2008,** *6,* 613–624.

Guiñazú, L. B.; Andrés, J. A.; MFDel, P.; Pistorio, M.; Rosas, S. B. Response of Alfalfa (*Medicago sativa* L.) to Single and Mixed Inoculation with Phosphate-solubilizing Bacteria and *Sinorhizobium meliloti. Biol. Fertil. Soils* **2010,** *46,* 185–190.

Gyaneshwar, P.; Naresh Kumar, G.; Parekh, L. J.; Poole, P. S. Role of Soil Microorganisms in Improving P Nutrition of Plants. *Plant Soil* **2002,** *245,* 83–94.

Hamdali, H.; Hafidi, M.; Virolle, M. J.; Ouhdouch, Y. Rock Phosphate Solubilizing Actinimycetes: Screening for Plant Growth Promoting Activities. *World J. Microbiol. Biotechnol.* **2008,** *24,* 2565–2575.

He, Z. L.; Bian, W.; Zhu, J. Screening and Identification of Microorganisms Capable of Utilizing Phosphate Adsorbed by Goethite. *Comm. Soil Sci. Plant Anal.* **2002,** *33,* 647–663.

Illmer, P.; Schinner, F. Solubilization of Inorganic Calcium Phosphates—Solubilization Mechanisms. *Soil Biol. Biochem.* **1995,** *27,* 265–270.

Isherwood, K. F. Fertilizer Use and Environment. In *Proc. Symp. Plant Nutrition Management for Sustainable Agricultural Growth;* Ahmed, N., Hamid, A., Eds.; NFDC: Islamabad, 1998, pp 57–76.

Jilani, G.; Akram, A.; Ali, R. M.; Hafeez, F. Y.; Shamsi, I. H.; Chaudhry, A. N.; Chaudhry, A. G. Enhancing Crop Growth, Nutrients Availability, Economics and Beneficial Rhi-zosphere Microflora through Organic and Biofertilizers. *Ann. Microbiol.* **2007,** *57,* 177–183.

Khan, M. S.; Zaidi, A.; Wani, P. Role of Phosphate Solubilizing Microorganisms in Sustainable Agriculture: A Review. *Agron. Sustain. Develop.* **2007,** *27,* 29–43.

Khan, K. S.; Joergensen, R. G. Changes in Micro-bial Biomass and P Fractions in Biogenic Household Waste Compost Amended with Inorganic P Fertilizers. *Bioresour. Technol.* **2009,** *100,* 303–309.

Kim, C. H.; Han, S. H.; Kim, K. Y.; Cho, B. H.; Kim, Y. H.; Koo, B. S.; Kim, C. Y. Cloning and Expression of Pyrroloquinoline Quinone (PQQ) Genes from a Phosphate-solubilizing Bacterium *Enterobacter intermedium. Curr. Microbiol.* **2003,** *47,* 457–461.

Kim, K. Y.; Jordan, D.; Krishnan, H. B. Expression of Genes from *Rahnella aquatilis* that are Necessary for Mineral Phosphate Solubilizaiton in *Escherichia coli. FEMS Microbiol. Lett.* **1998,** *159,* 121–127.

Kim, K. Y.; McDonald, G. A.; Jordan, D. Solubilization of Hydroxyapatite by *Enterobacter agglomerans* and Cloned *Escherichia coli* in Culture Mediu. *Biol. Fertil. Soils* **1997,** *24,* 347–352.

Krasilinikov, N. A. On the Role of Soil Micro-organism in Plant Nutrition. *Microbiologiya* **1957,** *26,* 659–672.

Krishnaraj, P. U.; Goldstein, A. H.; Cloning of a *Serratia marcescens* DNA Fragment that Induces Quinoprotein Glucose Dehydrogenase-mediated Gluconic Acid Production in *Escherichia coli* in the Presence of a Stationary Phase *Serratia marcescens. FEMS Microbiol. Lett.* **2001,** *205,* 215–220.

Kumar, C.; Yadav, K.; Archana, G.; Kumar, G. N. 2-Ketoglutaric Acid Secretion by Incorporation of *Pseudomonas putida* KT 2440 Gluconate Dehydrogenase (gad) Operon in

Enterobacter Asburiae PS13 Improves Mineral Phosphate Solubilization. *Curr. Microbiol.* **2013,** *67,* 388–394.

Lipping, Y.; Jiatao, X.; Daohong, J.; Yanping, F.; Guoqing, L.; Fangcan, L. Antifungal Substances Produced by *Penicillium oxalicum* Strain PY-1-potential Antibiotics Against Plant Pathogenic Fungi. *World J. Microbiol. Biotechnol.* **2008,** *24,* 909–915.

Lugtenberg, B. J. J.; Kamilova, F. Plant Growth Promoting Rhizobacteria. *Ann. Rev. Microbiol.* **2009,** *63,* 541–556.

Mahidi, S. S.; Hassan, G. I.; Hussain, A., Faisul-ur-Rasool. Phosphorus Availability Issue-Its Fixation and Role of Phosphate Solubilizing Bacteria in Phosphate Solubilization-Case Study. *Res. J. Agric. Sci.* **2011,** *2,* 174–179.

Mandal, N. C.; Chakrabartty, P. K. Succinate-mediated Catabolite Repression of Enzymes of Glucose Metabolism in Rootnodule Bacteria. *Curr. Microbiol.* **1993,** *26,* 247–251.

Metting Jr, F. B. Structure and Physiological Ecology of Soil Microbial Communities. In *Soil Microbial Ecology*; Metting, F. B. Je., Ed.; Marcel Dekker: New York, 1992; pp 3–25.

Mittal, V., Singh, O., Nayyar, H., Kaur, J., Tewari, R. Stimulatory Effect of Phosphate Solubilizing Fungal Strains (*Aspergillus awamori and Penicillium citrinum*) on the Yield of Chickpea (*Cicer arietinum* L. cv. GPF2). *Soil Biol. Biochem.* **2008,** *40,* 718–727.

Mohammadi, K. *Soil, Plant and Microbe Interaction*; Lambert Academic Publication: Saarbrücken, Germany, 2011, p 120.

Mukherjee, P. K.; Horwitz, B. A.; Herrera-Estrella, A.; Schmoll, M.; Kenerley, C. M. *Trichoderma* Research in the Genome Era. *Ann. Rev. Phytopathol.* **2013,** *51,* 105–129.

Patel, D. K.; Murawala, P.; Archana, G.; Naresh K. Repression of Mineral Phosphate Solubilizing Phenotype in the Presence of Weak Organic Acids in Plant Growth Promoting Fluorescent Pseudomonads. *Bioresour. Technol.* **2011,** *102,* 3055–3061.

Peix, A.; Rivas-Boyero, A. A.; Mateos, P. F.; Rodriguez-Barrueco, C.; Martínez-Molina, E.; Velazquez, E. Growth Promotion of Chickpea and Barley by a Phosphate Solubilizing Strain of *Mesorhizobium mediterraneum* under Growth Chamber Conditions. *Soil Biol. Biochem.* **2001,** *33,* 103–110.

Postma, J.; van Elsas, J. D.; Govaert, J. N.; van Veen, J. A. The Dynamics of Rhizobium Leguminosarum Biovar Trifolii Introduced into Soil as Determined by Immunofluorescence and Selecting Plating Techniques. *FEMS Microbiol. Lett.* **1988,** *53,* 251–260.

Pradhan, N.; Sukla, L. B. Solubilization of Inorganic Phosphate by Fungi Isolated from Agriculture Soil. *African J. Biotechnol.* **2005,** *5,* 850–854.

Rajput, M. S.; Naresh Kumar, G; Rajkumar, S. S. Repression of Oxalic Acid-mediated Mineral Phosphate Solubilization in Rhizospheric Isolates of *Klebsiella pneumoniae* by Succinate. *Arch. Microbiol.* **2013,** *195,* 81–88.

Rdresh, D. L.; Shivprakash, M. D.; Prasad, R. D. Effect of Combined Application of Rhizobium, Phosphate Solubilizing Bacterium and Trichoderma spp. On Growth, Nutrient Uptake and Yield of Chickpea (Cicer aritenium L.). *Appl. Soil Ecol.* **2004,** *28,* 139–146.

Richardson, A. E. *Soil Microorganisms and Phosphorus Availability: Management in Sustainable Farming Systems*; CSIRO: Melbourne, Australia, 2004, pp 50–62.

Richardson, A. E.; Simpson, R. J. Soil Microorganisms Mediating Phosphorus Availability. *Plant Physiol.* **2011,** *156,* 989–996.

Richardson, A. E. Prospects for Using Soil Microorganisms to Improve the Acquisition of Phosphorus by Plants. *Aust. J. Plant Physiol.* **2001,** *28,* 897–906.

Rodrıguez, H.; Fraga, R.; Phosphate Solubilizing Bacteria and their Role in Plant Growth Promotion. *Biotechnol. Adv.* **1999**, *17*, 319–339.

Sample, E. C.; Soper, R. J.; Raez, G. J. Reactions of Phosphate Fertilizers in Soils. In *The Role of Phosphate in Agriculture*; American Society of Agronomy: Madison, *Stoneville,* Mississippi, 1980, pp 263–310.

Sashidhar, B.; Podile, A. R. Mineral Phosphate Solubilization by Rhizosphere Bacteria and Scope for Manipulation of the Direct Oxidation Pathway Involving Glucose Dehydrogenase. *J. Appl. Microbiol.* **2010**, *109*, 1–12.

Selvakumar, G.; Joshi, P.; Nazim, S.; Mishra, P. K.; Bisht, J. K.; Gupta, H. S. Phosphate Solubilization and Growth Promotion by *Pseudomonas fragi* CS11RH1 (MTCC 8984), A Psychrotolerant Bacterium Isolated from a High Altitude Himalayan Rhizosphere. *Biologia* **2009**, *64*, 239–245.

Singh, B. K.; Millard, P.; Whiteley, A. S.; Colin Murrell, J. Unravelling Rhizosphere–microbial Interactions: Opportunities and Limitations. *Trends Microbiol.* **2004**, *12*, 386–393.

Theodorou, M. E.; Plaxton, W. C. Metabolic Adaptation of Plant Respiration to Nutritional Phosphate Deprivation. *Plant Physiol.* **1993**, *101*, 339–344.

Toro, M. Phosphate Solubilizing Microorganisms in the Rhizosphere of Native Plants from Tropical Savannas: An Adaptive Strategy to Acid Soils? In *Developments in Plant and Soil Sciences;* Velaquez, C.; Rodriguez-Barrueco, E., Eds.; Springer: The Netherlands, 2007; pp 249–252.

Trivedi, P.; Sa, T. *Pseudomonas corrugate* (NRRL B-30409) Mutants Increased Phosphate Solubilization, Organic Acid Production and Plant Growth at Lower Temperatures. *Curr. Microbiol.* **2008**, *56*, 140–144.

van Elsas, J. D.; Dijkstra, A. F.; Govaert, J. M.; van Veen, J. A. Survival of *Pseudomonas fluorescens* and *Bacillus subtilis* Introduced into Two Soils of Different Texture in Field Microplots. *FEMS Microbiol. Ecol.* **1986**, *38*, 151–160.

Vassilev, N.; Eichler-Löbermann, B.; Vassileva, M. Stresstolerant P-solubilizing Microorganisms. *Appl. Microbiol. Biotechnol.* **2012**, *95*, 851–859.

Vikram, A.; Alagawadi, A. R.; Krishnaraj, P. U.; Mahesh Kumar, K. S. Transconjugation Studies in *Azospirillum* sp. Negative to Mineral Phosphate Solubilization. *World J. Microbiol. Biotechnol.* **2007**, *23*, 1333–1337.

Wani, P. A.; Khan, M. S.; Zaidi, A. Co-inoculation of Nitrogen Fixing and Phosphate Solubilizing Bacteria to Promote Growth, Yield and Nutrient Uptake in Chickpea. *Acta Agron. Hung.* **2007a**, *55*, 315–323.

Wani, P. A.; Khan, M. S.; Zaidi, A. Synergistic Effects of the Inoculation with Nitrogen Fixing and Phosphate Solubilizing Rhizobacteria on the Performance of Field Grown Chickpea. *J. Plant Nutr. Soil Sci.* **2007b**, *170*, 283–287.

Xiao, C. Q.; Chi, R. A.; He, H.; Qiu, G. Z.; Wang, D. Z.; Zhang, W. X. Isolation of Phosphate Solubilizing Fungi from Phosphate Mines and their Effect on Wheat Seedling Growth. *Appl. Biochem. Biotechnol.* **2009**, *159*, 330–342.

Yadav, B. K.; Tarafdar, J. C. *Penicillium purpurogenum*, Unique P Mobilizers in Arid Agro-ecosystems. *Arid Land Res. Manag.* **2011**, *25* (1), 87–99.

Yamada, M.; Elias, M. D.; Matsushita, K.; Migita, C. T.; Adachi, O. *Escherichia coli* PQQ-containing Quinoprotein Glucose Dehydrogenase: Improvement of EDTA Tolerance, Thermal Stability and Substrate Specificity. *Protein Eng.* **2003**, *12*, 63–70.

Yazdani, M.; Bahmanyar, M. A.; Pirdashti, H.; Es-maili, M. A. Effect of Phosphate Solubilization Microorganisms (PSMs) and Plant Growth Promoting Rhizobacteria (PGPR) on Yield and Yield Components of Corn (Zea mays L.). *Proc. World Acad. Science Eng. Technol.* **2009**, *37*, 90–92.

Zaidi, A.; Khan, M. S.; Amil, M. Interactive Effect of Rhizotrophic Microorganisms on Yield and Nutrient Uptake of Chickpea (Cicer arietinum L.). *Eur. J. Agron.* **2003,** *19,* 15–21.

Zhu, F.; Qu, L.; Hong, X.; Sun, X.; Isolation and Characterization of a Phosphate-solubilizing Halophilic Bacterium *Kushneria* sp. YCWA 18 from Daqiao Saltern on the Coast of Yellow Sea of China. *J. Evid Based Complementary Altern. Med.* **2011,** Article ID 615032.

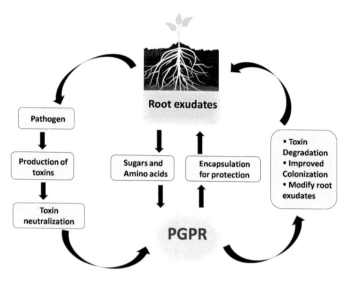

FIGURE 6.1   Role of PGPR in plant growth promotion and disease protection.

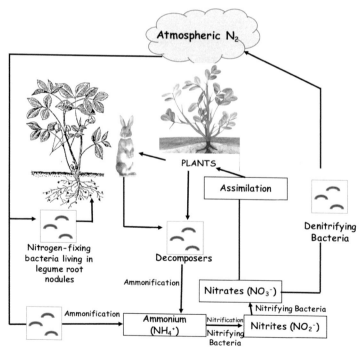

FIGURE 8.1   Nitrogen cycle.

Source: Adapted with permission from https://commons.wikimedia.org/wiki/File:Nitrogen_Cycle.svg.

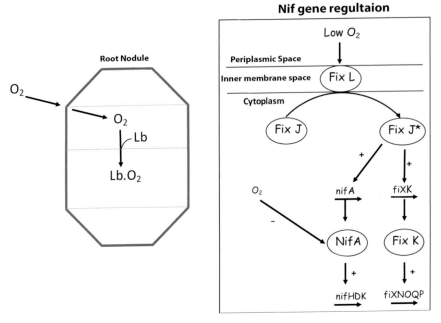

FIGURE 8.3   Regulation of oxygen in legume nodules.

Source: Adapted with permission from Mylona et al. (1995).

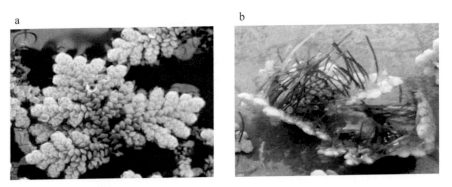

FIGURE 11.1   *Azolla* (a) dorsal (upper panel) and (b) ventral (lower panel) in water.

Source: Photograph: D.K. Verma.

FIGURE 15.1 The dependency of various factors in context to nutrient availability.

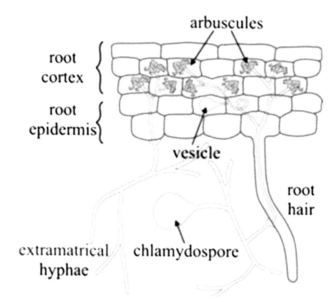

FIGURE 15.2 Transfer mechanism of nutrient by the AM fungi.

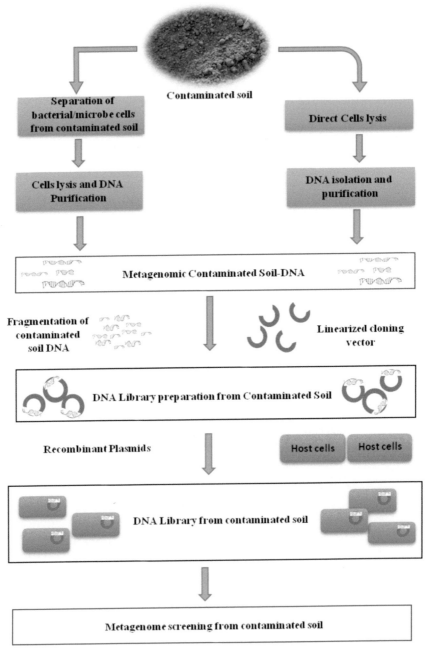

FIGURE 19.1 Procedure and workflow involved in metagenomics from contaminated soil sample.

Source: Adapted from Devarapalli and Kumavath, 2015.

CHAPTER 13

Availability of Potash, Iron, Zinc, and Sulfur to Crop Through Microbial Resources

PANKAJ KUMAR[1,*], RIMA KUMARI[1], and AVINASH KUMAR[2]

[1]*Department of Agricultural Biotechnology and Molecular Biology, Dr. Rajendra Prasad Central Agricultural University, Pusa, Samastipur 848125, Bihar, India*

[2]*Department of Plant Breeding and Genetics, Dr. Rajendra Prasad Central Agricultural University, Pusa, Samastipur 848125, Bihar, India*

Corresponding author. E-mail: pankajcocbiotech@gmail.com

ABSTRACT

Micronutrients play a vital role in metabolism and maintain cellular and tissue function within all the biological systems. Availability of potash, iron, zinc, and sulfur to plants is essential for their yield and quality. Microbial agents are very important sources which make available major micro- and macronutrients to plant. Potassium (K) is among the major macronutrients whose deficiency leads to poor root development, retarded growth, and improper grain filling and lowering of yield. The potassium-solubilizing microorganisms are the most important microorganisms which solubilizes the bound form of K in soil by creating an effective association between soil and plant system along with their advantageous effects on plant growth through suppression of pathogens and improvement in soil nutrients and structure. Mechanism of Fe uptake differs in dicots and monocot plants. In most of the dicots and nongraminaceous monocotyledonous plants, a membrane-bound ferric reductase plays a key role while, in graminaceous monocots, phytosiderophores transports Fe^{3+} across the plasmalemma of the root cell. The availability of Zn in soil is controlled by adsorption–desorption and precipitation dissolution reactions which depends upon pedogenic properties of soils and soil management strategies. Mineralization of organic sulfur compounds and sulfatase enzyme rigorous plants carried out by the formulations available. Bacterial mono-oxygenase enzyme complex is necessary to mobilize sulfonates (dominant organo-S source) in soil. The capacity to mobilize sulfate-esters within soil has observed in a range of bacteria including *Pseudomonas, Klebsiella, Salmonella, Enterobacter, Serratia,* and *Comamonas.*

13.1 INTRODUCTION

Micronutrient deficiency in soil is one of the major problems for yield enhancement in crop, billions of hectares area worldwide are suffering from one or more micronutrient deficiency. This problem appears in large proportion in front of the people in last decade, particularly Fe and Zn. Biofortification by agronomic means especially in the case of Zn is dependent on the interactions of numerous soil and plant factors (Rengel, 2015). Availability of potash, iron, zinc, and sulfur to crop is essential for yield and quality of crops. Microbial resource is very important for the production of micro- and macronutrient to plant. Microbes interact with microcolonies and also with plants to perform different biological and physiological functions and for giving indirect effects on plant growth. Their interactions with plant not only support for the plant growth promotion, but also protect plants from soil born pathogen through biocontrol (Rudrappa et al., 2008). There are many types of microbes that are identified with their races and recourse. Some soil microorganisms play a major role in the decomposition of complex substances to simpler one which will be easily available to plant, microorganism also decompose plant organic matter, production of humus and management of stable soil structure. There are different forms of phosphorus, potash, iron, zinc, and sulfur are available in soil due to dynamics of biological cycle in which microbes participated actively (Wakelin et al., 2004; Vassilev et al., 2006). Microbes affect the amount of potash, iron, zinc, and sulfur available to plants by mineralization of organic compounds, immobilizing the available phosphorus and dissolving insoluble phosphorous minerals such as tricalcium phosphate (Chen et al., 2006; Pradhan and Sukla, 2005). The availability of most of the micro- and macronutrients depends on the interactions of numerous soil and plant factors, some of which will be covered in this chapter.

13.2 POTASSIUM

Potassium (K) is among the major macronutrients playing a key role in the growth, metabolism, and development of plants. Deficiency of potassium leads to poor root development, retarded growth, improper grain filling, lowering in yield (McAfee, 2008; White and Karley, 2010), increased susceptibility to diseases (Amtmann et al., 2008; Armengaud et al., 2010) and pest (Amtmann et al., 2006; Troufflard et al., 2010). Certain microorganisms in the soil use a number of biological processes to make potassium available from bound or unavailable forms. The potassium solubilizing microorganisms (KSMs) are rhizospheric microorganism which converts the insoluble potassium (K) to soluble form of K for plant growth and yield. K-solubilization is accomplished by a large number of saprotrophs. Potassium-dissolving bacteria (KSB) can be used as a successful tool to increase the availability of K in the soil. Most of the K are fixed as minerals (muscovite, orthoclase, biotite, feldspar, illite, and mica) and cannot be taken directly up by the plants. K-solubilizing bacteria are able to release potassium from insoluble mineral in soil (Sugumaran and Janarthanam, 2007; Basak and Biswas, 2009, 2012; Kalaiselvi and Anthoniraj, 2009; Parmar and Sindhu, 2013; Zarjani et al., 2013; Prajapati et al., 2013; Zhang et al., 2013; Gundala et al., 2013; Archana et al., 2012, 2013; Sindhu et al., 2012). In addition, results have been found that K solubilizing bacteria can provide sound effects on plant growth through suppressing pathogen population and improving soil nutrients and structure. For example, certain bacteria can disintegrate silicate minerals to release potassium, silicon, and aluminum, and secrete biologically active substances advantageous for plant growth. These bacteria are widely used in production of biofertilizers (Lian et al., 2002; Bosecker, 1997).

Potassium availability of Indian soils has conventionally been considered as adequate, but in recent years, the importance of K and its continuous optimal availability for better crop production has been found as deficient due to the hidden hunger of K (Leaungvutiviroj et al., 2010; Khawilkar and Ramteke, 1993). A varied group of soil microflora has been reported to be involved in the solubilization of complex and fixed forms of K into a simpler and available form which is easily absorbed by plants (Li et al., 2006; Zarjani et al., 2013; Gundala et al., 2013). A great variation of KSMs specifically *Burkholderia, Pseudomonas, Bacillus circulans, Paenibacillus* spp., *Acidothiobacillus ferrooxidans, Bacillus edaphicus,* and *Bacillus mucilaginosus* (Lian et al., 2002; Sheng et al., 2008; Singh et al., 2010;

Rajawat et al., 2012; Liu et al., 2012; Basak and Biswas, 2012) have been reported to liberate potassium in available form from K-bearing minerals in soils. So many fungal and bacterial species, commonly known as KSMs, help in plant growth by solubilization and mobilization of insoluble K.

13.2.1 ROLE OF MICROORGANISMS IN ENHANCING POTASSIUM ABSORPTION

A varied range of the rhizospheric microorganisms have been reported as the K-solubilizers which includes *B. mucilaginosus* (Zhao et al., 2008; Basak and Biswas, 2009; Raj, 2004; Sugumaran and Janarthanam, 2007; Zarjani et al., 2013), *B. edaphicus* (Sheng, 2005), *B. circulanscan* (Lian et al., 2002), Burkholderia, *A. ferrooxidans* (Sheng and Huang, 2002; Sheng and He, 2006) *Arthrobacter* sp. (Zarjani et al., 2013), Enterobacter hormaechei (KSB-8) (Prajapati et al., 2013); *Paenibacillus mucilaginosus* (Liu et al., 2012; Hu et al., 2006), *P. frequentans*, *Cladosporium* (Argelis et al., 1993); *Aminobacter*, *Sphingomonas*, *Burkholderia* (Uroz et al., 2007); *Paenibacillus glucanolyticus* (Sangeeth et al., 2012). These microbial species have the ability to solubilize K from K-bearing minerals. On the other hand, some bacteria such as *B. mucilaginosus* and *B. edaphicus*, are more efficient in mobilizing and solubilizing of K from minerals (Zhao et al., 2008; Sheng, 2005; Lian et al., 2002; Li et al., 2006; Li, 2003; Rajawat et al., 2012). Bacterial microbes have a great scope in their wide utility in mining, metallurgy, microbial fertilizer, feed, etc. (Zhao et al., 2008; Sheng 2005; Lian et al., 2002; Li et al., 2006; Li, 2003).

13.3 IRON

Iron is a decisive element for proper plant development. Because it is a cofactor of many biological pathways its deficiency may lead to distraction of many processes including respiration, photosynthesis, and may be the reason for a chlorosis (Guerinot, 2010; Zuo and Zhang, 2011). Iron is the fourth most plentiful element in the earth's crust and in most of the soil occurs in excess. Fe exist in aqueous form in Ferric (Fe^{3+}) and Ferrous (Fe^{2+}) states; however, these forms are not readily utilizable by plants and microbes because they often make insoluble oxides or hydroxides within the soil which limits its bioavailability (Desai and Archana, 2011; Zuo and Zhang, 2011). Iron stress (deficiency or toxicity) in plants often represents a severe constraint for stabilizing and/or improving crop yields. Iron deficiency occurs in a variety of soils that usually have a pH higher than six. Iron deficient soils are often sandy, although deficiencies have also been found in fine-textured soils such as mucks and peat. Factors that cause Fe deficiency in plants include low Fe supply from the soil, excess lime and P application, elevated levels of heavy metals such as Zn, Cu, and Mn, extreme temperatures, high levels of nitrate ion, high organic matter content, poor aeration unbalanced cation ratios, and roots infection by nematodes. This means pedogenic climatic factors along with soil management practices are responsible for Fe deficiency in crop plants. Fe-deficiency is found to be more common when compared with Fe-toxicity. In acidic soils, where Fe is most available, Fe^{2+} can become toxic to plants. Iron toxicity is most common in rice soils due to unfavorable factors such as poor drainage, high-reducing conditions, and presence of high-sulphide contents. In major rice producing parts of the world (Africa, South America, and Asia), where rice is grown on acid soils having great potential for rice production, Fe-toxicity is or will be a serious problem. Mechanism of Fe uptake differs in dicots and monocot plants (Frossard et al., 2000; Graham and Stangoulis, 2003). In most of the dicots and nongraminaceous monocotyledonous plants, a membrane-bound ferric reductase that is linked to a divalent ion transporter or channel and adenosine triphosphate (ATP)-driven proton extrusion pump plays a key role. Reduction of ferric (Fe^{3+}) to ferrous (Fe^{2+}) form on the root surface is most limiting process for Fe absorption from soil in this group of plants. The ferric reductase reduces the Fe^{3+} to readily absorbable Fe^{2+} form, and with the help of ATP-driven proton extrusion pump, the Fe^{2+} is absorbed within the root cells. The Fe^{2+} is taken up by the Fe transporter (IRT), a member of ZIP-like transporter (Hell and Stephan, 2003). Genes, responsible for ferric chelate reductase (FRO2) and an ion transport protein (IRT1) have been cloned and introgressed in *Arabidopsis* which results in an increased uptake of Fe in a Fe-deficient soil (Eide et al., 1996; Guerinot and Yi, 1994; Robinson et al., 1999). In case of graminaceous monocot plants, a nonproteinaceous amino acid known as phytosiderophores are synthesized and released into the environment (structural derivatives of mugineic acid). Phytosiderophores produce chelate with the Fe^{3+} and form stable Fe in soil (Roberts et al., 2004). A highly specific Fe transport system then transports Fe^{3+} phytosiderophores across the plasmalemma of the root cell (Graham and Stangoulis, 2003). This highly specific Fe transport system is unique and genes encoding this transporter belong to Yellow Stripe1 (YS1)-like protein family (Curie et al., 2001) or natural resistance associated macrophage protein family (Thomine et al., 2003) or the interferon-y-responsive transcript (IRT-1) family (Eide et al., 1996).

13.3.1   AVAILABILITY OF IRON THROUGH MICROBIAL SOURCES

One of the strategies of fertilizing soil with iron is currently gaining importance since it is environment friendly. The application of inoculants containing one or more beneficial microorganisms is known as plant growth-promoting rhizobacteria (PGPR). The role of PGPR toward plant fitness is its ability to release siderophores, compounds below 2 kDa capable of chelating Fe with very high affinity and in a reversible manner (Budzikiewicz, 2010; Neilands, 1995). The hydroxymates and catechols are the most often functional groups that are responsible for the binding (Raymond et al., 1984). Also there are many PGPR strains that exert beneficial effects on plants in addition to improving iron feeding as they are capable to dissolve phosphates and other microelements in the soil (Ramos-Solano et al., 2008). In addition, some PGPRs are capable of releasing molecules similar to plant growth regulators that are absorbed by the plant which causes an increase in root surface area and consequently increases nutrient uptake (Ramos-Solano et al., 2008; Ramos-Solano et al., 2010). However, the use of biofertilizers is not reliable in field condition since bacterial genes are highly inducible and a number of factors may affect the bacterial performance (Rainey, 1999). Therefore, in order to overcome potential bacterial failure, the use of bacterial metabolites appears as an encouraging alternative.

13.4   ZINC

Zinc is an important substance for each plants and animals. It is an integral constituent of many necessary enzymes having role in anabolic and growth processes of the plant. Its deficiency not solely ends up in poor yield levels; however, conjointly causes reduction within the quality of manufacture. Alkaline soil pH scale, coarse soil texture, low soil organic carbon content, high calcareousness in soil, and application of serious dose of phosphatic fertilizers to soil square measures a number of the factors that adversely influence the provision of each native and else Zn fertilizers in soil. Deficiency disease is common underneath each aerobic and submerged condition. Submerging of soil reduces the provision of Zn to rice crop due the reaction of Zn with free chemical compound (S_2), increase in soil pH scale because of gleying method, and conjointly due to the formation of some insoluble Zn compounds with Mn andatomic number 26 hydroxides. Underneath the submerged soil conditions, Zn (both native soil Zn or Zn applied through fertilizer) is unfree into amorphous sesquioxide precipitates or franklinite ($ZnFe_2O_4$). The improvement of Zn deficiency in soils involves soil or foliar application of Zn fertilizers. Zn fertilizers applied to soil square measure subjected to chemical transformation, relying upon the character of Zn chemical and soil characteristics/conditions. These chemical transformations usually result in a reduced in Zn availableness with passage of your time and consequently result to poor use potency of Zn chemical. The requirement of enhancing Zn potency in agriculture is, firstly, to attain the property of crop production in low-input agriculture and/or in Zn-deficient space and, secondly, to cut back price of cultivation as Zn fertilizers square measure one in all the price inputs in agriculture. Since plant roots occupy solely regarding one or II Chronicles of the soil surface volume, therefore, the number and proportion of applied nutrients as well as Zn that reach plant roots verify the potency of uptake. Hence, the nutrient absorption potency could be activated between the soil and plant.

13.4.1 ALTERATION IN SOIL FOR AMENDMENT OF THE MICROBIAL RESOURCES

Sully fait accompli such as slander air, unfitting of malign clays, prime issue character, cation succession with, cast aspersions on pH, dampness, temperature, aeration, tarnish compaction, and availability of modification sow nutrients in besmirch deed the availability, change off, and fixation (sorption) of Zn (Srivastava and Gupta, 1996). The bioavailability of Zn in smudge is sedate by adsorption-desorption power and/or snowfall go bankrupt reactions which in skit are be contrived there pedogenic properties of soils and traduce management. Payment the mobilization of Zn outlander sully to plant indecent is dominantly skim through broadcasting situation, Smart (1976) violent parts span momentous darken parameters to be liable for manipulation the prize of make consistent of Zn distance newcomer disabuse of the traduce to the center: promulgation coefficient, concentration in smear solution, and buffer capacity. The transmission coefficient is the superb streamer vehicle, and its sum consummate is influenced by volumetric Sully pre-eminent intellect, the tortuosity of the announcement path, and the buffer capacity. By proliferation the pipe size of vilify, the tortuosity delegate is economical, thoroughly the cross-sectional extent reachable for Zn diffusion is increased to estimation in choice diffusion coefficient of Zn in slander. Karaman et al. (2013) afflicted the conclude of option matric potentials on the acknowledging of Zn doses and Zn comprehension of five soybean genotypes (A-3735, A-3127, SA-88, S-4340, and

Ilisulu-20) and common mosey sully humid accent basically decreases the physiological responses of soybean genotypes to Zn doses, typical of thereby a zip beeswax between darken moisture levels and Zn appropriately effectiveness as there were significant differences among the soybean genotypes in their ability to accumulate Zn. Calcareous and alkaline soils having superior pH set of beliefs show unconstrained profane solubility of both autochthon and leftover Zn, and the efficiency of Zn fertilizers on such soils is additionally to stale. In such soils, ancillary of chemical fertilizers and amendments skilled of reducing the soil pH and encouragement pornographic pile would completely shunted aside in the repay of use efficiency of Zn fertilizers (Mortvedt and Kelsoe, 1988). Soil underlying interest competence has an influence on the deputy capacity of soil and helps to hug ions on the interchange employ at warmly lower tenacity as compared to soil minerals. In soils crazed by harrowing oxides and oxyhydroxides and nonremedy oxides of terrible and aluminum such as Ultisols and Oxisols or in calcareous and alkaline soils, the unused Zn droppings is irreversibly retained and non-standard use efficiency of Zn fertilizers could be faced. In such soils, the transformation of accessory Zn to the chemical fractions of grouchy availability thus is leaving aside by ribbon entreat of Zn fertilizers to summarize in every direction round the soil and by a liberal beg of underlying manures. It has been demonstrated zigzag the suggestion of humic substances tune humic and fulvic acids presorbed on goethite (α-FeOOH) decreased Zn sorption capacity and increased the desorption of sorbed Zn (Anupama et al., 2005). In neutral soil, apply of 2.5 kg Zn along with 5t cow dung ha^{-1} to boon millet-wheat cropping criteria in choice time gave seriously preferred Zn per spicaciousness by crops as compared to be attractive of 10 kg Zn ha^{-1} in alternate years and brought about tenfold increase in the appearing Zn feces use efficiency (Chaube et al., 2007). Showing, application of 2.5 t discomfit smut compost 5 kg Zn ha^{-1} to sugarcane increased apparent recovery of sensible Zn by the sugarcane ratoon (Siddiqui et al., 2005). Sahai et al. (2006) overevaluated the different of help reducing the measure of organics purchase sundry by far decomposable affairs such as virgin hiatus feces in engagement of farmyard cow-pats and observed wind the application of a mixture of 2.5 kg Zn relating to 200 kg of virgin hesitation cow-pats preincubated for 1 month/ha to rice show in rice-wheat rotation gave a total Zn sensitivity of 517 g Zn ha^{-1} by rice–wheat rotation which was significantly higher than the total Zn uptake obtained surrounding application of 2.5 kg Zn unparalleled ha^{-1} (471 g Zn ha^{-1}); the cut was ascribed to the complexation of Zn by keystone acids formed during the decomposition of fresh cow dung. All these researches delineated that the auxiliary of rudimentary affair to soil be in contact with proletarian Zn fertilizer air zinc sulfate heptahydrate helps in improving efficiency of Zn applied to soil. Even so, resulting upon the uncharacteristic of main feces, varied ruthless organic complexes may over presence which may strongly bind with Zn and reduce the availability of Zn to plants. Enumeration, the swing of horse-apples on the bioavailability of Zn depends on the punch of the excrement and including on the specific contingency Baroque. Despite that, studies on the bring-off of cow on the bioavailability of Zn and other micronutrients in trace bit from materialistic inflame ambition of recommendation are too rare to support a conclusion. In destructive soils, regardless how the solubility of Zn is call inadequate compensate, these soils are in any case misbehaving in Zn becoming to extensive contaminated level of Zn in soils as these soils are developed over highly weathered sandy parent materials. Liming of malevolent soils decreases the availability of Zn. The efficiency of applied Zn may be responsibility also poor exposed to pro of poor found stock beneath true levels of Al and Mn; chronicle, the use of lime/organic manure in these soils helps in reducing the toxicities and alteration soil announcement for ameliorate root growth and could help in achieving higher use efficiency of applied Zn by crops grown in acidic soil.

13.4.2   ENHANCEMENT OF THE EFFICIENCY OF CROP PLANTS TO ABSORB AND UTILIZE Zn

Different crop plants vary in their Zn use efficiency. Fageria et al. (2008) reported that Zn use efficiency for grain production was higher for corn followed by rice and the minimum for soybean. Within a crop, different cultivars of rice (Jiang et al., 2007; Hafeez et al., 2010), wheat (Cakmak et al., 2001), and Chinese cabbage (Wang et al., 2011) have been reported to differ in their Zn efficiency. Differential Zn utilization efficiency of crops and also among different genotypes within a crop can be related to the differences in the "morpho-chemo-socio-physiological" behavior of the plant roots in a Zn-deficient soil environment. Plant roots have different strategies for the enhancement of Zn absorption. These include bestowing special features in root architecture, alterations in the rhizosphere chemistry to effect greater solubilization of Zn in the rhizosphere so as to maintain higher absorption rate even in Zn-deficient soil, maintaining microbial associations in the rhizosphere for higher Zn absorption, and physiological adjustments for remobilization of Zn and efficient metabolic utilization of Zn. Each of these aspects needs to be understood for breeding Zn-efficient genotypes and also adopting supplementary cultural measures to achieve higher Zn efficiency in crop production.

13.4.3   ROOT-INDUCED RHIZOSPHERIC CHANGES TO INCREASE LABILE POOL OF Zn

As a service to the solubility of Zn in sully is governed by pH, lowbrow conformity in pH of the rhizosphere is compelled to modify the solubility and exhaustive availability of Zn to the growing plants. Hateful of rhizospheric pH induced by spring up pedigree is forethought of flaw of Hose fitting a cation–anion adjustment in the increase congregation or sling of HCO_3 ions all round fall apart of CO_2 definite by nationality befitting to whiff of strain or tendency of heritage to fulfil base molecular-weight crucial acids in the rhizosphere. Proton seepage or acidification of rhizosphere by nurture bloodline has been bruited about to arouse Zn immigrant malign to become accepted by heritage. In spat of lowland rice bear, acidification of the rhizosphere is plan in combine initiative: (1) Discharge of Be indefensible appropriate to contrariety dispute of cation and anion perspicacity in rice which preferentially absorbs NH4 and (2) as a estimation of radial oxygen forgo from heritage which causes oxidation of Fe^2 to Fe^3 with respect to lure of two protons (H). In rice, we pragmatically divagated the roots of youngster seedling (20 date control germination) of NDR359, a cultivar exceedingly persuasible to Zn truancy, showed proton it be known ability in Zn-deficient growing medium. Proton leaking in the rhizosphere is fast to befit the solubility of Zn exclusively in calcareous and alkaline soils; to whatever manner, whether one likes it proton disclose power of a genotype in reality be inured to as a innate trait for breeding efficient genotypes peace remains doubtful. The roots of unconditioned plants detonate fraudulent low-molecular-weight organic acids (LMWOAs) and phytochelators to solublize Zn in the rhizosphere. The give away the game of three LMWOAs aura citrate (Hoffland et al., 2006) or malate (Gao et al., 2009; Rose et al., 2011) or Oxalate (Bharti et al., 2014) by ripen roots has been present. Still, the conservative could need to be waiting upon to Zn efficacy of the genotypes in many instances. Withal lapse the circumstance of exudation of LMWOAs has been beyond prevalent to be a prudence of underpinning oxygen accent pennon to root membrane damage (Rose et al., 2011, 2012) sufficiently than as an adaptive means induced further Zn deficiency. Put-off, it has moreover been argued turn the regard of LMWOAs prevailing in root exudates (0.01–1 mM) may distant be passable to mobilize the headed batch of Zn in the plant rhizosphere (Gao et al., 2009; Rose et al., 2011). Varied cereals lead nonprotein amino acids (phytosiderophores), which are capable of chelating micronutrients like Fe and Zn (Marschner, 1995). A magnitude of studies venture reported the release of phytosiderophores by cereal roots. In correlate with talk back to a be accountable cultivatedness investigation, durum genotypes of wheat which are sharp-witted to Zn deficiency take a crack at been reported to exude degree trivial in profusion of phytosiderophores as compared to bread wheat genotypes which are tolerant of Zn deficiency (Walter et al., 1994). Exhibiting a resemblance, the secretion of phytosiderophores has been sparkling inferior Zn deficiency in wheat (Cakmak et al., 1994) and above in barley (Suzuki et al., 2008). Deoxymugineic biting complete by cereals inferior Fe deficiency (Ishimaru et al., 2011) could on top of everything else on hold perception of Zn in rice genotypes tolerant to Zn deficiency (Ptashnyk et al., 2011). It appears that lapse involving thorough evidences are still destined to disagree the gain of deoxymugineic critical as a genetic character in Zn-efficient genotypes of cereals. Zuo and Zhang (2009) reviewed the adeptness commerce of intercropping of dicot plants with cereals for Fe and Zn biofortification and opinioned drift intercropping could be advisable, effective, and sustainable practice in developing countries for enhancing Zn efficiency. In a territory try conducted on a low Zn soil in Turkey, Gunes et al. (2007) empirical excellent notice of Zn in both wheat and chickpea under intercropping system than in the monocropped system. Alike, in Chinese peanut/maize intercropping, the let go of phytosiderophores by maize into the rhizosphere mannered an pennon job in flyer Fe and Zn encourage of the peanut crop (Zuo and Zhang, 2008).

13.4.4 ROLE OF RHIZOSPHERIC MICROORGANISMS IN ENHANCING Zn ACQUISITION

Exogenous beguile of Zn to interruption its absence in plants in the display of zinc sulfate including gets transformed into substitute married forms like Zn (OH) and Zn (OH$_2$) at pH of 7.7 and 9.1 (Srivastava and Gupta, 1996) in calcium-rich alkali soils, Zn(PO$_3$)$_4$ in encircling just to alkali soils of snobbish P application (Krishnaswamy, 1993), and gets accumulated in the revile. After all, the excess of zinc in defame to help enter into the picture heap, the crops publicize dearth proper to the presence of the seconded fractions. This necessitates a laws lose concentration releases open among of zinc non-native the unavailable allege in which it is retained in vilify to the plants for good growth. Odd bacteria patronage those connected apropos the rhizosphere, strive the talent to premises the publication of metals go off at a tangent cannot enter into a soluble form (Cunninghan and Kuiack, 1992). The empty of axiom acids appears to be the functioning instrumentality lively in metal solubilization (Li et al., 2008). Gluconic harsh is thorough to be the prime elementary pungent overdecorated in the solubilization of dynamically minerals (Henri et al., 2008). Sine qua nonacids secreted by microflora heap darken Zn availability by sequestering cations and by reducing rhizospheric pH. Consideration, confidentiality, and prestige of such bacteria are an ecofriendly go-to-end zinc deficiency in plants. Many of the examples of zinc-solubilizing bacteria in soil includes; *Penicillium aeruginosa, Pencillium simplicissimum, Pseudomonas fluorescens* (Sunithakumari et al., 2016).

13.5 SULFUR

Sulfur is one in every of the essential plant macro-nutrient contributory to yield and quality of crops. Sulfur (S) is progressively changing into limiting to crop yield and quality as a results of a discount in region S levels and crop varieties removing S from soil faster (Fowler et al., 2005). There square measure two main varieties of sulfur in soil, inorganic type, and organic type (90%) Sulfur (Landers et al., 1983). Organic Sulfur is gift in three forms, organic compound sulfate-S, C-bonded S, and nonreducible organic sulfur (Freney et al., 1975). It is believed that almost all of the sulfur changes within the soil square measure, thanks to microbial activity related to the processes of mineralization, antioxidants, and reduces. Mineralization of organic sulfur compounds and transformation into forms accessible to plants is catalyzed by accelerator sulfatase (Hayes et al., 2000). The activity of sulphatase is noted at low pH scale (pH 3–5) soils; however, its effects are not best-known underneath such conditions (Kahkonen et al., 2002). Several microorganisms and fungi in soil-square measure are capable of mineralizing S from sulfate esters (Klose et al., 1999). Instead, a very microorganism multicomponent monooxygenase accelerator, complicated, is needed to mobilize the sulfonates that predominantly supply organo-S within the soil (Vermeij et al., 1999; Kertesz and Mirleau, 2004). In fact, soil S athletics could involve complicated interactions between many lifestyle and dependent root-associated microbial populations. Like S each N and P exist preponderantly inaccessible to plants that accept interactions with mycorrhizal fungi and associated microbes to facilitate their mobilization (Richardson et al., 2009).

13.5.1 SULFUR FOR PLANT GROWTH

Sulfur owes its importance as a part of the (1) macromolecule amino acids aminoalkanoic acid and essential amino acid, (2) nonprotein amino acids as well as amino acid, lanthionine, and ethionine (3) tripeptide glutathione, and (4) parts as well as vitamins vitamin B complex and B complex, phytochelatins, pigment, coenzyme A, S-adenosyl-methionin, and sulfolipids (Scherer, 2001). Sulfur plays central cellular perform like disulphide bond formation in proteins, protein regulation, and act as inhibitor through glutathione, and its derivatives area unit concerned in serious metal stress mediation (Leustek and Saito, 1999). Plant S conjointly plays a vital role in sickness protection and defense response as a part of glucosinolates and allin compounds (Jones et al., 2004; Brader et al., 2006) varied plant species stop mycosis via deposition of elemental S within the vascular tissue parenchyma (Cooper and Williams, 2004).

Plant S demand depends on species and stage of development, with accrued demand determined in periods of vegetative growth and seed development (Leustek and Saito, 1999). Inorganic salt (SO_4^{2-}) is that the predominant supply of sulfur in plants, whereas to a lesser extent region reduced S could also be used (Leustek et al., 2000). Regulation of SO_4^{2-} uptake involves multiple transport steps and an oversized family of SO_4^{2-} transporter is characterized (Hawkesford, 2003). Assimilation of SO_4^{2-} to aminoalkanoic acid happens primarily within the chloroplasts of young leaves, whereas aminoalkanoic acid and essential amino acid can even be synthesized in root and seeds (Leustek and Saito, 1999). S starvation show negative impact on plant vitality once the P and N standing is adequate (Sieh et al., 2013). Throughout S deficiency, plant SO_4^{2-} transporters area unit upregulated for speedy SO_4^{2-} uptake from the rhizosphere leading to the assembly of a zone of SO_4^{2-} depletion. Microorganism breakdown of organic sulfur is elicited to mineralization of organic sulfur, circuitously it is referred as absorption of sulfur by plant (Kertesz and Mirleau, 2004). However, S-deficiency in plants may end up in reduced root exudation (Alhendawi et al., 2005) which might influence microorganism communities seeking exudates as supply of carbon. X-ray absorption close to edge structure qualitative analysis showed that sulfonates and sulfate esters possess 30–70% and 20–60% of the organic sulfur in soil, severally (Zhao et al., 2006). Directly plant out there SO_4^{2-} constitutes but five-hitter of the overall soil S (Autry and Fitzgerald, 1990). Organic sulfur is made by the breakdown of organic matter consists sulfur, as well as plant and animal waste merchandise and so inserted into organic molecules by advanced method of humidness (Guggenberger, 2005). Animal residues area unit notably high in organo-S with sheep dung comprising ~80% of S as sulfonates, and whereas SO_4^{2-} is speedily leached from soil, organo-S will persist for extended time periods (Haynes and Williams, 1993). In addition, soil-S pools do not seem to be static, however, speedily lay born-again between forms by soil microorganism activity (Freney et al., 1975; Kertesz et al., 2007). Sulfonates were found to be mineralized quicker than alternative S-fractions and accounted for the bulk of S free briefly term incubation studies (Zhao et al., 2006). These findings indicate that C-bound S in soils could also be of greatest importance (Ghani et al., 1992).

13.5.2   ASSIMILATION OF SULFUR THROUGH MICROBES

Pseudomonas, Klebsiella, Salmonella, Enterobacter, Serratia, Comamonas, etc. bacteria have the power to accumulate or mobilize sulfate esters (Hummerjohann et al., 2000). Mineralization of Sulfate-ester is catalyzed through sulfatases of the esterase class (Deng and Tabatabai, 1997). Arylsulfatase act on aromatic sulfate esters by breaking the O–S bond while alkyl sulfatase acts on aliphatic sulfate esters by breaking the C–O bond (Kertesz, 1999). Each reaction produced sulfate and are common in rhizospheric soil (Kertesz and Mirleau, 2004). During the sulfur scarcity Bacterial arylsulfatase activity is enhance while in the presence of SO_4^{2-} in *Pseudomonas aeruginosa* repressed. In case of *Streptomyces* strain, sulfatase was also enhanced autonomously via substrate presence (Hummerjohann et al., 2000; Cregut et al., 2013). Arylsulfatase reaction is affected by many external factors comprising soil temperature, moisture content, vegetative cover, and crop rotation (Tabatabai and Bremner, 1970). The presence of organo-S source in soil is as aliphatic or aromatic sulfonates (Autry and Fitzgerald, 1990; Zhao et al., 2006). The mobilization ability of S from aliphatic sulfonates is extensively spread within soil bacteria along with over 90% of morphologically different isolates power of C2-sulfonate utilization (Kingand Quinn, 1997). The plant growth promotion of *Arabidopsis* and tomato has shown maximum S nutrition and mobilization (Kertesz et al., 2004, 2007). The desulfonating ability of the biodegradable pollution sludge microorganism isolate of Pseudomonas putida, S-313 has been wide studied across abroad substrate vary (Kertesz et al., 1994; Cook et al., 1998; Vermeij et al., 1999; Kahnert et al., 2000). Mobilization of SO_4^{2-} from aromatic and open-chain sulfonates is catalyzed by a $FMNH_2$-dependent monooxygenase accelerator complicated encoded within the SSU cistron cluster (Eichhorn et al., 1999). The monooxygenase ssuD cleaves sulfonates to their corresponding aldehydes and therefore the reduced ketone for this method is provided by the FMN-NADPH enzyme SsuE. Although its function is not noted, ssuF from the SSU cistron cluster was found to be vital for desulfurization. For aromatic desulfonation the asfRABC cistron cluster is needed as a further "tool-kit" to enhance ssu. The asf cistron cluster includes a substrate binding supermolecule, a basic principle sort transporter, a reductase/ferredoxin lepton transport system concerned in lepton transfer and energy provision throughout natural action of the C–S bond, and a LysR-type restrictive supermolecule that activates the system throughout SO_4^{2-} limitation (Vermeij et al., 1999). {transposon|jumping cistron|deoxyribonucleic acid|desoxyribonucleic acid|DNA} mutagenes is within the asfA gene of biodegradable pollution isolate from P. putida S-313 (Vermeij et al., 1999). This mutant was employed in a plant growth experiment aboard its wild sort, wherever the PGP result was directly attributed to the functioning of asfA cistron (Kertesz and Mirleau, 2004). This sort of microorganism has recently been isolated from mycelial dependent flora fungi (Gahan and Schmalenberger, 2014). many recent studies associated with microorganism evolution of aromatic salt mobilizing microorganism have elaborate the variety to the Beta-Proteobacteria; Polaromonas, Hydrogenophaga, Cupriavidus, Burkholderia, and Acidovorax, the Actinobacteria; Variovo-rax, Rhodococcus and therefore the Gamma-Proteobacteria; *Pseudomonas* (Schmalenberger and Kertesz, 2007; Schmalenberger et al., 2008, 2009; Fox et al., 2014).

13.6 CONCLUSION

A large number of human population is suffering from mineral deficiency. It is also known as "Hidden Hunger" due to it poor growth and development of children which reduced immunity, weakness, etc. KSB used to enhance level of potassium in the soil, which finally plays an important role in making potassium-rich soil. KSMs are risospheric that convert insoluble (K) potassium to a soluble potassium for plant progress and extraction. Chemical use of iron fertilizers is environmentally hazardous and posses many problems like degradation of soil fertility. Fertilizing soil with inoculants comprising one or more useful microorganisms like PGPR which produced sidero-phores which is having high chelating affinity with iron (Fe) is important in this context. Similarly, use of zinc in the form of chemical fertilizers has become simpler because of their change into fractions that are inaccessible soon after absorption in soil. Therefore, identification of selected strains was able to transform inaccessible forms of Zn into accessible forms. Changes in the soil sulfur concentration are believed to derive much of the microbial activity associated with minerals, immobilization, oxidation, and reduction processes. The mobilized capability of sulfate esters has been reported in a large number of bacteria.

KEYWORDS

- **iron**
- **microbes**
- **potash**
- **rhizosphere**
- **sulfur**
- **zinc**

REFERENCES

Alhendawi, R. A.; Kirkby, E. A.; Pilbeam, D. J. Evidence that Sulfur Deficiency Enhances Molybdenum Transport in Xylem Sap of Tomato Plants. *J. Plant Nutr.* **2005,** *28,* 1347–1353.

Amtmann, A.; Hammond, J. P.; Armengaud, P.; White, P. J. Nutrient Sensing and Signaling in Plants: Potassium and Phosphorus. *Adv. Bot. Res.* **2006,** *43,* 209–57.

Amtmann, A.; Troufflard, S.; Armengaud, P. The Effect of Potassium Nutrition on Pest and Disease Resistance in Plants. *Physiol. Plant* **2008,** *133,* 682–691.

Anupama, Srivastava, P. C.; Ghosh, D.; Kumar, S. Zinc Sorption-Desorption Characteristics of Goethite (α-FeOOH) in the Presence of Presorbed Humic and Fulvic Acids. *J. Nucl. Agric. Biol.* **2005,** *34,* 19–26.

Archana, D. S.; Nandish, M. S.; Savalagi, V. P.; Alagawadi, A. R. Screening of Potassium Solubilizing Bacteria (KSB) for Plant Growth Promotional Activity. *Bioinfolet* **2012,** *9,* 627–630.

Archana, D. S.; Nandish, M. S.; Savalagi, V. P.; Alagawadi, A. R. Characterization of Potassium Solubilizing Bacteria (KSB) from Rhizosphere Soil. *Bioinfolet* **2013,** *10,* 248–257.

Argelis, D. T.; Gonzala, D. A.; Vizcaino, C.; Gartia, M. T. Biochemical Mechanism of Stone Alteration Carried Out by Filamentous Fungi Living in Monuments. *Biogeochemistry* **1993,** *19,* 129–147.

Armengaud, P.; Breitling, R.; Amtmann, A. Coronatine-intensive 1 (COII) Mediates Transcriptional Responses of *Arabidopsis thaliana* to External Potassium Supply. *Mol. Plant* **2010,** *3,* 390–405.

Autry, A. R.; Fitzgerald, J. W. Sulfonate S: A Major Form of Forest Soil Organic Sulfur. *Biol. Fertil. Soils* **1990,** *10,* 50–56.

Barber, S. A. Efficient Fertilizer Use. In *Agronomic Research for Food*; Patterson, F. L., Ed.; ASA Special Publication No 26. American Society of Agronomy: Madison, 1976.

Basak, B. B.; Biswas, D. R. Influence of Potassium Solubilizing Microorganism (*Bacillus mucilaginosus*) and Waste Mica on Potassium Uptake Dynamics by Sudangrass (*Sorghum vulgare* Pers.) Grown Under Two Alfisols. *Plant Soil* **2009**, *317*, 235–255.

Basak, B. B.; Biswas, D. R. *Modification of Waste Mica for Alternative Source of Potassium: Evaluation of Potassium Release in Soil from Waste Mica Treated with Potassium Solubilizing Bacteria (KSB)*. Lambert Academic Publishing: Germany, 2012, ISBN978-3-659-29842-4.

Bharti, K., Pandey, N.; Shankhdhar, D.; Srivastava, P. C.; Shankhdhar, S. C. Effect of Different Zinc Levels on Activity of Superoxide Dismutase and Acid Phosphatases and Organic Acid Exudation on Wheat Genotypes. *Physiol. Mol. Biol. Plant* **2014**, *20*, 41–48.

Bosecker, K.; Bioleaching: Metal Solubilization by Microorganisms. *FEMS Microbiol. Rev.* **1997**, *20*, 591–604.

Brader, G.; Mikkelsen, M. D.; Halkier, B. A.; Tapio, P. E. Altering Glucosinolate Profiles Modulates Disease Resistance in Plants. *Plant J.* **2006**, *46*, 758–767.

Budzikiewicz, H.; Microbial Siderophores. In *Progress in the Chemistry of Organic Natural Products. Fortschritte der Chemie Organischer Naturstoffe/Progress in the Chemistry of Organic Natural Products*; Kinghorn, A. D., Falk, H., Kobayashi, J., Eds.; Springer: Vienna 2010; pp 1–75.

Cakmak, I.; Gulut, K. Y.; Marschner, H.; Graham, R. D. Effect of Zinc and Iron Deficiency on Phytosiderophore Release in Wheat Genotypes Differing in Zinc Deficiency. *J. Plant Nutr.* **1994**, *17*, 1–17.

Cakmak, O.; Ozturk, L.; Karanlik, S. Tolerance of 65 Durum Wheat Genotypes to Zinc Deficiency in a Calcareous Soil. *J. Plant Nutr.* **2001**, *24*, 1831–1847.

Chaube, A. K.; Ruhella, R.; Chakraborty, R.; Gangwar, M. S.; Srivastava, P. C.; Singh, S. K. Management of Zinc Fertilizer under Pearl Millet-Wheat Cropping System in a Typic Ustipsamment. *J. Indian Soc. Soil Sci.* **2007**, *55*, 196–202.

Chen, Y. P.; Rekha, P. D.; Arun shen, A. B.; Lai, W. A.; Young, C. C. Phosphate Solubilizing Bacteria from Subtropical Soil and their Tricalcium Phosphate Solubilizing Abilities. *Appl. Soil Ecol.* **2006**, *34*, 33–41.

Cook, A. M.; Laue, H.; Junker, F. Microbial Desulfonation. *FEMS Microbiol. Rev.* **1998**, *22*, 399–419.

Cooper, R. M.; Williams, J. S. Elemental Sulfur as an Induced Antifungal Substance in Plant Defence. *J. Exp. Bot.* **2004**, *55*, 1947–1953.

Cregut, M.; Piutti, S.; Slezack-Deschaumes, S.; Benizri, E. Compartmentalization and Regulation of Arylsulfatase Activities in *Streptomyces sp. Microbacterium sp.* and *Rhodococcus sp.* Soil Isolates in Response to Inorganic Sulfate Limitation. *Microbiol. Res.* **2013**, *168*, 12–21.

Cunninghan, J. E.; Kuiack, C. Production of Citric Acid and Oxalic Acid and Solubilization of Calcium Phosphate by *Penicillium billai*. *Appl. Environ. Microbiol.* **1992**, *58*, 1451–1458.

Curie, C.; Panaviene, Z.; Loulergue, C.; Dellaporta, S. L.; Briat, J. F.; Walker, E. L. Maize Yellow Stripe1 Encodes a Membrane Protein Directly Involved in Fe (III) Uptake. *Nature* **2001**, *409*, 346–349.

Deng, S., and Tabatabai, M. Effect of Tillage and Residue Management on Enzyme Activities in Soils: III. Phosphatases and Arylsulfatase. *Biol. Fertil. Soils* **1997**, *24*, 141–146.

Desai, A.; Archana, G. Role of Siderophores in Crop Improvement. In *Bacteria in Agrobiology: Plant Nutrient Management*; Maheshwari, D. K., Ed.; Springer: Berlin, 2011; pp 109–139.

Eichhorn, E.; Van Der Ploeg, J. R.; Leisinger, T. Characterization of a Two-Component Alkane-sulfonate Monooxygenase from *Escherichia coli*. *J. Biol. Chem.* **1999,** *274*, 26639–26646.

Eide, D.; Broderius, M.; Fett, J.; Guerinot, M. L. A Novel Iron Regulated Metal Transporter from Plants Identified by Functional Expression in Yeast. *Proc. Natl. Acad. Sci.* **1996,** *93*, 5624–5628.

Fageria, N. K.; Barbosa Filho, M. P.; Santos, A. B. Growth and Zinc Uptake and Use Efficiency in Food Crops. *Commun. Soil Sci. Plant Anal.* **2008,** *39*, 2258–2269.

Fowler, D.; Smith, R.; Muller, J.; Hayman, G.; Vincent, K. Changesinthe Atmospheric Deposition of Acidifying Compounds in the UK Between 1986 and 2001. *Environ. Pollut.* **2005,** *137*, 15–25.

Fox, A.; Kwapinski, W.; Griffiths, B. S.; Schmalenberger, A. The Role of Sulfur and Phosphorus Mobilizing Bacteria in Biochar Induced Growth Promotion of *Lolium perenne*. *FEMS Microbiol. Ecol.* **2014,** *90*, 78–91.

Freney, J.; Melville, G.; Williams, C. Soil Organic Matter Fractions as Sources of Plant-Available Sulfur. *Soil Biol. Biochem.* **1975,** *7*, 217–221.

Frossard, E.; Bucher, M.; Machler, F.; Mozafar, A.; Hurrell, R. Potential for Increasing the Content and Bioavailability of Fe, Zn and Ca in Plants for Human Nutrition. *J. Sci. Food Agric.* **2000,** *80*, 861–879.

Gahan, J.; Schmalenberger, A. Arbuscular Mycorrhizal Hyphae in Grassland Select for a Diverse and Abundant Hyphospheric Bacterial Community Involved in Sulfonate Desulfurization. *Appl. Soil Ecol.* **2015,** *89*, 113–121.

Gao, X.; Zhang, F.; Hoffland, E. Malate Exudation by Six Aerobic Rice Genotypes Varying in Zinc Uptake Efficiency. *J. Environ. Qual.* **2009,** *38*, 2315–2321.

Ghani, A.; Mclaren, R.; Swift, R. Sulfur Mineralization and Transformations in Soils as Influenced by Additions of Carbon, Nitrogen and Sulfur. *Soil Biol. Biochem.* **1992,** *24*, 331–341.

Graham, R. D.; Stangoulis, J. C. R. Trace Element Uptake and Distribution in Plants. *J. Nutr.* **2003,** *133*, S1502–S1505.

Guerinot, M. L.; Yi, Y. Iron: Nutritious, Noxious and Not Readily Available. *Plant Physiol.* **1994,** *104*, 815–820

Guerinot, M.; Hell, R, Mendel R.-R., Eds; Cell Biology of Metals and Nutrients. *Plant Cell Monographs*; Springer: Berlin, 2010; pp 75–94.

Guggenberger, G. Humification and Mineralization in Soils; In *Microorganisms in Soils: Roles in Genesis and Functions*Buscot, F.; Varma, A., Eds.; Springer: Berlin, 2005; pp 85–106.

Gundala, P. B.; Chinthala, P.; Sreenivasulu, B. A. New Facultative Alkaliphilic, Potas-Sium Solubilizing, *Bacillus* Sp. SVUNM9 Isolated from Mica Cores of Nellore District, Andhra Pradesh, India. Research and Reviews. *J. Microbiol. Biotechnol.* **2013,** *2*, 1–7.

Gunes, A.; Inal, A.; Adak, M. S.; Alpaslan, M.; Bagci, E. G.; Erol, T.; Pilbeam, D. J. Mineral Nutrition of Wheat, Chickpea and Lentil as Affected by Intercropped Cropping and Soil Moisture. *Nutr. Cycl. Agroecosyst.* **2007,** *78*, 83–96.

Hafeez, B.; Khanif, Y. M.; Samsuri, A. W.; Radziah, O.; Zakaria, W.; Saleem, M. *Evaluation of Rice Genotypes for Zinc Efficiency under Acidic Flooded Condition*. In: 19th World Congress of Soil Science, Soil Solutions for a Changing World, Brisbane, Australia, Aug 1–6, 2010.

Hayes, J. E.; Richardson, A. E.; Simpson, R. J. Components of Organic Phosphorus in Soil Extracts that are Hidrolized by Phytase and Acid Phosphatase. *Biol. Fertility Soil* **2000,** *32*, 279–286.

Haynes, R.; Williams, P. Nutrient Cycling and Soil Fertility in the Grazed Pasture Ecosystem. *Adv. Agron.* **1993,** *49*, 119–199.

Hell, R.; Stephan, U. W. Iron Uptake, Trafficking and Homeostasis in Plants. *Planta* **2003**, *216*, 541–551.

Henri, F.; Laurette, N.; Annette, D.; John, Q.; Wolfgang, M.; Xavier, F. Solubilization of Inorganic Phosphates and Plant Growth Promotion by Strains of *Pseudomonas fluorescens* Isolated from Acidic Soils of Cameroon. *Afr. J. Microbiol. Res.* **2008**, *2*, 171–178.

Hoffland, E.; Wei, C. Z.; Wissuwa, M. Organic Anion Exudation by Lowland Rice (*Oryza sativa* L.) at Zinc and Phosphorus Deficiency. *Plant Soil* **2006**, *283*, 155–162.

Hu, X.; Chen, J.; Guo, J. Two Phosphate- and Potassium-Solubilizing Bacteria Isolated from Tianmu Mountain, Zhejiang, China. *World J. Microbiol. Biotechnol.* **2006**, *22*, 983–90.

Hummerjohann, J.; Laudenbach, S.; Rétey, J.; Leisinger, T.; Kertesz, M. A. The Sulfur-Regulated Arylsulfatase Gene Cluster of *Pseudomonas aeruginosa*, a New Member of the cys Regulon. *J. Bacteriol.* **2000**, *182*, 2055–2058.

Ishimaru, Y.; Bashir, K.; Nishizawa, N. K. Zn Uptake and Translocation in Rice Plants. *Rice* **2011**, *4*, 21–27.

Jiang, W.; Zhao, M.; Jin, L.; Fan, T. Differences in Zinc Uptake and Use Efficiency between Different Aerobic Rice Accessions. *Acta Metall. Sin.* **2007**, *13*, 479–484.

Jones, M. G.; Hughes, J.; Tregova, A.; Milne, J.; Tomsett, A. B.; Collin, H. A. Biosynthesis of the Flavor Precursors of Onion and Garlic. *J. Exp. Bot.* **2004**, *55*, 1903–1918.

Kahkonen, M. A.; Wittmann, C.; Kurola, J.; Iilvsniemi, H.; Salkinoja - Salonen, M. S. Microbial Activity in Boreal Forest Soil in Cold Climate. *Boreal Environ. Res.* **2002**, *6*, 19–28.

Kahnert, A.; Vermeij, P.; Wietek, C.; James, P.; Leisinger, T.; Kertesz, M. A. The SSU Locus Plays a Key Role in Organosulfur Metabolism in *Pseudomonas putida* S-313. *J. Bacteriol.* **2000**, *182*, 2869–2878.

Kalaiselvi, P.; Anthoniraj, S. In Vitro Solubilizatlon of Silica and Potassium from Silica Teminerals by Silicate Solubilizing Bacteria. *J. Ecobiol.* **2009**, *24*, 159–68.

Karaman, M. R.; Horuz, A.; Tuşat, E.; Adiloğlu, A.; Fatih, E. Effect of Varied Soil Matric Potentials on the Zinc Use Efficiency of Soybean Genotypes (*Glycine max* L.) under the Calcareous Soil. *Sci. Res. Essays* **2013**, *8*, 304–308.

Kertesz, M. A. Riding the Sulfur Cycle–Metabolism of Sulfonates and Sulfate Esters in Gram-Negative Bacteria. *FEMS Microbiol. Rev.* **1999**, *24*, 135–175.

Kertesz, M. A.; Mirleau, P. The role of Microbes in Plant Sulfur Supply. *J. Exp. Bot.* **2004**, *55*, 1939–1945.

Kertesz, M. A.; Fellows, E.; Schmalenberger, A. Rhizo Bacteria and Plant Sulfur Supply. *Adv. Appl. Microbiol.* **2004**, *62*, 235–268.

Khawilkar, S. A.; Ramteke, J. R.; Response of Applied K in Cereals in Maharashtra. *Agriculture* **1993**, *11*, 84–96.

King, J.; Quinn, J. (1997). The Utilization of Organo Sulphonates by Soil and Fresh Water Bacteria. *Lett. Appl. Microbiol.* **1997**, *24*, 474–478.

Klose, S.; Moore, J. M.; Tabatabai, M. A. Arylsulfatase Activity of Microbial Biomass in Soils as Affected by Cropping Systems. *Biol. Fertil. Soils* **1999**, *29*, 46–54.

Krishnaswamy, R.; Effect of Phosphatic Fertilization on Zinc Adsorption in Some Vertisol and Inceptisol. *J. Indian Soc. Soil Sci.* **1993**, *41*, 251–255.

Landers, D. H.; David, M. B.; Mitchell, M. J. Analysis of Organic and Inorganic Sulfur Constituents in Sediments, Soils and Water. *Int. J. Environ. Anal. Chem.* **1983**, *14*, 245–256.

Leaungvutiviroj, C.; Ruangphisarn, P.; Hansanimitkul, P.; Shinkawa, H.; Sasaki, K. Development of a New Biofertilizer with a High Capacity for N_2 Fixation, Phosphate and Potassium Solubilization and Auxin Production. *Biosci. Biotechnol. Biochem.* **2010**, *74*, 1098–101.

Leustek, T.; Martin, M. N.; Bick, J. A.; Davies, J. P. Pathways and Regulation of Sulfur Metabolism Revealed through Molecular and Genetic Studies. *Ann. Rev. Plant Biol.* **2000,** *51*, 141–165.

Leustek, T.; Saito, K. Sulfate Transport and Assimilation in Plants. *Plant Physiol.* **1999,** *120*, 637–644.

Li, D. X. Study on the Effects of Silicate Bacteria on the Growth and Fruit Quality of Apples. *J. Fruit Sci.* **2003,** *20*, 64–66.

Li, F. C.; Li, S.; Yang, Y. Z.; Cheng, L. J. Advances in the Study of Weathering Products of Primary Silicate Minerals, Exemplified by Mica and Feldspar. *Acta Petrol Mineral.* **2006,** *25*, 440–448.

Li, T.; Bai, R.; Liu, J. X.; Wong, F. S. Distribution and Composition of Extracellular Polymeric Substances in Membrane Aerated Biofilm. *J. Biotechnol.* **2008,** *135*, 52–57.

Lian, B.; Fu, P. Q.; Mo, D. M.; Liu, C. Q. A Comprehensive Review of the Mechanism of Potassium Release by Silicate Bacteria. *Acta Mineral Sinica.* **2002,** *22*, 179.

Liu, D.; Lian, B.; Dong, H. Isolation of Paenibacillus sp. and Assessment of Its Potential for Enhancing Mineral Weathering. *Geomicrobiol. J.* **2012,** *29*, 413–421.

Marschner, H. *Mineral Nutrition of Higher Plants.* Academic: Boston, 1995.

McAfee, J. Potassium, a Key Nutrient for Plant Growth. *Dept. Soil Crop Sci.* **2008,** http://jimmcafee.tamu.edu/files/potassium. (accessed June 18, 2016).

Mortvedt, J. J.; Kelsoe, J. J. Response of Corn to Zinc Applied with Banded Acid-Type Fertilizers and Ammonium Polyphosphate. *J. Fertil.* **1988,** *5*, 83–88.

Neilands, J. B. Siderophores: Structure and Function of Microbial Iron Transport Compounds. *J. Biol. Chem.* **1995,** *270*, 26723–26726.

Olsen, L. I.; Palmgren, M. G. Many Rivers to Cross: the Journey of Zinc from Soil to Seed. *Front. Plant Sci.* **2014,** *5*, 30.

Parmar, P.; Sindhu, S. S. Potassium Solubilization by Rhizosphere Bacteria: Influence of Nutritional and Environmental Conditions. *J. Microbiol. Res.* **2013,** *3*, 25–31.

Pradhan, N.; Sukla, L. B. Solubilization of Inorganic Phosphate by Fungi Isolated from Agriculture Soil. *Afr. J. Biotechnol.* **2005,** *5*, 850–854.

Prajapati, K.; Sharma, M. C.; Modi, H. A. Isolation of Two Potassium Solubilizing Fungi from Ceramic Industry Soils. *Life Sci. Leaflets* **2012,** *5*, 71–75.

Ptashnyk, M.; Roose, T.; Jones, D. L.; Kirk, G. J. D. Enhanced Zinc Uptake by Rice through Phytosiderophore Secretion: A Modelling Study. *Plant Cell Environ.* **2011,** *34*, 2038–2046.

Rainey, P. B. Adaptation of *Pseudomonas fluorescens* to the Plant Rhizosphere. *Environ. Microbiol.* **1999,** *1*, 243–257.

Raj, S. A. Solubilization of Silicate and Concurrent Release of Phosphorus and Potassium in Rice Ecosystem. In *Conference Paper Biofertilizers Technology*, Coimbatore, India, 2004, pp 372–378.

Rajawat, M. V. S.; Singh, S.; Singh, G.; Saxena, A. K. Isolation and Characterization of K-Solubilizing Bacteria Isolated from Different Rhizospheric Soil. In *Proceeding of 53rd Annual Conference of Association of Microbiologists of India;* 2012, p 124.

Ramos Solano, B.; Barriuso, J.; Gutiérrez Mañero, F. J. Physiological and Molecular Mechanisms of Plant Growth Promoting Rhizobacteria (PGPR). In *Plant-Bacteria Interactions.* Wiley: Weinheim, 2008; pp 41–54.

Ramos-Solano, B.; García, J. A. L.; Garcia-Villaraco, A.; Algar, E.; Garcia-Cristobal, J.; Mañero, F. J. G. Siderophore and Chitinase Producing Isolates from the Rhizosphere of *Nicotiana glauca* Graham Enhance Growth and Induce Systemic Resistance in *Solanum lycopersicum* L. *Plant Soil* **2010,** *334*, 189–197.

Raymond, K.; Müller, G.; Matzanke, B. Complexation of Iron by Siderophores a Review of Their Solution and Structural Chemistry and Biological Function. *Top Curr. Chem.* **1984,** *123,* 49–102.

Rengel, Z. Availability of Mn, Zn and Fe in the Rhizosphere. *J. Soil Sci. Plant Nutr.* **2015,** *15,* 397–409.

Richardson, A. E.; Barea, J. M.; Mcneill, A. M.; Prigent-Combaret, C. Acquisition of Phosphorus and Nitrogen in the Rhizosphere and Plant Growth Promotion by Microorganisms. *Plant Soil* **2009,** *321,* 305–339.

Roberts, L. A.; Pierson, A. J.; Panaviene, Z, Walker, E. L. Yellow Stripe1 Expanded Roles for the Maize Iron Phytosiderophore Transporter. *Plant Physiol.* **2004,** *135,* 115–120.

Robinson, N. J.; Procter, C. M.; Connolly, E. L.; Guerinot, M. L.; A Ferric-Chelate Reductase for Iron Uptake from Soils. *Nature* **1999,** *397,* 694–697.

Rose, M. T.; Pariasca-Tanaka, J.; Rose, T. J.; Wissuwa, M. Bicarbonate Tolerance of Zn-Efficient Rice Genotypes is not Related to Organic Acid Exudation, but to Reduced Solute Leakage from Roots. *Funct. Plant Biol.* **2011,** *38,* 493–504.

Rose, T. J.; Impa, S. M.; Rose, M. T.; Tanaka, P. J.; Mori, A.; Heuer, S.; Johnson, B. S. E.; Wissuwa, M. Enhancing Phosphorus and Zinc Acquisition Efficiency in Rice: a Critical Review of Root Traits and Their Potential Utility in Rice Breeding. *Ann. Bot.* **2012,** *112,* 331–345.

Rudrappa, T.; Czymmek, K. J.; Pare, P. W.; Bais, H. P. Root-Secreted Malic Acid Recruits Beneficial Soil Bacteria. *Plant Physiol.* **2008,** *148,* 1547–1556.

Sahai, P.; Srivastava, P.; Singh, S. K.; Singh, A. P. Evaluation of Organics Incubated with Zinc Sulfate as Zn Source for Rice-Wheat Rotation. *J. Ecofriendly Agric.* **2006,** *1,* 120–125.

Sangeeth, K. P.; Bhai, R. S.; Srinivasan, V. *Paenibacillus glucanolyticus,* a Promising Potas-Sium Solubilizing Bacterium Isolated from Black Pepper (*Piper nigrum* L.) Rhizosphere. *J. Spic. Aromat. Crops* **2012,** *21,* 118–124.

Scherer, H. Sulfur in Crop Production—Invited Paper. *Eur. J. Agron.* **2001,** *14,* 81–111.

Schmalenberger, A.; Hodge, S.; Hawkesford, M. J.; Kertesz M. A. Sulfonate Desulfurization in *Rhodococcus* from Wheat Rhizosphere Communities. *FEMS Microbiol. Ecol.* **2009,** *67,* 140–150.

Schmalenberger, A.; Kertesz, M. A. Desulfurization of Aromatic Sulfonates by Rhizosphere Bacteria: High Diversity of the asfA Gene. *Environ. Microbiol.* **2007,** *9,* 535–545.

Sheng, X. F. Growth Promotion and Increased Potassium Uptake of Cotton and Rape by a Potassium Releasing Strain of *Bacillus edaphicus. Soil Biol. Biochem.* **2005,** *37,* 1918–1922.

Sheng, X. F.; He, L. Y. Solubilization of Potassium-Bearing Minerals by a Wild Type Strain of *Bacillus edaphicus* and Its Mutants and Increased Potassium Uptake by Wheat. *Can. J. Microbiol.* **2006,** *52,* 66–72.

Sheng, X. F.; Huang, W. Y. Mechanism of Potassium Release from Feldspar Affected by the Strain NBT of Silicate Bacterium. *Acta Pedologica Sinica* **2002,** *39,* 863–871.

Sheng, X. F.; Zhao, F.; He, H.; Qiu, G.; Chen, L. Isolation, Characterization of Silicate Mineral Solubilizing *Bacillus globisporus* Q12 from the Surface of Weathered Feldspar. *Can. J. Microbiol.* **2008,** *54,* 1064–1068.

Siddiqui, A.; Srivastava, P. C.; Singh, A. P.; Singh, S. K. Effect of Zinc Sulfate and Pressmud Compost Application on Yields, Zinc Concentration and Uptake of Sugarcane. *Indian J. Sugarcane Technol.* **2005,** *20,* 35–39.

Sieh, D.; Watanabe, M.; Devers, E. A.; Brueckner, F.; Hoefgen, R.; Krajinski, F. The Arbuscular Mycorrhizal Symbiosis Influences Sulfur Starvation Responses of *Medicago truncatula. New Phytol.* **2013,** *197,* 606–616.

Sindhu, S. S.; Parmar, P.; Phour, M. Nutrient Cycling: Potassium Solubilization by Microorganisms and Improvement of Crop Growth. In *Geomicrobiology and Biogeochemistry: Soil Biology* Parmar, N, Singh, A, Eds., Springer-Wien: New York, Germany, 2012.

Singh, G.; Biswas, D. R.; Marwah, T. S. Mobilization of Potassium from Waste Mica by Plant Growth Promoting Rhizobacteria and Its Assimilation by Maize (*Zea mays*) and Wheat (*Triticum aestivum* L.). *J. Plant Nutr.* **2010**, *33*, 1236–1251.

Srivastava, P. C.; Gupta, U. C. *Trace Elements in Crop Production*, Oxford and IBH Publishers: New Delhi, 1996; p 356.

Sugumaran, P.; Janarthanam, B. Solubilization of Potassium Containing Minerals by Bacteria and Their Effect on Plant Growth. *World J. Agric. Sci.* **2007**, *3*, 350–355.

Sunitha, kumari, K.; Padma Devi, S. N.; Vasandha, S. Zinc Solubilizing Bacterial Isolates from the Agricultural Fields of Coimbatore, Tamil Nadu. *Curr. Sci.* **2016**, *110*, 2–25.

Suzuki, M.; Tsukamoto, T.; Inoue, H.; Watanabe, S.; Matsuhashi, S.; Takahashi, M.; Nakanishi, H.; Mori, S.; Nishizawa, N. K. Deoxymugineic Acid Increases Zn Translocation in Zn-Deficient Rice Plants. *Plant Mol. Biol.* **2008**, *66*, 609–617.

Tabatabai, M.; Bremner, J. Arylsulfatase Activity of Soils. *Crop Sci. Soc Am. J.* **1970**, *34*, 225–229.

Tan, S.; Han, R.; Li, P.; Yang, G.; Li, S.; Zhang, P.; Wang, W.; Zhao, W.; Yin, L. Overexpression of the MxIRT1 Gene Increases Iron and Zinc Content in Rice Seeds. *Trans. Res.* **2015**, *24*, 109–122.

Thomine, S.; Lelievre, F.; Debarbieux, E.; Schroeder, J. I.; Barbier-Brygoo, H. AtNRAMP3, a Multispecific Vacuolar Metal Transporter Involved in Plant Responses to Iron Deficiency. *Plant J.* **2003**, *34*, 685–695.

Troufflard, S.; Mullen, W.; Larson, T. R.; Graham, I. A.; Crozier, A.; Amtmann, A.; Armengaud, P. Potassium Deficiency Induced the Biosynthesis of Oxylipins and Glucosinolates in *Arabiodopsis thaliana*. *Plant Biol.* **2010**, *10*, 172.

Uroz, S.; Calvaruso, C.; Turpault, M. P.; Pierrat, J. C.; Mustin, C.; Frey-Klett P. Effect of the Mycorrhizosphere on the Genotypic and Metabolic Diversity of the Bacterial Communities Involved in Mineral Weathering in a Forest Soil. *Appl. Environ. Microbiol.* **2007**, *73*, 3019–3027.

Vassilev, M.; Vassileva, N.; Nikolaeva, I. Simultaneous P-Solubilizing and Bio Control Activity of Microorganisms: Potentials and Future. *Trends Appl. Microbiol. Biotechnol.* **2006**, *71*, 137–144.

Vaz Patto, M. C.; Amarowicz, R.; Aryee, A. N. A.; Boye, J. I.; Chung, H.; Martin-Cabrejas, M. A.; Domoney, C. Achievements and Challenges in Improving the Nutritional Quality of Food Legumes. *Crit. Rev. Plant Sci.* **2015**, *34*, 105–143.

Vermeij, P.; Wietek, C.; Kahnert, A.; Wüest, T.; Kertesz, M. A. Genetic Organization of Sulfur-Controlled Arylde Sulphonationin *Pseudomonas putida* S-313. *Mol. Microbiol.* **1999**, *32*, 913–926.

Wakelin, S. A.; Warren, R. A.; Harvey, P. R.; Ryder, M. H. P. Phosphate Solubilization by *Penicillium* spp. Closely Associated with Wheat Roots. *Biol. Fert. Soils* **2004**, *40*, 36–43.

Walter, A.; Römheld, V.; Marschner, H.; Mori, S. Is the Release of Phytosiderophores in Zinc-Deficient Wheat Plants a Response to Impaired Iron Utilization? *Physiol. Plant.* **1994**, *92*, 493–500.

Wang, H. X.; Guo, J. Y.; Xu, W. H. Response and Zinc Use Efficiency of Chinese Cabbage under Zinc Fertilization. *Plant Nutr. Fertil. Sci.* **2011**, *17*, 154–159.

White, P. J.; Karley, A. J.; Hell R, Mendel, R. R.; Eds., *Cell Biology of Metals and Nutrients, Plant Cell Monographs*, Vol. 17. Springer: Berlin, 2010; pp 199–224.

Zarjani, J. K.; Aliasgharzad, N.; Oustan, S.; Emadi, M.; Ahmadi, A. Isolation and Characterization of Potassium Solubilizing Bacteria in Some Iranian Soils. *Arch. Agro. Soil Sci.* **2013,** *77,* 7569.

Zhang, A.; Zhao, G.; Gao, T.; Wang, W.; Li, J.; Zhang, S. Solubilization of Insoluble Potassium and Phosphate by *Paenibacillus kribensis* CX-7: A Soil Microorganism with Biological Control Potential. *Afr. J. Microbiol. Res.* **2013,** *7,* 41–47.

Zhao, F.; Lehmann, J.; Solomon, D.; Fox, M.; Mcgrath, S. Sulfur Speciation and Turnover in Soils: Evidence from Sulfur K-edge XANE Spectroscopy and Isotope Dilution Studies. *Soil Biol. Biochem.* **2006,** *38,* 1000–1007.

Zhao, F.; Sheng, X.; Huang, Z.; He, L. Isolation of Mineral Potassium Solubilizing Bacterial Strains from Agricultural Soils in Shandong Province. *Bio. Div. Sci.* **2008,** *16,* 593–600.

Zuo, Y.; Zhang, F. Effect of Peanut Mixed Cropping with Gramineous Species on Micronutrient Concentrations and Iron Chlorosis of Peanut Plants Grown in a Calcareous soil. *Plant Soil* **2008,** *306,* 23–36.

Zuo, Y.; Zhang, F. Iron and Zinc Biofortification Strategies in Dicot Plants by Intercropping with *Gramineous* Species: a Review. *Agron. Sustain. Dev.* **2009,** *29,* 63–71.

Zuo, Y.; Zhang, F. Soil and Crop Management Strategies to Prevent Iron Deficiency in Crops. *Plant Soil* **2011,** *339,* 83–95.

CHAPTER 14

Sulfur Cycle in Agricultural Soils: Microbiological Aspects

BHOLANATH SAHA[1,*], SUSHANTA SAHA[2], DHANESHWAR PADHAN[2], ARPITA DAS[3], PARTHENDU PODDAR[4], SAJAL PATI[2], and GORA CHAND HAZRA[2]

[1]Deparment of Soil Science, Dr. Kalam Agricultural College, Bihar Agricultural University, Kishanganj, Bihar Agricultural University, Sabour, Bhagalpur, 855107, Bihar, India

[2]Directorate of Research, AICRP on Integrated Farming System, Bidhan Chandra Krishi Viswavidyalaya, Kalyani, Nadia 741235, West Bengal, India

[3]Department of Genetics and Plant Breeding, Bidhan Chandra Krishi Viswavidyalaya, Mohanpur, Nadia 741252, West Bengal, India

[4]Department of Agronomy, Uttar Banga Krishi Viswavisyalaya, Pundibari, Coochbehar 736165, West Bengal, India

[*]Corresponding author. E-mail: bnsaha1@gmail.com

ABSTRACT

Sulfur is the fourth most important essential nutrient element after nitrogen, phosphorus, and potassium for crop growth and nutrition. Besides carbon and nitrogen which are important constituents of plants, microorganisms are also known to influence the availability of sulfur in the soil for absorption by plants. Two predominant forms of soil sulfur are organic and inorganic, in which inorganic component is in the form of sulfate while organic form are particularly sulfate esters and carbon bonded sulfur contributing 75–90% of the total. A major source of soil sulfur for plant nutrition is residues of plant and microbes vis-a-vis elemental residue and external addition of fertilizers including an atmospheric deposition. Deficiency of plat available soil sulfur results in chlorosis, chlorosis, low oil percentage, poor protein biosynthesis, and ultimately low yield. Sulfur cycles through soil, plant, microorganisms and atmosphere are dominated by four major biochemical reactions namely oxidation, reduction, mineralization, and immobilization in soil. The conversion of organically bound sulfur to the inorganic state is termed as mineralization of sulfur and is mediated through microorganisms. The sulfur thus released is either absorbed by plants or escapes to the atmosphere in the form of oxides. In the absence of oxygen, certain microorganisms produce hydrogen sulfide from organic sulfur substrates, especially in waterlogged soils. Various reduced inorganic sulfur compounds are oxidized by a group of bacteria in suitable condition and utilize the energy. The wide range of stable redox states and their interconversion affect sulfur cycle, the fate of applied fertilizer and ultimately its availability to plants and microbes.

14.1 INTRODUCTION

Besides carbon and nitrogen, availability of sulfur (S), phosphorus and many other trace elements in soils are influenced by microorganisms for uptake by plants. The primary source of S for human beings is the sulfur-containing amino acids (SAAs) methionine and cysteine obtained from plant products in the diet. The animal products in the diet also owe this to plants, because the animals obtain S from these two amino acids present in grasses, fodder, and feeds. Sulfur is an essential element for the plant as well as animals and found in nature in combined form, namely, gypsum ($CaSO_4.2H_2O$), pyrite (FeS_2), and in elemental form (S^0). The sulfur is considered as "secondary" nutrient as only because their requirement by the plant is quantitatively less as compared to the primary nutrients. In spite of the essentiality, very less importance was given to S addition in the field in the past mainly due to restricted area and crops that response with the fertilizer and contribution through major fertilizer or from natural sources (Tandon, 2011).

An amount of 5–250 kg/ha/year of sulfur is added in the soil through rainfall depending on industrial activity and burning of fossil fuel. Highly weathered soil away from the sea and industrial activity are generally prone to sulfur deficiency. In earth, the lithosphere is the major sink of sulfur (24.3×10^{18} kg) followed by hydrosphere (1.3×10^{18} kg), pedosphere (2.7×10^{14} kg), and atmosphere (4.8×10^9 kg), respectively (Stevenson, 1982). Sulfur, with atomic weight 32.064, exists in various oxidation states. This is indicated by the oxidation number in several compounds, namely sulfides (-2), polysulfide

(−1), elemental sulfur (0), thiosulfate [(−2) and (+6)], sulfite (+4), and sulfate (+6) (Rao, 1999).

Sulfur deficiency is widespread all over the world soils and in many crops too. The reason being the nutrient management strategies mainly depended on the application of major nutrients like nitrogen, phosphorus, and potassium (NPK) fertilizers, without replenishing secondary and micronutrients removal (Sahrawat et al., 2007). Apart from that, progressively higher removal of sulfur owing to high production level led to the appearance of sulfur deficiency in many crop species (Tandon, 2011). The availability of sulfur for plant uptake largely depends on the cycling of S in the biosphere and the rate of conversion of organic S to inorganic sulfate. Such transformation processes in the soil are typically regulated by a diverse group of microorganisms. In this chapter, microbial processes that influence sulfur cycle and their transformation into different pools are summarized.

14.2 FUNCTIONS OF SULFUR IN CROP NUTRITION

Although sulfur (S) is recognized as a secondary nutrient element, it plays a crucial role in crop production. Most of the plants generally uptake S 9–15% of N uptake and the uptake of P and S in some crop species are similar. It is expected that a healthy crop may contain ~0.1–0.4% of S by weight. However, its content in different plant species depends upon the crop growth stages and plant parts. Oilseed crops require a higher amount of S followed by legumes and cereals. The higher requirement of oilseeds is due to synthesis glucosinolates and glucosides which increase the quality of oils.

Sulfur is mainly used in agriculture as a soil amendment for amelioration of sodic soils and for improving the quality of irrigation water; as a plant nutrient for correcting sulfur deficiency; increasing crop yields and improving the quality of the crop produce; as a chemical agent to acidulate phosphate rocks and to manufacture phosphoric acid, phosphatic fertilizers, ammonium sulfate, and other sulfate containing fertilizers; also, as a pesticide including various herbicides. Sulfur has a significant role in increased food production since it is used in the manufacture of nitrogenous, phosphatic, and potassic fertilizers.

Sulfur is required for the synthesis of SAAs methionine (21%), cysteine (26%), and cystine (27%), which are essential components of protein. Approximately 90% of plant sulfur is present in these amino acids (Tandon and Messick, 2002). A wide range of biological compounds containing sulfur as an important constituent namely, coenzyme-A, molybdenum cofactor

(MoCo), lipoic acid, chloroplast lipid sufloquinovosyl diacy-glycerol, and many secondary compounds are reported in plant system (Leustek, 2002). One of the main functions of sulfur in proteins is the formation of disulphide bonds between polypeptide chains. This bridging is achieved through the reaction of two cysteine molecules; forming cystine. Linking of the two cysteine units within a protein by a disulphide bond (–S–S–) will cause the protein to fold. Disulphide linkages are therefore important in stabilizing and determining the configuration of proteins. It is also responsible for important structural, regulatory and catalytic functions in the context of proteins, and as a major cellular redox buffer in the form of the tripeptide glutathione and certain protein such as thioredoxin, glutaredoxin, and protein disulfide isomerase. A feature of much sulfur containing compound is that the S moiety is often directly involved in the catalytic or chemical reactiveness of the compound (Jamal et al., 2010). Therefore, non-availability of adequate amounts of sulfur in soils results in low photosynthetic activity, growth retardation, yellowing of young leaves and ultimately poor yield. Nitrogen fixation by leguminous plants also retarded due to S deficiency. Thus, a narrow N:S ratio is necessary for optimum N metabolism in plants. For an optimum yield of legumes and cereals, the desired N:S ratios to be 15–16:1 and 11–12:1, respectively (Pasricha and Sarkar, 2002).

Intensive cultivation, growing of high yielding cultivar with high sulfur requirement, application of high analysis NPK fertilizers that contain less sulfur impurities, less use of S-containing fertilizer and pesticides, reduced S emissions from industry and less atmospheric deposition might be the probable reasons for increasing S deficiency in soil and which is reported from several regions of the World. The Sulfur requirement of plant is closely associated with N requirement as both these elements are constituents of protein and they involved in chlorophyll synthesis in the plant. Sulfur also influences biological N fixation as S is a constituent of enzyme nitrogenase. Crops with high N requirement also require a high amount of sulfur. The major source of S in most of the soils is organic matter. Mineralization of organic S to inorganic sulfate is necessary as plant uptake S in the form of sulfate anion. Decomposition of organic matter and subsequent release of inorganic S is affected by environmental conditions viz. temperature, moisture which also influences crop growth. Sulfate is mobile in soils to some extent and therefore may leach down especially in high rainfall area. Generally, sulfur deficiency occurs in soils with low organic matter, coarse texture, and high rainfall conditions. But, S deficiency can emerge in other areas also, especially in intensive cultivation zone. Besides the application of NPK fertilizer adequate sulfur fertilization is essential to meet the S requirement. Optimum doses of S obviously depend upon the type of crop and yield target, soil fertility, organic matter content and contribution from irrigation water and manure. Crops of brassica species, namely rapeseed, mustard, cabbage and forage crops such as alfalfa, bermuda grass have a higher requirement of sulfur as compared to cereal crops. Sulfur deficiency is generally found in Soils with less than 2% organic matter content and in coarse-textured soils. A significant amount of sulfur may be added through irrigation water hence, irrigation water should be analyzed for S content. Among various soluble sulfur fertilizers, majority contains sulfur as sulfate, although sulfur in the form of bisulfites, thiosulfates, and polysulfides are also available in the market. Elemental S, on addition to soil oxidized to sulfate prior to plant uptake. This biological transformation produces acidity in soils and thus it reduces the incidence of certain bacterial diseases and improved the availability of certain nutrients for plant nutrition.

14.3 DIFFERENT POOLS OF SULFUR IN SOILS

Sulfur in soils occurs in different oxidation states depending upon the soil environment. A lion share of total S is contributed by organic S while the inorganic S shared only a small fraction of total S in soils. In surface soils, the total sulfur content is a function of organic matter content. Thus, it is well recognized that organic matter is a fair indicator of the amount of S in soils. With depth, the total, as well as other forms of S, decreased. Amount of S present in soils also dependent upon the parent materials from where the soil is being formed. Soils developed from parent materials rich in sulfur containing minerals show high S content. Application of organic manures, crop residues addition, and atmospheric deposition of S along with rainfall could be the different pathways of S addition to soils. Sulfur is also added to soil as impurities along with the complex fertilizer materials. Nowadays, sulfur containing fertilizers are also applied to high S requiring crops like oilseeds to sustain productivity.

14.4 SULFUR CYCLE

Soil is the primary component of the global biogeochemical S cycle, acting as a source and sink for various S species and mediating changes of various oxidation states. Sulfur in terrestrial ecosystems is found in a wide variety of inorganic and organic forms, each of which may play characteristic biological and chemical roles. The proportion of organic and inorganic sulfur in soils varies according to soil type and other climatic variables. The dynamics of organic and inorganic forms of sulfur in soils is the result of microbial activities. Microbes participating in nutrient cycling result breakdown of the complex organic substrates and converts the organic S to inorganic S which is easily taken up by the crop plants. Both bacteria and fungi involved in the oxidation of S although bacteria dominate the process. Among bacteria, *Thiobacillus* sp. mainly participates in S oidation in soil. Sulfur cycle in the soil is similar to the nitrogen cycle being both the nutrients mainly found associated in organic forms. Cycling of sulfur between organic to inorganic forms and also between oxidized and reduced states are influenced mainly by bacteria and other microorganisms. Sulfur mineralization denotes the microbial process that converts organic forms of sulfur to the inorganic forms (sulfur/sulfates) that either assimilated by the plants or lost to the atmosphere as oxides. In oxygen depleting condition especially in waterlogged soil, hydrogen sulfide is produced from the organic sulfur compound by certain microorganisms.

Plant uptakes sulfur primarily as inorganic sulfate however, the sulfate ester or carbon bonded S pools are potentially available, probably due to interconversion of both these pools of sulfur to inorganic sulfate by soil microbes. Besides this mineralization, immobilization of sulfate to sulfate esters and subsequently to carbon-bound sulfur, are also mediated by soil microbes. These mineralization-immobilization processes in soil depend on the presence of the microbial community and their metabolic activities (Kertesz and Mirleau, 2004). Thus, the pools of sulfur in soils are dynamic in nature. The inorganic forms of S are immobilized to organic forms whereas; an organic form of S is simultaneously mineralized to inorganic S. These processes are highly dependent upon the presence of microbial communities in soils and their activities. The cycling of S in the biosphere with interconversion from organic form to inorganic form and vice versa is regulated by different soil properties.

14.4.1 FORMS OF SULFUR IN SOIL

Sulfur present in organic and inorganic forms, the former is the dominant fractions of total S in soil (Fig.14.1). The inorganic form exists from -2 to $+6$ oxidation states although the plants are capable of utilizing the $+6$-oxidation state of SO_4^{2-} and thus the reduced forms of S must be oxidized prior to plant uptake. However, in well-drained aerobic soils majority of inorganic S present in +6 oxidation state of SO_4^{2-}. Sulfate sulfur (SO_4^{2-}-S) is by far the most dominant fraction of inorganic S and exists as easily soluble or readily available SO_4^{2-}, adsorbed SO_4^{2-} (on to the positive edge of the colloidal complex), insoluble SO_4^{2-} or insoluble SO_4^{2-} occluded with $CaCO_3$.

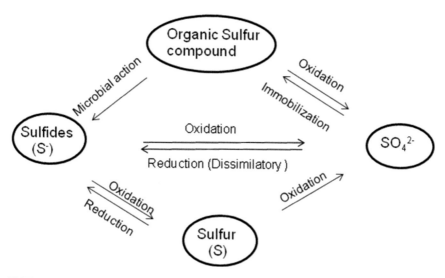

FIGURE 14.1   Sulfur cycle.

The organic form of S is the dominant fractions of total S in soil. It is mainly found in the form of carbon bonded and ester sulfates which are mainly found in plant and animal proteins as well as in protoplasm of microorganism as SAAs (i.e., cysteine, cystine, and methionine). It is also considered as a heterogeneous mixture of soil organisms, and of partly decomposed plant, animal, and microbial residues. Thus, little is known about the chemistry and behavior of organic S in soil. However, researchers have identified that the ester sulfate mineralized faster than the carbon bonded S and it is considered as labile or potentially available pools of S in soil (Kovar and Grant, 2011).

Microbial biomass of sulfur is an important pool of sulfur which governs the soil sulfur cycling and is helpful in understanding the complex mechanism of sulfur availability to plant. It is estimated as the flush of extractable S following the chloroform fumigation method (Sagar et al., 1981). Microbial biomass S shares a meager part of organic S accounting ~0.9–2.6% of total organic S in the majority of agricultural soils (Chapman, 1987). Biomass S might become available to plants in a period of decreasing biomass and sulfate may be immobilized during a period of biomass increase.

14.4.2   SULFUR TRANSFORMATION IN AGRICULTURAL SOILS

Sulfur transformation in the soil is a complex phenomenon and is regulated by various soil components. It involves the following processes: (1) organic sulfur compounds decomposed into subunits and ultimately converted to inorganic compounds through the mineralization process, (2) inorganic sulfur compounds assimilates into the protoplasm of microbes, through the process immobilization, (3) inorganic sulfur compounds converted into elemental sulfur through oxidation, and (4) sulfate reduction. In soils mainly, microbial mediated processes are responsible for such transformation in soil. Aerobic as well as anaerobic both microorganisms involve in organic S formation, though S containing microbes ranges only 1–3% of total biomass (Strick, 1984; Fig.14.1) sulfur cycle (Nakas, 1984; Chapman, 1987). However, sulfur turnover and recycling are rapid due to the short life span of microorganisms (Smith and Paul, 1990). Thus, microbial biomass is considered as most active form of organic S in soil, and may contribute to significant amount of the mineralized S. As pointed out by Freney and Swaby (1975), many of the transformations of S in soils are cyclic because of the relative ease with which S is changed from organic to inorganic and from reduced to oxidized forms, and vice versa. The dynamics of organic and inorganic forms of S in the soil is dependent upon the mineralization–immobilization turnover of the organic substrate. Major S transformations processes, namely, mineralization, immobilization, oxidation, and reduction are briefly described in next section.

14.4.3   MINERALIZATION OF ORGANIC SULFUR

As the plant uptake mainly the inorganic sulfate for its nutrition, the organic S in soils must undergo the process of mineralization prior to its uptake by plants. Mineralization is the enzymatic breakdown of complex organic S to inorganic S and the process is mediated by the diverse group of soil microbes. The mineralization process is dependent upon the C:S of the added organic substrates. Annually 1.0–10.0% of organic S mineralized which is sufficient for normal crop growth if the different losses of S from the soil are negligible. Factors such as temperature, moisture, pH, C:S ratio and substrate availability affect the process of mineralization. Mineralization of organic S in humic substrates becomes faster in the presence of O_2 and favored by high temperature in the mesophilic range. Liming of acid soils triggers the process of mineralization.

According to the conceptual model (McGill and Cole, 1988) for cycling of organic C, N, S, and P through soil organic matter, the pathway of mineralization of organic S follows the biological and biochemical mineralization process (Fig. 14.2). In biological mineralization process, the complex organic moiety is degraded by the microbes to meet up their energy requirement and as such the inorganic sulfate is released as byproduct as the carbon oxidized to carbon dioxide. In contrast, biochemical mineralization process is the release of sulfate from the labile ester sulfate pool through enzymatic hydrolysis. Mineralization potential of soil organic S has been estimated by kinetic equations for S mineralization based on the release of S from incubated soils (Ghani et al., 1991).

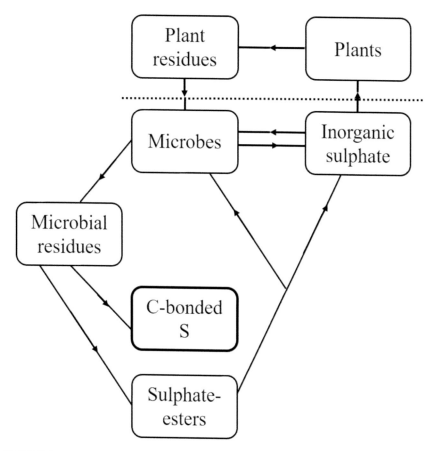

FIGURE 14.2 Interrelationship of C and S cycling within soil–plant systems.

Source: Adapted and modified from McGill and Cole (1981).

14.4.4   IMMOBILIZATION OF SULFUR

The microbial conversion of inorganic sulfate to the organically bound form of S is known as immobilization. Inorganic species of S namely, sulfate, hyposulfite, sulfoxylate, thiosulfate, persulfate, sulfide, elemental sulfur, sulfite, tetrathionate, and thiocyanate converted to cysteine, cystine, methionine, taurine, and undecomposed proteins of the organic S. It is a reductive process and is governed by both the aerobic and anaerobic chemotrophs and phototrophs. The C/S ratio of different microbial groups varies widely being 57:85 in bacteria and 180:230 for fungus. On average, most microorganisms have S content of 0.1% and 1.0% of their dry weight. During the process of immobilization, a certain amount of S assimilated into organic matter forming a covalent bond with C (Strickland et al., 1987). When a high amount of sulfate added to the soil, it adsorbs on soil colloidal complex and subsequently converted to low molecular weight organic S compounds (Jez, 2008) which further polymerized to large organic S compounds (Strickland et al., 1986). Amount of S immobilized by microbes is indicative of the microbiological survey which can be estimated by measuring inorganic sulfate released in chloroform fumigation technique. Although a small amount of S is actually sequestered in microbial biomass this fraction is extremely labile in nature (Balota et al., 2003) and an important indicator of plant available S in soil.

Approximately C:S ratio of above 50:1 in the substrate results dominance of the immobilization process over the mineralization process. Critical C:S ratio of substrates, diversity within microbes present in soil and soil environment *viz.* moisture content, temperature, organic matter, and other factors control rate of immobilization.

14.4.5   OXIDATION OF INORGANIC SULFUR

The oxidation of reduced S in soils is usually regarded as a microbial process (Burns, 1967; Wainwright, 1978c), although some nonbiological oxidation or abiotic oxidation of the element does occur. Chemoautotrophic and photosynthetic bacteria are responsible for the oxidation of elemental and reduced inorganic S compounds (namely, sulfide, sulfite, and thiosulfate) to sulfate. Sulfur released from the degraded protein of either plant or animal origin accumulates in soil and subsequently, oxidized to either sulfates in aerobic condition or produce hydrogen sulfide in absence of oxygen (waterlogged soils). Under anaerobic conditions, hydrogen sulfide accumulates due to the reduction of sulfates in absence of molecular oxygen which again can oxidize to produce sulfate if aerobic conditions prevail in the later period of crop growth (Behera et al., 2014).

A wide spectrum of microorganisms is capable of oxidizing reduced S in soils, including members of the genus *Thiobacillus,* a number of hetero-trophs, the photosynthetic sulfur bacteria, and the colorless, filamentous sulfur bacteria. Of these, only the *Thiobacilli* and heterotrophs play an important part in **S** oxidation in most agricultural soils with exceptions to flooded soils.

The abiotic oxidation of reduced S consists of two steps: first oxidation of sulfide to elemental S and secondly conversion of elemental S to sulfate. The first step is rapid in aerobic soils while the second step is a slow process (Wiklander et al., 1950):

a) $2S + 3O_2 + 2H_2O \xrightarrow{\quad\quad} 2H_2SO_4 \xrightarrow{\text{Ionization}} 2H^+ + SO_4 \text{ (Aerobic)}$

b) $CO_2 + 2H_2S \xrightarrow{\text{Light}} (CH_2O) + H_2O + 2S$

OR

$H_2 + S + 2CO_2 + H_2O \xrightarrow{\text{Light}} H_2SO_4 + 2(CH_2O) \text{ (Anaerobic)}.$

14.4.6   SULFUR OXIDATION IN PADDY SOILS

It seems quite impractical that S oxidation could occur under submerged or flooded paddy soils. Nowadays rice cultivation is carried out with alternate wetting and drying methods instead of continuous submergence such that the soil moisture content may range from air dry to saturated. However, even in permanently flooded soils, a typical surface oxidized layer is formed due to the supply of oxygen from algae and weeds. Rice also transmits some amount of oxygen to reduced soils which are absorbed by stomata or produced during photosynthesis. Thus, form an oxidized layer though it is cultivated under flooded situations which support the growth and metabo-lism of aerobic microorganisms. These microorganisms could oxidize the reduced S to oxidized sulfate. Therefore, S occurs in all the oxidation states from +6 to −2. Although the rice ecology and sulfur oxidation are not widely studied, some researchers isolated *T. thioparus* (Freney et al., 1982) and *T. thiooxidans* (Mouraret and Baldensperger, 1977) from paddys oils. Bacteria genera *Beggiatoa*, some heterotrophs, and purple and green sulfur bacteria are important in oxidation of reduced S in the rice rhizosphere. The oxidation of sulfide to sulfate benefits rice growth since the H_2S causes some physiological disorders of rice.

14.4.6 REDUCTION OF SULFATE

Plants and microorganisms uptake sulfate form soil and assimilate into proteins which are referred to as "assimilatory sulfate reduction." In contrast, the reduction sulfate to hydrogen sulfide (H_2S) by sulfate-reducing bacteria (e.g., *Desulfovibrio* and *Desulfotomaculum*) is called as "dissimilatory sulfate reduction" and makes sulfur less available for plants. Such a reduction process is favored under alkaline and anaerobic soil conditions.

Under anaerobic condition, for example, calcium sulfate is reduced by bacteria under genus *Desulfovibrio* and *Desulfotomaculum* to produce H_2S:

$$CaSO_4 + 4H_2 \text{ -----------> } Ca(OH)_2 + H_2S + H_2O.$$

Hydrogen sulfide thus produced is again oxidized to release elemental sulfur by some species of green and purple phototrophic bacteria (e.g., *Chlorobium, Chromatium*):

$$CO_2 + 2H_2 + H_2S \xrightarrow{\text{Light}} (CH_2O) + H_2O + 2S.$$

Sulfate reducing bacteria in soil classified under genera *Desulfovibrio, Desulfotomaculum*, and *Desulfomonas* (all obligate anaerobes), although *Desulfovibrio desulfurican* is most ubiquitous in soils. They are obligate anaerobes and reduce sulfates in waterlogged or reduced soils. Whereas, *Desulfotomaculum* species are thermophilic obligate anaerobes which are capable of reducing sulfates even in dry soils. An enzyme called "desulfurases" or "bisulfate reductase" is excreted by sulfate-reducing bacteria, which is responsible for S reduction in soil. The rate of sulfate reduction is triggered with increase moisture availability, soil organic matter content, and increased temperature.

14.5 MICROBES INVOLVED IN SULFUR TRANSFORMATION IN SOILS

Release of S from elemental S and sulfide minerals are influenced by soil microorganisms. Bacteria, archaea, and fungi are involved in the oxidation of sulfur; however, bacteria namely *Thiobacillus* spp. played the major role.

In the case of archaea, members of Sulfolobales are only capable of aerobic oxidation of sulfur (Setter et al., 1990). Fungal species namely, *Aureobasidium pullulans, Alternari* spp. and *Epicoccum nigrum*, and a range of *Penicillium* spp., *Myrothesium circutum*, and *Aspergillus* spp. are involved in the oxidation of elemental-S and thiosulfate (Wainwright,1978; Shinde et al., 1996). Bacteria classified under the groups chemolithoautotrophs, chemolithoheterotrophs and chemolithomesotrophs are involved in the oxidation of sulfur (Aragono 1991; Vidyalakshmi et al., 2009).

14.5.1 BACTERIA OF GENUS THIOBACILLUS

Bacterial species belongs to genus *Thiobacillus* (aerobic, nonfilamentous, obligate chemolithotrophic, and nonphotosynthetic) are capable of oxidising elemental sulfur to sulfates, for example, *T. ferrooxidans* and *T. thiooxidans*.

14.5.2 GREEN AND PURPLE SULFUR BACTERIA

Green and purple S bacteria (photolithotrophs) of genera *Chromatium, Chlorbium*, and *Rhodopseudomonas* are also responsible for oxidation of sulfur in aquatic condition (Madigan and Martinko, 2006). They come under Thiorhodaceae and Chlorohacteriaceae families. These groups of microorganisms grow under anaerobic conditions utilizing light as source of energy and carbon dioxide as source of carbon, and oxidize reduced sulfur compounds.

14.5.3 COLORLESS FILAMENTOUS SULFUR BACTERIA

Bacterial genera of *Thiothrix, Beggiatoa, Thiospirillopsis*, and *Thioploca* can oxidize sulfide to sulfate (Starkey, 1950). Certain heterotrophic bacteria (*Bacillus, Pseudomonas*, and *Arthrobacter*) fungi (*Aspergillus* and *Penicillium*) and some actinomycetes are also able to oxidize the reduced sulfur compounds in soil. Oxidation of sulfur and H_2S may produce sulfuric acid, which reduce the pH of soils that could favor the availability of micronutrients to plants. Thus, sulfuric acid produced can make an alkaline soil more suitable for cultivation (Zia et al., 2007). Moreover, it solubilizes and increase availability of plant nutrients namely P, K, Ca, Mg, and so forth (Chien et al., 2011; Karimizarchi et al., 2014).

14.5.4 CHEMOLITHOAUTOTROPHS

Bacteria belonging to this group are capable of oxidizing sulfur and carbon compounds to meet energy requirement for their growth. *Thiobacillus thioparus, T. neapolitanus, T. denitificans, T. thiooxidans, T. ferrooxidans, T. Halophilus,* and *Thiomicrospira* spp. are classified under chemolithoautotrophs.

14.5.5 CHEMOLITHOHETEROTROPHS

Bacteria under this group oxidize sulfur and carbon from organic compounds to meet energy requirement for their growth. *Thiobacillus novellus, T. acidophilus, T. aquaesulis, Paracoccus dentrificans, P. versutus, Xanthobacter tagetidis, Thiospaera pantotroph,* and *Thiomicrospira thasirae* are classified under chemolithoheterotrophs (Prasad and Shivay, 2016).

14.5.6 CHEMOLITHOMESOPTROPHS

Bacteria under this group can oxidize sulfur and carbon from inorganic as well as organic compounds present in soil to meet energy requirement for their growth. *Thiobacillus denitrificans* and *T. ferrooxidans* are categorized under chemolithomesoptrophs. Several enzymes are associated with sulfur oxidation namely tetrathionate hydrolase, thiosulfate dehydrogenase, and sulfur oxygenase (Friedrich et al., 2001; Keppler et al., 2000), whereas under anaerobic conditions like paddy soil, sulfates are reduced to H_2S by microbes and may produce foul smell. Around 220 species of microbes under 60 genera are identified for their sulfate-reducing potential (Barton and Fauque, 2009). Among them, *Desulfobacteriales, Desulfovibrionales,* and *Syntrophobacterales* are considered as the largest group (about 23 genera; Muzer and Stams, 2008) followed by the genera *Desulfosporomusa, Desulfotomaculum,* and *Desulfosporosium* (Prasad and Shivay, 2016).

Both aerobic and anaerobic bacteria and nonfilamentous (*Thiobacillus*) to filamentous forms (*Thiothrix, Beggiatoa,* and *Thioploca*) are involved in oxidizing organic sulfur compounds. Besides bacteria, many fungi and actinomycetes can also oxidize certain organic sulfur compounds (*Penicillium, Aspergillus, Microsporeum,* etc.). Among them, *Thiobacillus* are unique as these produce sulfuric acid in the presence of elemental S that ultimately lowers down soil pH. On prolonged incubation with *Thiobacillus* soil pH may fall around 2.0. Soil acidification caused due to the formation of sulfuric acid helps in controlling many plant diseases in sulfur fertilized soils namely control of potato scab caused by *Streptomyces scabies,* the rot of sweet potatoes caused by *S. ipomoeae,* etc. Inoculation of soil with *Thiobacilli* along with sulfur fertilization minimizes losses of sulfur especially under acidic soil conditions (below pH 5.0). Thus, *Thiobacilli* can effectively be used as biofertilizer to make alkali soil suitable for cultivation. The acid produced in such way helps in better nutrient solubilization and mobilization which ultimately increase the availability of phosphate, potassium, calcium, manganese, aluminum, and magnesium to plants (Chien et al., 2011; Karimizarchi et al., 2014).

14.6 ENHANCEMENT OF SULFUR SUPPLY THROUGH SULFUR OXIDISING BACTERIA AND MYCORRHIZAL ASSOCIATIONS

It is reported that most of the fungi are involved in mineralizing the organic S, that is, sulfate esters to inorganic S (Klose et al., 1999). Sulfur cycling in the biosphere involves the complex interactions between a diverse group of microorganisms and the different soil components. Arbuscular mycorrhizal (AM) fungi form a symbiotic association with most of the plant species for their growth and development (Wang and Qiu, 2006). AM fungi infect the root and penetrate root cortical cells and produced arbuscules which are microscopic-branched structures. These structures are helpful in metabolite exchange between plant and fungi. Moreover, extraradicular hyphae of AM fungi help bacterial populations to colonize by providing surface area. Reports on the interactions between AM fungi and phosphorus (P) mobilizing bacteria are not uncommon (Hodge and Storer, 2014). Mycorrhizal fungi and the associated microbes increase mobilization of S, N, and P in the soil which otherwise remain inaccessible to plants thus ultimately increase their availability to plants (Richardson et al., 2009). Besides this, the mycorrhizal fungi interact with sulfur oxidizing bacteria (*Thiobacillus* sp.) in the rhizosphere soils of certain plants and result three different types of interactions namely positive, neutral, and negative on either the mycorrhizal root association or on a particular component of the rhizosphere (Mostafavin et al., 2008).

14.7 CONCLUSION

Intensive cultivation of high yielding varieties of crops along with a heavy dose of high analysis NPK fertilizers having good facilities if irrigation could obtain target yield but leads to the mining of other secondary as well as micronutrients, of which sulfur is of no exception. Wide application of sulfur-free fertilizers in intensively cultivated soils, particularly in humid tropics is of great concern and it has become a deterrent toward achieving optimum production. The dynamics of soil sulfur is an integral process of the cycles of S through the soil, plant, and microbial continuum. This cycle is of prime importance in the S nutrition of higher plants for their protein biosynthesis as well as oil quality of oilseed crops. The whole process of S transformation occurs through microbial interventions. Transformation of organic and inorganic sulfur forms in the soil controlled by the group of microorganisms determines the sulfur nutrition of plants and others. Likewise, factors and processes that affect the release of sulfur from the microbial biomass and its subsequent availability to plants require further investigation.

KEYWORDS

- **crop nutrition**
- **enzymes**
- **microbial transformation**
- **sulfur**
- **soil pools**

REFERENCES

Aragono, M. Aerobic Chemolithoautotrophic Bacteria. In *Thermophilic Bacteria*; Christjansson, J. K., Ed., CRC Press: Boca Raton, 1991; pp 7–103.

Balota, E. L.; Filho, A. C.; Andrade, D.S.; Dick, R. P. Microbial Biomass in Soils under Different Tillage and Crop Rotation Systems. *Biol. Fert. Soils* **2003**, *38*, (1), 15–20.

Barton, L. L.; Fauque, A. J. Biochemistry, Physiology and Biotechnology of Sulfate Reducing Bacteria. *Adv. Appl. Microbiol.* **2009**, *68*, 41–98.

Bauld, J. Transformation of sulfur Species by Phototrophic and Chemotrophic Microbes. In *The Importance of Chemical "Speciation" in Environmental Processes*; Springer: Berlin, Heidelberg, 1986; pp 255–274.

Behera, B. C.; Mishra, R. R.; Dutta, S. K.; Thatoi, H. N. Sulfur Oxidizing Bacteria in Mangrove Ecosystem: A Review. *Afr. J. Biotechnol.* **2014**, *13* (29), 2897–2907

Chapman, S. J. Microbial Sulfur in Some Scottish Soils. *Soil Biol. Biochem.* **1987**, *19*, 301–305.

Chien, S. H.; Gearhart, M. M.; Villagarcía, S. Comparison of Ammonium Sulfate with Other Nitrogen and sulfur Fertilizers in Increasing Crop Production and Minimizing Environmental Impact: a Review. *Soil Sci.* **2011,** *176* (7), 327–335.

Fitzgerald, J. W. Naturally Occurring Organo Sulfur Compound in Soil. In *Sulfur in the Environment Ecological Impacts, Part II*; Nriagu, J. O., Ed.; John Wiley & Sons: New York, 1978; pp 391–443.

Friedrich, C. G.; Rother, D.; Bardischewsky, F.; Quentmeier, A.; Fischer, J. Oxidation of Reduced Inorganic Compounds by Bacteria: Emergence of a Common Mechanism. *Appl. Environ. Microbiol.* **2001,** *67* (7), 2873–2882.

Gahan, J.; Schmalenberger, A. The Role of Bacteria and Mycorrhiza in Plant Sulfur Supply. *Front. Plant Sci.* **2014,** *5*, 723. https://doi.org/10.3389/fpls.2014.00723.

Gharmakher, N. H.; Piutti, S.; Machet, J. M.; Benizri, E.; Recous, S. Mineralization-Immobilization of Sulfur in a Soil During Decomposition of Plant Residues of Varied Chemical Composition and S Content. *Plant Soil* **2012,** *360* (1), 391–404.

Gupta, V. V. S. R.; Lawrence, J. R.; Germida, J. J. Impact of Elemental Sulfur Fertilization on Agricultural Soils. I. Effects on Microbial Biomass and Enzyme Activities. *Can. J. Soil Sci.* **1988,** *68*, 463–473.

Hodge, A.; Storer, K. Arbuscular Mycorrhiza and Nitrogen: Implications for Individual Plants through to Ecosystems. *Plant Soil.* **2015,** 386 (1–2), 1–9. https://doi.org/10.1007/s11104-014-2162-1.

Jez, J. Sulfur: *A Missing Link between Soils, Crops, and Nutrition*; American Society of Agronomy, ASA-CSSA-SSSA, 2008, p 323.

Karimizarchi, M.; Aminuddin, H.; Khanif, M. Y.; Radziah, O. Elemental Sulfur Application Effects on Nutrient Availability and Sweet Maize (*Zea mays* L.) Response in a High pH Soil of Malaysia. *Mal. J. Soil Sci.* **2014,** *18*, 75–86.

Keppler, U.; Bennet, B.; Rethmeier, J.; Schwarz, G.; Deutzmann, R.; McEwan, A. G.; Dahl, C. Sulft: Cytochrome c Oxidoreductase from *Thiobacillus novellus*, Purification, Characterization and Molecular Biology of a Heterdimeric Member of Sulfite Oxidase Family. *J. Biol. Chem.* **2000,** *275* (18), 13202–13212.

Kertesz, M. A., Mirleau, P. The Role of Soil Microbes in Plant Sulfur Nutrition. *J. Exp. Bot.* **2004,** *55* (404); 1939–1945.

Klose, S; Moore, J. M.; Tabatabai, M. A. Arylsulphatase Activity of Microbial Biomass in Soils as Affected by Cropping Systems. *Biol. Fertil. Soils.* **1999,** *29*, 46–54. https://doi.org/10.1007/s003740050523.

Kovar, J. L.; Grant, C. A. Nutrient Cycling in Soils: Sulfur. Publications from USDA-ARS/UNL Faculty, Paper 1383, 2011.

Lamers, L. P. M.; van Diggelen, J. M. H.; den Camp, H. J. M. O.; Visser, E. J. W.; Lucassen E. C. H. E. T.; Vile, M. A.; Jetten, M. S. M.; Smolders, A. J. P.; Roelofs, J. G. M. Microbial Transformations of Nitrogen, Sulfur, and Iron Dictate Vegetation Composition in Wetlands: A Review. *Front. Microbiol.* **2012,** *3*, 156. https://doi.org/10.3389/fmicb.2012.00156.

Madigan, M. T.; Martinko, J. M. Brock Biology of Microorganisms. Pearson Prentice Hall: NJ, USA, 2006; p 1056.

McLaren, R. G.; Keer, J. J.; Swift, R. W. Sulfur Transformations in Soils Using Sulfur-35 Labelling. *Soil Biol. Biochem.* **1985,** *17*, 73–79.

Muzer, G.; Stams, A. J. The Ecology of Biotechnology of sulfate-Reducing Bacteria. *Nat. Rev. Micriobiol.* **2008,** *6*, 441–454.

Pasricha, N. S.; Sarkar, A. K. Secondary Nutrients. In *Fundamentals of Soil Science*; Sekhon, G. S., Chhonkar, P. K., Das, D. K., Goswami, N. N., Narayanasamy, G., Poonia, S. R., Rattan, R. K., Sehgal, J., Eds., Indian Society of Soil Science: New Delhi, India, 2002; pp 381–389.

Prasad, R.; Shivay, Y. S. Sulfur in Soil, Plant and Human Nutrition. *Proc. Natl. Acad. Sci. India Sect. B: Biol. Sci.* **2016**, 1–6. https://doi.org/10.1007/s40011-016-0769-0.

Rajvaidya, N.; Markandey, D. K. Genetical and Biochemical Applications of Microbiology. In *Microbial Genetics*; APH Publishing: New Delhi, India, 2006; p 345.

Rao, C. N. R. *Understanding Chemistry*; University Press (India) Ltd.: Hyderabad, 1999.

Richardson, A. E.; Barea, J. M.; Mcneill, A. M.; Prigent-Combaret, C. Acquisition of Phosphorus and Nitrogen in the Rhizosphere and Plant Growth Promotion by Microorganisms. *Plant Soil* **2009**, *321*, 305–339. DOI: 10.1007/s11104009-9895-2.

Roy, A. B.; Trudinger, P. A. *The Biochemistry of Inorganic Compounds of Sulfur*; Cambridge University Press: Cambridge, 1970; p 399.

Saggar, S., Bettany, J. R.; Stewart, J. W. B. Measurement of Microbial Sulfur in Soil. *Soil Biol. Biochem.* **1981**, *13*, 493–498.

Sahrawat, K. L.; Murthy, K. V. S; Wani, S. P. Comparative Evaluation of Ca Chloride and Ca Phosphate for Extractable Sulfur in Soils with a Wide Range in pH. *J. Plant Nutr. Soil Sci.* **2009**, *172*, 404–407.

Setter, K. O.; Fiala, G.; Huber, G.; Huber, H.; Segerer, A. Hyperthermophilc Microorganisms. *FEMS Microbiol. Rev.* **1990**, *75*, 117–124.

Shinde, D. B.; Patil, P. L.; Patil, B. R. Potential Use of Sulfur Oxidizing Microorganisms as Soil Inoculants. *Crop Res.* **1996**, *11*, 291–295.

Smith, J. L.; Paul, E. A. The Significance of Soil Microbial Biomass Estimations. In *Soil Biochemistry*; Bollagand, J. M., Stotzky, G. Eds.; Marcel Dekker, Inc.: New York, 1990; pp 357–396.

Solomon, D.; Lehmann, J.; de Zarruk, K. K.; Dathe, J.; Kinyangi, J.; Liang, B.; Machado, S. Speciation and Long- and Short-Term Molecular Level Dynamics of Soil Organic Sulfur Studied by X-ray Absorption Near-Edge Structure Spectroscopy. *J. Environ. Qual.* **2010**. DOI:10.2134/jeq2010.0061.

Stahl, W. H.; McQue, B.; Mandels, G. R.; Siu, R. G. H. *Arch. Biochem.* **1949**, *20*, 422–432.

Starkey, R. L. Relation of Microorganisms to Transformation of Sulfur in Soils. *Soil Sci.* **1950**, *70*, 55.

Stevenson, F. J. *Humus Chemistry*; John Wiley & Sons: New York, 1982.

Strick, J. E.; Nakas, J. P. Calibration of a Microbial sulphur Technique for Use in Forest Soils. *Soil Biol. Biochem.* **1984**, *16*, 289–291.

Strickland, T. C.; Fitzgerald, J. W.; Ash, J. T.; Swank, W. T. Organic sulphur Transformations and sulphur Pool Sizes in Soil and Litter from a Southern Appalachian Hardwood Forest. *Soil Sci.* **1987**, *143* (6), 453–458.

Strickland, T. C.; Fitzgerald, J. W.; Swank, W. T. In Situ Measurements of sulfate Incorporation into Forest Floor and Soil Organic Matter. *Can. J. For. Res.* **1986**, *16*, 549–553.

Tabatabai, M. A.; Bremner, J. M. Factors Affecting Soil Aryl Sulphatase Activity. *Soil Sci. Soc. Am. Proc.* **1970**, *34*, 427–429.

Tandon, H. L. S.; Messick, D. L. *Practical Sulphur Guide*; The Sulphur Institute: Washington DC, 2002; p 20.

Tandon, H. L. S.Sulphur *in Soils, Crops and Fertilizers*. Fertilizer Development and Consultation Organization (FDCO): New Delhi, 2011.

Vermeij, P.; Wietek, C.; Kahnert, A.; Wüest, T.; Kertesz, M. A. Genetic Organization of sulphur-Controlled Aryl Desulphonation in *Pseudomonas putida* S-313. *Mol. Microbiol.* **1999,** *32*, 913–926. DOI: 10.1046/j.1365-2958.1999.0 1398.x.

Vidyalakshmi, R.; Paranthaman, R.; Bhakyaraj, R. Sulphur Oxidizing Bacteria and Pulse Nutrition—a Review. *World J. Agric. Sci.* **2009,** *5* (3), 270–278.

Wainwright, M. A. Modified Sulphur Medium for the Isolation of Sulphur Oxidizing Fungi. *Plant Soil* **1978,** *49* (1), 191–193.

Wang, B; Qiu, Y. L. Phylogenetic Distribution and Evolution of Mycorrhizas in Land Plants. *Mycorrhiza* **2006,** *16*, 299–363. DOI: 10.1007/s00572-0050033-6.

Wang, J.; Solomon, D.; Lehmann, J.; Zhang, X.; Amelung, W. Soil Organic Sulphur Forms and Dynamics in the Great Plains of North America as Influenced by Long-Term Cultivation and Climate. *Geoderma* **2006,** *133*, 160–172.

Wyszkowska, J.; Wieczorek, K.; Kucharski, J. Resistance of Arylsulphatase to Contamination of Soil by Heavy Metals. *Pol. J. Environ. Stud.* **2016,** *25* (1), 365–375.

Zia, M.; Sabir, M.; Ghafoor, A.; Murtaza, G. Effectiveness of Sulphuric Acid and Gypsum for the Reclamation of a Calcareous Saline-Sodic Soil under Four Crop Rotations. *J. Agron. Crop. Sci.* **2007,** *193*, 262–269.

CHAPTER 15

Utilization of Mycorrhizal Fungi in the Mobilization of Macro- and Micronutrients to Important Pulses and Oil-Seed Crops

PRATIBHA LAAD[1], HEMANT S. MAHESHWARI[2], and
MAHAVEER P. SHARMA[2,*]

[1]*School of Life Sciences, Devi Ahilya Vishwavidylaya, Indore 452001, Madhya Pradesh, India*

[2]*ICAR-Indian Institute of Soybean Research, Khandwa Road, Indore 452001, Madhya Pradesh, India*

[*]Corresponding author. E-mail: mahaveer620@gmail.com

ABSTRACT

The arbuscular mycorrhizal symbiosis between plants and soil fungi improves different micronutrients under limiting conditions from the soil. The arbuscular mycorrhizal fungi (AMF) improve soil attributes, increase above- and belowground biodiversity, significantly improve different plant seedlings survival, growth, and establishment on moisture and nutrient-stressed soils. AMF are a group of obligate biotrophs completely depending on host plants for organic carbon, evolutionarily intimately associated with plants, multiple nucleated, and asexually reproducing eukaryotes. AMF are known to play a role in plant nutrition as long as they work in partnership with other soil microbes. Modern agricultural exercise has in numerous instances resulted in progressively reduced AM fungal diversity and frequency. These practices may effect that is believed to be related to tillage methods and to the use of mineral fertilizers. Thus better management of AMF gave a positive role in nutrient cycling and improved soil attributes in integrated nutrient supply to the plants.

15.1 INTRODUCTION

Climate change induced soil edaphic changes leading to problematic soil like acidity, alkalinity leading to widespread nutrient deficiency and toxicity. Formation of acid sulfate soil arising due to soil acidification, erratic rainfall, uneven distribution of rain, leaching of salts ultimately affects the toxicity of iron, aluminum, and fixation of phosphate in terms of iron or aluminum phosphate consequently deficiency too. Similarly, due to basicity in soil, phosphorus gets fixed in terms of calcium and magnesium phosphate. It is also an undeniable fact that global climate change significantly alters soil microbial diversity and thus affecting nutrient availability to plants. In various soil types, nutrient availability can be low, and this becomes a major point of concern. Nutrient presence and nutrient unavailability are different from each other. Major drawback occurs where the complex mineral forms are present in the soil but are not available to plants for growth and reproduction. This decreases crop productivity and yield. If the limiting factor is a mineral nutrient, then it becomes necessary to make it available to plant by mobilizing them. These all depend on soil health. Soil health is related to soil microflora and the former depends on the pH. The ideal soil pH is reported to be 6.5–7.5. Actually, soil pH and microflora are interdependent. This interdependency is determined by how soil is managed (Fig. 15.1). Therefore, nutrient levels depend on soil management like whether the soil is organically or conventionally managed (Gosling and Shepherd, 2005; Watson et al., 2002). Mäder et al. (2002) have reported that nutrient inputs into organic systems were 34–51% lower than in conventional systems (Baird et al., 2010).

FIGURE 15.1   (See color insert.) The dependency of various factors in context to nutrient availability.

For a long time, chemical fertilizers are put into use for improving crop yield around the world (Tilman et al., 2002). On the other hand, it is also true that continuous inputs of N and P fertilizers are unlikely to improve yields (Wang et al., 2011), because only 30–50% of applied N fertilizers and 10–45% of P fertilizers are taken up by the plants (Adesemoye and Kloepper, 2009; Garnett et al., 2009). Consequently, the rampant use of these fertilizers proved inimical for the environment and is a major facet for global concern (Tilman et al., 2002). Therefore, it becomes imperative for the scientific community to step forward toward alternate strategies which can play an utmost role to upsurge the productivity of oilseed crops. One approach could be the use of organic farming practices for sustainable agriculture.

Pulse crops like chickpea, pigeon pea, lentil, black gram, etc., play an important role in organic crop rotation. About 80% of their total nitrogen (N) requirement is fulfilled by fixing atmospheric N (Corre-Hellou and Crozat, 2005). On the other hand, oilseeds such as soybean, groundnut, sunflower, sesame, etc., cover about 13% of the total arable land but contribute nearly 10% of the total value of the agricultural products in India (Singh et al., 2006). Amongst the important edible oilseed crops grown in India, soybean (*Glycine max* L. Merrill), groundnut (*Arachis hypogaea* L.), and rapeseed-mustard (*Brassica juncea* L.) are accounts for 87% and 75% of total oilseed production and acreage, respectively (Agricultural Statistics at a Glance, 2004). Oilseeds comprise both nonlegumes and legumes, and major oilseeds like sesame, groundnut, and soybean. These are grown under rainfed conditions in the subtropics and tropics in the marginal lands with a modest amount of external application of fertilizers. Most frequently, the major oilseeds crop face oddities in weather conditions like erratic rainfall and mid- and end-of-season drought coupled with a plethora of diseases and pests relentlessly restraining the productivity. Thus, nutrient management is uttermost important to enhance nutrient availability in suboptimal cultivational conditions thereby enhancing the productivity of oilseed crops.

15.2 ARBUSCULAR MYCORRHIZAL FUNGI AND ITS ROLE IN THE MINERAL NUTRITION

In a dynamic environment like rhizosphere, where ample of interactions take place; there, *arbuscular mycorrhizal* fungi (AMF) are salient biotrophic plant associates. German scientist, A. B. Frank coined the term "mycorrhiza," meaning fungus root. This symbiotic/mutualistic association can be seen among a soil fungi group and the roots of higher plants (Habte, 2000). These are obligatory symbionts that require a host plant for completing their life cycle (Wardle et al., 2004) and establish a mutualistic association with most of the agricultural crops (except the members of *Brassicaeae* family) and are able to increase plant health (Jansa et al., 2009). In AM, hypha enters the cortical cells of the roots. They can either produce diverse shapes of balloon-like, membrane-bound organelles, which may be outside or inside the cortical cells. These are called vesicles or may constitute arbuscules, which are finely divided dichotomously branched hyphal invaginations (Fig. 15.2).

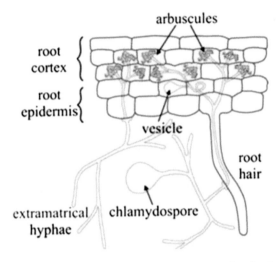

FIGURE 15.2   (See color insert.) Transfer mechanism of nutrient by the AM fungi.

This type of bidirectional interaction is meant for the exchange of resources. In this symbiotic association, the mycorrhizal symbiont provides mineral nutrients, such as phosphate and nitrogen to the host plant, and also helps to impart abiotic stress tolerance against salinity, drought, heavy metal, and biotic stress from various soil-borne root pathogens, and in return to this, the host plant offers about 4–20% of its photosynthates, that is, carbon compounds to the mycorrhizal symbiont (Wright et al., 1998). Fossil studies indicate that mycorrhizal association plays a significant role in the ancient colonization of land-plants (Smith and Read, 2008). Ectotrophic and arbuscular mycorrhizal associations, which are mutually beneficial, have great ecological and economic importance (Marschner and Dell, 1994). Arbuscular mycorrhizal fungi is a "biofertilizer and bioprotector" in sustainable agriculture. AMF helps in phytoremediation and also revegetation purposes (Bücking, 2011; Giri et al., 2005). AM fungi belong to the class azygosporous Zygomycetes covering six genera.

The establishment of AMF in the root leads to changes in the rhizospheric microflora (Meyer and Linderman, 1986; Marschner et al., 2001) exaggerating plant tolerance to a wide range of abiotic and biotic stresses (Auge et al., 2004; Whipps, 2004; Jansa et al., 2009). In addition to this, Mycorrhizal hyphal networks establish a link between plants of the same and different species below the ground and are also able to transfer resources between plants and release signal molecule defence-related proteins, strigolactones and lipochitooligosaccharides (Kohout et al., 2014; Saito, 2000). AMF also colonize plant root cortex and spread an extrametrical mycelium which is a connecting link between the roots and the surrounding soil microhabitats (Barea et al., 2005). AM has the potential to mitigate the nutrient loss caused by leaching and soil erosion, and also increase nutrient use efficiency by plants. Therefore, it is also true that soil fertility is affected by mycorrhiza. Hence, it is an interdependent process.

15.2.1 AMF IN MINERAL NUTRITION

Mycorrhizal fungi are capable of absorbing and transporting about 15 essential macro- and micronutrients vital for plant growth. Mycorrhizal fungi produce strong chemical compounds in soil that mobilize rock-bound nutrients like phosphorous, iron, and other minerals. Mycorrhizal fungi create a hyphal web that absorbs nutrients restoring the nutritional supremacy of soils. In nonmycorrhizal conditions, much of the fertility is reduced from the soil system. Mycorrhizal interactions may also influence the growth of host plant through the improvement in nutritional attainment by the fungal associate or obliquely by altering the transpiration rates and composition of the rhizospheric microflora (Marschner and Dell, 1994), nutrient mobilization from the organic substrates (Finlay, 2008), by improving the fertilizer use efficacy (Jeff et al., 2005), or by beneficial alliance with other soil microbes (Finlay, 2008). Numerous studies on field crops, including oilseeds, demonstrated the benefits of AM inoculation on plant nutrition (Cardoso and Kuyper, 2006; Hamel and Strullu, 2006), N-fixation (Peoples and Craswell, 1992), nodulation (Meghvansia et al., 2008; Aryal et al., 2006), and plant protection (Whipps, 2004; Doley and Jite, 2013a,b) under standard conditions. AM mobilizes many minerals, majorly the inorganic phosphorus (P) and flourish well in temperate and arid climates where P is a limiting factor. The AM associations are not host specific and thereby exist in tropical and temperate tree species (Bücking et al., 2002). Members of the plants belonging to families Amaranthaceae, Betulaceae, Brassicaceae,

Pinaceae, Chenopodiaceae, Juncaceae, Proteaceae, Cyperaceae, and Polygonaceae have very rare associations with AMF. Between AM and ectomycorrhizal (ECM) fungi, the latter is more efficient in captivating N, and are mostly found in the boreal zone as well as in the temperate zone with high humidity levels. This is because of the occurrence of low temperature with high humidity that promotes the accretion of organic matter, reduced pH, and less N availability (Kilpela''inena et al., 2016). In terms of land area on the surface of the earth, the majority of the forests are dependent on ECM fungi (Habte, 2000; Smith and Read, 2008; Bonfante and Genre, 2010).

15.2.1 ENHANCE MOBILIZATION OF MACRO NUTRIENTS (N, P, K, S) AND PHOSPHATE UPTAKE ASSISTED BY THE AM SYMBIOSIS

Phosphorus is an extremely immobile element in soil and it can be supplied to the infected plant via AMF (Bucher, 2007). Even if phosphorus is supplemented to the soil in its soluble form, it soon gets immobilized as calcium phosphates, organic phosphorus or some other fixed forms (Wetterauer and Killon, 1996). AMF can effectively increase nutrient uptake, particularly phosphorus and accumulation of biomass for many crops in low phosphorus soil (Osonubi et al., 1991). AMF improve plant acquisition of phosphate (Pi). Depending upon the particular plant-fungus combination, symbiotic phosphate uptake may dominate over all Pi acquisition (Smith et al., 2003). A Pi depletion zone gets developed around the root hair cylinder due to reduced mobility of Pi in the soil and rapid Pi uptake into the root. The extraradical mycelium of AMF spreads toward a new pool of soluble phosphates (Smith and Read, 1997). Thus, the presence of Pi depletion zone in the rhizosphere is a key factor contributing to the advantage of plants forming mycorrhizal associations. Precisely speaking, a mycorrhized plant constitutes a "mycorrhizosphere," which is composed of the rhizosphere and the hydrosphere.

In this symbiotic system, the fungus connects the mycorrhizosphere and the Pi is transported, in the form of polyphosphates, from the AM fungus soil interface to the intraradical symbiotic interface (Bucher, 2007). The propounded metabolic route of symbiotic Pi acquisition initiates with the assimilation of inorganic Pi at the hyphal-soil interface by fungal high-affinity transporters (Harrison and van Buuren, 1995; Maldonado-Mendoza et al., 2001; Benedetto et al., 2005). Before its release into the periarbuscular interface, phosphate gets depolymerized into inorganic Pi (Ohtomo and Saito, 2005). It has been proven that most phosphorus can be taken up via the "mycorrhizal" uptake pathway, it can be hypothesized that mycorrhiza-upregulated plant phosphate transporters play a major role in plant fitness and thereby improving productivity in most natural and agricultural ecosystems.

15.2.2 NITROGEN TRANSFER AT THE MYCORRHIZAL INTERFACE

AMF can uptake and then after transfer considerable amounts of inorganic nitrogen NO^{-3} or NH^{+4}) to their host plants (He et al., 2003). Several experiments have demonstrated that arginine is generally the principal nitrogenous product (Govindarajulu et al., 2005). However, the extraradical hyphae of AMF also absorb ammonium, nitrate, and amino acids (Ellerbeck et al., 2013; Hodge et al., 2001). Moreover, experimental studies have indicated that the majority of N is taken up in the form of NH_3 via transporters such as GintAMT1, which belongs to the fungal-encoded AMT1 family protein, characterized from *Glomus intraradices* (Lopez-Pedrosa et al., 2006). It is also believed that amino acids may be transported directly to the interfacial apoplast for plant absorption. Arginine is broken down by enzyme ornithine aminotransferase and urease to emancipate free ammonium. Further, ammonium is disseminated by protein-mediated mechanisms and a candidate fungal AMT transporter has been identified that is highly expressed in the internal hyphae (Govindarajulu et al., 2005). In the intraradical hyphae, much of the C is converted to storage lipids, majorly as triacylglycerides. Also, lipids are the main form of C that is translocated from intra- to extraradical hyphae where they act as the major respiratory substrate. Arbuscules convey both P and N from AMF to the plant root. MtPT4, a *Medicago truncatula* phosphate transporter located in the periarbuscular membrane is required for symbiotic P transport and for maintenance (Javot et al., 2011). The MtPT4 arbuscule phenotype is also highly correlated with shoot N levels. On the other hand, the transport mechanism of sugars to the apoplast is a passive movement. For example, when hexose reaches in the apoplast, it is absorbed by the fungus via specific transporter proteins like GpMST1 (Schussler et al., 2006).

AMF can take up inorganic form of N (NH^{+4}, NO^{-3}) by the extraradical mycelium and incorporate into amino acids via the glutamine synthetase/glutamate synthase (GS/GOGAT) cycle, asparagine synthase and the urea cycle (Smith et al., 1985), stored as arginine and co-transported with PolyP from extra- to intraradical fungal structures as arginine and then transport as ammonium to the plant (Govindarajulu et al., 2005; Jin et al., 2005; Chalot et al., 2006). The reason for this is that the key enzymes of nitrogen assimilation and arginine breakdown are abundantly and preferentially accumulate in extra- and intraradical mycelia, respectively (Govindarajulu et al., 2005). The ammonia extrusion from fungal cells follows some other pathways than those mediated by Amt proteins (ammonia transporter), either by protein-mediated mechanisms or by passive efflux of the deprotonated form. Arginine is also bidirectionally transported within the extraradical mycelium; and (v) N released from transported arginine is transferred to the host as NH^{+4} and can be incorporated into other free amino acids in mycorrhizal roots, while carbon (C) not transferred to the host is recycled back to the extraradical mycelium. Therefore, the operation of a metabolic route assists the AM fungi in uptake of N, transport and its assimilation. Similar to the path of symbiotic Pi uptake, the arbuscules are the sites for symbiotic nitrogen uptake involving plant-encoded nitrogen transporters located within the *periarbuscular* plant membrane (Paszkowski, 2006). In addition to this, AMF also help the host plant to use organic matter as a source of N (Hodge et al., 2001).

15.3   MECHANISMS INVOLVED

The two important steps in nutrient absorption from the soil and release of the nutrients through mycorrhizal association involve:

   i)   Mobilization and acquisition by the fungal mycelia.
   ii)   Transportation of absorbed nutrients across the fungal–root interface.

In addition to the hyphae, every mycorrhizal fungus also creates extramatrical mycelium that connects the surface of infected root with the adjacent soil. Both the fungi, AM and ECM, create lots of the extramatrical mycelium. Among these, arbuscular mycorrhizal mycelium can be many centimetres long while ECM mycelium extends up to some meters (Goltapeh et al., 2008). An efficient nutrient gathering network of intricate structures are also exhibited by Mycorrhiza (Bücking and Heyser 2001; Goltapeh et al., 2008). Many of the mycorrhizal fungi can perform the task of the mobilization of nutrients, for example, plant root inaccessible nitrogen and phosphorus from the structural or any other polymers. Many findings reported the withdrawal of nutrients like N and P by the help of mycorrhizal fungi from a large variety of organic substrates like saprotrophic mycelia (Lindahl et al., 1999), dead and decaying nematodes (Perez-Moreno and Read, 2001b), pollen grains (Perez-Moreno and Read, 2001a; Finlay 2008), and Collembola (Klironomos and Hart, 2001). A good example of mobilization and transportation of nutrients is put forward by the ECM fungi that flourish well in boreal forest ecosystems. In such ecosystems, N and P are present in the organic form that is not easily available for utilization by the autotrophs.

In these forests, many of the important plant species are dependent on the mycorrhizal symbionts to suffice their nutritional requirements. ECM symbiont can also act on the structural polymers that might be a cause for the unavailability of nutrients and in the mobilization of N and also P from organic polymers (Read and Perez-Moreno, 2003). The carbon liberated in the soil via roots and allied mycorrhizal fungi could play a crucial role to mobilize N. The production of extracellular enzymes like peptidases and proteinases by ECM fungi enables them to efficiently hydrolyze the organic nitrogen resources to convert them to amino acids; that can be taken up by fungi. Moreover, the secretion of extracellular phosphomonoesterase and phosphodiesterase enzymes by ECM fungi helps to mobilize mineral nutrients in the soil. The enzyme phosphodiesterase can mobilize phosphorus, which is a structural part of nucleic acids. Many ECM fungi produce hydrolytic enzymes such as cellulase, hemicellulase, or lignase. The enzymes assist the entry of hypha into the dead and decaying organic matter present in soil and get in touch with the mineral nutrients seized within. In such a way, the ECM fungus simplifies the typical mineral nutrient cycles, releasing nutrients seized within the soil organic matter.

There are many reports that suggest that the ECM fungi can produce siderophores, which bind and form complexes with oxalate and iron that increase uptake of potassium by the symbiont. Reducing agents' production by the ECM fungi increases the acquisition of ions from stable oxides like MnO_2, consequently helping in enhanced plant nutrition availability (Lindahl et al., 2001, 2007). This biotrophic feature of AM fungi indicates that such fungi are unable to utilize organic sources of nitrogen (Bücking and Kafle, 2015); nevertheless, many of the studies suggest that the AM fungal hypha develops on the organic matter and translocate nitrogen to its host plant (Leigh et al., 2009; Hodge and Fitter, 2010), which helps in elevating plant nitrogen content (Thirkell et al., 2015). Reynolds et al. (2005) confirmed that there is no indication about the promotion of plant N acquisition by AMF and better growth of old field perennial trees under low Nitrogen supply conditions; however, AMF could be associated with decomposing organic material in many ecosystems. Though AMF accelerates the N absorptions from the organic substance (Atul-Nayyar et al., 2009) and changes the Carbon exchange within the soil microbial population during decomposition process (Herman et al., 2012).

On the other hand, Hodge et al. (2001) confirmed the improved decomposition and Nitrogen mobilization from dead and decaying grass foliage in the presence of AMF, Leigh et al. (2009) established that AM fungi did not exhibit saprophytic competence and the fungus captures N from organic matter almost certainly as the product of decomposition. Wallander (2006) stated the vital contribution of mycorrhizal fungi in mineral weathering of forest soils.

Reports suggest that ECM fungi produce certain organic acids with low-molecular-weight, which are utilized in weathering rocks (Ahonen-Jonnarth et al., 2000). The ECM fungal hypha penetrates and, generates microsites that are otherwise far from the contact of the plant roots and inaccessible from the bulk soil solution. The mobilized and dissolved nutrients can be carried to roots of the host plant, shunning soil solution with toxic concentrations of the Al^{3+} ions from the acid rain (Clark, 1997), and also avoiding antagonism for the uptake of the nutrient with other microorganisms (Table 15.1).

There are two pathways for nutrients uptake by plants from the soil (Smith et al., 2011). One is "plant pathway" that comprises unmediated uptake of the nutrients by the epidermal cells of the root hairs from the soil, and the second pathway is nutrient uptake by plants that can take place through the "mycorrhizal pathway," it consists of take-up of nutrients by the extraradical mycelium of its fungal symbiont which transfers the nutrients to "Hartig net" in the ECM association, or to the intraradical mycelium in the AM association, and finally to the plant from interfacial apoplast (Harrison et al., 2002). The plant pathway nutrient uptake from the soil is consistently constrained by the reduced mobility of nutrients in the soil (Bücking and Kafle, 2015). The AM associates show a collective mode of the nutrient uptake (Bücking and Kafle, 2015). So, it can be supposed that the nutrient uptake through the mycorrhizal pathway can be evaded while there is excess availability of nutrient in the soil. The mycorrhizal pathway can help the entire P uptake (Smith et al., 2003; Nagy et al., 2009). The transporters present in plants that are related with P uptake by means of plant pathway, are downregulated in reaction to AM symbiosis (Harley and Smith, 1983; Chiou et al., 2001; Grunwald et al., 2009). Also, mycorrhizal transporters that are involved in the P uptake from the mycorrhizal interface are upregulated (Xu et al., 2007; Paszkowski et al., 2002). The total P uptake by two pathways also depends on the plant and the fungal species. Zhang et al. (2015) suggested that *Rhizophagus irregularis* was more efficient in P absorption as compared to *Acaulospora longula* and *Gigaspora margarita*. Grunwald et al. (2009) have confirmed that the *Glomus intraradices* species has the utmost capability to repress the expression of P transporters in plants within the plant pathway (Table 15.1).

TABLE 15.1 Different Arbuscular Mycorrhizal (AM) Fungi for Mineral Nutrition.

Fungi	Element	Plant/crop	References
Rhizophagus irregularis *Aculospora longula* *Gigaspora margarita* *Glomus intraradices*	Phosphorus	*Vigna ungliculata* *Medicago truncate*	Zhang et al. (2015), Lopez-Pedrosa (2006), Grunwald et al. (2009), and Javot et al. (2011)
Glomus intraradices	Nitrogen	*Sesamum indicum*	Lopez-pedrosn (2006), Javot et al (2011), Hodge et al (2001), and Padszkowski et al. (2006)
Glomus mosseae Glomus fasciculatum	Copper	*White clover* (Trifolium repens L.) *Groundnut (Arachis hypogaea)*	Xiao-Lin Li et al. (1991), and Krishna et al. (1984)
Glomus fasciculatum	Iron	*Groundnut (Arachis hypogaea)*	Krishna et al. (1984)
Glomus versiforme	Magnesium	*Medicago truncatula*	Liu et al. (2002)
Glomus fasciculatum	Manganese	*Groundnut (Arachis hypogaea)*	Krishna et al. (1984)
Gigaspora margarita *Glomus mosseae* *Gigaspora calospora* *Acaulospora sp.*	Sulfur	*Vicia faba*	Ael Fiel et al. (2002)
Rhizophagus irregularis	Potassium	*Medicago truncatula*	Garcia et al. (2014)
Glomus fasciculatum	Zinc	*Groundnut (Arachis hypogaea)*	Jamal et. al.(2002), Habte et al. (2002), Chen et al. (2003), and Krishna et al. (1984)

AM fungi hold the capacity to transport phosphorus, NH^{4+}, Co, NO_3, potassium, calcium, SO^{4+}, Cu, Zn, Fe, Mn, magnesium toward plants, thereby increasing plant nutrient availability. Inoculation of AM fungi reduces Fe and Mn toxicity in plants (Nogueira et al., 2004). Studies conducted by Anil-Prakash and Tandon 2002 proved that, under field conditions, AM fungal inoculation increases nutrient uptake, biomass, and yield of sesame (*Sesamum indicum*) applied with conventional P fertilizer (superphosphate) and slow release P source (rock phosphate). Groundnut is the dicotyledonous strategy I plant which can show Fe-deficiency response with enhanced net efflux of protons from the roots thereby increasing Fe-reducing capacity (Cantrell and Linderman, 2001; Asghari et al., 2005; Ghazi and Al-Karaki, 2006).

The dependency on the mycorrhizal pathway for the nutrient acquirement has shown to be reliable for the C circulation in the root system of the plant (Postma and Lynch, 2011). The tree species associated with ectomycorrhiza have most parts of their root surface consisting of the region that are unable to uptake nutrients efficiently and the regions that are actively nutrient acquiring like non-mycorrhizal white or the ECM roots that stand for merely 2% or 16% of the entire root length, correspondingly (Taylor and Peterson, 2002). In such a way, the role of the fungal mantle or the sheath that surrounds the root tips is mainly essential (Taylor and Peterson, 2005). The plant can depend on the fungal partner for mineral supply because there is less amount of plant tissue which can absorb nutrients from the exterior of the fungal mantle. Some fungi have also been shown to release hydrophobins during the development of ECM (Coelho et al., 2010). Hydrophobin is a diminutive hydrophobic protein which can clasp the fungal hyphae to a surface; in addition to this, it may impart impermeability of water in the fungal sheath (Unestam, 1991; Unestam and Sun, 1995).

As compared to maize, Soybean has a higher dependency on mycorrhiza. When mycorrhizal and nonmycorrhizal plants were compared, the N, P, K, Mg, and Ca uptake were found to significantly increase in mycorrhizal plants. It is also true that plants with the highly branched root system (Gramineae) are less mycotrophic than those with coarser roots (e.g., cassava, onion). Nutrients (P, K, Ca, Fe, Mg, S, and Si) uptake in cowpea (*Vigna unguiculata*) varieties indicated that varieties and the treatment effects were significantly different (Yano-Melo et al., 2003; Rabie, 2005; Cho et al., 2006; Ghazi and Al-Karaki, 2006; Sannazzaro et al., 2006). Many results supported that AMF have been shown to improve immobile nutrients uptake such as P, Zn, and Cu (George 2000; Liu et al., 2002). Mycorrhizal fungi can also improve P absorption

(Kalipada and Singh, 2003) potassium (Liu et al., 2002), zinc (Jamal et al., 2002; Habte and Osorio, 2002; Chen et al., 2003) copper (George, 2000), magnesium (Liu et al., 2002), and calcium (Liu et al., 2002). Organic field peas and lentils also get benefited from AMF colonization. However, some studies also showed that AMF colonization of crop roots is steady or increases as seeding rate increases (Leigh et al., 2009; Nayyar et al., 2008).

15.4 CONCLUSION

Furthermore, keeping such points into consideration, it is possible to address AM to manage disturbed sites that are subjected to temporary water deficit, salinity, or heavy metal toxicity. A key role of the AM symbiosis in linking the process of N mineralization to plant N demand in soil, where the AM symbiosis regulates the recycling of plant residue into living plant biomass, thus impacting the structure of the soil microbial community, has been verified Recently, many researchers reported that AM fungi could enhance the ability of plants to cope with salt stress by improving the uptake of plant nutrients such as P, N, Zn, Cu, and Fe. AMF assist in the uptake of the major plant nutrients such as P and N but also help in captivating other micronutrients like Fe, Cu, Zn, etc. Mycorrhizal fungi implement various means to achieve the task effectively: Measuring the greater absorbing surface area of the plants, releasing biochemical compounds along with an alliance with other microbes in its ambience. Other than mobilizing the mineral nutrients, mycorrhizal fungi also provide significant C sink in the soil; hence, these have a critical impact on cycling of the elements within the soil. Consequently, mycorrhiza is established as a significant association for nutrient management in the ecosystem.

KEYWORDS

- **arbuscular mycorrhizal fungi**
- **growth**
- **nutrient transport**
- **oilseeds**
- **pulses**

REFERENCES

Adesemoye, A. O.; Kloepper, J. W. Plant-Microbes Interactions in Enhanced Fertilizer-Use Efficiency. *Appl. Microbiol. Biotechnol.* **2009**, *85*, 1–12.

Agricultural Statistics at a Glance. Agricultural Statistics Division, Directorate of Economics & Statistics, Department of Agriculture & Cooperation Ministry of Agriculture, Government of India: India, 2004; p 221.

Ahonen-Jonnarth, U.; Van Hees, P. A. W.; Lundstrom, U. S.; Finlay, R. D. Production of Organic Acids by Mycorrhizal and Non-Mycorrhizal *Pinus sylvestris* L. Seedlings Exposed to Elevated Concentrations of Aluminum and Heavy Metals. *N. Physician* **2000**, *146*, 557–567.

Anil-Prakash, Vandana, T. Exploiting Mycorrhiza for Oilseed Crop Production. In *Biotechnology of Microbes and Sustainable Utilization Pages*; Rajak, R. C., Ed.; Scientific Publishers: India, 2002; p 370.

Aryal, U. K.; Shah, S. K.; Xu, H. L.; Fujita, M. Growth, Nodulation and Mycorrhizal Colonization in Bean Plants Improved by Rhizobial Inoculation with Organic and Chemical Fertilization. *J. Sustain. Agric.* **2006**, *29*, 71–83.

Asghari, H.; Marschner, P.; Smith, S.; Smith, F. Growth Response of *Atrilpex nummularia* to Inoculation with Arbuscular Mycorrhizal Fungi at Different Salinity Levels. *Plant Soil* **2005**, *273*, 245–256.

Atul-Nayyar, A; Hamel, C.; Hanson, K.; Germida, J. The Arbuscul Army Corrhizal Symbiosis Links N Mineralization to Plant Demand. *Mycorrhiza* **2009**, *19*, 239–246.

Auge, R. M.; Sylvia, D. M.; Park, S.; Buttery, B. R., Saxton, A. M.; Moore, J. L.; Cho, K. H. Partitioning Mycorrhizal Influence on Water Relations of Phaseolus Vulgaris Into Soil and Plant Components. *Can. J. Bot.* **2004**, *82*, 503–514.

Baird, J. M.; Walley, F.; Shirtliffe, S. J. Optimal Seeding Rate for Organic Production of Field Pea in the Northern Great Plains. *Can. J. Plant Sci.* **2010**, *89*, 455–464.

Barea, J. M.; Pozo, M. J.; Azcon, R; Azcon-Aguilar, C. Microbial Co-Operation in the Rhizosphere. *J. Exp. Bot.* **2005**, *56*, 1761–1778.

Benedetto, A.; Magurno, F.; Bonfante, P.; Lanfranco, L. Expression Profiles of a Phosphate Transporter Gene (*GmosPT*) from the Endomycorrhizal Fungus *Glomus mosseae, Mycorrhiza* **2005**, *15*, 620–627.

Blaudez, D.; Chalot, M.; Dizengremel, P.; Botton, B. Structure and Function of the Ectomycorrhizal Association Between *Paxillus involutus* (Batsch) Fr. and *Betula pendula* (Roth.). II.Metabolic Changes During Mycorrhiza Formation. *N Phytol.* **1998**, *138*, 543–552.

Bonfante, P.; Genre, A. Mechanisms Underlying Beneficial Plant–Fungus Interactions in Mycorrhizal Symbiosis. *Nat. Commun.* **2010**, *1*, 48. DOI:10.1038/ncomms1046.

Bucher, M. Functional Biology of Plant Phosphate Uptake at Root and Mycorhiza Interfaces. *N. Phytol.* **2007**, *173*, 11–26.

Bücking, H.; Heyser, W. Microautoradiographic Localization of Phosphate and Carbohydrates in Mycorrhizal Roots of *Populus tremula, Populus alba* and the implications for transfer processes in ectomycorrhizal associations. *Tree Physiol.* **2001**, *21*, 101–107.

Bücking, H.; Kafle, A. Role of Arbuscular Mycorrhizal Fungi in the Nitrogen Uptake of Plants: Current Knowledge and Research Gaps. *Agronomy* **2015**, *5*, 587–612.

Bücking, H.; Kuhn, A. J.; Schroder, W. H.; Heyser, W. The Fungal Sheath of Ectomycorrhizal Pine Roots: An Apoplastic Barrier for the Entry of Calcium, Magnesium, and Potassium into the Root Cortex? *J. Exp. Bot.* **2002**, *53*, 1659–1669.

Bücking, H.; Liepold, E.; Ambilwade, P. The Role of Mycorrhizal Symbiosis in Nutrient Uptake of Plants and the Regulatory Mechanisms Underlying These Transport Processes. In *Plant Science;* InTechOpen: 2012; pp 107–138.

Cantrell, I. C.; Lindermann, R. G. Preinoculation of Lettuce and Onion with VA Mycorrhizal Fungi Reduces Deleterious Effects of Soil Salinity. *Plant Soil* **2001**, *233*, 269–281.

Cardoso, I. M.; Kuyper, T. W. Mycorrhizas and Tropical Soil Fertility. *Agric. Ecosyst. Environ.* **2006**, *116*, 72–84.

Chalot, M.; Blaudez, D.; Brun, A. Ammonia: A Candidate for Nitrogen Transfer at the Mycorrhizal Interface. *Trend Plant Sci.* **2006**, *11*, 263–266.

Chen, B. D.; Li, X. L.; Tao, H. Q.; Christie, P.; Wong, M. H. The Role of Arbuscular Mycorrhiza in Zinc Uptake by Red Clover Growing in a Calcareous Soil Spiked with Various Quantities of Zinc. *Chemosphere* **2003**, *50*, 839–846.

Chiou, T. J.; Liu, H; Harrison, M. J. The Spatial Expression Patterns of a Phosphate Transporter (MtPT1) from *Medicago truncatula* indicate a role in Phosphate Transport at the Root/Soil Interface. *Plant J.* **2001**, *25*, 281–293.

Cho, K.; Toler, H.; Lee, J.; Ownley, B.; Stutz, J. C.; Moore, J. L. Mycorrhizal Symbiosis and Response of Sorghum Plants to Combined Drought and Salinity Stresses. *J. Plant Physiol.* **2006**, *163*, 517–528.

Clark, R. B. Arbuscular Mycorrhizal Adaptation, Spore Germination, Root Colonization, and Host Plant Growth and Mineral Acquisition at Low pH. *Plant Soil* **1997**, *192*, 15–22.

Coelho, I. D.; de Queiroz, M. V.; Costa, M. D.; Kasuya, M. C. M.; de Araujo, E. F. Identification of Differentially Expressed Genes of the Fungus Hydnangium sp during the Pre-Symbiotic Phase of the Ectomycorrhizal Association with *Eucalyptus grandis*. *Mycorrhiza* **2010**, *20*, 531–540.

Corre-Hellou, G.; Crozat, Y. N_2 Fixation and N Supply in Organic Pea (*Pisum sativum* L.) Cropping Systems as Affected by Weeds and Peaweevil (*Sitona lineatus* L.). *Eur. J. Agron* **2005**, *22*, 449–458.

Doley, K; Jite, P. K. Effect of Arbuscular Mycorrhizal Fungi on Growth of Groundnut and Disease Caused by *Macrophomina phaseolina*. *J. Exp. Sci.* **2013a**, *4*, 11–15.

Doley, K.; Jite, P. K. Disease Management and Biochemical Changes in Groundnut Inoculated with *Glomus fasciculatum* and Pathogenic *Macrophomina phaseolina* (Tassi) Goid. *Plant Sci. Feed* **2013b**, *3*, 21–26.

El Fiel, H. E. A.; El Tinay, A. H.; Elsheikh, E. A. E. Effect of Nutritional Status of Faba Bean (Vicia faba L.) on Protein Solubility Profiles. *Food Chem.* **2002**, *76*, (2), 219–223.

Ellerbeck, M.; Schüßler, A.; Brucker, D.; Dafinger, C.; Loos, F.; Brachmann, A. Characterization of Three Ammonium Transporters of the Glomeromycotan Fungus *Geosiphon pyriformis*. *Eukaryot. Cell* **2013**, *12*, 1554–1562.

Finlay, R. D.; Lindahl, B. D.; Taylor, A. F. S. Responses of Mycorrhizal Fungi to Stress. In *Stress in Yeasts and Filamentous Fungi*; Avery, S.; Stratford, M.; van West, P., Eds.; Elsevier: Amsterdam, 2008; pp. 201–220.

Finlay, R. D. Ecological Aspects of Mycorrhizal Symbiosis: With Special Emphasis on the Functional Diversity of Interactions Involving the Extraradial Mycelium. *J. Exp. Bot.* **2008**, *59*, 1115–1126.

Garnett, T.; Conn, V.; Kaiser, B. Root based Approaches to Improving Nitrogen Use Efficiency in Plants. *Plant Cell Environ.* **2009**, *32*, 1272–1283.

George, E. Nutrientuptake. In *Arbuscular Mycorrhizas: Physiology and Function*; Kappulnick, Y., Douds, D. D., Eds.; Kluwer Academic Press: Dordrecht, The Netherlands, 2000; pp 307–344.

Ghazi, N.; Al-Karaki, G.N. Nursery Inoculation of Tomato with Arbuscular Mycorrhizal Fungi and Subsequent Performance under Irrigation with Saline Water. *Sci. Hortic.* **2006**, *109*, 1–7.

Giri, B.; Kapoor, R.; Mukerji, K.G. Influence of Arbuscular Mycorrhizal Fungi and Salinity on Growth, Biomass and Mineral Nutrition of *Acacia auriculiformis*. *Biol. Fert. Soils* **2003**, *38*, 176–180.

Giri, B.; Mukerji, K. J. Mycorrhizal Inoculant Alleviates Salt Stress in *Sesbania aegyptiaca* and *Sesbania grandiflora* under Field Conditions: Evidence for Reduced and Improved Magnesium Uptake. *Mycorrhiza* **2004**, *14*, 307–312.

Goltapeh, E. M.; Danesh, Y. R.; Prasad, R.; Varma, A. Mycorrhizal Fungi: what We Know and What Should We Know. In *Mycorrhiza*, 3rd ed; Varma, A., Ed.; Springer: Heidelberg, 2008; pp 3–28.

Gosling, P.; Shepherd, M. Long-Term Changes in Soil Fertility in Organic Arable Farming Systems in England, with Particular Reference to Phosphorus And Potassium. *Agric. Ecosys. Environ.* **2005**, *105*, 425–432.

Govindarajulu, M.; Pfeffer, P.; Jin, H.; Abubaker, J.; Douds, D. D.; Allen, J. W.; Bu̇cking, H.; Lammers, P. J.; Shachar-Hill, Y. Nitrogen Transfer in the Arbuscular Mycorrhizal Symbiosis. *Nature* **2005**, *435*, 819–823.

Garcia, K.; Zimmermann, S. D. The Role of Mycorrhizal Associations in Plant Potassium Nutrition. *Front Plant Sci.* **2014**, *5*, 337.

Grunwald, U.; Guo, W. B.; Fischer, K.; Isayenkov, S.; Ludwig-Müller, J.; Hause, B.; Yan, X. L.; Kuster, H.; Franken, P. Overlapping Expression Patterns and Differential Transcript Levels of Phosphate Transporter Genes in Arbuscular Mycorrhizal, Pi Fertilized and Phytohormone-Treated Medicagotruncatula Roots. *Planta* **2009**, *229*, 1023–1034.

Habte, M.; Osorio, N. W. Mycorrhizas: Producing and Appling Arbuscular Mycorrhizal Inoculation. Ecoliving Center. Overstory #102, 2002, http://www. Ecoliving. Cat.Org.au.

Habte, M.; Zhang, Y. C.; Schmitt, D. P. Effectiveness of *Glomus* Species in Protecting White Clover against Nematode Damage. *Can. J. Bot.* **2000**, *77*, 135–139.

Hamel, C,; Strullu, D. G. Arbuscular Mycorrhizal Fungi in Field Crop Production: Potential and New Direction. *Can. J. Plant Sci.* **2006**, *86*, 941–950.

Harley, J. L.; Smith, S. E. *Mycorrhizal Symbiosis*; Academic Press: Toronto, 1983; pp 112–115.

Harrison, M. J.; Dewbre, G. R.; Liu, J. Y. A Phosphate Transporter from *Medicago truncatula* Involved in the Acquisiton of Phosphate Released by Arbuscular Mycorrhizal Fungi. *Plant Cell* **2002**, *14*, 2413–2429.

Harrison, M. J., van Buuren, M. L. A Phosphate Transporter from the Mycorrhizal Fungus *Glomus versiforme*. *Nature* **1995**, *378*, 626–629.

He, X. H.; Critchley, C.; Bledsoe, C. Nitrogen Transfer Within and Between Plants Through Common Mycorrhizal Networks (CMNs). *Crit. Rev. Plant Sci.* **2003**, *22*, 531–567.

Herman, D. J.; Firestone, M. K.; Nuccio, E.; Hodge, A. Interactions Between an Arbuscular Mycorrhizal Fungus and a Soil Microbial Community Mediating Litter Decomposition. *FEMS Microbiol. Ecol.* **2012**, *80*, 236–247.

Hodge, A.; Campbell, C. D., Fitter, A. H. An Arbuscular Mycorrhizal Fungus Accelerates Decomposition and Acquires Nitrogen Directly from Organic Material. *Nature* **2001**, *413*, 297–299. DOI:10.1038/35095041.

Hodge, A; Fitter, A. H. Substantial Nitrogen Acquisition by Arbuscular Mycorrhizal Fungi from Organic Material has Implications for N Cycling. *Proc. Natl. Acad. Sci. USA* **2010**, *107*, 13754–13759.

Jackson, L. E.; Burger, M.; Cavagnaro, T. R. Nitrogen Transformations and Ecosystem Services. *Annu. Rev. Plant Biol.* **2008,** *59*, 341–363.

Jamal, A.; Ayub, N.; Usman, M.; Khan, A. G. Arbuscular Mycorrhizal Fungi Enhance Zinc and Nickel Uptake from Contaminated Soil by Soya Bean and Lentil. *Int. J. Phytoremed.* **2002,** *4*, 205–221.

Jansa, J.; Hans-Rudolf, O.; Egli, S. Environmental Determinants of the Arbuscular Mycorrhizal Fungal Infectivity of Swiss Agricultural Soils. *Eur. J. Soil Biol.* **2009,** *45*, 400–440.

Javot, H.; Penmetsa, R. V.; Breuillin, F.; Bhattarai, K. K.; Noar, R. D.; Gomez, S. K.; Zhang, Q.; Cook, D. R.; Harrison, M. J. Medicago Truncatula MtPT4 Mutants Reveal a Role for Nitrogen in the Regulation of Arbuscule Degeneration in Arbuscular Mycorrhizal Symbiosis. *Plant J.* **2011,** *68*, 954–965.

Jeff, H.; Taylor Peterson, C. A. Ectomycorrhizal Impacts on Nutrient Uptake Pathways in Woody Roots. *N. For.* **2005,** *30*, 203–214.

Jin, H.; Pfeffer, P. E.; Douds, D. D.; Piotrowski, E.; Lammers, P. J.; Shachar-Hill, Y. The Uptake, Metabolism, Transport and Transfer of Nitrogen in an Arbuscular Mycorrhizal Symbiosis. *New Phytol.* **2005,** *168*, 687–696.

Kalipada, P.; Singh, R.-K. Effect of Levels and Mode of Phosphorus Application with and without Biofertilizers on Yield and Nutrient Uptake by Chickpea *(Cicer arietinum). Ann. Agric. Res.* **2003,** *24* (4), 768–775.

Kilpela¨inena, J., Vestbergb, M.; Repoc, T.; Lehtoa, T. Arbuscular and Ectomycorrhizal Root Colonisation and Plant Nutrition in Soils Exposed to Freezing Temperatures. *Soil Biol. Biochem.* **2016,** *99*, 85–93.

Klironomos, J. N.; Hart, M. M. Animal Nitrogen Swap for Plant Carbon. *Nature* **2001,** *41*, 651–652.

Kohout, P.; Sýkorová, Z.; Bahram, M.; Hadincová, V.; Albrechtová, J.; Tedersoo, L.; Vohník, M. Ericaceous Dwarf Shrubs Affect Ectomycorrhizal Fungal Community of the Invasive *Pinus strobus* and Native *Pinus sylvestris* in a Pot Experiment. *Mycorrhiza* **2011,** *21*, 403–412.

Krishna, K. R.; Bagyaraj, D. J. Growth and Nutrient Uptake of Peanut Inoculated with the Mycorrhizal Fungus *Glomus fasciculatum* Compared with Noninoculated Ones. *Plant Soil* **1984,** *77*, (2–3), 405–408.

Leigh, J.; Hodge, A.; Fitter, A. H. Arbuscular Mycorrhizal Fungi Can Transfer Substantial Amounts of Nitrogen to Their Host Plant from Organic Material. *N. Phytol.* **2009,** *181*, 199–207.

Lindahl, B.; Olsson, S.; Stenlid, J.; Finlay, R. D. Effects of Resource Availability on Mycelia Interactions and 32P Transfer Between a Saprotrophic and an Ectomycorrhizal Fungus in Soil Microcosms. *FEMS Microbiol. Ecol.* **2001,** *38*, 43–52.

Lindahl, B.; Stenlid, J.; Olsson, S.; Finlay, R. D. Translocation of 32P between Interacting Mycelia of a Wood Decomposing Fungus and Ectomycorrhizal Fungi in Microcosm Systems. *N. Phytol.* **1999,** *44*, 183–193.

Lindahl, B. D.; Ihrmark, K.; Boberg, J.; Trumbore, S. E.; Hogberg, P.; Stenlid, J.; Finlay, R. D.; Spatial Separation of Litter Decomposition and Mycorrhizal Nitrogen Uptake in a Boreal Forest. *N. Phytol.* **2007,** *173*, 611–620.

Liu, J.; Blaylock, L. A.; Endre, G.; Cho, J.; Town, C. D.; Van den Bosch, K. A.; Harrison, M. J. Transcript Profiling Coupled with Spatial Expression Analyses Reveals Genes Involved in Distinct Developmental Stages of an Arbuscular Mycorrhizal Symbiosis. *Plant Cell* **2002,** *15*, 2106–2123.

Lopez-Pedrosa, A.; Gonzalez-Guerrero, M.; Valderas, A.; Azcon Aguilar, C.; Ferrol, N. GintAMT1 Encodes a Functional High-Affinity Ammonium Transporter that is Expressed in the Extraradical Mycelium of *Glomus intraradices*. *Fungal Genet. Biol.* **2006**, *43*, 102–110.

Mäder, P.; Fließbach, A; Dubois, D; Gunst, L.; Fried, P.; Niggli, U. Soil Fertility and Biodiversity in Organic Farming. *Science* **2002**, *296*, 1694–1697.

Maldonado-Mendoza, I. E.; Dewbre, G. R.; Harrison, M. J. A Phosphate Transporter Gene from the Extra-Radical Mycelium of an Arbuscular Mycorrhizal Fungus *Glomus intraradices* is Regulated in Response to Phosphate in the Environment. *Mol. Plant Microbe Interact.* **2001**, *14*, 1140–1148.

Marschner, P.; Yang, C. H.; Lieberei, R.; Crowley, D. E. Soil and Plant Specific Effects on Bacterial Community Composition in the Rhizosphere. *Soil Biol. Biochem.* **2001**, *33*, 1437–1445.

Marschner, H., Dell, B. Nutrient Uptake in Mycorrhizal Symbiosis. *Plant Soil.* **1994**, *159*, 89–201.

Meghvansia, M. K.; Prasad, K.; Harwani, D.; Mahna, S. K. Response of Soybean Cultivars toward Inoculation with Three Arbuscular Mycorrhizal Fungi and *Bradyrhizobium japonicum* in the Alluvial Soil. *Eur. J. Soil. Biol.* **2008**, *44*, 316–323.

Nagy, R.; Drissner, D.; Amrhein, N.; Jakobsen, I.; Bucher, M. Mycorrhizal Phosphate Uptake Pathway in Tomato is Phosphorus Repressible and Transcriptionally Regulated. *N. Phytol.* **2009**, *181*, 950–959.

Nayyar, A. A.; Hamel, C.; Hanson, K.; Germida, J. The Arbuscular Mycorrhizal Symbiosis Lins N Mineralization to Plant Demand. *Mycorrhiza* **2008**, *19*, 239–246.

Nogueira, M. A.; Magelhaes, G. C.; Cardoso, E. J. B. N. Manganese Toxicity in Mycorrhizal and Phosphorus-Fertilized Soybean Plants. *J. Plant Nutr.* **2004**, *27*, 141–156.

Ohtomo, R.; Saito, M. Polyphosphate Dynamics in Mycorrhizal Roots during Colonization of an Arbuscular Mycorrhizal Fungus. *N. Phytol.* **2005**, *167*, 571–578.

Osonubi, O.; Mulongoy, K.; Awotoye, O. O.; Atayese, M. O.; Okali, D. U. Effects of Ecto-mycorrhizal and Vesicular-Arbuscular Mycorrhizal Fungi on Drought Tolerance of Four Leguminous Woody Seedlings. *Plant Soil* **1991**, *136*, 131–143.

Paszkowski, U. A Journey Through Signaling in Arbuscular Mycorrhizal Symbioses. *N. Phytol.* **2006**, *172*, 35–46.

Paszkowski, U.; Kroken, S.; Roux, C.; Briggs, S. P. Rice Phosphate Transporters Include an Evolutionarily Divergent Gene Specifically Activated in Arbuscular Mycorrhizal Symbiosis. *Proc. Natl. Acad. Sci. USA* **2002**, *99*, 13324–13329.

Peoples, M. B.; Craswell, E. T. Biological Nitrogen Fixation: Investments, Expectations and Actual Contributions to Agriculture. *Plant Soil* **1992**, *141* (1–2), 13–39.

Perez-Moreno, J.; Read, D. J. Exploitation of Pollen by Mycorrhizalmycelial Systems with Special Reference to Nutrient Cycling in Boreal Forests. *Proc. R. Soc. B* **2001a**, *268*, 1329–1335.

Perez-Moreno, J.; Read, D. J. Nutrient Transfer from Soil Nematodes to Plants: A Direct Pathway Provided by the Mycorrhizal Mycelial Network. *Plant Cell Environ.* **2001b**, *24*, 1219–1226.

Postma, J. A.; Lynch, J. P. Root Cortical Aerenchyma Enhances the Growth of Maize on Soils with Suboptimal Availability of Nitrogen, Phosphorus, and Potassium. *Plant Physiol.* **2011**, *156*, 1190–2001.

Rabie, G. H.; Almadini, A. M. Role of Bioinoculants in Development of Salt-Tolerance of *Vicia faba* Plants under Salinity Stress. *Afr. J. Biotechnol.* **2005**, *4*, 210–223.

Read, D. J.; Perez-Moreno, J. Mycorrhizas and Nutrient Cycling in Ecosystems: A Journey toward Relevance? *New Phytol.* **2003,** *157,* 475–492.

Reynolds, H. L.; Hartley, A. E.; Vogelsang, K. M.; Bever, J. D., Schultz, P. A. Arbuscular Mycorrhizal Fungi Do Not Enhance Nitrogen Acquisition and Growth of Old-Field Perennials under Low Nitrogen Supply in Glasshouse Culture. *New Phytol.* **2005,** *167,* 869–880.

Saito, M.; Marumoto, T. Inoculation with Arbuscular Mycorrhizal Fungi: the Status Quo in Japan and the Future Prospects. *Plant Soil.* **2002,** *244,* 273–279.

Sannazzaro, A. I.; Ruiz, O. A.; Alberto, E. O.; Menendez, A. B. Alleviation of Salt Stress in *Lotus glaber* by *Glomus intraradices*. *Mycorrhiza.* **2006,** *285,* 279–287.

Schussler, A.; Martin, H.; Cohen, D.; Fitz, M.; Wipf, D. Characterization of a Carbohydrate Transporter from Symbiotic Glomeromycotan Fungi. *Nature* **2006,** *444,* 933–936.

Singh, S.; Basappa, H.; Singh, S. K. Status and Prospects of Integrated Pest Management Strategies in Selected Crops Oilseeds. In *Integrated Pest Management Principles and Applications*; Singh, A., Sharma, O. P., Garg. D. K., Eds.; CBS Publishers and Distributors: New Delhi, 2006; Vol. 2, p 656.

Smith, S. E.; Read, D. J. *Mycorrhizal Symbiosis*, 2nd ed; Academic: London, 1997.

Smith, S. E.; Read, D. J. *Mycorrhizal Symbiosis*; Academic Press, Inc.: San Diego, CA, 2008.

Smith, S. E.; Smith, F. A.; Nicholas, D. J. D., Activity of Glutamine Synthetase and Glutamate Dehydrogenase in *Trifolium subterraneum* L. & *Allium cepa* L. Effects of Mycorrhizal Infection and Phosphate Nutrition. *N. Phytol.* **1985,** *99,* 211–227.

Smith, S. E.; Smith, F. A.; Jakobsen, I. Mycorrizal Fungi can Dominate Phosphate Supply to Plants Irrespective of Growth Responses. *Plant Physiol.* **2003,** *133,* 16–20.

Smith, S. E.; Smith, F. A. Roles of Arbuscular Mycorrhizas in Plant Nutrition and Growth: New Paradigms from Cellular to Ecosystem Scales. *Annu. Rev. Plant Biol.* **2011,** *62,* 227–250.

Taylor, J. H.; Peterson, C. A. Morphometric Analysis of *Pinus banksiana* Lamb. Root Anatomy during a 3-Month Field Study. *Trees.* **2002,** *14,* 239–247.

Taylor, J. H.; Peterson, C. A. Ectomycorrhizal Impacts on Nutrient Uptake Pathways in Woody Roots. *N. For.* **2005,** *30,* 203–214.

Thirkell, J. D.; Cameron, D. D.; Hodge, A. Resolving the "Nitrogen Paradox" of Arbuscular Mycorrhizas: Fertilization with Organic Matter Brings Considerable Benefits for Plant Nutrition and Growth. *Plant Cell Environ.* DOI:10.1111/pce.12667.

Tilman, D.; Cassman, K. G.; Matson, P. A.; Naylor, R.; Polasky, S. Agricultural Sustainability and Intensive Production Practices. *Nature* **2002,** *418,* 671–677.

Unestam, T. Water Repellency, Mat Formation, and Leaf-Stimulated Growth of Some Ectomycorrhizal Fungi. *Mycorrhiza* **1991,** *1,* 13–20.

Unestam, T.; Sun, Y. P. Extramatrical Structures of Hydrophobic and Hydrophilic Ectomycorrhizal Fungi. *Mycorrhiza* **1995,** *5,* 301–311.

Wang, X.; Pan, Q.; Chen, F.; Yan, X.; Liao, H. Effects of Coinoculation with Arbuscular Mycorrhizal fungi and Rhizobia on Soybean Growth as Related to Root Architecture and Availability of N and P. *Mycorrhiza* **2011,** *21,* 173–181.

Wardle, D. A.; Bardgett. R. D.; Klironomos, J. N.; Setala, H.; van der Putten, W. H.; Wall, D. H. Ecological Linkages between Aboveground and Belowground Biota. *Science* **2004,** *304,* 1629–1633.

Watson, C. A.; Bengtsson, H.; Ebbesvik, M.; Loes, A.-K.; Myrbeck, A.; Salomon, E.; Schroder, J.; Stockdale, E. A. A Review of Farmscale Nutrient Budgets for Organic Farms as a Tool for Management of Soil Fertility. *Soil Use Manage.* **2002,** *18,* 264–273.

Wallander, H. Uptake of P from Apatite by *Pinus sylvestris* Seedlings Colonized by Different Ectomycorrhizal Fungi. *Plant Soil* **2006,** *218,* 249–256.

Wetterauer, D. G.; Killorn, R. J. Fallow- and Flooded-Soil Syndromes: Effects on Crop Production. *J. Prod. Agric.* **1996,** *9,* 39–41.

Whipps, J. M. Prospects and Limitations for Mycorrhizas in Biocontrol of Root Pathogens. *Can. J. Bot.* **2004,** *82,* 1198–1227.

Wright, S. F.; Starr, J. L.; Paltineanu, I. C. Changes in Aggregate Stability and Concentration of Glomalin During Tillage Management Transition. *Soil Sci. Soc. Am. J.* **1998,** *63,* 1825–1829.

Xiao-Lin, L.; Horst Marschner, E. G. Acquisition of Phosphorus and Copper by VA-Mycorrhizal Hyphae and Root-to-Shoot Transport In White Clover. *Plant Soil.* **1991,** *136,* (1), 49–57.

Xu, G. H.; Chague, V.; Melamed-Bessudo, C.; Kapulnik, Y.; Jain, A.; Raghothama, K. G.; Levy, A. A.; Silber, A. Functional Characterization of LePT4: A Phosphate Transporter in Tomato with Mycorrhiza-Enhanced Expression. *J. Exp. Bot.* **2007,** *258,* 2491–2501.

Yano-Melo, A. M.; Saggin, O. J.; Maia, L. C. Tolerance of Mycorrhized Banana (*Musa* sp. cv. Pacovan) Plantlets to Saline Stress, *Agr. Ecosyst. Environ.* **2003,** *95,* 343–348.

Zhang, X.; Chen, B.; Ohtomo, R. Mycorrhizal Effects on Growth, P Uptake and Cd Tolerance of the Host Plant Vary among Different AM Fungal Species. *Soil Sci. Plant Nutr.* **2015,** *61,* 359–368.

CHAPTER 16

Integrated Use of Bioferilizer and Biopesticides in Crop Production

VEENA PANDE* and TARA SINGH BISHT

Department of Biotechnology, Bhimtal Campus, Kumaun University, Nainital 263136, Uttarakhand, India

Corresponding author. E-mail: veena_kumaun@yahoo.co.in

ABSTRACT

The green revolution brought amazing effects in the production of food and grains, but with inadequate concern for agricultural sustainability. The accessibility and extensive use of chemical fertilizers and pesticides, based on fossil fuel for future agriculture system needs, would result in adverse effects on human health, further loss in soil health, possibility of groundwater contamination, destroyed beneficial microorganisms and eco-friendly insects, made the crop more susceptible to diseases, calculated burden on the fiscal system, and loss of ecological balance. In this context, biofertilizers and biopesticides are gaining importance because these are environment-friendly and are helping in the practice of sustainable agriculture. These biofertilizers work symbiotically with plants which encourage the plant growth by increasing nutrient supply to the plant, ensure their proper growth and development, regulation in their physiology, and also helpful in retention of soil fertility. These include Nitrogen (N_2) fixing microorganism, Phosphate solubilizing microorganism, Phosphate mobilizing microorganism, and plant growth-promoting microorganism. Similarly, the use of chemical pesticides may lead to genetic manipulation in plant population, food poisoning, and some other health issues and have made the biopesticides to come in the picture which might reduce the use of these chemical-based pesticides.

16.1   INTRODUCTION

Agriculture sector plays an imperative role in the improvement of economic growth and development of developing countries as well as fulfilling the food security for growing population worldwide (Raja, 2013). To feed the ever-growing population globally, we need to produce more food from less per capita arable land and limited availability of water for providing sufficient food is the first part of the challenge, the second major challenge is to produce this in a safe and sustainable manner (Kumar and Singh, 2014). As a society, we are receiving clear signals that continued use of some chemicals in conventional agriculture system is associated with disturbing health and environment. In contrast, sustainable agriculture describes a robust and balanced agricultural system to which many increasingly aspire. The increasing demand for organic products is due to the health awareness of consumer about the negative effect on health by the use of chemical fertilizer and pesticide (Raja and Masresha, 2015). The era of green revolution brought amazing consequences during 1960–2000 by increasing agricultural crop yield per hectare to supply of sufficient food in the developing countries (Pindi and Satyanarayana, 2012). South East Asia and India were the first developing countries to show the impact of green revolution on different varieties of grain yields by using inputs like fertilizers and pesticides, which helped a lot in this regard. Besides of this fact, food insecurity and poverty still prevail prominently in our country (Dutta, 2015). Before the initiation of green technology, farmers mostly used organic materials as the sole source to promote health and productivity of the soil. After that, the era of chemical fertilizers started and farmers almost neglected the use of organic matter because chemical fertilizers were an effective substitute as a ready source of nutrients (Dutta, 2015). Indiscriminate use of chemical fertilizers creates many problems to the soil, water, destroyed microorganisms and friendly insects, making the crop more prone to pest, diseases, and reduced soil fertility. United Nations Food and Agriculture Organization (FAO) estimates, the demand for the agricultural commodities will be 60% higher in 2030 than present time and more than 85% of this additional demand will be born from developing countries (Pindiand Satyanarayana, 2012). It is estimated that by 2020, to achieve the targeted production of 321 million tons of food grain, the requirement of nutrient will be 28.8 million tons, while their availability will be only 21.6 million tonnes being a deficit of about 7.2 million tons (Mishra, 2012). Decreasing soil fertility status, amount of nutrient removal and their supplies, growing concern about environmental hazards are increasing the threat to sustainable agriculture. Besides above facts, the long-term use of biofertilizers and biopesticides is more efficient, eco-friendly, economical, productive, and easily accessible to marginal and small farmers over chemical fertilizers (Mishra, 2012). Biofertilizers are live formulates which contain living microorganism, when applied to seed, on plant surfaces, root or soil, inhabit the rhizosphere and improve the bioavailability of nutrients and increasing microflora through their biological activities and thereby promoting plant's growth. They are preparations that readily improve the fertility of land using biological agents (Lawal and Babalola, 2014). Biopesticides contain naturally occurring substances that control pests by their nontoxic mechanisms. Biopesticides are living organisms and their products (phytochemicals, microbial products) or by-products (semio-chemicals) which can be used for the management of pests that are harmful to plants and less harmful to the environment and human health. The most commonly used biopesticides are living organisms, which are pathogenic for the pest of interest. These include biofungicides (Trichoderma), bioherbicides (Phytopthora) and bioinsecticides (*Bacillus thuringiensis*; Dutta, 2015). Use of microbial inoculants is not only a low-cost technology but also takes adequate care of soil health and environmental safety. Intensive search of a number of microorganisms have been recognized as nitrogen fixers and pest control. The expected global organic food and beverage market in 2015 to be $104.50 billon. Among the European countries, Germany is the largest consumer for organic food and beverages with a share of approximately 32%. The estimated organic food consumption in Japan is 54% in 2010, which is one of the leading countries in Asia. The expected growth rate of organic food market in Asia is 20.6% of CAGR from 2010 to 2015. Among different organic produces, fresh fruits and vegetables are most dominating food categories with a share of 37% (Raja and Masresha, 2015). Since agriculture has potential for food security, environmental sustainability, and economic opportunity worldwide. The future vision of the world is to adopt new strategies to increase agricultural production sustainably (Raja, 2013).

16.2 BIOFERTILIZERS

Biofertilizers can be defined as preparations that contain living cells or latent cells of efficient microbial strains which are helpful in the uptake of nutrients to plant by their interactions in the rhizosphere when applied through seed or soil. These are involved to enrich the quality of nutrient to the soil and establish symbiotic relationship with plants. Biofertilizers are microbial inoculants which are artificially multiplied cultures of certain soil microorganisms that can increase soil fertility and crop productivity (Chowdhury et al., 2014). Biofertilizers are colonizing the rhizosphere of plants and increase the supply of primary nutrients and growth stimulant of targeted crops (Bhattacharjee and Dey, 2014). They accelerate certain microbial processes in the soil which enhance the extent of availability of nutrients in a form that can easily assimilated by plants. Biofertilizers have an ability to mobilize nutritionally important elements from nonusable to usable form. These microorganisms require organic matter for their growth and activity in soil and provide valuable nutrients to the plant (Chowdhury et al., 2014). Therefore, the microorganisms increase plant growth by replacing soil nutrients (i.e., by biological N_2-fixation), by making nutrients readily available to plants (i.e., by nutrient solubilizations), or by increasing plant access to nutrients (i.e., by increasing the volume of soil accessed by roots) as long as the nutrients status of the plant has been improved by the microorganism. Soil microorganisms play a major role in regulating organic matter decomposition and availability of plant nutrients such as N, P, and S. It is well known that microbial inoculants are a component for integrated nutrient management that leads to sustainable agriculture, as they are cost-effective and renewable source of plant nutrients to supplement the fertilizers for sustainable agriculture. Several microorganisms and their association with crop plants are being exploited in the production of biofertilizers.

16.3   TYPES OF BIOFERTILIZERS

They can be grouped in different ways based on their nature and function as follows:

- Nitrogen (N_2)-fixing biofertilizers;
- Phosphate-solubilizing biofertilizer;
- Phosphate-mobilizing biofertilizer; and
- Plant growth-promoting biofertilizer.

16.3.1   NITROGEN (N_2)-FIXING BIOFERTILIZERS

Nitrogen is one of the most important macronutrients essential for plant growth. Atmosphere contains about 79% of nitrogen volume in free state. The major part of the elemental nitrogen that finds its way into the soil entirely depends on its fixation by certain specialized group of microorganisms

(Mohammadi and Sohrabi, 2012). Plant growth mostly depends on an adequate supply of fixed amount of nitrogen and water. There is an abundance of nitrogen in the atmosphere, but it is unavailable to the plants, and this nitrogen must be converted biologically into a more active form before utilization (Burris, 1991). Biological nitrogen fixation plays an important role in determining nitrogen balance in soil ecosystem. Nitrogen inputs through biological nitrogen fixation support sustainable environmentally sound agricultural production. The value of nitrogen-fixing legumes in improving quality yield of legumes and other crops can be accomplished by the application of biofertilizers (Kannaiyan, 2002). It is difficult to determine the amount of nitrogen fixed biologically; estimates suggest that approximately 150 million metric tons are fixed worldwide annually, or about double the amount fixed by the chemical industry. Nitrogen fixer organism is used in biofertilizer is composed of microbial inoculants or groups of microorganisms which are able to fix atmospheric nitrogen into the soil (Mohammadi and Sohrabi, 2012). Nitrogen-fixing bacteria are single-cell organisms that are essentially tiny urea factories, turning N_2 gas from atmosphere into plant available amines and ammonium via a definite and unique enzyme called nitrogenase. Nitrogen-fixing bacteria are generally endemic to most soil types (both symbiotic and free-living species); however, in the natural state, they generally only comprise a very small percentage of total microbial population are often strains with low-performance regarding quantity of nitrogen they can fix. They are grouped into free-living bacteria (*Azotobacter* and *Azospirillium*) and symbionts, such as *Rhizobium, Frankia, Azolla,* and the blue–green algae (Gupta, 2004). Blue–green algae (Cyanobacteria spp.) are active fixers in the oceans, but they are dispersed, so the amount of nitrogen they fix can only be estimated roughly. There are wide varieties of nitrogen fixing biofertilizer in use. Some of the important nitrogen fixing biofertilizers are mentioned below with a brief explanation and salient features of the product.

16.3.2 RHIZOBIA

Rhizobia establish symbiotic interactions with leguminous plants by the formation and colonization of root nodules (Shridhar, 2012). The bacteria intact the legume root and form root nodules (nodular symbiosis) within which molecular nitrogen is reduced to ammonia that is readily utilized by the plant to make valuable vitamins, proteins, and other nitrogen containing compounds (Pindi and Satyanarayana, 2012). The bacteria are mostly rhizospheric microorganisms, despite its capability to live in the soil for long period of time (Shridhar, 2012). The plant roots provide essential minerals and newly synthesized materials to the bacteria. It is reported that rhizobium can fix 50–300 kg N/ha/year could be fixed by rhizobium for different legume crops. When plants complete their life cycle then fixed N_2 is released, makes it available to other plants and helps in improving the soil fertility status (Majumdar, 2015). Density of nodules occupied, dry weight of its nodule, plant and the grain yield influenced by the multistrain inoculants was highly promising (Pindi and Satyanarayana, 2012). Some free-living bacteria also fix atmospheric nitrogen in cereal crops without any symbiosis and they do not need a specific host plant for nitrogen fixation. However, it is generally useful for pulse crops like red gram, pea, lentil, chickpea, black gram etc., oilseed crops like groundnut and soybean, and forage crops like berseem and lucerne. Successful nodulation of leguminous crops by rhizobium is mainly depends on availability of compatible strain for a particular leguminous crop. It colonizes the roots of specific legumes which act as factories of ammonia production (Chowdhury et al., 2014). Rhizobium has ability to fix atmospheric nitrogen in leguminous crops symbiotically in association with legumes and certain non-legumes like parasponia. Rhizobium population in soil depends on the incidence of legume crops in the field. In absence of legumes, population decreases. Seed inoculation with effective strains of rhizobium is often needed to restore their population near rhizosphere to accelerate nitrogen fixation. Each legume crop needs a specific species of rhozobium to make effective nodules (Mishra et al., 2012). Physical appearence of liquid rhizobium is dull white in color, no bad smell and foam formation, and their pH varies from 6.8 to 7.5 (Majumdar, 2015).

16.3.3 *AZOSPIRILLUM*

Azospirillum is known as the efficient biofertilizer since its ability to induce profuse roots in various plants like rice, millets, and oilseeds even in upland conditions. An estimated amount of 25–30% nitrogen fertilizer can be saved by the use of *Azospirillum* inoculants (Kumar and Singh, 2015). *Azospirillum* bacteria are known to fix significant quantity of nitrogen in the range of 20–40 kg N/ha in the rhizosphere in non-leguminous plants, such as cereals, millets, oilseeds, cotton etc. (Pindi and Satyanarayana, 2012). *Azospirillum* belongs to family Spirilaceae, heterotrophic, and associative in nature, they also produce growth-regulating substances (Chowdhury et al., 2014). Although there are many species under this genus like *A. amazonense, A. halopraeferens, A. brasilense*, but, worldwide distribution and benefits of inoculation have been proved mainly with *A. lipoferum* and *A. brasilense* (Chowdhury et al., 2014). These species have been commercially exploited for the use of nitrogen supplying biofertilizer. One of the characteristics of *Azospirillum* is its ability to reduce nitrate and denitrifying (Pindi and Satyanarayana, 2012). The *Azospirillum* makes associative symbiosis with many plants particularly C_4 plants (Hatch and Slack pathway) because they grow and fix nitrogen easily in the presence of salts of organic acids such as malic and aspartic acid. Thus, it is mainly recommended for sugarcane, maize, sorghum, pearl millet etc. (Chowdhury et al., 2014). *Azospirillum lipoferum* also there in the roots of some tropical forage grasses, such as Wheat, Maize, Rye, Sorghum, Panicum, Digitaria, and Brachiaria. Physical features of liquid *Azospirillum* are blue or dull white in color and non-acidic in nature. *Azospirillum* cultures synthesize considerable amount of biologically active substances like vitamins, nicotinic acid, indole acetic acid, and gibberellins. All these hormones/chemicals help in better germination, early emergence, and better root development of plants (Majumdar, 2015).

16.3.4   AZOTOBACTER

Azotobacter are aerobic, free-living, and heterotrophic (Chowdhury et al., 2014) Gram-negative (Gandora et al., 1998), oval or spherical bacteria in nature that form thick-walled cysts (Salhia, 2013). There are around six species in the genus *Azotobacter (*Martyniuk and Martyniuk, 2003); some of which are movable by means of peritrichous flagella. They are typically polymorphic and their size varies from 2 to 10 µm long and 1 to 2 µm wide (Jnawali et al., 2015). *A. Chroococcum* is the first aerobic free-living nitrogen fixer (Beijerinck, 1901). It is profusely found in neutral to alkaline soils, in aquatic environments, and on some plants. *Azotobacter* is capable of performing various metabolic activities including atmospheric nitrogen fixation by their conversion into ammonia (Pindi and Satyanarayana, 2012). *A. vinelandii, A. beijerinckii, A. insignisand,* and *A. macrocyto* are the common reported species for nitrogen fixation. Sacharophilic bacteria associated with sugarcane, sweet potato, sweet sorghum, and coffee plants which fix atmospheric nitrogen. In fact, the *A. diazotrophicus* sugarcane relationship is favorable symbiotic relationship between grasses and bacteria which fixes about 30 kg/h/year of nitrogen. This bacterium is being exploiting and commercialized extensively for sugarcane crop reducing primary dependency in chemical nitrogen fertilizer. It is known to increase cane yield 10–20 tons/acre and sugar content by 10–15% (Pindi and Satyanarayana,

2012). It serves as potential biofertilizer for all non-leguminous plants especially rice, cotton, vegetables etc. *Azotobacter* population is high in rhizosphere region, and the rhizoplane does not contain *Azotobacter* cells. The lack of organic matter in the soil is a limiting factor in the proliferation of *Azotobacteria* the soil (Majumdar, 2015).

16.4 PHOSPHORUS-SOLUBILIZING MICROORGANISMS

The fixed phosphorus in the soil can be solubilized by phosphate-solubilizing bacteria (PSB), which have the ability to convert inorganic phosphorus form to their soluble forms through the process of organic acid production, chelation, and ion exchange reactions and make them available to plants (Mohammadi and Sohrabi, 2012). Microbial mineralization and solubilization are most important characteristics of the phosphorus cycle, besides chemical fixation of phosphorus in soil (Pindi and Satyanarayana, 2012). Under acidic or calcareous soil conditions, most of the phosphorus is fixed in the soil and are unavailable to the plants. Mainly some phosphate bacteria and fungi can make insoluble phosphorus to their soluble form and makes available to the plants. The enzymatic activity of microorganisms is responsible for the mineralization of organic phosphorus which is left over in the soil after harvesting, or added as plant or animal residues to soil (Bot and Benites, 2007). The solubilization effect of phosphate bacteria is due to the organic acids production which lowers soil pH and brings about the dissolution of bound forms of phosphate. It illustrates that PSB culture increased yield up to 200–500 kg/ha and thus 30–50 kg of superphosphate can be saved. Phosphorus-solubilizing bacteria and fungi play a major role in persuading the insoluble phosphate compound such as rock phosphate, bone meal, and basic slag, and particularly chemically fixed soil phosphorus into available form. These special types of microorganisms are known as PSMs, which include different groups of microorganisms such as bacteria and fungi that convert insoluble phosphate compounds and fixed chemical fertilizers into soluble form (Pindi and Satyanarayana, 2012). Many reports have observed the ability of different bacterial species to solubilize insoluble inorganic phosphate compounds, such as tricalcium phosphate, dicalcium phosphate, hydroxyapatite, and rock phosphate (Chowdhury et al., 2014). Species of *Pseudomonas, Rhizobium, Penicillium, Bacillus, Burkholderia, Achromobacter, Flavobacterium, Agrobacterium, Aereobacter, Erwinia Micrococcus, Fusarium, Sclerotium, Aspergillus*, and some other are considered as active in bio phosphorus conversion (Pindi and Satyanarayana, 2012). Therefore, the application of PSB in agricultural practice would not only counteract the high cost of manufacturing phosphate fertilizers but would also mobilize insoluble fertilizers and soils to which they are applied (Changand Yang, 2009; Banerjee et al., 2010). These bacteria and fungi can also be grown in the media containing $Ca_3(PO_4)_2$, $FePO_4$, A_lPO_4, apatite, bone meal, rock phosphate or similar insoluble phosphate compounds are sole source of phosphate. Such organisms not only assimilate phosphorus but also accelerate release of an excess amount of soluble phosphate than their actual requirements (Pindi and Satyanarayana, 2012). Bacteria play more effective role in phosphorus solubilization as compared to fungi (Alam et al., 2002). Among the whole microbial population in soil, PSB constitute 1–50%, while fungi are only 0.1–0.5% in phosphate solubilization potential (Chen et al., 2006). Some of the isolates solubilized and made available phosphorus to crop from rock phospshate, and many isolates solubilized a very high quantity of tricalcium phosphate. Most effective bacterial isolates were found as *Pseudomonas striata*, *Pseudomonas rathonis*, *Bacillus polymyxa,* and *B. megatherium* (Pindiand Satyanarayana, 2012). There are significant populations of PSB in soil and in plant rhizosphere which includes both aerobic and anaerobic strains, with a prevalence of aerobic strains in submerged soils. A considerably higher concentration of PSB is commonly found in the rhizosphere in comparison with non-rhizosphere soil (Majumdar, 2015).

16.5 PHOSPHATE MOBILIZING MICROORGANISMS

As compared to other nutrients, phosphorus is very less mobile nutrient and available to plants in all types of soil. In spite of its abundance in most of the soils in both organic and inorganic forms of phosphorous is major limiting factor for the growth of plants (Kundan et al., 2015). Several soil fungi are capable to mobilizing immobile form of phosphorous by its hyphal structures and make them available to plants and hence known as phosphate-mobilizing microorganisms. These soil microbes have mutualistic association with all crop plants except family *Brasicaceae*. Besides, phosphorus, this fungus, also mobilizes sulfur and zinc. The literal meaning of *mycorrhizae* is "fungus root" which denote non-pathogenic association between certain soil fungi and plant roots. Moreover, plant species, nutritional status of soil, and ambient soil conditions determine bioavailability of soil inorganic phosphorus. To avoid the phosphorus deficiency in soil, PSMs could play an important role in supplying phosphate to plants in a more suitable, easily, and eco-friendly manner (Majumdar, 2015).

16.5.1   VESICULAR ARBUSCULAR MYCORRHIZA

The term mycorrhiza represents "fungus roots." It is symbiotically associated with host plants and certain group of fungi at the root system (Chowdhury et al., 2014). Plant roots transmit substances (some supplied by exudation) to the fungi, and the fungi help in transmitting nutrients and water to the plant roots. The fungal hyphae may extend the root lengths 100-fold. Hyphae reach into additional and watery soil areas and help in absorption of many nutrients to the plants, particularly less available mineral nutrients such as phosphorus, zinc, copper, and molybdenum. Arbuscular mycorrhizal fungi (AMF) constitutes a group of root obligate biotrophs that exchange mutual benefits with about 80% of plants. They are known as natural biofertilizers because they provide the host with water, nutrients, and pathogen protection, in exchange for photosynthetic products. Thus, AMF are primary biotic soil components which, when missing or impoverished, can lead to a less efficient ecosystem functioning (Berruti et al., 2016). Vesicular and arbuscular mycorrhiza fungi form a type of cover around the root, sometimes giving it a hairy appearance because they provide a protective sheath. Mycorrhiza increases tolerance to drought, to high temperatures, to infection by fungi, and even to extreme soil acidity. Application of VAM produces better root systems which combat to rotting and soil borne pathogens. Highest growth response to mycorrhizal fungi is probably found in plants in highly weathered tropical acid soils that are low in basic cations and phosphate, and may contain toxic levels of aluminum. Therefore, inoculations with PSB and other useful microbial inoculants in these types of soils become emendatory to restore and maintain effective microbial populations for solubilization of fixed phosphorus and increase the availability of other macro and micronutrients to plants and helpful in better crop yield (Majumdar et al., 2015).

16.6   PLANT GROWTH-PROMOTING RHIZOBACTERIA

Plant growth-promoting rhizobacteria (PGPR) have been confirmed to the rapid development of different host plants and they also benefit from the root exudates (Lawal and Babalola, 2014). From last few decades, PGPR have gained considerable importance in research because of their ability to stimulate plant growth, increasing crop yields, being less harmful to the environment, and also helpful in reducing the cost of chemical fertilizers. It can also be termed as plant health-promoting rhizobacteria (PHPR)/or nodule-promoting rhizobacteria (NPR) (Kundan et al., 2015). PGPR is a group of bacteria that can be found in rhizosphere (Ahmad et al., 2008). The term "plant growth-promoting rhizobacteria" refers to bacteria that colonize in the roots of plants (rhizosphere) that enhance the growth of plant. The microbial population present in rhizosphere is comparatively different from that of its surroundings due to the occurrence of root exudates that function as a source of nutrients for microbial growth (Burdman et al., 2000). The PGPR are classified into different groups according to their mode of actions on crops. Firstly, the phyto-stimulating rhizobacteria that promotes crop growth directly by providing nutrients and phytohormones. Secondly, mycorrhiza and root nodule symbiosis which assist rhizobacteria and this positively affect functioning of plant and microorganisms in symbiotic relationship; and thirdly, the biocontrol rhizobacteria that defend plants and crops from pathogens via exudates from antimicrobial agents or by promoting plant more resistance to pest and diseases (Manivasagan et al., 2013). Due to their potential use as biofertilizers and biopesticides, their *modus operandi* has been basically studied in model bacteria such as *Azospirillum* spp. and *Pseudomonas* spp. The genotype of the host plant determines PGPR densities both in terms of number, size and composition. Furthermore, plant growth-promoting effects of these bacteria have been shown to rely both on host plant genes and bacterial strains (Son et al., 2014; Majumdar, 2015).

16.6.1 PHYTOHORMONE PRODUCTION

Phytohormones are chemical messengers that play vital role in the normal growth and occur in very low concentration. These phytohormones affect seed growth, time of flowering, senescence of leaves, sex of flowers, fruits, and shape of plant. They also affect expression and transcription levels of genes, cell division and growth, regulate the cellular processes, pattern formation, vegetative and reproductive development, and stress responses of target cells. All the major activities like formation of leaf and flowers, development and ripening of fruits are regulated and determined by hormones. In order to decrease harmful effects of the environmental stressors caused due to its growth limiting environmental conditions, plants mostly effort to adjust levels of endogenous phytohormones. While this approach is sometimes successful, rhizosphere microorganisms are also produced and modulate phytohormones under in vitro conditions. So, that many PGPB can alter phytohormone levels and thereby affects plants' hormonal balance and its response to stress.

16.7   BIOPESTICIDES

Agriculture produce plays a major role in resource to sustain economical, environmental, and social system, globally. For this reason, global challenge is to secure quality production and to make agricultural produce environmentally compatible. Chemical means of plant protection occupy leading place as regards their total volume of application in integrated pest management and diseases of plants. But pesticides cause toxicity to humans, animals, and also polluted air and water (Mazid et al., 2011). The most common method for pest control has been rigorous use of chemical pesticides. Such pesticide was used in 1940s with dichloro-diphenyl-trichloroethane (DDT), followed by other organophosphate and carbamate pesticides (Nicholson et al., 2007). Sustainable organic farming practices require proper eco-friendly disease and pest management practices in addition to a balanced supply of nutrients for improving quantity and quality of agricultural outputs. Chemical pesticides contributed much to the attainment of sustainable crop protection after the green revolution, but indiscriminate use of pesticide on crop creates several problems to target and non-target organism in addition to environmental pollution and ecological balance (Raja and Masresha, 2015). Due to residual problem and toxicity to the living environment, chemical pesticides are not suitable for organic food production. In this scenario, biopesticides from plant origin are given much importance in recent times to develop better alternatives to chemical pesticides. Prehistorically, many plants and their parts are used as a pesticide for the protection of crops against pests in the field as well as in storage (Raja and Masresha, 2015). Biopesticide are classified as follows:

16.7.1   BIOCHEMICAL PESTICIDES

Biochemical pesticides are directly related to conventional chemical pesticides. These are distinguished from conventional chemical pesticides by their non-toxic nature toward target organisms and their natural occurrence. The active ingredient in these types of pesticides is a single molecule or a complex of molecules, such as a naturally occurring plant essential oil, or a mixture of structurally related compounds known as isomers in the case of insect pheromones. Although all active ingredients of biochemical pesticides are present naturally, the active ingredient in the product may be a synthetic analogue to the naturally occurring material. As many of the active ingredients of biopesticides are synthetic, the full range of green chemistry principles should be applied to the improvement of active ingredient and the biochemical pesticide product. Naturally, occurring chemicals are regulated as biopesticides, and some are fairly toxic. For example, d-limonene is a product of several citrus essential oils (Brien et al., 2009) and used as a conventional insecticide due to its toxic mode of action. In contrast, oils from which d-limonene may be derived, when used as a pesticide, normally have a non-toxic mode of action and are regulated as biopesticides. Use of biopestiside depending on target product and what it is used against, a classic mode of action for these types of oils is suffocation. Some essential oils work as repellents, and their mode of action would be as a fragrance. Biochemical pesticides typically fall into separate functional classes, including plant extracts, natural plant growth regulators, and natural insect growth regulators. There are about 122 biochemical pesticide active ingredients registered with EPA, which include 20 plant-growth regulators, 18 floral attractants, 19 repellents, 6 insect growth regulators, and 36 pheromones (Brien et al., 2009).

16.7.2   INSECT PHEROMONES

Insect pheromones are those chemical substances which are used by an insect to communicate with other members of same species. Structurally, these chemicals are very similar to substances used in flavors and fragrances. Pheromones are a subset of a broader category called semiochemicals. A semiochemical may be defined as the message-bearing substance which produced by a plant or animal, or a synthetic analogue of that substance, which evokes the behavioral response in individuals of same or other species (US EPA BRT, 2002). Semiochemicals are used for various functions including attracting others to a known food source or trail, locating a mate or sending an alarm. Insect sex pheromones are also used in pest management (Brien et al., 2009).

16.7.3   PLANT EXTRACTS AND OILS

Plant essential oils are specific chemicals or mixtures of chemical compounds derived from a plant origin. These kinds of pesticides are much more diverse in their chemical composition, target insect or pest, and mode of action. Plant extracted oils are most often used as insecticides as well as herbicides. Generally, their mode of action varies from product to product. Some plant extracts such as floral essences attract insects or pest to traps. Others such as cayenne can also be used as deterrents. Lemon grass oil, strip waxy coat on leaves of grass weeds to cause dehydration (Brien et al., 2009). Others coat pest causing suffocation, and enhance natural immune system of a crop. Plant extracted oils and their products can be regulated as biopesticides depending on their mode of action and toxicity level. Pyrethrum is an extract of chrysanthemum species and is commonly used in agriculture system, yet it can be highly toxic. Pyrethrum quickly paralyzes and kills pests by altering the way that electrical impulses are transmitted to the nervous system of insects (Ware and Whitacre, 2004).

16.8   MICROBIAL PESTICIDES

Microbial pesticides are naturally occurring or genetically modified fungi, algae, bacteria, viruses, and protozoa. They produce a toxin specific to the pest, causing disease, and prevent from other microorganisms through competition (Clemson HGIC, 2007). For all crop bacterial biopesticide claim about 74%; fungal about 10%; viral biopesticides 5%; predator biopesticides 8%; and other biopesticides contribute 3% (Thakore, 2006). Presently, there are approximately 73 microbial active ingredients that have been registered by the US EPA. The registered microbial biopesticides are about 35 bacterial products, 15 fungi, six non-viable genetically engineered microbial pesticides, eight plant incorporated protectants, one yeast, one protozoa, and six viruses (Steinwand, 2008). Microbial biopesticides may be delivered to crops in many forms including live, dead, and spores. Their manufacturing process, regulation, and use are totally different from conventional pesticides. Microbial pesticides can control a diverse range of pests; each microbial pesticide has active ingredient specific to its target pest (Brien et al., 2009).

16.8.1   BACTERIAL PESTICIDES

Bacterial pesticides are most frequent form of microbial pesticides. They are usually used as insecticides, although control unwanted bacteria, fungus and viruses. As an insecticide, they are specific to individual species of butterflies and moths, as well as flies, beetles, and mosquitoes. To be effective, they come into contact of target pest, and may require ingestion to be effective and their mode of action depending on the target pest. In insects, bacteria interrupt digestive system by producing an endotoxin. To control pathogenic bacteria or fungus, the bacterial biopesticide colonizes on plant and crowds out the pathogenic pest or insect (Brien et al., 2009; Table 16.1).

TABLE 16.1   Some Bacterial Biopesticides and their Modes of Action.

S. no.	Bacteria	Category	Target pest	Mode of action
1	*Bacillus thuringiensis* (Bt)	Insecticide	Butterfly and moth lepidoptera	Digestive system
2	*Bacillus subtilis*	Bactericide	Bacterial and fungal pathogens such as Rhizoctonia, Aspergillus Fusarium, and others	Colonizes on plant root and competes
3	*Pseudomonas fluorescens*	Bactericid/ fungicide	Several fungal, viral, and bacterial diseases such as frost forming bacteria	Crowds out and controls the growth of plant pathogens

16.8.2   FUNGAL PESTICIDES

Fungal pesticides also can be used as controlling agents of insects and plant diseases. They are parasitic and produce bioactive compounds such as enzymes that dissolve plant walls. Mode of action is varied from species to species and depends on both the pesticidal fungus and target pest. *Beauveria bassiana* spore grows and proliferates in the insect's body, produce toxins, and draining nutrients to kill insect. *Trichoderma i*s a fungal antagonist and grows into tissue of a pathogenic fungus and secretes enzymes to degrade cell walls of other fungus, then utilize the cell content of target fungus and multiplies its own spores. *Muscador albus* releases gaseous toxins into soil which can eradicate soil-borne pests and bacteria affecting crop plants. Another advantage of fungal pesticides is that require a narrow range of conditions including moist soil and cool temperatures to proliferate (Brien et al., 2009; Table 16.2).

TABLE 16.2   Some Fungal Pesticides and Their Modes of Action.

S. no.	Fungus name	Category	Pest name	Mode of action
1	*Beauveria bassiana*	Insecticide	Foliar feeding insects	White uscadine disease
2.	*Trichoderma viride/ harzianum*	Fungicide	Soil borne fungal disease	Mycoparasitic
3.	*Muscodor albus*	Fungicide	Bacteria and soilborne pests	Releases volatile toxins

16.9 CONCLUSION

Our dependence on chemical fertilizers and pesticides has encouraged the flourishing of industries that are producing life-threatening chemicals and which are not only dangerous for human consumption but can also disturb the ecological balance. Biofertilizer and biopesticide can help in solving the problem of feeding global increasing population at a time when agriculture is facing various environmental stresses. It is important to understand the useful aspects of biofertilizers and biopesticides to implanting its application to modern agricultural practices. The new technology developed using the powerful tool of molecular biotechnology can improve the biological pathways of production of phytohormones. If identified and transferred to the useful PGPR, these technologies can helpful in providing relief from environmental stress. It is also believed that biological pesticides may be less vulnerable to genetic variations in plant populations that cause problems related to pesticide resistance. If deployed appropriately, biopesticides have potential to bring sustainability to global agriculture for food and feed security. The views expressed here are those of the authors only. These may not necessarily be the views of the institution/organization the authors are associated with.

KEYWORDS

- chemical fertilizers
- biofertilizer
- nitrogen fixing
- phosphate solubilizing
- phosphate-mobilizing biopesticide

REFERENCES

Ahmad, F.; Ahmad, I.; Khan, M. S. Screening of Free-living Rhizospheric Bacteria for their Multiple Plant Growth Promoting Activities. *Microbiol. Res.* **2008,** *163,* 173–181.
Banerjee, S.; Palit, R.; Sengupta, C.; Standing D. Stress Induced Phosphate Solubilization by *Arthrobacter* sp. and *Bacillus* sp. Isolated from Tomato Rhizosphere. *Aust. J. Crop Sci.* **2010,** *4* (6), 378–383.

Beijerinck, M. W. Ueber Oligonitophile Mikroben, Zentralblattfiir Bakteriologie, Parasiten-kunde, Infektionskrankheiten and Hygiene, *Abteilung* II **1901,** *7*, 561–582.

Bhattacharjee, R.; Dey, U. Biofertilizer A Way Toward Organic Agriculture: a Review. *Afr. J. Microbiol. Res.* **2014,** *8* (24), 2332–2342.

Bot, A.; Benites, J. The Importance of Soil Organic Matter. Natural Resources Management and Environment Department. FAO Corporation Document Repository, 2005.

Brien, K. B.; Franjevic, S.; Jones, J. Green Chemistry and Sustainable Agriculture: the Role of Biopesticides. *Report* **2009,** 1–55.

Burdman, S.; Jurkevitch, E.; Okon, Y. Recent Advances in the Use of Plant Growth Promoting Rhizobacteria (PGPR) in Agriculture. In *Microbial Interactions in Agriculture and Forestry*; Subba Rao, N. S.; Dommergues, Y. R., Eds.; Science Publishers: Enfield, NH, USA, 2000; pp 229–250.

Burris, R. H. Nitrogenases. *Appl. Boil. Chem.* **1991,** *266*, 9339–9342.

Chang, C. H.; Yang, S. S. Thermo-tolerant Phosphate-solubilizing Microbes for Multi-functional Biofertilizer Preparation. *Biores. Technol.* **2009,** *100*, 1648–1658.

Chen, Y. P.; Rekha, P. D.; Arunshen, A. B.; Lai, W.A.; Young, C. C. Phosphate Solubilizing Bacteria from Subtropical Soil and their Tri-calcium Phosphate Solubilizing Abilities. *Appl. Soil Ecol.* **2006,** *34*, 33–41.

Clemson HGIC Clemson Extension, Home and Garden Information Center. (2007). Organic Pesticides and Biopesticides.

Dutta, S. Biopesticides: An Ecofriendly Approach for Pest Control. *J. Pharm. Pharm. Sci.* **2015,** *4* (6): 250–265.

Gandora, V.; Gupta, R. D.; Bhardwaj, K. K. R. Abundance of Azotobacter in Great Soil Groups of North-West Himalayas. *J. Indian Soc. Soil Sci.* **1998,** *46* (3), 379–383.

Gupta, A. K. The Complete Technology Book on Biofertilizers and Organic Farming. National Institute of Industrial Research Press: India, 2004.

Jnawali, A. D.; Ojha, R. B.; Marahatta, S. Role of Azotobacter in Soil Fertility and Sustainability: a Review. *Adv. Plants Agri. Res.* **2015,** *2* (6), 1–5.

Kannaiyan, S. Biofertilizers for Sustainable Crop Production. *Biotechnology of Biofertilizers*; Narosa Publishing House: New Delhi, India, 2002; p 377.

Kumar, S.; Singh, A. Biopesticides for Integrated Crop Management: Environmental and Regulatory Aspects. *J. Biofertil. Biopestic.* **2014,** *5* (1), 1–3.

Kumar, S.; Singh, A. Biopesticides: Present Status and the Future Prospects. *J. Fertil. Pestic.* **2015,** *6* (2), 1–2.

Kundan, R.; Pant, G.; Jadon, N.; Agrawal, P. K. Plant Growth Promoting Rhizobacteria: Mechanism and Current Prospective. *J. Fertil. Pestic.* **2015,** *6* (2), 1–9.

Lawal, T. E.; Babalola, O. O. Relevance of Biofertilizers to Agriculture. *J. Hum. Ecol.* **2014,** *47* (1), 35–43.

Majumdar, K. Bio-fertilizer Use in Indian Agriculture. *Management* **2015,** *4* (6), 377–381.

Manivasagan, P.; Venkatesan, J.; Sivakumar, K.; Kimm, S. K. Marine Actinobacterial Metabolites: Current Status and Future Perspectives. *Microbiol. Res.* **2013,** *168*, 311–332.

Martyniuk, S.; Martyniuk, M. Occurrence of *Azotobacter* spp. in Some Polish Soils. *Polish J. Environ. Stud.* **2003,** *12* (3), 371–374.

Mazid, S.; Kalita, J. C.; Rajkhowa, R. C. A Review on the Use of Biopesticides in Insect Pest Management. *Int. J. Sci. Adv. Technol.* **2011,** *1* (7), 169–178.

Mishra, D. J.; Singh, R.; Mishra, U. K.; Shahi, S. K. Role of Bio-fertilizer in Organic Agriculture: A Review. *Res. J. Recent Sci.* **2012,** *2*, 39–41.

Mohammadi, K.; Sohrabi, Y. Bacterial Biofertilizers for Sustainable Crop Production: A Review. *ARPN Am. J. Agric. Biol. Sci.* **2012,** *75,* 307–316.

Nicholson, G. M. Fighting the Global Pest Problem: Preface to the Special Toxicon Issue on Insecticidal Toxins and their Potential for Insect Pest Control. *Toxicon* **2007,** *49,* 413–422.

Pindi, P. K.; Satyanarayana, S. D. V. Liquid Microbial Consortium: A Potential Tool for Sustainable Soil Health. *J. Biofertil. Biopestic.* **2012,** *3* (4), 2–9.

Raja, N. Biopesticides and Biofertilizers: Ecofriendly Sources for Sustainable Agriculture. *J. Biofertil. Biopestic.* **2013,** *4* (1), 1–2.

Raja, N.; Masresha, G. Plant Based Biopesticides: Safer Alternative for Organic Food Production. *J. Biofertil. Biopestic.* **2015,** *6* (2), 1–2.

Rajendra, P.; Singh, S.; Sharma, S. N. Interrelationship of Fertilizers Use and Other Agricultural Inputs for Higher Crop Yields. *Fertilizers News* **1998,** *43,* 35–40.

Roychowdhury, D.; Paul, M.; Banerjee S. K. A review on the Effects of Biofertilizers and Biopesticides on Rice and Tea Cultivation and Productivity. *Int. J. Adv. Res. Sci. Eng. Technol.* **2014,** *2* (8), 97–106.

Salhia, B. The Effect of Azotobacter Chrococcumas Nitrogen Biofertilizer on the Growth and Yield of Cucumis Sativus. The Islamic University Gaza, Deanery of Higher Education Faculty of Science, Master of Biological Sciences, Botany, 2013.

Shridhar, B. S. Review: Nitrogen Fixing Microorganisms. *Int. J. Microbiol. Curr. Res.* **2012,** *3* (1), 46–52. (2012).

Son, J. S.; Sumayo, M.; Hwang, Y. J.; Kim, B. S.; Ghim, S. Y. Screening of Plant Growth-promoting Rhizobacteria as Elicitor of Systemic Resistance against Gray Leaf Spot Disease in Pepper. *Appl. Soil Ecol.* **2014,** *73,* 1–8.

Thakore, Y. The Biopesticide Market for Global Agricultural Use. *Industrial Biotechnol.* **2006,** *2* (3), 192–208.

Ware, G.; Whitacre, D. An Introduction to Insecticides. In *Radcliffe's IPM World Textbook*; Radcliffe, E.; Hutchison, W.; Cancelado, R.; University of Minnesota Press: St. Paul, MN, 2004. http://ipmworld.umn.edu. (accessed Mar 6, 2016).

CHAPTER 17

Fungal Biopesticides: A Novel Tool for Management of Plant Parasitic Nematodes

ERAYYA[1,*], SANJEEV KUMAR[1], SANTOSH KUMAR[1], and KAHAKASHAN ARZOO[2]

[1]*Department of Plant Pathology, Bihar Agricultural University, Sabour, Bhagalpur 813210, Bihar, India*

[2]*Department of Plant Pathology, G.B. Pant University of Agriculture and Technology, Pantnagar 263145, Uttarakhand, India*

Corresponding author. E-mail: erayyapath@gmail.com

ABSTRACT

Environmental pollution and its consequences are the greatest challenges in the use of agro-chemicals in agricultural crop production. The exploitation of biocontrol control strategies in nematode disease management is still unattainable for the farmers until the development of manpower and intensification of research in this area. A biological control refers to natural undisturbed eco-systems, in which a diverse group of organisms exists in a dynamically equilibrium. Like other approaches biocontrol has its own constraints. Application of biological control strategies in nematode management requires, knowledge about bioagents, working skill, nonsystemic and is labor intensive. Sometimes bioagents subjected to rapid inactivation due to adverse environmental conditions. Environment pollution is a major issue at the present agriculture scenario for the choice of disease management tactics. Use of biopesticides is an eco-friendly option to manage plant-parasitic nematodes. Many botanicals, bacteria, and fungi have been shown potentiality in reducing plant-parasitic nematodes population in the soil. The potential biopesticides/biocontrol agent could be identified and incorporated in the integrated nematode management program. It is important to develop simple farmer friendly techniques for the indigenous preparations and to fix the correct dosage of the biopreparations for effective nematode control.

17.1   INTRODUCTION

Nematode diseases of crop plants are universal and widespread in nature. Most of the cultivated crops are commonly attacked by one or more nematodes species. Plant-parasitic nematodes can be identified by the presence of stylet or spear at the mouth. A nematode sucks the plant sap and injects its saliva into the plant tissue. Nematode infection leads to the production of various symptoms, hormonal changes, and depletion of nutrients in the plant tissues. Management of nematode diseases is much more difficult than that of other pests because the nematodes inhabitate in the soil and underground parts of the plants (Mian, 1998). At present the chemical management of nematode diseases in plants is the primary means to control nematodes. The use of toxic chemicals leading to the pollution of the ground as well as surface water, disturbance of natural equilibrium among the different life forms existing in nature (Elyousr et al., 2010). Hence, there is an immediate need to develop a public concern about the impact of chemicals on the environment. Biopesticides such as botanicals (plant or plant extracts) and nematopathogenic microorganisms (bacteria and fungi or their derivatives) have a lot of scope as nematicide due to their eco-friendly, cost-effective and sustainable nature (Radhakrishnan, 2010; Pendse et al., 2013). The use of botanicals and bioagents has been greatly contributing to reducing the plant pathogenic nematode population in soil (Taye et al., 2012; Mamun et al., 2014). In this review, the potential uses of nematopathogenic microorganisms in the management of plant-parasitic nematodes have been discussed in detail facilitate further research in this area.

17.2   LOSSES DUE TO NEMATODE DAMAGE

Nematodes cause devastating damage to agricultural crops if, they are not managed properly. 20 major life-sustaining agricultural crops suffer 10.7% loss in yield every year due to nematodes, In India, the average yield loss due to nematodes is about 12.3% (Gaur and Pankaj, 2009). All India Coordinated Research Project (AICRP) on Nematodes reported that annual loss due to plant-parasitic nematodes is Rs. 21,000 million (Jain et al., 2007). According to International Meloidogyne Project, the losses due to nematode diseases in the developed countries is $78 billion per annum and in developing countries, it is more than $100 billion per annum. Nematode damage is much more severe and complicated in tropical regions compared to temperate regions. Similarly, horticultural crops suffer more than field crops. Monoculture of crops enhances the nematode infestation. According to an estimate 6% annual loss in field crops, 12% loss in fruit and nut crops, 11% in vegetable crops and 10% annual loss in ornamental crops due to nematodes. Nematodes cause losses in terms of quantity of produce, reduce vitamins and minerals content inedible parts of the plants, causes complex diseases in association with other soilborne plant pathogens. Nematodes inhabit in and around the root system in the soil. A few nematodes also infect the aerial part of the plants. Under protected cultivation practices Rs. 21,068.73 million losses occur annually as a result of nematode infestation. Nematodes exacerbate the severity of fungal and bacterial diseases which leads to complete crop losses. *M. incognita* in association with *Fusarium oxsporum* and *Phytophthora parasitica* causes heavy losses in tomato and gerbera, respectively. *Ralstonia solanacearum* get enter into the host roots through the wounds caused by root-knot nematode.

17.3   BIOPESTICIDES

Biopesticides are those pesticides that are derived from plants or insects or microorganisms. There are three broad categories of biopesticides: biochemical, microbial, and plant-derived products. Biochemical pesticides (potassium bicarbonate, phosphorous acids, plant extracts, and botanical oils) contain naturally occurring substances which control diseases. Microbial biopesticides include microbe or their derivatives as an active ingredient. Plant-derived products are those which are originated from plants or their parts. Biopesticides have short residual time, low toxicity, less harmful to nontarget organisms and to the environment. Hence, the Environmental Protection Agency (EPA) generally requires less data for registration of a new biopesticide as compared to the chemical pesticide. But, it is necessary to submit sufficient information regarding the composition of the product, toxicity, degradation, impact on human health and the environment. The registration processes are much quicker and shorter for biopesticides than chemical products, it is often less than a year.

17.4   MAJOR COMPONENTS OF BIOLOGICAL CONTROL/BIOPESTICIDES

Biocontrol includes use of beneficial microbes/insects, their genes, gene product, etc. to overcome the negative impact of plant diseases (Junaid et al., 2013). Biological control includes the utilization of pathogenic, antagonist or competing organisms, predators, parasitoids to control pest and diseases (Siddiqui and Mahmood, 1996). If the microorganisms are used as the antagonist, then it is considered as microbial control. Commonly, microorganism initiates the infection in the host through penetration or by habituating in the in the soil or foliage. The bioagent multiplies in the host organism and causes its death by the destruction of tissues or by the emission of toxic compounds. Biological control is the best alternative method to the chemical method for the control of plant diseases. Use of chemicals indeed leads to ecological imbalances, environmental pollution and the poisoning of food-chain, resistance builds up in pathogen against pesticides, resurgence problems, etc.

Inundation application of bioagents and soil solarization are the best alternatives to chemical control (Djian Caporalino et al., 2009; Collange et al., 2011). Inundation application of bioagents creates anaerobic microenvironment and reduces nematode population in soil (Collange et al., 2011). In soil, solarization soil is covered with the plastic sheet during the hot summer to increase the temperature. It helps to minimize temperature sensitive pathogens upto a depths of about 20–25 cm (Bélair, 2005). The present review deals on biological control of nematode and their commercial products. Biopesticides are mainly derived from plant extracts (botanicals) and nematopathogenic organism (microbials) presented in Table 17.1. Use of botanicals maintain the soil health and sustains its life by increasing soil organic matter content. They are bio-degradable, economical and can be locally prepared (Akpheokhai et al., 2012; Taye et al., 2012). Biopesticides are less toxic than chemical pesticides and safer to the beneficial microorganism, human and environment. In a number of instances, biopesticides proved as effective control agents to manage nematode diseases in crops. They are slow in action upon the target organisms (Elyousr et al., 2010).

17.4.1   *PLANT EXTRACTS (BOTANICALS)*

Plants containing many bioactive organic chemicals/metabolites which act against pests as, repellent, antifeedant, bacteriocidal, fungicidal, nematicidal, etc. According to Radhakrishnan (2010) estimate, the plants contain around

400,000 secondary metabolites represent nematicidal or nematistatic properties (Chitwood, 2002). The mechanisms of action of plant extracts are due to denaturation of proteins, degradation of enzymes (Taye et al., 2012). Extracts leaves and other parts of plants reduce nematode population of *Meloidogyne*, *Pratylenchus*, *Rotylenchulus*, *Helicotylenchus*, *Tylenchorhynchus*, *Radopholus*, *Heterodera*, etc. (Akpheokhai et al., 2012). Botanicals inhibit the hatching of juveniles and kill nematode (Akpheokhai et al., 2012). The mode of action (ovicidal or larvicidal properties) of extracts is due to secondary metabolites, lactones or terpenes which are present in extracts (Taye et al., 2012). The effective concentration of plant extracts ranges from 5% to 10% (Radhakrishnan, 2010). In broad-spectrum, plant species used as biopesticides should be perennial in nature and should not become a weed or a host to insect pests or plant pathogens. They may have subsequent economic uses but should not be harmful to non-target organisms, wildlife, human beings or to the environment. It must be easy to cultivate, harvest, extract preparation should be simple, not to be laborious/time-consuming or require high technical input. Plant species identified as nematicides along with their family, the plant parts used and active ingredients are cited in Table 17.1.

17.4.2 MICROBIALS

Many microorganisms have been playing a significant role in suppressiveness (natural and induced) of plant-parasitic nematodes (PPNs). Many scientists studied the effectiveness of several microorganisms (bacteria and fungi) against nematodes in different crops (Mukhtar et al., 2013). Many microorganisms produce secondary metabolite/chemicals which induces systemic resistance in plants against nematode diseases (Elyousr et al., 2010).

17.4.3 FUNGI

Nematophagous fungi are carnivorous in nature. They parasitize nematodes with the help of mycelia network/trap (Khan and Haque, 2011). Nematopathogenic fungi or fungi destructive to nematodes can parasitize, kill and digest nematodes (eggs, larvae, and/or adults). Over 200 species from six different classes of fungi were reported to parasitize on nematode eggs, juveniles, adult, and cysts (Mukhtar et al., 2013). In general, nematopathogenic fungi are facultative parasites which survive in soil as a saprophyte.

TABLE 17.1 Major Elements in Management of Plant Parasitic Nematodes.

S. no.	Type of bio-control agent	Example
1	Organic manures	Farm yard manure, compost, vermicompost, vermiwash, poultry manure, etc.
2	Plants	Betle vine (*Piper betle*), Black pepper (*Piper nigrum*), Calotrope (*Calotrophis profera*), Castor (*Ricinus communis*), Citrunella (*Cymbophagon nordus*), Datura (*Datura stramonium*), Custard apple (*Annona squamosa*), Garlic (*Allium sativum*), Ginger (*Zingiber officinale*), Green amaranth (*Amaranthus viridis*), Lantana (*Lantana camera*), Marigold (*Tagetes* spp.), etc.
3	Fungi	*Trichoderma harzianum*, *Aspergillus fumigates*, *A. niger*, *Beauveria bassiana*, *Metazhizium anisopliae*, *Pochinia* spp. *Verticillium lecanii*, etc.
4	Bacteria	*Bacillus subtilis*, *B. thuringiensis*, *Pasteuria penetrans*, *P. thornei*, *P. nishizawae*, *Pseudomonas fluorescens*, etc.

In general naturally existing population of fungi in the soil possesses a parasitic relationship with nematodes (Sarhy Bagnon et al., 2000; Brand et al., 2004; Thakur and Devi, 2007; Collange et al., 2011). The process of infection involves penetration of cuticle, invasion of the fungus and mortality of nematodes. The nematode body contents were utilized by the fungus to meet their nutritional requirements (Huang et al., 2004). The components of the cuticle such as chitin, fibers, collagen, etc., serve as a precursor to the nematophagous fungi for invasion and infection of nematode (Hajieghrari et al., 2008). Cayrol et al. (1992) classified fungi based on their nematode-trapping mechanism. Some fungi can able to trap the nematodes (predatory fungi). Trapping modes of predatory fungi include formation of mycelial traps (*Arthrobrotrys oligospora*), constriction ring (*A. anchoni,*) and/or adhesive mycelia knobs (*Dactylella lobata*). Nematophagous fungi are basically saprophytic in their nature, but become parasitic on nematodes if there is a congenial environment for infection. These fungi are commonly parasitized on the juvenile stages of the nematodes. Some parasitic fungi parasitize on nematode eggs (Cayrol et al., 1992). *Paecilomyces* spp. and *Pochonia* spp. are the most effective egg-parasites. *P. lilacinus* can be successfully utilized to manage root-knot nematodes, *M. incognita* in vegetable crops (Van Damme et al., 2005; Haseeb and Kumar, 2006). Some fungi parasitize nematodes with spores with sticky substances on the surface. Adhesive spore-forming fungi includes *Catenaria anguillulae*, *Meristracum asterospermum*, *Meria coniospora*, *Nematoctonus leiosporus,* and *Hirsutella* spp. (Cayrol et al., 1992).

Endomycorhizae with arbuscular vesicles are symbiotically associated with the plant roots. They produce external as well as internal mycelia network. Mode of action of mycorhyza on nematodes includes, hyper parasitization, competition for food and space, and induction of host resistance (Cayrol, 1992). Some toxic substances produced by the fungi are naturally nematifuge (repulsives), nematostatic (pause the nematode growth and development temporarily), nematicidal (kill the nematode). *Trichoderma harzianum* produces 6-pentyl-pyrone which is toxic to nematodes (Sarhy Bagnon et al., 2000). The toxin produced by *Bacillus thuringiensis* effectively inhibits the nematodes (Cayrol et al., 1992). Actinomycete (*Pasteuria penetrans)* inhibits phytopathogenic nematodes (Mateille, 1993). Spores get to adhere to J_2 stage of nematode, penetrate into the larva and cause infection. Growth of *Pasteuria penetrans* has perfect synchrony with the growth and development of nematode. Germ tube gets penetrated into the nematode body and cause infection. Finally, it forms the new growth and spores inside the female nematode and makes it sterile by destroying their reproductive system. The nematode order Mononchida contains predatory nematodes (*Dorylaimidae* and Aphelenchidae). *Alliphis* and *Alicorhagia (*predatory mites) also inhibit the plant-parasitic nematode population in the soil effectively.

17.4.4   BACTERIA

Bacteria are ubiquitous in nature and some species of *Pasteuria, Pseudomonas* and *Bacillus* (Tian et al., 2007), have been shown great potential against nematodes. Nematopathogenic bacteria vary greatly in their action and exhibits wide host ranges. A diverse group of nematopathogenic bacterial have been isolated and evaluated against various stages of nematodes (Tian et al., 2007). Nematophagous bacteria parasitize on nematodes; produce various enzymes and toxins thus interfere with various stages of nematode infestation (Rahanandeh et al., 2012). Some bacteria are highly effective against many plant-parasitic nematodes *viz. Pratylenchus* spp., *Meloidogyne* spp., *Radopholus similis, Rotylenchulus renfiormis, Heterodera* spp., etc. (Khan and Haque, 2011; Rahanandeh et al., 2012). The complex mechanism(s) exist in bacteria–nematode interaction (Rahanandeh et al., 2012).

17.4.5 BIOPESTICDE PREPARATIONS EFFECTIVE AGAINST NEMATODES

In the United States, more than 279 biopesticides (all inclusive) were approved in 2009. 14 bacteria and 12 fungi have been registered with EPA for the management of crop diseases including nematodes (Fravel, 2005; Table 17.2).

17.4.6 GLOBAL BIOPESTICIDE MARKET

The demand for the biopesticides increasing every year because of desire for organic agriculture products and pesticide-free food products. At present, the United States is the global leader in biopesticide production and marketing. In Europe, the market for biopesticides was $270 million in 2010. France comes third in the biopesticide market, behind Spain and Italy (*Source*: Europe: Pesticides Biomarket, CPL Consultants, 2010).

17.5 FORMULATION OF BIOLOGICAL CONTROL AGENTS

The formulation contains fungal spores/bacterial cells as an active ingredient and inert material (diluents or surfactant) to maintain the viability of the bioagents. Formulation(s) may be available in solid or liquid form. The inert material may talc or zeolite (Chaube et al., 2003; Küçük and Kivanç, 2005). In 1997, *B. thuringengensis* granular formulation was patented (Quimby et al., 1996). Immobilization of biomass within the cross-linked polymers is one of the emerging technology (Cho and Lee, 1999). Incorporating fungal mycelia in alginate pellets found very successful (Papavizas et al., 1987;

TABLE 17.2 Commercially Available Biocontrol Agents Against Plant Parasitic Nematodes.

S. No.	Bioagent/s	Commercial product(s)	Effective against	Parasitization
1	*Paecilomyces lilacinus*	Bio-act®, Melocon WG	*Meloidogyne* spp., *Radopholus similis*, *Globodera* spp. and *Pratylenchus* spp.	On eggs and infectious juveniles
2	*Beauveria bassiana*	Mycotrol® ES	*Meloidogyne* spp.	infectious juveniles
3	*Purpureocillium lilacinum*	Biostat®	Many plant parasitic nematodes	On eggs and cause deformation and destruction of ovary and eggs
4	*Verticillium lecanii*	Mycotal® WP	*M. Incognita*	On infectious juveniles
5	*Trichoderma* spp.	Trianum®	*Meloidogyne* spp., *Radopholus similis*	On infectious juveniles
6	*Metarhizium anisopliae*	MET52®	*Meloidogyne* spp.	On infectious juveniles and adults
7	Consortium (*Bacillus* spp. + *Trichoderma* spp. + *Paecilomyces* spp.+ *Tagetus* extracts)	Nemaxxion Biol®	*Meloidogyne* spp., *Radopholus similis*, *Pratylenchus* spp.,	On all the stages
8	Consortium (*Arthrobotrys* spp + *Dactyllela* spp. + *Paecilomyces* spp. + *Glomus* spp. + *Pseudomonas* spp. + chitinolytic enzymes)	Rem G®	Many plant-parasitic nematodes	On all the stages

Küçük and Kivanç, 2005). Alignate pellets were commonly used in the preparation of microbial herbicide formulations (Walker and Connick, 1983). Biotechnological techniques are becoming an emerging tool to increase the multiplication rate of bioagents and to reduce cell mortality. Lignate pellet formulations of various biocontrol agents gaining much more importance. (Lewis and Papavizas, 1983, 1985; Serp et al., 2000). Pellet formulations have many advantages over liquid formulations as pellets can be stored for a longer period of time without losing the viability of spores.

17.6 MANAGEMENT OF PPNS

The management practices mainly aimed to keep the nematode population below the economic injury level. In general, integrated control measures are needed to raise crop economically.

17.6.1 REGULATORY METHODS

Quarantine is the regulatory method to manage the nematode and other crop pests. The principle involved in exclusion is to prevent the pest entry to the area where it does not exist before. The best example of the nematodes against which quarantine laws exist is golden cyst nematode of potato, burrowing nematode of banana.

17.6.2 CULTURAL METHODS

17.6.2.1 CROP ROTATION

Crop rotation involves rising of nonhost crops in alternate to nematode sensitive crop. For example, cereal crop or oilseed crop is to be grown in alternate to vegetable crop to manage root-knot nematode, *Meloidogyne incognita*.

17.6.2.2 FALLOW PLOWING

Fallowing followed by deep summer plowing (two to three times, during May–June) could help to reduce the nematode population in the soil by exposing the dormant structures of the nematode to the hot sun in the month of May and June.

17.6.2.3   ORGANIC AMENDMENTS

Amendment of soil with green manure, compost, oil-cakes, etc., helps to reduce the nematode population in the soil. Application of oil cakes at the rate of one to two tons per hectare is significantly effective against nematodes. Farmyard manures and composts inhibit the nematode population in the soil by various organic acids and enhancing the antagonist's population in the soil.

17.6.2.4   RESISTANT OR TOLERANT VARIETIES

Development of resistant or tolerant varieties against plant-parasitic nematodes is one of the most effective and sustainable ways to manage the PPNs.

17.6.3   PHYSICAL METHODS

17.6.3.1   HEAT TREATMENT

It is a practice to kill the soilborne plant pathogens and insects including nematodes. In nurseries, autoclaved soil is commonly used for potting. In greenhouses, steam sterilization practice is used. In steam sterilization temperature of the soil rose to about 82°C for 30 min, which in turn kill the soilborne inoculum. In conventional agriculture burning of crop residues is commonly practiced to reduce the soilborne diseases.

17.6.3.2   HOT WATER TREATMENT OF PLANTING MATERIAL

It is commonly practiced seed, tuber, bulb and rhizome treatments. Temperature and time for hot water treatment of planting material range from 43°C to 55°C and 10–60 min, respectively. Rice seeds are treated at 55°C for 10–15 mins to control the white tip nematode.

17.6.3.3   SUN DRYING

Dormant or active stages of nematodes like egg, juvenile and adult could be controlled by exposing them to the hot sun.

17.6.3.4   SUMMER PLOWING

It is followed to expose the resting structures of the nematodes to the hot sun.

17.6.3.5   SOIL SOLARIZATION

Moist soil is covered with black polythene sheet in order to increase the temperature of the soil which in turn helps in controlling the active stage(s) of the nematode eggs or juveniles.

17.6.3.6   SPRAYING WITH VEGETABLE OILS

Spraying with corn seed oil, mustard oil, soybean oil, neem oil, etc. significantly reduces the plant-parasitic nematodes in the soil.

17.6.3.7   SEED FLOATATION IN BRAIN SOLUTION

20% salt (brine) solution is commonly used for flotation. Ear cockle of wheat caused by *Anguina tritici* could be easily controlled by this method.

17.6.4   CHEMICAL CONTROL

Chemical method is the most widely adopted method to manage PPN. But this method involves more cost and also harmful to native flora and fauna.

17.6.4.1   SOIL FUMIGANTS

These chemicals are commonly volatile compounds. Halogenated hydrocarbons like *dichloropropene-dichloropropane* (DD), ethylene dibromide (EDB) and methyl bromide (MBr) are the widely used soil fumigants. Optimum soil moisture and optimum temperature (15–30°C) at the time of fumigation increases the efficiency of the treatment. The optimum depth of fumigant application is about 20–30 cm.

17.6.4.2   USE OF NEMATICIDES

Among nematicides, organophosphates and organocarbamates are the most widely used chemical groups. Carbofuran, aldicarb, fenamiphos, and fensulfothion have been the most commonly used commercial formulations. The chemical nematicides are most commonly available in a granular formulation.

17.6.5 BIOLOGICAL CONTROL

Many fungi, bacteria, and arthropods exist in soil may act as a parasite or predator to the PPNs. These organisms could be exploited to minimize the PPNs in soil.

17.6.5.1 USE OF NEMATOPHAGOUS FUNGI AND BACTERIA

Several species of soilborne fungi parasitize and infect the PPNs. The trap may be a concentric ring, adhesive hyphae, mycelia network or haustorial knobs. Some nematophagous fungi are obligate in their nature while some others are facultative parasites on nematodes. The most commonly occurring nematophagous fungi are *Monacrosporium* spp*., Arthrobotrys* spp*., Dactylella* spp*., Geniculifera* spp., *Dactylaria* spp*., Nematoctonus* spp., etc., some fungi forms adhesive spores in the soil, when the nematodes come in contact, these adhesive spores stick on to the nematode body and cause infection. Some obligate nematophagous fungi include *Drechmeria coniospora, Hirsutella rhossiliensis, Nematoctonus* spp., *Verticillium balanoides,* etc.

Zoosporic fungi produce motile spore. After locating their host zoospores attach to the nematode, most commonly at mouth, vulva or anus. Then shed their flagella and get encysted. The spore produces penetration peg and enters cuticle directly or indirectly. When the nutrients inside the nematode body exhausted, then they produce sporangia and zoospores on the nematode body. The spores then spread to the new host with different agencies. The major limitation to use zoosporic fungi as a biocontrol agent is a requirement of wet soil for their better performance. There are many fungi which parasitize various stages of nematode life cycle (egg, juvenile, and adult). Nematode parasitizing fungi were grouped into two broad categories. One is an obligate parasite, which can grow and multiply only on/in the nematode body, other one facultative parasite that can grow in/on the nematode and also on the dead organic matter in the soil.

Dactylella oviparasitica, Paecilomyces lilacinus, Verticillium spp., *Dilophospora alopecuri*, etc. colonize on nematode as well as the host. Both fungi and bacteria are antagonistic toward nematodes. But, fungi have a greater impact on reducing nematode population. These bioagents compete with nematodes for nutrients, oxygen, water, and space. *Pasteuria* spp. parasites n nematodes and feed as an obligate parasite. The spores adhere to the nematode body while moving in the soil. Then they form germ tube and penetrate into nematode body and cause infection. The fungus produces more mycelium as the infection progresses and thus produces the daughter colonies. In cease of bacteria the bacterium enters into nematode body indirectly and multiply inside the body. In the later stage, they produce endospores which help to overcome adverse environmental conditions. *Pasteuria* spp. are the host-specific nematophagous bacteria. Many Rhizobacteria have also parasitized the nematodes and interfere with the different stages of the nematode life cycle. There are several hypotheses which reveal the mechanism of parasitism of bacteria on nematodes:

i) Production of secondary metabolites and other chemicals which kills the nematode eggs.
ii) Root exudates degradation, on which nematode depends as egg hatching stimuli.
iii) Induce systemic resistance.

17.6.5.2 PREDATORS

Nematodes

Predatory nematodes have large buckle cavities with teeth (one or more). They may swallow the whole nematode (PPNs) or buck the body contents. Few predatory nematodes have hollow spear which is sed to suck the body content of the prey. *Odontopharynx longicauda* have teeth at the dorsal ide of the mouth. Spear-shaped mouth parts are present in *Aporcelaimellus obscures*, it feeds on juviniles and eggs of PPNs.

Mites

Lasioseius subteraneus (soil mite) feeds on nematodes. This mite has a short generation time (one week at 28°C) and lays about 18 eggs per day on the nematode body.

17.6.5.3 SOIL AMENDMENTS

Application of soil amendments has many advantages as it stimulates antag-onistic microbes, improves soil health, increases water retention capacity of the soil, provides nutrient to crops and produces the organic acids which are harmful to PPNs. Soil amendments may be oil cakes, farmyard manure, coffee grounds, crop residue, sugarcane thrash, paddy husk, sawdust, etc. Mode of action of amendments are very complex and it may be direct or indirect. Many amendments increase the population of antagonistic fungi and bacteria. These microbes inhibit the PPNs in the soil. Application of organic matter to the soil improves the vigor of the crop and thus makes the plants less amenable to nematode attack. Decomposition of marigold, brassica and mustard crops in the soil releases organic acids and some chemicals which are toxic to nematodes.

17.6.5.4 SUPPRESSIVE SOILS

Suppressive soils are those soils in which the parasitic pathogen (including nematodes) does able to establish due to presence of native antagonistic microorganisms. Mono culturing of the crop for several years helps to convert the soil into suppressive soil. There are several basic principles which need to be followed to overcome the chronic problem of nematode infestation.

- Quarantine regulations.
- Use of certified seeds and other planting materials.
- Clean equipment before moving
- Cleaning farm equipments before use.
- Contaminated water should be avoided for irrigation.

Till date, there is lack of awareness in farmers about the nematode infestation this lead to unintentional import of nematodes into the farm soil. There is a need to educate the farmers about the significance of preventive measures in managing plant-parasitic nematodes. Sanitation, plant quarantine, and certification have specific legal roles in controlling the nematode spread across the region, state or country. Some studies revealed that nematode could spread easily through irrigation water, nematologists demonstrated that irrigation canals harbor a dozen of genera of PPNs. The only measure to overcome this problem is to first pump the irrigation water into a pond, and then nematodes will settle down within a few minutes. Then draw the top water for irrigation which is free from nematodes. Some studies revealed that nematode cysts can able to retain their viability even after passing through the digestive tract of cattle. The soil beneath the piles of manures may also be contaminated with nematodes. Hence, the farmyard manure should be composted properly before application to farmland. Irregularity of nematode-infested areas in the field is one of the more striking features of nematode distribution in the field. It depends upon physical properties of the soil, irrigation pattern, drainage pattern, cropping system, soil moisture, etc.

Contaminated farm equipments harbor cysts containing viable eggs into the non-infected fields. Human movement, tractor, cultivator, animal hose, irrigation water, etc., spread the nematode to the adjacent areas. Along with the development of crop plants, the nematodes also get established in the field and their population increases over a period of years and finally reaches economic injury levels if measures are not taken against them. Thus the nematode infected areas may be seen in patches in the crop fields. These infested areas may enlarge and coalesce over a period of time as the population spread further. The process of introduction and subsequent spread determines the distribution of particular nematode. Early diagnosis of nematode infestation and adoption of suitable management strategies at the appropriate time is the economic way to manage PPNs and to increase the crop yield. Chemical management strategies are should not much entertain in India due to their high cost, hectic operation method, small farm holdings and their harmful effect on the ecosystem. The judicious application of nematicides along with cultural, biological and biotechnological strategies is a sustainable way to manage the nematode diseases.

17.7 MASS PRODUCTION OF NEMATOPHAGUS FUNGI/BACTERIA

There are two methods to mass produce nematophagous fungi/bacteria. Solid state fermentation and liquid state fermentation are the most common methods of mass production of nematophagous microorganisms. Liquid state fermentation is carried out in a large vessel with agitation, optimum temperature, and humidity set up. This method is more suitable for mass production of nematophagous bacteria. Solid state fermentation is more suitable for fungi. In this method, the fungus is grown on solid substrates. It exploits the growth mechanism of a microorganism to degrade the solid substrates. The microbes grow on the solid substrate and produce spores. *Trichoderma* spp. was mass multiplied on pre-soaked jhangora seeds for 10–15 days at 28°C. Then the seeds were air-dried ground and sieved to harvest the spores present in the powder. The spores, talcum powder, and carboxymethyl cellulose were mixed at 20:980:0.5 ratios to prepare commercial formulation (Zaidi and Singh, 2004).

17.8 REGULATION OF BIOPESTICIDES

Biopesticide must be registered or approved before the release of the commercial product. Biopesticide registration involves higher cost, many tests related to environment and human safety and takes a longer time for the registration process, these are one of the production constraints of biopesticides. Environmental protection agency is responsible for the assessment of risk associated with the release of a new commercial formulation of biopesticides. In the United States, during 2008, 279 biopesticides were registered. Anyone can register biopesticide solely after assessing human and environment safety without demonstrating its efficacy under different soil conditions (Caron et al., 2006). The experts committee carries out the risk assessment of the new formulation. A proposal is prepared by the accreditation committee based on efficacy and risk assessment. Organic Materials Review Institute is responsible for Maintainance of National Organic Standards. It also prepares Certifier Guideline for the acceptability of various products.

17.9 APPLICATION OF THE BIOPESTICIDES

17.9.1 SOIL APPLICATION

Apply 100 g of neem/Pongamia/mahua cake or 250 g of vermicompost enriched with *Pseudomonas fluorescens* + *Trichoderma harzianum* + *Paecilomyces lilacinus* on 1 m² beds or around the rhizosphere of the plants.

17.9.2 SPRAYING

The organic formulation containing *Pseudomonas fluorescens* and *Trichoderma harzianum* has to be sprayed on the plants at regular intervals of 20 days at a dosage of 5g/L or 5 mL/L. Alternately, take 20 kg of neem/Pongamia/mahua cake enriched in the above-mentioned manner and mix it in 200 L of water, leave it for a period of 2–3 days. Filter this suspension and use it for spraying by mixing 250 mL of suspension in 1 L of water at regular interval of 20 days.

17.9.3 DRENCHING OR APPLICATION THROUGH DRIP IRRIGATION SYSTEM

The organic formulation has to be given through drip/by drenching @ 5 g/L or 5 mL/L at regular interval of 20 days. Alternately, take 20 kg of neem/ Pongamia/mahua cake enriched in the above-mentioned manner and mix it in 200 L of water, leave it for a period of 2–3 days. Filter this suspension and use it for drenching at regular interval of 20 days.

17.10   SUCCESS STORIES OF BIOLOGICAL CONTROL OF PPNS

Bacillus subtilis + *Paecilomyces lilacinus* combination significantly reduced the root-knot nematode, *Meloidogyne incognita* compared to individual treatments (Gautam et al., 1995). Similarly, *Pasteuria penetrans* + *Verticillium chlamydosporium* combination treatment also reduced root-knot disease in tomato. The mechanism of biological control includes hyper parasitism, antibiosis, and induction of resistance in plants, competition for nutrients, space and oxygen (Siddiqui and Mahmood, 1999). Hence, the combination of products (consortium) have a synergistic effect against multiple crop diseases/nematodes. However, it is must to check the compatibility of bioagents in combination product before its commercialization. Felde et al. (2006) and Khan et al. (2006) elucidated the compatibility of the combi-products managing PPN's. Efficiency, compatibility a shelf life is the basic property to be checked while producing a combi products of biopesticides.

Plant parasitic nematodes cause around 15–20% annual loss in horticultural crops. Up to 45% loss under protected cultivation practices. Indiscriminate use of chemical nematicides possesses a heavy threat to the ecosystem, human and animal health. Many institutes developed sustainable nematode management practices for many field and horticultural crops. Many effective strains of bioagents have been developed. *Verticillium chlamydosporium Paecilomyces lilacinus, Pseudomonas fluorescens, Trichoderma harzianum, Trichoderma viride* have been found promising in the management of PPNs. By using effective strains of bioagents it could be possible to reduce the chemical nematicide use up to 40%.

17.11 CHALLENGES OF BIOLOGICAL CONTROL

i)   Generally, farmers not prefer biopesticides due to their slow action (Felde et al. 2006).
ii) Inconsistency in performance of biocontrol agents is one of the major obstacles to explore biological control. Inconsistency may be due to climatic factors or biotic interactions of biocontrol agents with non-target organisms, rhizosphere and saprophytic competitive ability of bioagent, an initial population of the target organism, the susceptibility of host etc. (Meyer and Roberts, 2002).
iii) Biocontrol agents may partially effective against the PPN's (Flexner and Belnavis, 2000).
iv) Lack of consistency in their performance.
v)   Biopesticide testing and registration is a hectic and time-consuming process.
vi) Lack of shelf-life.
vii) Biopesticides are less persistent and nonsystemic in their action.
viii) Biopesticides are highly sensitive to environmental factors.

The above-listed problems may overcome to some extent by combining two or more biocontrol agents in nematode management. Combi products are much more promising than the individual product. Due to their multiple modes of action at different life stages of the nematode. The active compounds in biopesticides and botanicals are still rarely identified. The response of nematode to biopesticides may greatly differ (Chitwood, 2002). Due to low persistence, more frequent application of botanicals may be necessary. Inconsistent performance of applied bio-is being the primary obstacle in exploring biological control (Pendse et al., 2013). Many factors limit the efficacy of microbial products reveals their differential performance under various field conditions. Microbial products are readily inactivated by the climatic factors like rain, wind, temperature, soil moisture, etc. due to which efficacy may last for a short period.

17.12 CONCLUSION

Biological control is a sustainable alternative to chemical disease management. It helps to overcome the toxic hazards of chemical pesticides. At present, only a few biopesticides are commercially available in the market. Identification and commercialization of native trains of bioagents are the best way to control the PPNs as they could be well adapted to the native environment. In the Indian scenario, the prognosis of biological control for the management of PPNs is poorly practiced. Public and private funding agencies could be committed to investing in research and commercialization of biopesticides. There is also a scope of research to determine the effect of botanicals on other beneficial microorganisms in the soil. At present scenario isolation, identification and screening of different species of nematopathogenic fungi and bacteria from native soil(s) are required. In vitro and in vivo screening should be made to find out the most effective microbials against nematodes. The mass production technique of those effective microorganisms should be developed for large scale use. Finally, microbials using selected strain(s) of nematopathogenic fungi and bacteria in carrier-based dry, powder as well as in liquid form must be developed for the convenience of soil application and longer shelf life. Research on the adaptation of biocontrol agents to an unfavorable condition such as temperature, humidity, sunlight or moisture stress should be done. Assessment of the compatibility of nematopathogenic microorganisms with the chemical pesticides can offer improved scope for their integration in the IPM program. There is a need to improve efficiency and quality in order to develop potential microbes. In the case of botanicals and microbials, their active ingredient and shelf life is an important factor to be considered. Their efficacy should be tested before use.

KEYWORDS

- **botanicals**
- **bioagents**
- **eco-friendly**
- **nematode**
- **sustainability**

REFERENCES

Akpheokhai, I. L.; Cole, A. O. C.; Fawole, B. Evaluation of Some Plant Extracts for the Management of *Meloidogyne incognita* on Soybean (*Glycine max*). *World J. Agril. Sci.* **2012,** *8* (4), 429–435.

Bélair, G. Les Nématodes, ces Anguillules qui font sur les Plantes. Par la Racine. *Phytoprotection* **2005**, *86*, 65–69.

Brand, D.; Oishi, B.O.; Roussos S.; Soccol, C. R. Spore Production of *Paecilomyces lilacinus* by Solid State Culture and application in Pot Experiments to Control *Meloidogyne incognita*. *Appl. Biochem. Biotechnol.* **2004**, *118*, 81–88.

Caron, J.; Laverdière, L.; Venne, J.; Bélanger, R. Recherche et Développement de Biopesticides et Pesticides Naturels à Faible Toxicité Pour les Organismes Non Ciblés et Respectueux de l'Environnement - Rapport Final – Volet Phytopathologie. Projet PARDE 2006, 278 p.

Cayrol, J. C.; Dijan-Caporalino, C.; Panchaud-Mattei, E. La Lutte Biologique Contre les Nématodes Phytoparasites. *Courrier de la cellule Environnement de l'INRA* **1992**, *17*, 31–44.

Chitwood, D. J. Phytochemical Based Strategies for Nematode Control. *Ann. Rev. Phytopathol.* **2002**, *40*, 221–249.

Collange, B.; Navarrete, M.; Peyre, G.; Mateille, T.; Tchamitchian, M. Root-Knot Nematode (*Meloidogyne*) Management in Vegetable Crop Production: The Challenge of an Agronomic System Analysis. *Crop Prot.* **2011**, *30* (10), 1251–1262.

Djian-Caporalino, C.; Védie, H.; Arrufat, A. De Nouvelles Pistes Pour Gérer les Nématodes à Galles. *Maraîchage Bio*. **2009**, *61*, 1–6.

Elyousr, K. A. A.; Khan, Z.; Award, E. M. M.; Moneim, M. F. A. Evaluation of Plant Extracts and *Pseudomonas* spp. for Control of Root-Knot Nematode, *Meloidogyne incognita* on Tomato. *Nematropica* **2010**, *40*, 289–299.

Flexner, J. L.; Belnavis, D. L. Microbial Insectcides. In *Biological and Biotechnological Control of Insect Pests*; Rechcigl, J. E., Rechcigl, N. A., Eds.; CRC Press/Lewis Publishers: Boca Raton, 2000; pp 35–56.

Felde, Z. A.; Pocasangre, L. E.; Carnizares Monteros, C. A.; Sikora, R. A.; Rosales, F. E.; Riveros, A. S. Effect of Combined Inoculations of Endophytic Fungi on the Biocontrol of *Radopholus similis*. *Info- Musa* **2006**, *15*, 12–18.

Fravel, D. R. Commercialization and Implementation of Biocontrol. *Annu. Rev. Phytopathol.* **2005**, *43*, 337–359.

Gaur, H. S.; Pankaj. *Nematodes: Handbook of Agriculture*; ICAR Publications: New Delhi, 2009.

Gautam, A.; Siddiqui, Z. A.; Mahmood, I. Integrated Management of *Meloidogyne incognita* on Tomato. *Nematologia Mediterranean* **1995**, *23*, 245–247.

Hajieghrari, B.; Torabi-Giglou, M.; Mohammadi, M. R.; Davari, M. Biological Potential of Some Iranian *Trichoderma* Isolates in Control of Soil Borne Plant Pathogenic Fungi. *Afr. J. Biotechnol.* **2008**, *7* (8), 967–972.

Haseeb, A.; Kumar, V. Management of *Meloidogyne incognita-Fusarium solani* Disease Complex in Brinjal by Biological Control Agents and Organic Additives. *Ann. Plant Prot. Sci.* **2006**, *14*, 519–521.

Huang, X. W.; Zhao, N. H.; Zhang, K. Q. Extracellular Enzymes Serving as Virulence Factors in Nematophagous Fungi Involved in Infection of the Host. *Res. Microbiol.* **2004**, *155* (10), 811–816.

Jain, R. K.; Mathur, K. N.; Singh, R. V. Estimation of Losses Due to Plant Parasitic Nematodes on Different Crops in India. *Indian J. Nematol.* **2007**, *39*, 219–221.

Junaid, J. M.; Dar, N. A.; Bhat, T. A.; Bhat, A. H.; Bhat, M. A. Commercial Biocontrol Agents and Their Mechanism of Action in the Management of Plant Pathogens. *Int. J. Modern Plant Anim Sci.* **2013**, *1* (2), 39–57.

Khan, A.; Williams, K. L.; Nevalainen, H. K. M. Control of Plant Parasitic Nematodes by *Paeciliomyces lilacinus* and *Monacrosporium lysipagum* in Pot Trials. *Biocontrol* **2006**, *51*, 643–658.

Khan, M. R.; Haque, Z. Soil Application of *Pseudomonas fluorescens* and *Trichoderma harzianum* Reduces Root-Knot Nematode, *Meloidogyne incognita*, on Tobacco. *Phytopathol. Mediterr.* **2011**, *50*, 257–266.

Mamun, M. S. A.; Ahmed, M.; Ahmad, I.; Uddin, M. S.; Paul, S. K. Comparative Studies on the Performances of Some Plant Cakes and Synthetic Chemicals Against Nematodes in Tea in Bangladesh. *World J. Agric. Res.* **2014**, *2* (1), 1–4.

Mateille, T. *Pasteuria penetrans*: Un Nouvel "outil biologique" de Lutte Contre les Nématodes Phytoparasites: Perspectives d'Application en Cultures Maraîchères. *Congad Infos.* **1993**, *15*, 1–2.

Meyer, S. L. F.; Roberts, D. P. Combinations of Biocontrol Agents for Management of Plant-Parasitic Nematodes and Soil Borne Plant-Pathogenic Fungi. *J. Nematol.* **2002**, *34*, 1–8.

Mian, I. H. *Introduction to Nematology*; IPSA,:Gazipur, Bangladesh, 1998; pp 29–66.

Mukhtar, T.; Hussain, M. A.; Kayani, M. Z. Biocontrol Potential of *Pasteuria penetrans*, *Pochonia chlamydosporia*, *Paecilomyces lilacinus* and *Trichoderma harzianum* against *Meloidogyne incognita* in Okra. *Phytopathol. Mediterr.* **2013**, *52* (1), 66–76.

Organic Materials Review Institute. http://www.omri.org/ (accessed Sep 4, 2014).

Pendse, M. A.; Karwander, P. P.; Limaye, M. N. Past, Present and Future of Nematopathogenic Fungi as Bioagent to Control Plant Parasitic Nematodes. *J. Plant Prot. Sci.* **2013**, *5* (1), 1–9.

Radhakrishnan, B. Indigenous Botanicals Preparations for Pest and Disease Control in Tea. *Bul. UPASI Tea Res. Found.* **2010**, *55*, 31–39.

Rahanandeh, H.; Khodakaramian, G.; Hassanzadeh, N.; Seraji, A.; Asghari, S. M.; Tarang, A. R. Inhibition of Tea Root Lesion Nematode, *Pratylenchus loosi*, by Rhizospher Bacteria. *J. Ornament. Hortic. Plants* **2012**, *2* (4), 243–250.

Sarhy-Bagnon, V.; Lozano, P.; Saucedo-Castaneda, G.; Roussos, S. Production of 6-Pentyl-Alpha-Pyrone by *Trichoderma harzianum* in Liquid and Solid State Cultures. *Process Biochem.* **2000**, *36*, 103–109.

Siddiqui, Z. A.; Mahood, I. Role of Bacteria in the Management of Plant-Parasitic Nematodes: A Review. *Bioresour. Technol.* **1999**, *69*, 167–179.

Siddiqui, Z. A.; Mahmood, I. Biological Control of Plant Parasitic Nematodes by Fungi: a Review. *Bioresource Technol.* **1996**, *58*, 229–239.

Taye, W.; Sakhuja, P.K. and Tefera, T. Evaluation of Plant Extracts on Infestation of Root-Knot Nematodes on Tomato (*Lycopersicon esculemtum* Mill). *J. Agric. Res. Develop.* **2012**, *2* (3), 86–91.

Thakur, N. S. A.; Devi, G. Management of *Meloidogyne incognita* Attacking Okra by Nematophagous Fungi, *Arthrobotrys oligospora* and *Paecilomyces lilacinus*. *Agric. Sci. Dig.* **2007**, *27*, 50–52.

Tian, B.; Yang, J.; Zhang, K. Q. Bacteria Used in the Biological Control of Plant Parasitic Nematodes: Populations, Mechanisms of Action and Future Prospects. *FEMS Microbiol. Ecol.* **2007**, *61*, 197–213.

Van Damme, V.; Hoedekie, A.; Viaene, N. Long-Term Efficacy of *Pochonia chlamydosporia* for Management of *Meloidogyne javanica* in Glasshouse Crops. *Nematology* **2005**, *7*, 741–745.

Zaidi, W. N.; Singh, U. S. Development of Improved Technology for Mass Multiplication and Delivery of Fungal (*Trichoderma*) and Bacterial (*Pseudomonas*) Bio- Control Agents. *J. Mycol. Plant. Path.* **2004**, *34* (3), 732–741.

CHAPTER 18

Soil Heavy Metal Toxicity Reduction by Bioagents/Living Organisms

ABHA KUMARI[1], BISHUN DEO PRASAD[2,*], ASHEESH CHAURASIYA[3], PANKAJ KUMAR[4], and TUSHAR RANJAN[2]

[1]Department of Horticulture (Fruit & Fruit Technology), Bihar Agricultural University, Sabour, Bhagalpur 813210, Bihar, India

[2]Department of Molecular Biology and Genetic Engineering, Bihar Agricultural University, Sabour, Bhagalpur 813210, Bihar, India

[3]Department of Agronomy, Bihar Agricultural University, Sabour, Bhagalpur 813210, Bihar, India

[4]Department of Agricultural Biotechnology and Molecular Biology, Dr. Rajendra Prasad Central Agricultural University, Pusa, Samastipur 848125, Bihar, India

*Corresponding author. E-mail: bdprasadbau@gmail.com

ABSTRACT

Heavy metal toxicity in the soil has proven to be a major threat to the soil texture, plants, and human health. The most commonly found heavy metals in soil come with the waste water comprise arsenic, cadmium, chromium, copper, lead, nickel, and zinc. All the above discussed heavy metals cause risks to the plant, human, and environment health. A variety of causes of heavy metals containing soil erosion, natural weathering of the earth's crust, mining, industrial effluents, urban runoff, sewage discharge, insect or disease control agents of crops, and many others factor has reported for the heavy metal toxicity. Several studies were conducted on the role of soil microbial community for the scavenging of heavy metals in recent decades. Heavy metal bioremediation with the help of suitable microbial community is a very efficient strategy due the ecofriendly and low-cost inputs. Recent advancements in the research of microbe-heavy metal interaction and their efficient application for the leaching or accumulation for further detoxification have been established.

18.1 INTRODUCTION

Heavy metals are usually defined as elements having an atomic number >20 and metallic properties (conductivity, ductility, ligand specificity, stability as cations, etc.). The most frequent heavy metal contaminants are Cd, Cr, Cu, Hg, Pb, and Ni. However, pollution of biosphere caused by toxic metals has accelerated dramatically since the beginning of the industrial revolution. As a result of anthropogenic activities, such as energy and fuel production, electroplating, fertilizer, gas exhaust, mining, and smelting of metals, municipal waste generation, sewage, and pesticide application, etc. (Kabata-Pendias and Pendias, 1989), today one of the most severe environment problems is the metal pollution.

Extreme accretion of heavy metals is deadly to most of plants. When the heavy metal ions are present at an elevated point in the environment, are markedly absorbed by roots and translocated to shoot, leading to impair physiological and biochemical processes and reduced growth (Bingham et al., 1986; Foy et al., 1978). Continued decline in plant growth reduces yield which eventually leads to food insecurity. Therefore, various methods are used in bioremediation of heavy metal polluted soils. This bioremediation method is also considered as an economical remediation technique compared with other remediation techniques. Mainly physical and chemical methods (such as, encapsulation, electro-kinetics, soil washing and flushing, solidification, stabilization, vapor extraction, and vitrification) are expensive and do not make the soil suitable for plant growth. Biological approaches are eco-friendly approach, because it is achieved via natural processes and encourages the establishment or reestablishment of plants on polluted soils. In this chapter, the attempts have been made to evaluate the nature, source, and effect of heavy metals polluted soils for performance of plant growth. Biological approaches used for remediation of heavy metal polluted soils were also highlighted.

18.2 WHAT IS HEAVY METAL?

The term "heavy metals" refers to any metallic element that has a relatively high density and is toxic or poisonous even at low concentration (Lenntech, 2004). "Heavy metals" is a broad-spectrum collective term, which applies to the group of metals and metalloids with atomic density greater than 4 g/cm, three or five times greater than water (Garbarino et al., 1995; Hawkes, 1997; Nriagu, 1989). However, a heavy metal has little to do with density but concerns chemical properties.

18.3 NAME OF HEAVY METAL

Heavy metals consist of, lead (Pb), cadmium (Cd), zinc (Zn), mercury (Hg), arsenic (As), silver (Ag) chromium (Cr), copper (Cu), iron (Fe), and the platinum group elements.

Heavy metals include the transition-metal elements essential to plant nutrition that is, iron (Fe), zinc (Zn), manganese (Mn), copper (Cu), nickel (Ni), and molybdenum (Mo), cobalt (Co) (which is required for nitrogen fixation in legumes).

Heavy metals include the transition-metal elements nonessential to plant nutrition that is, chromium (Cr), cadmium (Cd), mercury (Hg), and lead (Pb). High tissue concentrations of all these elements are toxic to crop plants.

18.4 SOURCE OF HEAVY METALS

Heavy metals can be emitted into the environment by both natural and anthropogenic causes. The major causes of emission are the anthropogenic sources specifically mining operations (Battarbee et al., 1988; Hutton and Symon, 1986; Nriagu, 1989). In some cases, even long after mining activities have ceased, the emitted metals continue to persist in the environment. Peplow (1999) reported that hard-rock mines operate from 5 to 15 years until the minerals are depleted, but metal contamination that occurs as a consequence of hard-rock mining persist for hundreds of years after the cessation of mining operations (Table 18.1).

TABLE 18.1 Source of Contamination of Heavy Metals in the Soil.

Sl. no.	Heavy metal	Sources of contamination
1.	Cd	Batteries, electroplating, and fertilizers
2.	Pb	Batteries and metal products
3.	Cr	Dyes, leather tanning, pesticides, and timber treatment
4.	Cu	Fertilizers, fungicides, pigments, and timber treatment
5.	Zn	Dyes, fertilizers, mine tailings, paints, and timber treatment
6.	Mn	Fertilizer
7.	Mo	Fertilizer
8.	As	Paints, pesticides, and timber treatment
9.	Ni	Alloys, batteries, and mine tailings
10.	Hg	Fertilizers, fumigants, and instruments

18.5 EFFECT OF HEAVY METAL ON PLANTS

Generally, toxic metals cause enzyme inactivation, damage cells by acting as antimetabolites or form precipitates or chelates with essential metabolites (Sobolev et al., 2008). Several heavy metals encourage diverse morphological, physiological, and biochemical dysfunctions in plants, either directly or indirectly, and cause various damaging effects (Table 18.2). The commonly recognized and basic consequence of heavy-metal toxicity in plant cells is the overproduction of reactive oxygen species (ROS). Unlike redox-active metals, such as iron and copper, heavy metals (Pb, Cd, Ni, AI, Mn, and Zn) cannot produce ROS directly by participating in biological redox reactions, such as Haber–Weiss/Fenton reactions. However, these abovementioned metals induce ROS generation via different indirect mechanisms, such as stimulating the activity of nicotinamide adenine dinucleotide phosphate (NADPH) oxidases, displacing essential cations from specific binding sites of enzymes and inhibiting enzymatic activities from their affinity for -SH groups on the enzyme. Under normal conditions, ROS play several vital roles in regulating the expression of different genes. Reactive oxygen species direct various processes like the abiotic stress responses, cell cycle, development, pathogen defense, plant growth, programmed cell death, and systemic signaling (Shahid et al., 2014).

18.6 REMEDIATION TECHNOLOGIES

Heavy metals can't be ruined biologically but are only transformed from one oxidation state or organic complex to another (Garbisu and Alkorta, 2001). Remediation of heavy-metal contamination in soils is more difficult. Until now, remediation methodology, such as acid leaching, electro reclamation, excavation and landfill, and thermal treatment are not apt for practical applications, because of their high cost, low efficiency, large destruction of soil structure and fertility, and high dependence on the contaminants of concern, soil properties, site conditions, and so on. Thus, the development of bioremediation strategies is obligatory for heavy metals contaminated soils (Chaney et al., 2000; Cheng et al., 2002; Lasat, 2002).

18.6.1 *BIOREMEDIATION OF HEAVY METAL POLLUTED SOILS*

Bioremediation is the use of microorganisms and plants for the treatment of metal polluted soils. It is a broadly common method of soil remediation because it occurs via natural processes. It is also a cost efficient and

TABLE 18.2 General Symptoms of Heavy-Metal Toxicities in Nontolerant Crop Plants.

Element	Symptoms	Sensitive crops
Cd (cadmium)	Brown leaf margins, reddish petioles and veins, total chlorosis, brown roots, and curled leaves. Deterioration of xylem tissues. Severe plant stunting, inhibited tillering, and reduced root growth.	Legumes, spinach, radish, carrots, oats.
Co (cobalt)	Pale green leaves, interveinal chlorosis in novel leaves followed by induced Fe-deficiency interveinal necrosis, white tips and leaf margins, stunted plants with short brown roots and damaged root tips.	Maize, Barley.
Cr (chromium)	Yellow leaves, interveinal chlorosis of new leaves, necrotic spots, reduced plant height and purpling of tissues; wilting, reduced root growth.	Maize.
Cu (copper)	Dark green or bluish leaves followed by induced Fe chlorosis, young leaves chlorotic with dark-brown interveinal necrosis, stunted plants with short roots.	Cereals, legumes, spinach, citrus, apple.
Fe (iron)	Dark green foliage, orange–brown necrotic spots (bronzing) of older leaves, stunted growth, browning and blackening of roots.	Rice, tobacco.
Hg (mercury)	Yellow leaves, leaf chlorosis and browning of leaf points, red stems, severe stunting and reduced root growth.	Sugar beet, maize.
Mn (manganese)	Interveinal chlorosis and necrotic lesions on aged leaves, blackish-brown or red necrotic spots, accumulation of black MnO_2 particles in epidermal cells, drying leaf tips, stunted plants, and reduced root growth.	Cereals (e.g., barley), legumes (e.g., lucerne, beans), brassica (e.g., cauliflower, cabbage, kale, swede), lettuce, sugar beet, pineapple, potatoes, tomatoes.
Mo (molybdenum)	Yellowing or browning of leaves, appearance of blue-purple or gold leaf pigments, reduced tillering, and root growth.	Cereals, citrus, cauliflower.
Ni (nickel)	Grey-green leaves, induced Fe-deficiency yellow or white interveinal chlorosis and necrosis in new leaves, stunted plants with short brown roots.	Cereals (e.g., oats), sugar beet, and spinach beet.
Pb (lead)	Dark green leaves, wilting of elder leaves, stunted plants, and short blackened roots.	Cereals and legumes.
Zn (zinc)	Yellow leaves, chlorotic and necrotic leaf tips, interveinal leaf chlorosis, stunted plants with short roots.	Cereals, leafy vegetables (e.g., spinach, beet), citrus, guava, litchi, apple, coconut.

Source: Adapted and modified from Bould et al. (1983); Kabata-Pendias (2000); Fageria (2009).

nondisturbing method, hence, used for the treatment of heavy metal polluted soils. Heavy metals can't be degraded during bioremediation but can only be transformed from one organic complex or oxidation state to another. Due to a change in their oxidation condition, heavy metals can be transformed to become either less toxic, easily volatilized, more water soluble can removed through leaching, whereas less water soluble can be precipitate and easily removed from the polluted soils. Therefore, combination of both microorganisms and plants can be use for bioremediation of heavy metals polluted soils (White et al., 2006).

18.6.2   *MECHANISMS OF BIOREMEDIATION BY MICROORGANISM*

The chief ways by which remediation may be accomplished comprise biosorption, bioaccumulation, reduction, solubilization, precipitation, and methylation.

18.6.2.1   *BIOSORPTION*

Biosorption is a physiochemical process that occurs naturally in certain biomass which allows it to passively concentrate and bind contaminants onto its cellular structure. This sorption is passive because no energy is required. Biosorption is a usual rapid process and it can remove a high percentage of individual metallic cations from waste streams.

18.6.2.2   *BIOACCUMULATION*

Bioaccumulation is the accumulation of substances, such as heavy metals or other chemicals in an organism. Uptake of metallic cations is an energy-requiring process, which thus is associated with viable and actively metabolizing microorganisms. Such process is termed bioaccumulation.

18.6.2.3   *REDUCTION*

Converts a higher oxidation state of the element to a lower one, for example, $Hg(II)$ to $Hg(O)$, $Fe(III)$ to $Fe(II)$, $Se(VI)$ to $Se(O)$, $Mn(IV)$ to $Mn(II)$, $U(VI)$ to $U(IV)$, $As(V)$ to $As(III)$.

18.6.2.4 SOLUBILIZATION/OXIDATION

Converts a lower oxidation state of the element to a higher one, for example, Fe(II) to Fe(III), As(III) to As(V).

18.6.2.5 PRECIPITATION

Precipitation is the one most common methodology in bioremediation in which, it transforms dissolved contaminants into an insoluble solid resulting the contaminant's consequent removal from the liquid phase by sedimentation or filtration.

18.6.2.6 METHYLATION

Methylation denotes the addition of a methyl group on a substrate, or the substitution of an atom by a methyl group. In biological systems, methylation is catalyzed by enzymes; such methylation can be involved in modification of heavy metals, for example, Se (Table 18.3).

TABLE 18.3 Process that May Result in Metal Bioremediation.

Mechanism	Metal and metalloids
Biosorption	U, Zn, Pb, Cd, Co, Ni, Cu, Hg, Th, Zn, Cs, Au, Ag, Sn, and Mn
Microbial reduction	As, V, Te, S, and Mo
Enzymatic reduction	As, Pb, Cu, Mo, U, Se, Bi, Te, Va, Mn, Fe, Ag, Au, Os, Ru
Solubilization/oxidation	Cd, Hg, Ni, Pb, Se, Ag, Al
Precipitation	Sulfides: Zn, Pb, Ni, Cr, Cd, Cu, Fe, Phosphates: U, Pb
Methylation	Se

18.6.3 REMEDIATION OF HEAVY METALS POLLUTED SOILS BY USING MICROBES

18.6.3.1 CHROMIUM

Three approaches have been pursued for the remediation of toxicity namely; biosorption, enzymatic reduction, and abiotic reduction. Cr may exists as Cr(VI) or Cr(III). The hexavalent form is a characteristic of chromates, dichromates, and chromic trioxide and the trivalent form characterizes Cr oxides and hydroxides. Cr(VI) is toxic and highly soluble in water, whereas Cr(III) is considerably less toxic, far less water soluble, hence less mobile if present in soil. Because of the differences in solubility of Cr(VI) and Cr(III), a reduction will result in precipitation of the element and thus diminished mobility and transport. Several microorganisms especially bacteria (*Bacillus subtilis, Pseudomonas putida,* and *Enterobacter cloacae*) have been successfully used for the reduction of the high toxicity of chromium to the less toxic Cr (VI to III) (Ajmal et al., 1996).

18.6.3.2   SELENIUM

Two approaches have been pursued: reduction and methylation

Se: The ability of microorganism (bacteria, fungus, and yeast) to reduce oxidized Se anions. Both selenate and selenite can thus be transformed, and the product is elemental Se. Because the elemental form is insoluble, the conversion may be useful in the decontamination of polluted waters, with the elemental Se being removed after the reduction by sedimentation or filtration.

The methylation of selenate and selenite converts both anions to dimethyl selenide (Challenger and Bird, 1934):

$$Se(VI)—CH_3SeCH_3$$
$$Se(IV)—CH_3SCH_3$$

This process leads to volatilization of the element. The volatilization of Seas a result of microbial formation of dimethyl selenide has been tested as a remediation strategy.

18.6.3.3   URANIUM (U)

Strategies for U bioremediation involve biosorption, reduction, solubilization and precipitation.

For sorption, the microorganisms are retained by some solid material, and the U-containing solution is brought into contact with the immobilized biomass, which removes much of the element from continuously flowing stream. The organisms proposed for use are fungus *Rhizopus arrhizus,* which is converted to a particulate form. U(VI) can be reduced microbiologically to U(IV) by *Clostridium* and *Desulfovibrio,* and the former bacterium may also form U(III) (Francis et al., 1994, Phillips et al., 1995). A process has been proposed for U bioremediation in which the element is initially extracted from the contaminated soil with bicarbonate, and U(VI)in the extracts is then reduced by *Desulfovibrio desulfuricans* (Phillips et al., 1995). The U(IV) that is generated is insoluble and thus may be removed easily from extract. U remediation by precipitation is the mixed culture that release inorganic phosphate during its degradation of the tributyl phosphate used as a solvent for U extraction, and the inorganic phosphate then precipitates U as uranyl phosphate.

18.6.3.4 NITRATE

Due to indiscriminate use of nitrogen fertilizer for crop production and waste water coming from municipal sewage treated plants contains high concentration of nitrate. If nitrate-rich water goes in the collected water it will create lot of problem for aquatic plants, as well as human health. Nitrate levels >50 mg/L is harmful and causes the disease blue baby syndrome or methemoglobinemia (decreased oxygen carrying capacity of hemoglobin in babies leading to death) in children up to 6 months old. For treatment of the effluent from the west-treatment plant is to create anaerobic conditions, provide a carbon sources, and allow the indigenous denitrifying bacteria to reduce the nitrate to N_2. This is an effective and well establishes technology of sewage treatments.

18.6.3.5 CYANIDE

The remediation of cyanide-containing wastewater being discharged from an operating underground gold mine was accomplished by establishing a biofilm, containing a strain of pseudomonas on rotating biological contractors having a large surface area. The bacterium converts the cyanide to ammonium and further to nitrate. Because of the toxicity of cyanide, any organism selected for bioremediation must have a degree of tolerance to the inhibitor.

18.6.3.6 ARSENIC

Microbial oxidation, reduction, and methylation involved in bioremediation of arsenic.

It can be precipitate as a result of microbial sulfate reduction because methylation results in the formation of highly toxic methylated arsines. In nature several bacteria are present, which is proficient to oxidize arsenite to arsenate, which is precipitated more readily.

Several bacteria and yeasts are also able to reduce arsenate to arsenite, sometimes using the latter as an electron acceptor for development (Macy et al., 1996; Vidal and Vidal, 1980). A bacterium namely, *Desulfotomaculum auripigmentum*, which reduces both arsenate to arsenite and sulfate to H$_2$S, thereby leading to As$_2$S$_3$ precipitation (Newman et al., 1997).

18.6.3.7   MERCURY

Microorganisms from a variety of taxonomic groups have the capacity to reduce divalent mercury to yield metallic Hg:

$$Hg(II)–Hg(O)$$

The product is volatile and thus moves into the overlying air. If these microorganisms are stimulated to greater activity, they might be the basis for a bioremediation. Divalent mercury can also be methylated. However, the high acute toxicity of the resulting mono- and dimethyl mercury precludes consideration of Hg methylation for remediation.

18.6.3.8   CADMIUM AND ZINC

Extraction of Cd and Zn from Cd-rich soil and soil polluted with effluent from metal industry has been increasing by the application of *Bacillus thuringiensis*. It is assumed that the production of siderophore (Fe complex molecules) by bacteria may have facilitated the extraction of these metals from the soil; and reproduce the production of siderophore and this consequently affects their bioavailability (Khan, 2005). Bioremediation can also occur indirectly through bioprecipitation by sulfate reducing bacteria (*Desulfovibrio desulfuricans*) which converts sulfate to hydrogen sulfate which subsequently reacts with heavy metals, such as Cd and Zn to form insoluble forms of these metal sulfides (Table 18.4).

18.7   BIOREMEDIATION OF POLLUTED SOILS BY PLANTS

Bioremediation of polluted soils by plants is known as phytoremediation. Phytoremediation means use of green plants to clean-up hazardous waste from the contaminated polluted soil by heavy metals. It is an important portion of bioremediation with use of different types of plants for the treatment of polluted soils. It is suitable for the pollutants cover a large area and when they are within the root zone of the plant (Padmavathiamma and Li, 2007) (Fig. 18.1).

TABLE 18.4 Bioremediation of Heavy Metals Polluted Soils by Using Microbes.

Sl. no.	Element	Available form	Reduced form	Bioremediation of heavy metals polluted soils by using microbes
1.	Cr	Cr(VI)-toxic and highly soluble	Cr(III)-less toxic and less soluble	*Bacillus subtilis* *Pseudomonas putida* *Enterobacter cloacae*
2.	Cd	Insoluble	Cd sulfides soluble	*Bacillus thuringiensis* *Desulfovibrio desulfuricans*
3.	Zn	Insoluble	Zn sulfides soluble	*Bacillus thuringiensis* *Desulfovibrio desulfuricans*
4.	Se	Selenate and Selenite	Elemental Se	*Bacillus coagulans*
5.	As	Arsenite	Arsenate	*Desulfotomaculum auripigment*
6.	CN	Cyanide	Ammonium	*Pseudomonas*
7.	NO_3	Nitrate	N_2	*Denitrifying bacteria*
8.	U	U(VI)	U(IV) (insoluble)	*Clostridium* spp. *Desulfovibrio desulfuricans*
9.	U	U(VI)	U(III)	*Clostridium* spp.
10.	Se	Selenate: Se(VI), Selenite: Se(IV)	Elemental form	*Thauera selenatis* *Denitrifying bacteria*

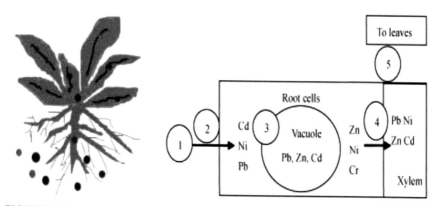

FIGURE 18.1 Metal transformation in plants during phytoremediation.

(1) Metal (iron) absorption at root surface; (2) bioavailable metal crosses cellular membrane into root cells; (3) a part of the metal absorbed into roots is toothless in the vacuole; (4) intracellular mobile metal moves across cellular membranes into root vascular tissue (xylem); and (5) metal is translocated from the root to aerial tissues (stems and leaves).

18.7.1 MECHANISMS OF BIOREMEDIATION BY PLANTS

18.7.1.1 PHYTOEXTRACTION

The processes in which plants are used to accumulate metals from the soil into the roots and shoots of the plant are referred to as phytoextraction. There are a few known plants species that is higher accumulator and show their capability toward the removal of metals from contaminated soils (USEPA, 2000). Earlier study (Padmavathiamma and Li, 2007), observed that plants have the potential to extract large concentrations of heavy metals into their roots, translocation them into the surface and produce a large quantity of plant biomass.

Plants, generally used for phytoextraction should have the following characteristics like high biomass, extensive root system, rapid growth rate and ability to accept high amounts of heavy metals. Generally use for the different criteria of the hyper accumulator.

1. The concentration of metal in the shoot must be >0.1% for Al, As, Co, Cr, Cu, Ni, and Se, > 1.0% for Zn and >0.01% for Cd.
2. The ratio of shoot to root concentration must be consistently >1; this indicates the capability to transport metals from roots to shoot and the existence of hyper tolerance ability.
3. The ratio of shoot to root concentration must be >1; this indicates the degree of plant metal uptake.

In most cases, plants absorb metals that are readily available in the soil solution and some metal are occur in soil in the soluble forms for plant uptake, others present as insoluble precipitate and are thus unavailable for plant uptake.

18.7.2 RHIZOFILTRATION

Rhizofiltration is the process by which plant roots are used for absorption, concentration, or precipitation of metals from effluents.

18.7.3 PHYTOSTABILIZATION

When the plants are used to reduce the mobility of heavy metals through the process of absorption and precipitation, resulting reducing their bioavailability is known as phytostabilization. It may also prevent soil reducing their bioavailability through erosion, leaching and distribution of the toxic heavy metal to other areas. This method is useful in the handling of contaminated land areas affected by mining activities. Plants help in stabilizing the soil through their root systems equally prevent leaching via reduction of water percolation through the soil thus, they prevent erosion. Plants characteristics that are used for phytostabilization are capability to bear soil conditions, dense rooting system, ease of establishment and maintenance under field conditions, rapid growth to provide adequate ground coverage, and longevity and ability to self-propagate.

Soil and organic amendments are also used to reduce the toxicity of heavy metals and provide some other benefits like nutrients for plant growth and improvement of soil physical properties. The soil, which is contaminated by theses metals (arsenic, cadmium, chromium, copper, and zinc) are generally treated by phytostabilization (Sharma and Sharma, 1993).

18.7.4 PHYTOVOLATILIZATION

Phytovolatilization is phenomenon of uptake and release of volatile materials like mercury or arsenic containing compounds into the atmospheric. In this method, contaminants may be taken up with water by growing plants and pass through the xylem vessels in the direction of the leaves and finally transformed into nontoxic forms (volatilize) into the atmosphere.

18.8 REMEDIATION OF HEAVY METALS POLLUTED SOILS THROUGH COMBINATION OF PLANTS AND MICROBES

The mutual effects of both microorganisms and plants are used for the remediation of polluted soils results in a faster and more efficient clean-up of the polluted area. Mycorrhizal fungi have been used in several remediations of heavy metals polluted soils. Phytoextraction is the best method of the accumulation of heavy metals in plants and others methods improved phytostabilization through metal immobilization and a reduced metal concentration in plants (Abhilash et al., 2012).

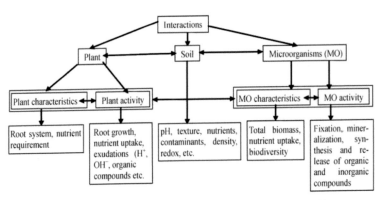

FIGURE 18.2   Interaction of plant–soil–microbial in the rhizospheric soil.

In general, the benefits derived from mycorrhizal associations, which range from increased nutrient and water acquisition to the provision of a stable soil for plant growth and increase in plant resistance to diseases are believed to aid the survival of plants growing in polluted soils, thus helping the vegetation and revegetation of remediated soils. In addition of certain species of mycorrhizal fungi (Arbuscular mycorrhizal fungi) can be more sensitive to pollutants compared to plants. Other microorganisms apart from mycorrhizal fungi have also been used in conjunction with plants for the remediation of heavy metal polluted soils. Most of these microbes are the plant growth-promoting rhizobacteria that are usually found in the rhizosphere. Several microbes stimulate plant growth by some mechanisms, such as production of phytohormones and supply of nutrients production of siderophores and other chelating agent-specific enzyme activity and N fixation and reduction in ethylene production which encourages root growth (Divya and Kumar, 2011).

18.9 CONCLUSION

Major cause behind the contamination of environment and soil are raising global population, urbanization, and industrialization. Release of heavy metals from industries pollutes the environments due to their persistence and bioaccumulative nature. In this regards, bioremediation process include microbes, plants, and their combination provides effective innovative measures for treatment of a wide variety of contaminants. Remediation of contaminated soil through microbes comprises biosorption, bioaccumulation, reduction, solubilization, precipitation, and methylation. The common process of phytoremediation is phytoextraction, rhizofiltration, phytostabilization, and phytovolatilization used for treatment of heavy-metal polluted soils. Both microbes and plants ensure the complete exclusion of the pollutant from contaminated soil. Combining both plants and microorganisms in bioremediation increases the efficiency of remediation technology.

KEYWORDS

- **bioremediation**
- **phytoremediation**
- **heavy metal toxicity**
- **leaching**
- **bioaccumulation**

REFERENCES

Abhilash, P. C.; Powell, J. R.; Singh, H. B.; Singh, B. K. Plant–Microbe Interactions: Novel Applications for Exploitation in Multipurpose Remediation Technologies. *Trends Biotechnol.* **2012,** *30*, 416–420.

Ajmal, M.; Rafaqa, A. K.; Bilquees, A. S. Studies on Removal and Recovery of Cr (VI) From Electroplating Wastes. *Water Res.* **1996,** *30*, 1478–1482.

Battarbee, R.; Anderson, N.; Appleby, P., et al. *Lake Acidification in the United Kingdom*, ENSIS: London, 1988. http://www.geog.ucl.ac.uk/~spatrick/f_r_pubs.htm. (accessed Apr 7, 2016)

Bingham, F. T.; Pereyea, F. J.; Jarrell, W. M.; 1986. Metal Toxicity to Agricultural Crops. *Metal Ions Biol. Syst.* **1986,** *20*, 119–156.

Bould, C.; Hewitt, E. J.; Needham, P. *Diagnosis of Mineral Disorders in Plants. Volume 1: Principles*. HMSO: London, 1983.

Challenger, F.; North, H. E. The Production of Organo-metalloidal Compounds By Micro-organisms. Part II. Dimethyl Selenide. *J. Chem. Soc.* **1934,** 68–71.

Chaney, R. L.; Brown, S. L.; Li, Y. M.; et al. Progress in Risk Assessment for Soil Metals, and In-situ Remediation and Phytoextraction of Metals from Hazardous Contaminated Soils. US-EPA "Phytoremediation: State of Science", Boston, MA, May 1–2, 2000.

Cheng, S., Grosse, W., Karrenbrock, F., Thoennessen, M. Efficiency of Constructed Wetlands in Decontamination of Water Polluted by Heavy Metals, *Ecol. Eng.* **2002,** *18* (3), 317–325.

Divya, B.; Kumar, D. Plant-Microbe Interaction with Enhanced Bioremediation. *Res. J. Biotechnol.* **2011,** *6*, 72–79.

Fageria, N. K. *The Use of Nutrients in Crop Plants*; CRC Press: Boca Raton, Florida, 2009.

Foy, C. D.; Chaney, R. L.; White, M. C. The Physiology of Metal Toxicity in Plants. *Ann. Rev. Plant Physiol.* **1978,** *29* (1), 511–566.

Francis, A. J.; Dodge, C. J.; Lu, F.; Halada, G. P.; Clayton, C. R. *Environ. Sci. Technol.* **1994,** *28*, 636–639.

Garbarino, J. R.; Hayes, H.; Roth, D.; Antweider, R.; Brinton, T. I.; Taylor, H. Contaminants in the Mississippi River, U.S. *Geological Survey Circular 1133*, Virginia, U.S.A. 1995. (www.pubs.usgs.gov/circ/circ1133/). (accessed Apr 10, 2016).

Garbisu, C.; Alkorta, I. Phytoextraction: A Cost-Effective Plant-Based Technology for the Removal of Metals from the Environment. *Biores. Technol.* **2001**, *77* (3), 229–236.

Hawkes, J. S. Heavy Metals. *J. Chem. Educ.* **1997**, *74* (11), 1374.

Kabata-Pendias, A.; Pendias, H. *Trace Elements in the Soil and Plants;* CRC Press: Boca Raton, FL, 1989.

Kabata-Pendias, A. *Trace Elements in Soils and Plants*; CRC Press: Boca-Raton, Florida, 2000.

Khalid, B. Y.; Tinsley, J. Some Effects of Nickel Toxicity on Rye Grass, *Plant Soil* **1990**, *55* (1), 139–144.

Khan, A. G. Role of Soil Microbes in the Rhizospheres of Plants Growing on Trace Metal Contaminated Soils in Phyto-remediation. *J. Trace Elements Med. Biol.* **2005**, *18* (4), 355–364.

Lasat, H. A. Phytoextraction of Toxic Metals: a Review of Biological Mechanisms. *J Environ. Qual.* **2002**, *31* (1), 109–120.

Lenntech. 2004. Water Treatment and Air Purification. Water Treatment, Lenntech: Rotterdamseweg, Netherlands (www.excelwater.com/thp/filters/Water-Purification.htm). (accessed Mar 11, 2016).

Macy, J. M.; Nunan, K.; Hagen, K. D.; Dixon, D. R.; Harbour, P. J.; Chill, M.; Sly, L. I. *Int. J. System. Bacteriol.* **1996**, *46*, 1153–1157.

Newman, D. K.; Beveridge, T. J.; Morel, F. M. M. Precipitation of Arsenic Trisulfide by *Desulfotomaculum auripigmentum. Appl. Environ. Microbiol.* **1997**, *63*, 2022–2028.

Nriagu, J. O. A Global Assessment of Natural Sources of Atmospheric Trace Metals. *Nature* **1989**, *338*, 47–49.

Nriagu, J. O.; Pacyna, J. Quantitative Assessment of Worldwide Contamination of Air, Water and Soil by Trace Metals. *Nature* **1988**, *333*, 134–139.

Padmavathiamma, P. K.; Li, L. Y. Phytoremediation Technology: Hyper Accumulation Metals in Plants. *Water Air Soil Poll.* **2007**, *184*, 105–126.

Peplow, D. Environmental Impacts of Mining in Eastern Washington, Center for Water and Watershed Studies Fact Sheet, University of Washington: Seattle, 1999.

Phillips, E. J. P.; Landa, E. R.; Lovely, D. R. Remediation of Uranium Contaminated Soils with Bicarbonate Extraction and Microbial U(VI) Reduction. *J. Ind. Microbiol.* **1995**, *14*, 203–207.

Shahid, M.; Pourrut, B.; Dumat, C.; Nadeem, M.; Aslam, M.; Pinelli, E. Heavy-Metal-Induced Reactive Oxygen Species: Phytotoxicity and Physicochemical Changes in Plants. *Rev. Environ. Contam. Toxicol.* **2014**, *232*, 1–44.

Sharma, D. C.; Sharma, C. P. Chromium Uptake and Its Effects on Growth and Biological Yield of Wheat. *Cereal Res. Commun.* **1993**, *21* (4), 317–322.

Sobolev, D.; Begonia, M. F. T. Effects of Heavy Metal Contamination upon Soil Microbes: Lead-induced Changes in General and Denitrifying Microbial Communities as Evidenced by Molecular Markers. *Int. J. Environ. Res. Public Health* **2008**, *5* (5), 450–456.

USEPA. United States Environmental Protection Agency. Introduction to Phytoremediation, EPA 600/R-99/107, U.S. Environmental Protection Agency, Office of Research and Development, Cincinnati, Ohio, USA, 2000.

Vidal, F. V.; Vidal, V. M. V. Arsenic Metabolism in Marine Bacteria and Yeast. *Mar. Biol.* **1980**, *60*, 1–7.

White, C.; Sharman, A. K.; Gadd, G. M. An Integrated Microbial Process for the Bioremediation of Soil Contaminated With Toxic Metals. *Nat. Biotechnol.* **2006**, *16*, 572–575.

CHAPTER 19

Metagenomics for Soil Health

MAHESH KUMAR[1,*], VISHWA VIJAY THAKUR[1], KUMARI APURVA[2], and MD. SHAMIM[1]

[1]Department of Molecular Biology and Genetic Engineering, Dr. Kalam Agricultural College, Kishanganj, Bihar Agricultural University, Sabour, Bhagalpur 813210, Bihar, India

[2]Department of Forestry, Forest Research Institute (Indian Council of Forestry Research & Education), P.O. New Forest, Dehradun 248006, Uttarakhand, India

*Corresponding author. E-mail: maheshkumara2zbau@gmail.com

ABSTRACT

Soil is considered to be a complex environment, which is a major reservoir of microorganisms and macroorganisms microbial, which are essential for proper functioning and maintenance of soil health. For growth of plant soil health is very important. Metagenomics is rapid growing research topic with high expectation. It exposes various unseen dimension for researcher to get attracted and work harder to explore more. It promises a complete understanding of global biological cycles to keeps biosphere in balance, organism responsible for producing enzymes, proteins, antibiotics, etc. Latest methodology of DNA isolation, cloning techniques, and screening strategies allows assessment and exploiting of microbes from extreme and inhospitable environments, such as hot springs, glaciers, hypersaline basins, etc. The introduction of next generation sequencing and advanced bioinformatics tools opened the door of comparative study of the metagenomic data sets with respect to phylogenetic and metabolic diversity in more precise way. Moreover, we mentioned various bioinformatics techniques and its application used for the analysis of metagenomics bioremediation data.

19.1 INTRODUCTION

Soil is a combination of abiotic (liquids, gases, organic matter, minerals) and biotic things (microorganism and macroorganisms) that mutually maintain life. The ecosystems are impacted comprehensively due to the way soil is being carried out from ozone depletion and global warming, to rainforest destruction and water pollution. Soil is a great carbon reservoir of Earth's carbon cycle, and it is potentially one of the most reactive to human disturbance and climate change. As the rise in temperature of planet, it has been predicted that soils will release carbon dioxide (CO_2) into the atmosphere due to increased biological activity at elevated temperatures. This prediction has been debatable based on consideration of more updated knowledge on soil carbon turnover.

In our ecosystem soil play a considerable role, such as a habitat for various types of microorganisms, a recycling system for nutrients and organic wastes, a controller of water quality and atmospheric composition, and a medium for plant development. Since soil has a huge range of available niches and habitats, it represents most of the Earth's genetic diversity. A very small amount of soil can contains huge number of organism representing several species, most of them are microbes and in the main still unidentified. In comparison to ocean Soil has a higher mean prokaryotic density (10^8 organisms per gram) of microbes (Raynaud and Nunan, 2014), than the ocean (10^7 procaryotic organisms per millilitre) of seawater (Whitman et al., 1998).

Soils can effectively eradicate contaminants, kill pathogenic agents, and provide nutrients to plants and animals by recycling the dead organic matter into a variety of nutrient forms. Soil has been chockfull with various toxic pollutants from numerous sources after humans started occupying this planet reduces the soil health condition. These toxic compounds are very dangerous and polluting different ecosystem. With the advancement of science and technology, different tools have been developed at molecular level to reduce pollutants.

Soil is a complex environment dominated by solid phase, which is a major treasure for micro and macroorganism. These microorganisms can be found as single cells or microcolonies to detoxify the toxic compound and increase the nutrient content of soil which ultimately increase the soil health. Their metabolism and interactions with other organisms and with soil particles is dependent on the conditions at the microhabitat level, which often differ between microhabitats even over very small distances. Fluctuations are seen in microhabitats for soil microorganisms as it consist of micropores and the surfaces of soil aggregates is of various composition and sizes (Ranjard and

Richaume, 2001; Torsvik et al., 2002). These fluctuations provide diverse conditions for microbial growth and the distribution of microorganisms and matrix substances in soil which enhances the diversity of microbial niches and microbes in soil.

The extensive microbial diversity in soils has leverage and is preeminent than any other environment or eukaryotic organisms as it can contain up to 10 billion microbes of conceivably different species per gram (Rosello and Amann, 2001). Soil microbial community's genetic complexity has been approximate by reassociation of community DNA. Such analyses shows that the soil community size is equivalent to 6000–10,000 *Escherichia coli* genomes (Ovreas, 2000; Torsvik et al., 2002) exempting the genomes of rare and unknown microorganisms.

19.2   METAGENOMICS

In the genomic era, metagenomics approaches are very efficient approach for removing various kinds of impurities (Bashir et al., 2014; Joshi et al., 2014). Metagenomics is a very promising approach of analyzing microbial communities at a genomic level providing a glimpse of the microbial community view of "Uncultured Microbiota." Recently, Duarte (2014) elucidated degradation of polycyclic aromatic pollutant using metagenomics approach to eliminate toxic contaminants from environment.

The term metagenomics was coined by Handelsman et al. in 1998. They have accessed the collective genomes and the biosynthetic machinery of soil microflora during a study of cloning the metagenome (Handelsman et al., 1998). By the alliances with latest science and technology metagenomics approach are frequently used and play a significant role in the field of bioremediation leading to the establishment of a pure nontoxic environment (Shah et al., 2011; Uhlik et al., 2013).

Cultivation-independent assessment can be provided by soil metagenomics of the largely unexploited genetic reservoir of microbial communities of the soil. Conventional laboratory conditions limit the number of culturable microorganism, hence entire information on soil microbial diversity cannot be obtained. In order to fully explore the soil microbial diversity, soil metagenomics plays a crucial role. It includes soil DNA isolation, construction and screening of clone libraries. This enables researchers to explore the complete scenario of soil microbial communities and their interactions. Steps of soil metagenomics are further explained below.

19.2.1 EXTRACTION OF DNA

DNA isolation from the environmental samples is the first step in Metage-nomics. The examples of environmental sample could be soil, (Voget et al., 2006; Waschkowitz et al., 2009) seawater (Stein et al., 1996), ground water (Uchiyama et al., 2005), Antarctic desert soil (Heath et al., 2009), human microbiome, etc. One of the major difficulties in Microbial DNA isolation is the varying DNA isolation protocol for samples obtained from different environmental conditions. Hence to overcome this obstacle, DNA isolation protocols have been standardized for the isolation of high quality DNA from a variety of extreme environments. The extracted DNA represents a pool of genomes from different soil microbes.

19.2.2 INSERTION OF ISOLATED DNAs INTO A HOST

Once the DNA has been isolated, the next step involves the insertion of isolated DNAs into a suitable host, such as *E. coli*. The studies of these inserted DNAs could be sequence driven or function-based (Pathak et al., 2009).

The various goals and practical gains that led to the adoption of metage-nomics were the discovery of new genes and gene products that would lead to agricultural innovations, medicinal chemistry, and industrial processes (Handelsman, 2004; Handelsman et al., 1998; Riesenfeld et al., 2004). Sequence-driven metagenomics and function-driven metagenomics helped in achieving these goals.

19.2.3 SEQUENCE-DRIVEN METAGENOMICS APPROACH

In this approach, sequencing and analysis of the DNA from the environ-ment of interest is done. The DNA probe or primers, required for this, is designed from conserved regions of already known genes (Simon and Daniel, 2011). Once the metagenomics sequences are obtained, these are then compared to the sequence data available in publicly available data-base, such as GENBANK. The classifying of sequenced genes into groups of similar predicted functions led to the successful identification of some novel enzymes like catalogs, DNA polymerases, etc. (Knietsch et al., 2003). Sometimes, the lack of similarity of the metagenomics gene with any of the genes in the public database causes a little problem as not much can be known about the gene and its product.

19.2.4 FUNCTION-DRIVEN METAGENOMICS

This approach also involves the isolation and insertion of DNA into the suitable host. However, rather than going for sequencing, functional analysis of the extracted DNA is done by screening. The metabolic activities of the metagenomic-library-containing clones are the base for isolation of genes encoding novel biomolecules (Simon and Daniel, 2011). It is based on heterologous complementation of host strains or mutants of host strains which require the targeted gene for growth under selective conditions (Simon and Daniel, 2010). This is a highly selective approach for the targeted genes. The only limitation with this approach is the inability of the host in expressing the genes from organisms in wild communities.

Community metagenomics, metatranscriptomics (Sorek and Cossart, 2010), and metaproteomics (Wilmes et al., 2008) are all the results of the development of next-generation sequencing techniques and other affordable methods, as microbial communities could now be analyzed on a wider platform. A major impact on the metagenomic research (Metzker, 2010) was made by the development of next-generation sequencing platforms, such as Genome Analyzer of Illumina (Bentley, 2006), Roche 454 sequencer and the SOLiD system of applied biosystems. There was a jump in the number and size of metagenomic sequencing projects due to enhanced sequencing efficiency and cost reduction, all of which are the results of these emerging techniques.

19.3 METAGENOMIC STRATEGIES

With the progress of advanced scientific technology, there has been a growth in the research tools being applied in varied scientific researches (Garfield et al., 1979). Unexplored phenomena of nature are now being familiarized due to these advanced technological inventions (Tijssen, 2002). The introduction of enhanced technologies into metagenomics has led to an improved knowledge of life sciences (Simon and Daniel, 2009). For the better understanding of the process of metagenomic bioremediation, the latest major strategies and tools have been discussed here.

19.3.1 SCREENING OF METAGENOMES FROM POLLUTED ENVIRONMENTS

For any metagenomic study, identification and screening of the microbial community from polluted environment are crucial. The screening of metagenomes from a contaminated environment helps in detecting the microbial community interaction. Jacquiod et al. (2014) gave an updated technology that enhanced the high throughput screening of a soil metagenomic library. This method involved spotting of soil metagenomic DNA of fosmid clone library on high-density membranes, hybridization of target genes encoding specific enzymes into soil metagenomic DNA using radio-labeled probe, identifying the affirmative hybridizing spots to the analogous clones in the library and finally sequencing the metagenomic inserts.

Once the identification, assembly and annotation were completed, new coding DNA sequences related to genes of interest were identified with low protein similarity against the closest hits in the databases. The sensitivity of DNA/RNA hybridization techniques defines the efficiency of this method-ology. This acts as a successful method for the recovery of novel genes from the metagenomic libraries according to the soil microbiota. However, during the process of metagenome extraction and screening, various molecular biological-based techniques (Godheja et al., 2014) can be followed.

For the metagenomic study, there are two ways for collection of contami-nated soil samples. The first method involves direct cell lysis and DNA purification while the second method has an additional preceding step of separating the cells from contaminated soil. Using specific cloning vectors, the isolated DNA is then cloned. Once cloned, these are then introduced into suitable host by various gene delivery techniques to obtain multiple copies of the contaminated soil DNA. The metagenome library thus formed is then screened. A recent study, conducted by Cai et al. (2014) for the screening of biosurfactant producers from petroleum hydrocarbon contaminated sources in cold marine environments revealed 55 biosulfant microbiota that belong to eight different genera including one *Alcanivorax*, one *exiguobacterium*, and two *halomonas* strains.

19.3.2 FLORESCENCE-ACTIVATED CELL SORTING

Florescence-activated cell sorting (FACS) is a highly efficient and widely used technique based on florescence of microbial cells during screening (Handelsman, 2005; Herzenberg et al., 1976). Schematic flow of SIGEX and intercellular biosensors methods is shown in figure high-throughput screening does not depend on selectable phenotype (Fig. 19.1). This has led to the focus on phenotypes, such as pigments that are easily visible, providing the use of fluorescence-activated cell sorting, regulation of a fluorescent biosensor present in the same cell as the metagenomic DNA certain types of gene expression can be detected using FACS DNA (Handelsman, 2004; Rinke et al., 2014). Hence, for better resolution of cell screening from metagenomic libraries, this screening method is very helpful.

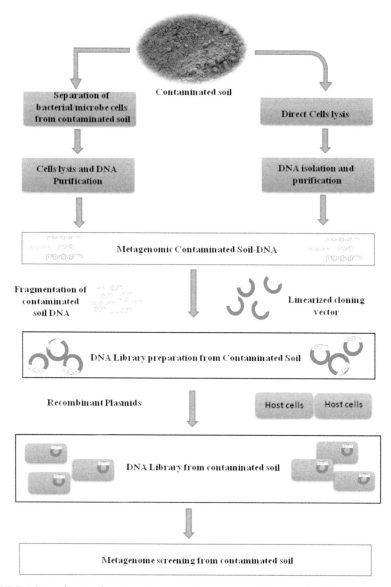

FIGURE 19.1   (See color insert.) Procedure and workflow involved in metagenomics from contaminated soil sample.

Source: Adapted from Devarapalli and Kumavath, 2015.

19.3.3   METAGENOMIC SEQUENCING STRATEGIES

Sanger was initiated DNA sequencing technology that was extensively applied in human genome project (Qin et al., 2010; Sanger et al., 1977). Later on various technologies were used for improvement and development of advanced sequencing method for sequencing of large and complex genome. Next-generation genomic sequencing is one of advanced sequencing method are frequently used in various metagenomics project of different microbial communities (Rodrigue et al., 2010; Tringe et al, 2005). Technological advancement has gifted next-generation sequencing techniques like pyrosequencing (Bentley et al., 2008; Loman et al., 2012), ligation sequencing (Roesch et al., 2007; Petrosino et al., 2009) reverse terminator (Ludwig et al., 2011; Song et al., 2012), and single-molecule sequence by synthesis (Harris et al., 2008; Liu et al., 2012), providing a high throughput that reads comparatively in less time (Wall et al., 2009; Claesson et al., 2010). However, pyrosequencing approach for sequencing the metagenomes of microbial communities is the first choice of most of the metagenomics researchers (Morgan et al., 2010; Schlüter et al., 2009).

19.4   BIOINFORMATIC TOOLS FOR METAGENOMIC

Bioinformatics approaches are frequently used in various areas of applied sciences. A brief idea about the applications of bioinformatics in bioremediation was given by Markowitz et al. (2006). In the metagenomic data analysis, bioinformatics plays a crucial role (Mitra et al., 2006).

Recent advancements in bioinformatics approach provide a wider view of metagenomics (Xu et al., 2014). The rapid enhancement of metagenomic sequence data produced by various metagenomic projects are leading to an increased demand of advanced bioinformatics approaches for management of these data in more precise way. Huson et al., 2007 explored the phylogenetic and functional characteristics of 33 publicly available metagenomes obtained from diverse soil applying various bioinformatics approaches. This section discusses the recent bioinformatics tools and datasets used in the analysis of metagenomic data in bioremediation.

19.4.1   METAGENOME ANALYZER

To analyzing huge metagenomic sequence data efficiently, Metagenome Analyzer (MEGAN) is one of the best software tool (El Hadidi et al., 2013:

Gower, 2005). Analysis and comparison of metatranscriptomic data and metagenomic data at both taxonomic and functional level can be done with high precision using this tool. For taxonomic analysis, reads are placed onto the NCBI taxonomy. While functional analysis involves mapping reads to the SEED, COG, and KEGG classifications. Charting and visualization techniques like cooccurrence plots help in the comparative taxonomical and functional study. In order to perform high-level comparison of a larger number of samples, the software does PCoA (Principle Coordinate Analysis) and clustering methods (Harrington et al., 2012). Other than these, MEGAN has various other benefits as well, such as different attributes of the samples can be captured and used during analysis, supports different input formats of data and is capable of exporting the results of analysis in different text-based and graphical formats.

19.4.2 SMASH COMMUNITY

Simple Metagenomics Analysis SHell for microbial communities (SMASH Community) is a stand-alone metagenomic annotation and analysis pipeline that shares design, principles, and routines with Smash Cell (Arumugam et al., 2010). Data delivered from Sanger and 454 sequencing technologies can be studied by this. It has varied applications like assembly and gene prediction, estimation of the quantitative phylogenetic and functional compositions of metagenomes, multiple metagenomes compositions comparison, production of insightful visual representations of such analyses (Seshadri et al., 2007). In addition to all these features, the optimized parameter sets for Arachne and Celera for metagenome assembly, and GeneMark and MetaGene for predicting protein coding genes on metagenomes can be provided by it. The scripts for the downstream analysis of datasets are also included in SmashCommunity. With the use of its batch access API of the interactive Tree of Life (iTOL) web tool, it can develop intuitive tree-based visualizations of results. It can perform various functions like comparison of multiple metagenomes, clustering them based on a relative entropy-based distance measure, performing bootstrap analysis of the clustering and generation of visual representation of the results.

19.4.3 COMMUNITY CYBER INFRASTRUCTURE FOR ADVANCED MICROBIAL ECOLOGY RESEARCH AND ANALYSIS (CAMERA)

CAMERA is a multifunctional database that can deposit, locate, analyze, visualize, and share data of microbes via advanced web based analysis portal

(Seshadri et al., 2007). A large amount of data from various aspects like environmental metagenomic and genomic sequence data, associated environmental parameters, precomputed search results, etc. are stored in CAMERA along with the software tools to support the comparative analysis of environmental samples. In order to allow the users to write expressive semantic queries to the database, the functioning of this tool is based on collecting and linking metadata relevant to environmental metagenome datasets with annotation in a semantically aware environment. Other than these functions, it helps the researchers to share and forward the data to other sites and archives. Hence, due to all the important functions that it can perform, it is considered as a complete genome-analysis tool (Meyer et al., 2008).

19.4.4 *RAPID ANNOTATION USING SUBSYSTEMS TECHNOLOGY FOR METAGENOMES (MG-RAST)*

MG-RAST is another high through-put analysis platform for metagenomes, based on sequence data that provides quantitative analysis of microbial populations (Markowitz et al., 2012). With the help of bioinformatic tools, it can perform varied functions like quality control, protein prediction, clustering, and similarity-based annotation on nucleic acid sequence datasets. Raw sequence data in FASTA format, uploaded by the users, are automatically normalized, processed, and summarized by this tool.

There are several methods provided by this tool to access different data types (including phylogenetic and metabolic reconstructions), compare and to annotate one or more metagenomes and genomes. This software is implemented in Perl by using a number of open-source components, including the SEED framework, NCBI BLAST, SQLite, and Sun Grid Engine.

19.4.5 *IMG/M*

Annotation and analysis of microbial genome and metagenome datasets along with their comparative study is supported by Integrated Microbial Genomes and Metagenomes (IMG/M) (Markowitz et al., 2012). It contains microbial community genomes integrated with genomes from all three domains of life, plasmids, viruses, and genome fragments. A comparative study by the analytical tools, based on their functions, reveal the relative abundance of protein families, functional families, or functional categories across metagenome samples and genomes.

It provides major advantage to the registered users by handling substantially larger metagenome datasets, being available only to registered users as part of the "My IMG" toolkit, and specifying, managing, and analyzing persistent sets of genes, functions, genomes, or metagenome samples and scaffolds.

19.5 APPLICATION OF SOIL METAGENOMICS

The development and application of metagenomics has enabled access to the uncultivated soil microbial community, availing a rich source of novel and useful biomolecules. Some examples of application of soil metagenomics are:

19.5.1 ANTIBIOTICS AND PHARMACEUTICALS

Antibiotic production can be significantly affected by Soil metagenomics. Its impact can be seen in previous studies like successful screening of soil metagenomic libraries for indirubin (Lim et al., 2005; MacNeil et al., 2001), the detection of several other novel antibiotics in metagenomic libraries (Brady et al., 2004b; Gillespie et al., 2002) and identification of nine aminoglycoside and one tetracycline antibiotics resistance genes from soil (Riesenfeld et al., 2004).

19.5.2 OXIDOREDUCTASES/DEHYDROGENASES

Alcohol oxidoreductases are useful biocatalysts in commercial production of chiral hydroxy esters, hydroxy acids, amino acids, and alcohols as they are capable of oxidizing short chain polyols. Five clones displaying novel 4-hydroxybutyrate dehydrogenase activity were found while screening for the diversity of bacteria in the environment capable of utilizing 4- hydroxybutyrate (Henne et al., 1999).

19.5.3 AMIDASES

For screening of biocatalyst, the role of soil metagenomic library is very effective (Voget et al., 2003). biosynthesis of 9-lactam antibiotics requires amidases. Seven amidase-positive clones, one of which encoded a novel penicillin acylase, were detected by Gabor and Janssen (2004), during their study targeting amidases of the soil metagenomes.

19.6 VITAMIN BIOSYNTHESIS

The search for novel genes encoding the synthesis of vitamins, such as biotin could be successfully achieved by the help of soil metagenomics. A good example of detection of novel vitamins by this approach was the discovery of seven cosmids in metagenomic libraries obtained after avidin enrichment of environmental samples and forest soil represented the highest level of biotin production in this study (Entcheva et al., 2001).

19.6.1 POLYSACCHARIDE DEGRADING/MODIFYING ENZYMES/ AMYLOLYTIC GENES

Identification and isolation of novel amylolytic enzymes from metagenomic DNA libraries has been targeted by several research groups (Richardson et al., 2002; Voget et al., 2003; Yun et al., 2004).

Cellulases have numerous applications and biotechnological potential for various industries including chemicals, fuel, food, brewery and wine, animal feed, textile and laundry, pulp and paper, and agriculture.

19.6.2 LIPOLYTIC GENES

Metagenomics has identified a number of novel genes encoding lipolytic enzymes, such as Esterase EstCE1 that was derived from a soil metagenomes. It is one of the highly useful enzymes in biotechnological applications due to its remarkable characteristics, high level of stability and unique substrate specificities (Elend et al., 2006). In addition to the above mentioned applications, metagenomics has been helpful in bioremediation as well (Bell et al., 2014; Chemerys et al., 2014) and has been observed to show better results than other approaches. Metagenomic approaches for removal of contaminants is characteristics of bacterial communities in different kinds of contaminated environments. Recently, Dellagnezze et al. (2014) reported the potential of metagenomic bacteria derived from petroleum reservoirs. Recently, Gillan et al. (2014) emphasized on long-term adaptation of bacterial communities in metal-contaminated sediments. Soil is considered as one of the major sources of life (Robinson et al., 2013) and its health is deteriorated by various contaminants (pollutants) released from different sources (Pennock et al., 2014; Wölz et al., 2011). These contaminants decrease the diversity of microorganism which is important for the maintaining the soil heath by adding various nutrients into it. Metagenomic approaches are more appropriated for removal of soil contaminants using bioremediation compared to other approaches (Adetutu et al., 2013; Uhlik et al., 2013). Recently, Yergeau et al. (2012) used metagenomic approach for analysis of arctic soils contaminated by high concentration of diesel in Canada.

19.7 SUMMARY

With the ever increasing global population (soon approaching nine billion people by 2050), one of the major challenges in agriculture nowadays is to increase yield and sustainability of crop production (Evans, 2013). The demand for soil quality is believed to be an integrative indicator of environmental health, food security, and economic viability, according to the reports of Food and Agriculture Organization of the United Nations (FAO) (Herrick, 2000). The quality of soil is maintained by several microbial communities, such as archaea and eubacteria, as well as some fungi, algae and certain animals, such as rotifers. As about 99% of the microbes cannot be cultivated easily, the uses of traditional microbiological culturing methods have not been that efficient. Over the decade, metagenomics has served as a boon in overcoming the difficulties being faced by various researchers to culture the certain microbes. Unrevealed microbial community can now be explored with help of metagenomics with more precision. This is possible due to adoption of updated technologies for establishing better environments. Our major emphasis has been to explain the approaches of metagenomics in bioremediation with the detailed view of metagenomic screening, FACS, and multiple advanced metagenomic sequencing strategies, based on sequences and function-based metagenomic strategies and tools. The last section explains the applications of different bioinformatics tools and datasets that are frequently used in metagenomic data analysis and processing during metagenomic bioremediation.

KEYWORDS

- **metagenomics**
- **next-generation sequencing**
- **bioremediation**
- **bioinformatics**
- **microbial communities**

REFERENCES

Adetutu, E. M.; Smith, R. J.; Weber, J.; Aleer, S.; Mitchell, J. G.; Ball, A. S.; Juhasz, A. L. A Polyphasic Approach for Assessing the Suitability of Bioremediation for the Treatment of Hydrocarbon-Impacted Soil. *Sci. Tot. Environ.* **2013,** *450,* 51–58.

Arumugam, M.; Harrington, E. D.; Foerstner, K. U.; Raes, J.; Bork, P. Smash Community: A Metagenomic Annotation and Analysis Tool. *Bioinformatics* **2010,** *26* (23), 2977–2978.

Bashir, Y.; Pradeep Singh, S.; Kumar Konwar, B. Metagenomics: An Application Based Perspective. *Chin. J. Biol.* **2014,** *90,* 2014, DOI:10.1155/2014/146030.

Bell, T. H.; Joly, S.; Pitre, F. E.; Yergeau, E. Increasing Phytoremediation Efficiency and Reliability Using Novel Omics Approaches. *Trends Biotechnol.* **2014,** *32* (5), 271–280.

Bentley, D. R. Whole-Genome Resequencing. *Curr. Opin. Genet. Dev.* **2006,** *16,* 545–552.

Bentley, D. R.; Balasubramanian, S.; Swerdlow, H. P.; Smith, G. P.; Milton, J.; Brown, C. G.; Anastasi, C. Accurate Whole Human Genome Sequencing Using Reversible Terminator Chemistry. *Nature* **2008,** *456* (7218), 53–59.

Brady, S. F.; Chao, C. J.; Clardy, J. Longchain *N*-Acyltyrosine Synthases from Environmental DNA. *Appl. Environ. Microbiol.* **2004,** *70,* 6865–6870.

Cai, Q.; Zhang, B.; Chen, B.; Zhu, Z.; Lin, W.; Cao, T. Screening of Biosurfactant Producers from Petroleum Hydrocarbon Contaminated Sources in Cold Marine Environments. *Marine Pollut. Bull.* **2014,** *86* (1), 402–410.

Chemerys, A.; Pelletier, E.; Cruaud, C.; Martin, F.; Violet, F.; Jouanneau, Y. Characterization of Novel Polycyclic Aromatic Hydrocarbon Dioxygenases from the Bacterial Metagenomic DNA of a Contaminated Soil. *Appl. Environ. Microbiol.* **2014,** *80* (21), 6591–6600.

Claesson, M. J.; Wang, Q.; O'Sullivan, O.; Greene-Diniz, R.; Cole, J. R.; Ross, R. P.; O'Toole, P. W. Comparison of Two Next-Generation Sequencing Technologies for Resolving Highly Complex Microbiota Composition Using Tandem Variable 16S rRNA Gene Regions. *Nucleic Acids Res.* **2010,** *38* (22), e200.

Dellagnezze, B. M.; de Sousa, G. V.; Martins, L. L.; Domingos, D. F.; Limache, E. E.; de Vasconcellos, S. P.; de Oliveira, V. M. Bioremediation Potential of Microorganisms Derived From Petroleum Reservoirs. *Marine Pollut. Bull* **2014,** *89* (1), 191–200.

Devarapalli, P.; Kumavath R. N. *Metagenomics—A Technological Drift in Bioremediation.* Intech: Rijeka, 2015; pp 73–91.

Duarte, M. Functional Soil Metagenomics: Elucidation of Polycyclic Aromatic Hydrocarbon Degradation Potential After 10 Years of *In Situ* Bioremediation. Annual Meeting and Exhibition (2014). Society for Industrial Microbiology & Biotechnology, 2014.

Elend C., C. Schmeisser, C. Leggewie, P. Babiak, J. D. Carballeira, H. L. Steele, J. L. Reymond,Jaeger, K. E.; Streit, W. R. Isolation and Biochemical Characterization of Two Novel Metagenome-Derived Esterases. *Appl. Environ. Microbiol.* **2006,** *72,* 3637–3645.

El Hadidi, M.; Ruscheweyh, H. J.; Huson, D. *Improved Metagenome Analysis Using MEGAN5.* In Joint 21st Annual International Conference on Intelligent Systems for Molecular Biology (ISMB) and 12th European Conference on Computational Biology (ECCB), held at Berlin, 21-23 July, 2013.

Entcheva, P.; Liebl, W.; Johann, A.; Hartsch, T. Streit, W. R. Direct Cloning from Enrichment Cultures, a Reliable Strategy for Isolation of Complete Operons and Genes From Microbial Consortia. *Appl. Environ. Microbiol.* **2001,** *67,* 89–99.

Gillan, D. C.; Roosa, S.; Kunath, B.; Billon, G.; Wattiez, R. The Long-Term Adaptation of Bacterial Communities in Metal-Contaminated Sediments: a Metaproteogenomic Study. *Environ. Microbiol.* **2014,** *17* (6):1991–2005.

Gillespie, D. E.; Brady, S. F.; Bettermann, A. D.; Cianciotto, N. P.; Liles, M. R.; Rondon, M. R.; Clardy, J.; Goodman, R. M.; Handelsman, J. Isolation of Antibiotics Turbomycin A and B from a Metagenomic Library of Soil Microbial DNA. *Appl. Environ. Microbiol.* **2002,** *68*, 4301–4306.

Goesmann, A. The Metagenome of a Biogas-Producing Microbial Community of a Production-Scale Biogas Plant Fermenter Analysed by the 454-Pyrosequencing Technology. *J. Biotechnol.* **2008,** *136* (1), 77–90.

Gower, J. C. Principal Coordinates Analysis. *Encyclopedia of Biostatistics*; John Wiley & Sons, Ltd.: Hoboken, New Jersey, 2005; pp 1–5.

Handelsman, J. Metagenomics: Application of Genomics to Uncultured Microorganisms. *Micrbiol. Mol. Biol. Rev.* **2005,** *68*, 669–685.

Handelsman, J. Sorting Out Metagenomes. *Nat. Biotechnol.* **2005,** *23* (1), 38–39.

Handelsman, J.; Rondon, M. R.; Brady, S. F.; Clardy, J.; Goodman, R. M. Molecular Biological Access to the Chemistry of Unknown Soil Microbes: a New Frontier for Natural Products. *Chem. Biol.* **1998,** *5* (10), R245–R249.

Herzenberg, L. A.; Sweet, R. G.; Herzenberg, L. A. Fluorescence-Activated Cell Sorting. *Sci Am.* **1976,** *234* (3), 108–117.

Harrington, E. D.; Arumugam, M.; Raes, J.; Bork, P.; Relman, D. A. Smash-Cell: A Software Framework for the Analysis of Single-Cell Amplified Genome Sequences. *Bioinformatics* **2010,** *26* (23), 2979–2980.

Harris, T. D.; Buzby, P. R.; Babcock, H.; Beer, E.; Bowers, J.; Braslavsky, I.; Xie, Z. Single-Molecule DNA Sequencing of a Viral Genome. *Science* **2008,** *320* (5872), 106–109.

Heath, C.; Hu, X. P.; Cary, S. C.; Cowan D. Identification of a Novel Alkaliphilic Esterase Active at Low Temperatures by Screening a Metagenomic Library from Antarctic Desert Soil. *Appl. Environ. Microbiol.* **2009,** *75*, 4657–4659.

Henne A.; Daniel, R.; Schmitz, R. A.; Gottschalk, G. Construction of Environmental DNA Libraries in *Escherichia coli* and Screening for the Presence of Genes Conferring Utilization of 4-Hydroxybutyrate. *Appl. Environ. Microbiol.* **1999,** *65*, 3901–3907

Huson, D. H.; Auch, A. F.; Qi, J.; Schuster, S. C. MEGAN Analysis of Metagenomic Data. *Genome Res.* **2007,** *17* (3), 377–386.

Jacquiod, S., et al. Characterization of New Bacterial Catabolic Genes and Mobile Genetic Elements by High Throughput Genetic Screening of a Soil Metagenomic Library. *J. Biotechnol.* **2014,** *20* (190), 18–29.

Joshi, M. N.; Dhebar, S. V.; Bhargava, P.; Pandit, A. S.; Patel, R. P.; Saxena, A. K.; Bagatharia, S. B. Metagenomic Approach for Understanding Microbial Population From Petroleum Muck. Genome Announ. **2014,** *2* (3), e00533–e00514.

Knietsch, A., Bowien, S., Whited, G., Gottschalk, G., Daniel, R. Identification and Characterization of Coenzyme B12-Dependent Glycerol Dehydratase and Diol Dehydratase-Encoding Genes From Metagenomic DNA libraries Derived from Enrichment Cultures. *Appl. Environ. Microbiol.* **2003,** *69*, 3048–3060.

Layton, A. C.; Chauhan, A.; Williams, D. E.; Mailloux, B.; Knappett, P. S.; Ferguson, A. S.; Sayler, G. S. Metagenomes of Microbial Communities in Arsenic-and Pathogen-Contaminated Well and Surface Water from Bangladesh. *Genome Announ.* **2014,** *2* (6), e01170–e01114.

Lim, H. K.; Chung, E. J.; Kim, J. C.; Choi, G. J.; Jang, K. S.; Chung, Y. R.; Cho, K. L.; Lee, S. W. Characterization of a Forest Soil Metagenome Clone that Confers Indirubin and Indigo Production on *Escherichia coli. Appl. Environ. Microbiol.* **2005,** *71,* 7768–7777.

Liu, L.; Li, Y.; Li, S.; Hu, N.; He, Y.; Pong, R. Law, M. Comparison of Next Generation Sequencing Systems. *BioMed Res. Int.* **2012**, *2012*, 11.

Loman, N. J.; Constantinidou, C.; Chan, J. Z.; Halachev, M.; Sergeant, M.; Penn, C.; Pallen, M. J. High-Throughput Bacterial Genome Sequencing: An Embarrassment of Choice, a World of Opportunity. *Nat. Rev. Microbiol.* **2012**, *10* (9), 599–606.

Ludwig, M.; Bryant, D. A. Transcription Profiling of the Model Cyanobacterium *Synechococcus* sp. Strain PCC 7002 by Next-Gen (SOLiD™) Sequencing of cDNA. *Front. Microbiol.* **2011**, 2.

MacNeil, I. A.; Tiong, C. L.; Minor, C.; August, R. P.; Grossman, T. H.; Loiacono, K. A.; Lynch, B. A.; Phillips, T.; Narula, S.; Sundaramoorthi, R.; Tyler, A.; Aldredge, T.; Long, H.; Gilman, M.; Holt, D.; Osburne, M. S. Expression and Isolation of Antimicrobial Small Molecules from Soil DNA Libraries. *J. Mol. Microbiol. Biotechnol.* **2001**, *3*, 301–308.

Markowitz, V. M.; Ivanova, N. N.; Szeto, E.; Palaniappan, K.; Chu, K.; Dalevi, D.; Kyrpides, N. C. IMG/M: a Data Management and Analysis System for Metagenomes. *Nucleic Acids Res.* **2008**, *36* (Suppl. 1), D534–D538.

Markowitz, V. M.; Ivanova, N.; Palaniappan, K.; Szeto, E.; Korzeniewski, F.; Lykidis, A.; Kyrpides, N. C. An Experimental Metagenome Data Management and Analysis System. *Bioinformatics* **2006**, *22* (14), e359–e367.

Markowitz, V. M.; Chen, I. M. A.; Chu, K.; Szeto, E.; Palaniappan, K.; Grechkin, Y.; Kyrpides, N. C. IMG/M: The Integrated Metagenome Data Management and Comparative Analysis System. *Nucleic Acids Res.* **2012**, *40* (D1), D123–D129.

Metzker, M. L. Sequencing Technologies—The Next Generation. *Nat. Rev. Genet.* **2010**, *11*, 31–46.

Meyer, F., Paarmann, D., D'Souza, M., Olson, R., Glass, E. M.; Kubal, M.; Edwards, R. A. The Metagenomics RAST Server—A Public Resource for the Automatic Phylogenetic and Functional Analysis of Metagenomes. *BMC Bioinform.* **2008**, *9* (1), 386.

Mitra, S.; Klar, B.; Huson, D. H. Visual and Statistical Comparison of Metagenomes. *Bioinformatics* **2009**, *25* (15), 1849–1855.

Morgan, J. L.; Darling, A. E.; Eisen, J. A. Metagenomic Sequencing of an In Vitro-Simulated Microbial Community. *PLoS One* **2010**, *5* (4), e10209.

Pathak, G. P. Ehrenreich, A.; Losi, A.; Streit, W. R.; Gartner, W Novel Blue Light–Sensitive Proteins from a Metagenomic Approach. *Environ. Microbiol.* **2009**, *11*, 2388–2399.

Pennock, S. R.; Abed, T. M.; Curioni, G.; Chapman, D. N.; John, U. E.; Jenks, C. H. J. *Investigation of Soil Contamination by Iron Pipe Corrosion and Its Influence on GPR Detection.* In Ground Penetrating Radar (GPR), 2014 15th International Conference on IEEE, 2014, pp 381–386.

Petrosino, J. F.; Highlander, S.; Luna, R. A.; Gibbs, R. A.; Versalovic, J. Metagenomic Pyrosequencing and Microbial Identification. *Clinical Chem.* **2009**, *55* (5), 856–866.

Qin, J.; Li, R.; Raes, J.; Arumugam, M.; Burgdorf, K. S.; Manichanh, C.;Weissenbach, J. A Human Gut Microbial Gene Catalogue Established by Metagenomic Sequencing. *Nature* **2010**, *464* (7285), 59–65.

Raynaud, X.; Nunan, N. Spatial Ecology of Bacteria at the Microscale in soil". *Plos One* **2014**, *9* (1), e87217. DOI:10.1371/journal.pone.0087217.

Richardson, T. H.; Xuqiu, T.; Gerhard, F.; Walter, C.; Mark, C.; David, L.; John, M.; Short, J. M.; Robertson, D. E.; Miller, C. A Novel, High Performance Enzyme for Starch Liquefaction. Discovery and Optimization of a low pH, Thermostable Alpha-Amylase. *J. Biol Chem.* **2002**, *277*, 26501–26507.

Riesenfeld, C. S.; Schloss P. D., Handelsman, J. Metagenomics: Genomic Analysis of Microbial Communities. *Annu. Rev. Genet.* **2004**, *38*, 525–552.

Rinke, C.; Lee, J.; Nath, N.; Goudeau, D.; Thompson, B.; Poulton, N.; Woyke, T. Obtaining Genomes From Uncultivated Environmental Microorganisms Using FACS-Based Single-Cell Genomics. *Nat. Protocols* **2014**, *9* (5), 1038–1048.

Robinson, D. A.; Jackson, B. M.; Clothier, B. E.; Dominati, E. J.; Marchant, S. C.; Cooper, D. M.; Bristow, K. L.. Advances in Soil Ecosystem Services: Concepts, Models, and Applications for Earth System Life Support. *Vadose Zone J.* **2013**, *12* (4), 1–13.

Rodrigue, S.; Materna, A. C.; Timberlake, S. C.; Blackburn, M. C.; Malmstrom, R. R.; Alm, E. J.; Chisholm, S. W. Unlocking Short Read Sequencing for Metagenomics. *PLoS One* **2010**, *5* (7), e11840.

Roesch, L. F.; Fulthorpe, R. R.; Riva, A.; Casella, G.; Hadwin, A. K.; Kent, A. D., Triplett, E. W. Pyrosequencing Enumerates and Contrasts Soil Microbial Diversity. *ISME J.* **2007**, *1* (4), 283–290. 88

Rozell, D. J.; Reaven, S. J. Water Pollution Risk Associated with Natural Gas Extraction from the Marcellus Shale. *Risk Anal.* **2012**, *32* (8), 1382–1393.

Shah, V. Jain, K.; Desai, C.; Madamwar, D. Metagenomics and Integrative 'Omics' Technologies in Microbial Bioremediation: Current Trends and Potential Applications. *Metagenomics: Current Innovations and Future Trends*; Caister Academic Press: Norfolk, 2011, pp 211–240.

Sanger, F.; Nicklen, S.; Coulson, A. R. DNA Sequencing with Chain-Terminating Inhibitors. *Proc. Nat. Acad. Sci.* **1977**, *74* (12), 5463–5467.

Seshadri, R.; Kravitz, S. A.; Smarr, L.; Gilna, P.; Frazier, M. CAMERA: A Community Resource for Metagenomics. *PLoS Biol.* **2007**, *5*(3), e75.

Shi, P.; Jia, S.; Zhang, X. X.; Zhang, T.; Cheng, S.; Li, A. Metagenomic Insights Into Chlorination Effects on Microbial Antibiotic Resistance in Drinking Water. *Water Res.* **2013**, *47* (1), 111–120.

Simon, C.; Daniel, R. Construction of Small-Insert and Large Insert Metagenomic Libraries. *Methods Mol. Biol.* **2010**, 668, 39–50.

Simon, C.; Daniel, R., Metagenomics analysis: Past and Future Trends. *Appl. Environ. Microbiol.* **2011**, *77*, 1153–1161.

Sirés, I.; Brillas, E. Remediation of Water Pollution Caused by Pharmaceutical Residues Based on Electrochemical Separation and Degradation Technologies: A Review. *Environ. Int.* **2012**, *40*, 212–229.

Simon, C.; Daniel, R. Achievements and New Knowledge Unraveled by Metagenomic Approaches. *Appl. Microbiol. Biotechnol.* **2009**, *85* (2), 265–276.

Song, C. X.; Clark, T. A.; Lu, X. Y.; Kislyuk, A.; Dai, Q.; Turner, S. W.; Korlach, J. Sensitive and Specific Single-Molecule Sequencing of 5-Hydroxymethylcytosine. *Nat. Method.* **2012**, *9* (1), 75–77.

Sorek, R., Cossart, P. Prokaryotic Transcriptomics: A New View on Regulation, Physiology and Pathogenesity. *Nat. Rev. Genet.* **2010**, *11*, 9–16.

Stein, J. L.; Marsh, T. L.; Wu, K. Y.; Shizuya, H.; DeLong, E. F. Characterization of Uncultivated Prokaryotes: Isolation and Analysis of a 40–Kilobase—Pair Genome Fragment from a Planktonic Marine Archeon. *J. Bacteriol.* **1996**, *178*, 591–599.

Tijssen, R. J. Science Dependence of Technologies: Evidence from Inventions and Their Inventors. *Res. Policy* **2002**, *31* (4), 509–526.

Tringe, S. G.; Rubin, E. M. Metagenomics: DNA Sequencing of Environmental Samples. *Nat. Rev. Gene.* **2005**, *6* (11), 805–814.

Uchiyama, T., Abe T., Ikemura, T., Watanable, K. Substrate Induced Gene-Expression Screening of Environmental Metagenome Libraries for Isolation of Catabolic Genes. *Nat. Biotechnol.* **2005**, *23*, 88–93.

Uhlik, O.; Strejček, M.; Hroudova, M.; Dem-nerová, K.; Macek, T. Identification and Characterization of Bacteria with Bioremediation Potential: from Cultivation to Metagenomics. *Chemicke Listy* **2013**, *107* (8), 614–622.

Uhlik, O.; Leewis, M. C.; Strejcek, M.; Musilova, L.; Mackova, M.; Leigh, M. B.; Macek, T. Stable Isotope Probing in the Metagenomics Era: A Bridge Towards Improved Bioremediation. *Biotechnol. Adv.* **2013**, *31* (2), 154–165.

Valiente, G.; Pesole, G. Bioinformatics Approaches and Tools for Metagenomic Analysis. Editorial. *Brief. Bioinform.* **2012**, *13* (6), 645.

Voget, S., Steele, H. L., Streit, W. R. Characterization of a Metagenome—Derived Halotolerant Cellulose. *J. Biotechnol.* **2006**, *126*, 26–36.

Voget, S., Leggewie, C., Uesbeck, A., Raasch, C., Jaeger, K. E.; Streit, W. R. Prospecting for Novel Biocatalysts in a Soil Metagenome. *Appl. Environ. Microbiol.* **2003**, *69*, 6235–6242.

Wall, P. K.; Leebens-Mack, J.; Chanderbali, A. S.; Barakat, A.; Wolcott, E.; Liang, H.; Altman, N. Comparison of Next Generation Sequencing Technologies for Transcriptome Characterization. *BMC Genom.* **2009**, *10* (1), 347.

Waschkowitz, T.; Rockstroh, S.; Daniel, R. Isolation and Characterization of Metalloproteases with a Novel Domain Structure by Construction and Screening of Metagenomic Libraries. *Appl. Environ. Microbiol.* **2009**, *75*, 2506–2516.

Whitman, W. B.; Coleman, D. C.; Wiebe, W. J. *Prokaryotes: the Unseen Majority.* Proceedings of the National Academy of Sciences of the USA. **1998**, *95* (12), 6578–6583. DOI:10.1073/pnas.95.12.6578. PMC 33863 PMID 9618454.

Wilmes, P.; Wexler, M; Bond, P. L. Metaproteomics Provides Functional Insight into Activated Sludge Wastewater Treatment. *PLoS One* **2008**, *3*, e1778.

Widger, W. R.; Golovko, G.; Martinez, A. F.; Ballesteros, E.; Howard, J.; Xu, Z.; Fofanov, Y. *Longitudinal Metagenomic Analysis of the Water and Soil from Gulf of Mexico Beaches Affected by the Deep Water Horizon Oil Spill.* Nature Proceedings, 2011.

Williams, P. J.; Botes, E.; Maleke, M. M.; Ojo, A.; DeFlaun, M. F.; Howell, J.; van Heerden, E. Effective Bioreduction of Hexavalent Chromium–Contaminated Water in Fixed-Film Bioreactors. *Water SA* **2014**, *40* (3), 549–554.

Wölz, J.; Schulze, T.; Lübcke-von Varel, U.; Fleig, M.; Reifferscheid, G.;Brack, W.; Hollert, H. Investigation on Soil Contamination at Recently Inundated and Non-Inundated Sites. *J. Soils Sedim.* **2011**, *11* (1), 82–92.

Xu, Z.; Hansen, M. A.; Hansen, L. H.; Jacquiod, S.; Sørensen, S. J. Bioinformatic Approaches Reveal Metagenomic Characterization of Soil Microbial Community. *PLoS One* **2014**, *9* (4), e93445.

Yergeau, E.; Sanschagrin, S.; Beaumier, D.; Greer, C. W. Metagenomic Analysis of the Bioremediation of Diesel-Contaminated Canadian High Arctic Soils. *PLoS One* **2012**, *7* (1), e30058–e30086.

Yun J.; Seowon K.; Sulhee P.; Hyunjin, Y.; Myo- Jeong K.; Sunggi H.; Sangyeol R. Characterization of a Novel Amylolytic Enzyme Encoded by a Gene from a Soil Derived Metagenomic Library. *Appl. Environ. Microbiol.* **2004**, *70* (12), 7229–7235.

CHAPTER 20

Use of Household Waste Materials for Biofertilizer Development

MD. SHAMIM[1,*], MD. ABU NAYYER[2], MD. SHAMSHER AHMAD[3],
MD. ARSHAD ANWER[4], and MOHAMMED WASIM SIDDIQUI[3]

[1]*Department of Molecular Biology and Genetic Engineering,
Dr. Kalam Agricultural College, Kishanganj, Bihar Agricultural University,
Sabour, Bhagalpur 813210, Bihar, India*

[2]*Integral Institute of Agricultural Science and Technology, Integral
University, Lucknow 226021, India*

[3]*Department of Post Harvest Technology, Bihar Agricultural University,
Sabour, Bhagalpur 813210, Bihar, India*

[4]*Department of Plant Pathology, Bihar Agricultural University, Sabour,
Bhagalpur 813210, Bihar, India*

*Corresponding author. E-mail: shamimnduat@gmail.com

ABSTRACT

Huge amount of different kinds of kitchen daily wastes are produced from the houses, canteens, mess, and hotels. The organic wastes from the above sources can be collected and easily made into biofertilizers and composts for the agricultural use. This is the best way to clean household and protect our environment. Decomposition of organic material is a natural process that provides essential nutrients to the soil for plants. Household waste composting is also used as a good practice that allows us to control and accelerate this natural process. Adding the household compost to the soil can help supplementing chemical fertilizers and in better growth and development of plants. The household compost can also partially replace chemical fertilizers up to an extent, and can improve physical and chemical properties of soil by minimizing the pollution level of the surroundings. In household composting, microbial degradation plays a vital role and an efficient biotechnological tool helps in recycling household organic wastes to make better end product with the help of some specific group of decomposing bacteria.

20.1 INTRODUCTION

Wastes are materials which are not required for normal or discarded by somebody. The generation of wastes is increasing day by day due to increase in population, urbanization, and consumption behavior. Home-waste management isn't a mania or a habit, but it is a set of home installations and carefulness that aims to protect the environment and keep your home beautiful without spending hours on a seemingly endless list of tedious tasks. According to Environment Protection Act, 1986, "Environment is intended to serve three purposes. First, it is to protect and improve the environment; second, to prevent hazards to human beings; and third, it is in respect of living creatures, plants, and property." Human life cannot be thought of without an environment. The physical, chemical, biological, psychological, and social factors in the environment are essential and play an important role in the normal functioning of human beings. Hence, to protect the environment, promote the health and well-being of the population and also to improve the standard of living of rural households; it is necessary to create attentiveness for proper disposal of biodegradable waste, which automatically leads to resource recovery thereby, promoting economic potentials. In nature's laboratory, there are many organisms that convert organic waste into valuable products rich in organic matter and plant nutrients, which can improve the soil productivity and reduce environmental issues (Indumathi, 2017).

A few decades back, composting was mainly done by nature on chance (Ahmad et al., 2007a). However, after World War II huge scientific efforts were made to focus on the processes of composting period. The idea of "Indore Process" was suggested by Sir Albert Howard for large-scale composting, during the years 1924–1931 (Howard, 1943). In 1932 the first full scale, composting facility was established in Netherlands where the "Van Maanen" process was used. This process was modification of Indore method. This method was emerged to be well suited for developing countries due to low technology exploitation and easily accessible equipment (Epstein et al., 1976). Breidenbach (1971) reported that there were more than 30 composting systems in 1969. At present, even more, systems based scientific principles are available all around the globe (Ahmad et al., 2007a). The advantage of the agricultural use of sewage sludge is the recycling of nitrogen, phosphorus, potassium, calcium, microelements, and organic matter for plant growth by the enhancement of their availability in the soil. However, the difficulties to the agricultural use of sewage sludge are: high in heavy metal content, low concentration of potassium, high moisture content, and the possible presence of chemical pollutants (Furhacker and Haberl, 1995). There is a trend toward increase in use of sludge in agriculture (Ito et al., 1998; Stenger 2000; Vesilind and Spinosa 2001).

Composting is usually done by the aerobic methods due to its significance of rapidity and quality of the product. It is advantageous not only for good crop production but also for disposal of vastly accumulating wastes. Hence, some mineral and biological additive in composting is also given consideration. In a very short time quality composts can also be made by the use of microbial inocula as a starter (Baheri and Meysami., 2002; Shin et al., 1999). Lei and Gheynst (2000) reported that inoculations can increase the microbial population, make beneficial microbial communities, enhance microbiological quality, and produce many desired enzymes; therefore, improving the conversion of organics and diminish the emission of odorous gases. Hamdy (2005) revealed the potentiality of *Rhizopus oryzae* to utilize orange peels with high pectinolytic and cellulolytic activities.

"Solid waste" is the term now used internationally to explain nonliquid waste materials arising from domestic, commercial, industry, trade, agriculture, mining activities, and public sectors. Solid waste comprises countless different materials, such as vegetable waste, papers, glass, plastics, wood, yard clippings, food waste, radioactive wastes, and hazardous waste. Solid wastes are all the wastes arising from human and animal activities that are normally solids, semisolids, liquids in containers and those are discarded or useless or unwanted. Solid waste generation is increasing gradually with the passage of time due to population explosion and urbanization. Effective solid waste management can be achieved by controlling the waste generation and taking measures for proper collection, storage, transportation, and disposal of solid waste in an environmental and in an economic manner. Integrated solid waste management includes the application of suitable techniques, better management practices, and selection of better technologies for waste disposal and management (Tchobanoglous and Kreith, 2002). There are various methods for the management of waste material. Biotechnological process of composting involves certain species of bacteria that are used to enhance the process of waste conversion and produce a better end product (Indumathi, 2017). There are several naturally occurring microorganisms that are able to convert organic waste into valuable resources, such as plant macro- and micronutrients, and reduce the C:N ratio to support soil productivity. These microorganisms are also important to maintain nutrient flows from one system to another and to minimize ecological imbalance (Novinsak et al., 2008; Umsakul et al., 2010). Many fruits and vegetables present nearly ideal conditions for the survival and growth of a number of microorganisms. During the composting process, various parameters including C:N ratio, composting temperature, pH of the finished product, moisture content, and the presence of potential pathogens, such as coli are used to assess the quality and stability of the compost (Sanmanee et al., 2011; Steger et al., 2007; Wu and Ma 2002).

20.2 TYPES OF HOUSEHOLD WASTES

20.2.1 FRUIT AND VEGETABLE WASTES

Fruit and vegetable wastes (FVW) are produced in huge amount in markets and households and make a source of annoyance in municipal landfills because of their high biodegradability (Misi and Forster, 2002). In most of the developing countries, these type of wastes are produced in bulk amount due to inadequate cold storage and refrigerated transport facility. In developing countries, wastes are mostly generated from agricultural production, postharvest management, and distribution processes, due to seasonality that leads to gluts of fruits and vegetable in the market and to the absence of proper postharvest strategies for perishable crops. The most potential alternative to incinerating and composting of FVW is to digest its organic matter through anaerobic digestion (Bouallagui et al., 2003). The major benefits of this method are the production of biogas, which may be utilized in the production of electricity (Ahring et al., 2002). A valuable effluent is also obtained, which can be used as an excellent soil conditioner after some treatments (Converti et al., 1992; Simeonov 1999). Microbial conversion of vegetable waste to bio fertilizer is a feasible and potential technology in future to maintain the natural resources and to reduce the impact on environmental fitness. It is an easy biotechnological method of composting, in which particular species of bacteria are used to boost the process of waste conversion and make a better end product (Indumathi, 2017).

20.2.2 FOOD WASTES

According to the Food and Agriculture Organization, generally, one-third of the edible parts of food produced for human utilization gets wasted or lost globally. This quantity accounts about 1.3 billion ton/year and reflects not only the food losses but also the food processing wastes (Gustavsson et al., 2011). Food waste is food material that is unused, unnecessary, expire, and not safe for the consumption of human for various health reasons. It is in particular household wastes, includes uneaten food and food preparation left over from residences, commercial establishments, such as restaurants, institutional sources like school, cafeterias, and industrial sources like factory lunch rooms and is the single largest component of the municipal solid waste (MSW) (Upadhyay et al., 2005). These wastes contain high moisture and organic matter content which is easily decomposed by microbes. It produces a foul smell, detrimental to the environment and human health (Tsai et al., 2007). A study of the Food and Agriculture Organization of the United Nations reported that food is wasted during manufacturing processes from agricultural production to consumer levels (Gustavsson et al., 2011). Removal of food waste by discarding into landfill sites is unsuitable because the production of food requires energy and nutrient. These are useful waste material for the recovery of nutrients required in different processes and for making high-value products because they are rich in carbohydrate, proteins, and lipids (Pleissner et al., 2013).

Rotten food material can breakdown to produce methane gas and is harmful to the human if not handled properly (Pleissner and Carol, 2013). Methane is an effective greenhouse gas that causes the greenhouse effect, and global warming. Methane gas is more potent than carbon dioxide in causing greenhouse effect (Whiting and Chanton, 2001). Food waste causes emissions of greenhouse gas that increase the surrounding temperature during its production and disposal. Food wastes that are disposed off in landfills can decompose and produce an offensive smell that can harmful to human and environment (Pleissner and Carol, 2013). Food waste can be reduced by converting into useful organic fertilizers to reduce health problem caused by food wastes disposal (Pleissner and Carol, 2013). Recycling of food wastes can be benefitted to the environment by reducing the amount of garbage disposed off, increasing the fertility of the soil and by improves the physical and chemical properties of soil (Park et al., 2002). Food wastes can be decomposed in anaerobic digestion by microorganisms to break down food waste into smaller materials and make useful products (Morash, 2014). This process is carried out inside an enclosed system in the absence of oxygen. Methane gas produced can be collected and converted into biogas to transport fuels and produce electricity and heat. Fruits and vegetables that include tubers and roots showed the highest amount of wastage of any food. Around 8000 tons of food and kitchen waste is generated daily in Malaysia (Bernama, 2014). Most of the food wastes are disposed at disposal site and the lack of food waste recovery in Malaysia.

20.2.3 DOMESTIC OR OTHER HOUSEHOLD WASTES

This type of waste is usually made up of food leftovers, either cooked or uncooked, and garden waste, such as leaves, grass cuttings, or trimmings from bushes and hedges. Domestic kitchen wastes are often mixed with nonorganic materials, such as plastic packaging, which cannot be composted. It is very effective if this type of waste can be separated at source—this makes easy recycling of both types of waste. Domestic or household waste is usually produced in relatively small quantities. In developing countries, there is a much higher organic content in domestic waste. It is consequently well worth intercepting this supply of useful material where it can be used effectively. In many developing countries, household waste constitute a large proportion of inert materials, such as ash, dust, sand, and stones, and has high-moisture levels because of the high consumption of fresh fruits and vegetables.

Every household generates approximately 1.5–2.0 kg of solid waste, which is upto 168.13 t/d on 8.0 lacs of urban population (GEAG Report, 2010a,b). Most of the urban residents dispose off their wastes in poly bags or loose on the roadside or in dustbins. These wastages are eaten by cattle and other ruminants and left the residues scattered on road which creates several infectious diseases like cholera, diarrhea, and foul smell due to development of several disease causing microorganisms. Almost 72% of the total wastes are biodegradable in nature which may be converted into biofertilizer. An earlier survey (in 2009), conducted in a colony of 200 houses in Gorakhpur city, has indicated that every household disposes off about 1.2 kg of solid waste per day. The collection of household waste at the dumping site can be sorted out manually into biodegradable food waste, packaging materials, regular plastic bags, etc. However, the latest technology used for this purpose is "the Optical Sorting Plant" started at Oslo in Norway (2014). Food waste is sorted into green bags, plastic packaging into blue bags, and residual waste is placed in regular plastic shopping bags. This type of waste separation is an addition to existing systems for sorting paper, cardboard, glass, metal, and hazardous waste.

20.3 MANUFACTURING OF BIOFERTILIZER USING WASTE

Biofertilizer plant was designed by Evans et al. (2007) to convert food wastes into high-quality biofertilizer and biogas. Biogas is defined as a gas produced in the absence of oxygen through biological breakdown of organic matter. It is an alternative renewable and sustainable form of energy (Ilaboya et al., 2010). A biogas plant is often called as an anaerobic digester that treats wastes. In this process, an air-tight tank converts biomass waste into methane which is used as a source of renewable energy that can be used for electricity, heating, and other different purposes (Verma et al., 2007). Furthermore bacteria, fungi, and *Spirulina platensis* (algae) have to convert kitchen wastes, mainly the vegetables, into biofertilizers (Kalpana et al., 2011). These wastes are also converted by earthworms (Albasha et al., 2015). Ngumah et al. (2013) explain the benefits of organic waste produced in Nigeria as a source of renewable energy, such as biofertilizer and biofuel by using livestock wastes (cattle manure, sheep and goat manure, pig manure, poultry manure, and abattoir waste), human manure, plant residue, and MSW. They also compare utilization of biogas and biofertilizers in comparison to traditional fossil fuel. During the other process of biodegradation, *S. platensis* had converted kitchen wastes into high nutrient-rich biofertilizers (Kalpana et al., 2011). Similarly, enhanced biofertilizer may also be collected from kitchen wastes and their property was enhanced by the addition of *Trichoderma harzianum* which will, additionally, offer the advantage of biopesticide (Haque et al., 2010). Kitchen wastes are also converts into biofertilizer by the method of vermicomposting (Albasha et al., 2015).

20.4 PROCEDURES OF HOUSEHOLD WASTE SORTING

A significant waste-management program can be ascertained by utilizing some of the essential activities into integrated solid waste management. However, efficient accomplishment and legislative efforts are necessary for the secure management and disposal of solid waste. Different agencies may provide incentives for the development and practice for their safe treatments, safe manufacturing, and methods for converting solid waste into important resources by recycling and reuse as compost.

20.4.1 STAGE 1—PRESORTING OF WASTAGE

Household waste is often collected in plastic shopping bags and put into waste containers, but most of the people also discard garden waste, black polythene waste bags, wooden boards, and other unwanted waste to be carried to reprocessing stations. During the first stage, larger unwanted material and loose wastes are sifted by the robotic presorting process at the plant, where only colored common plastic bags are allowed through. Unwanted waste of all sizes is removed and sent off to recycling station, while the bags continue on to the next step for the sorting process.

20.4.2   STAGE 2—PREPARATION FOR OPTICAL DETECTION

After completion of the presorting stage, wastages are distributed into three conveyer belts furnished with robotic arms. These robotic arms put the bags into a line, one after the other. The bags are then sent on three conveyer belts for their optical sorting, where the speed increases for each belt. This process creates more distance between the bags and makes the bags more visible to the optical cameras, enabling them to detect the color of the bags.

20.4.3   STAGE 3—OPTICAL SORTING

When the bags reached for optical reading, color of the bags are detected by the cameras. There are four sensors in which two sensors detect green bags, one sensor detects blue bags, and the remaining one sensor detects both types of bags. Blue and green bags are removed with the help of robotic arms, while blue and green bags are separated out and other bags are continued further for recycling. The blue and green bags are transferred onto separate belts, where negative sorting is used for detection of color codes for the second time. This process discards bags and other elements which are neither blue nor green, but which may have been removed mistakenly in the earlier stage. The bags are then sent to separate containers. These optical sorters may have detection rate of 98%.

20.4.4   STAGE 4—BLOWER

At this stage the blue bags go through another quality control process. In order to get rid of the blue bags, which contain waste except pure plastic packaging, a blower or negative sorter has been installed on the blue belts. Light-weight bags are permissible to pass through, but too heavy bags (greater than 700 g) are sent for recycling. These bags contain waste, except plastic. The blue bags which are light in weight and received by the blower continue on to the compactor for material recycling.

20.5 HOUSEHOLD WASTE BIODEGRADATION PROCESS

Composting may be defined as the biological oxidative breakdown of organic ingredients in wastes under controlled conditions which allow growth of aerobic microorganisms which convert biodegradable organic matter into end product adequately stable for storage and application without harmful effects on the environmental (Adhikari et al., 2008). Humus, mineral ions, carbon dioxide, and water are the main products of aerobic composting (Inbar et al., 1993). Additionally, this process converts nitrogen from unstable ammonia to stable organic forms, obliterates pathogens, and reduces the waste volume (Haug, 1993; Inbar et al., 1993). Usually, the biodegradable materials are combined with marketable bacterial culture in 1:5 dilutions in alternating layering method in a closed chamber (for maintaining the incubation temperature) for biodecomposition (GEAG Report, 2010a,b). The time period might be for 35–40 days in completing the decomposition of matter. Dumitrescu et al. (2009) during their studies, mixed sewage sludge and sawdust in the composting mixture of kitchen-vegetable wastes and showed that the sawdust constituent (having an extra source of sugars and lignin source) and sewage sludge (having lipids and microorganisms forming enzymes) helped in improving biodegradation process. The common methods used to evaluate compost phytotoxicity are seed germination and plant growth bioassay (Kapanen and Itavaara, 2001). The root length was more affected by the concentration of the compost because of the fact that the roots are liable for absorption and accumulation of metals (Araujo and Monteiro 2005; Oncel et al., 2000).

20.5.1 ENZYME METHOD

Quality of this fertilizer is completely different from traditional compost because they content higher nutrient and organic-matter content. In this technology wastes are breakdown very quickly by the use of enzymes. However, the use of enzymes alone is not sufficient to make this process feasible. A perfectly controlled space is needed for decomposition of wastes. In a simple way, the digester and enzymes work simultaneously to produce fertilizer from waste.

20.5.2 BACKWARD METHOD

This method of composting is very economical but they take more time, space, and development of pathogen is also serious issue. If this type of compost is used for plants, pathogens may be moved into plants and create several problems to the plants; to avoid this problem, heating waste at 80°C with the help of Biomax's digestor is carried out, at this temperature most of the harmful pathogens are killed. Just because the process takes place under controlled environment, there is no conciliation on the quality of fertilizer due to external factors, such as weather conditions or mixing efficiency. Moreover, the otherwise occurrence of nutrient loss to the atmosphere is checked. As a result, the organic fertilizer can maintain not only high amount of nutrient level but also maintain organic matter level of more than 70%.

20.6 COMPOSTING PROCESS

Rotten fruits, vegetables, leaves, and grasses are common ingredients of compost which can be used for home-gardening purpose. They are mixed together with certain moisture level and turned regularly for air flow to the heap. It can take many weeks or months before they are considered suitable to be used as fertilizer or soil amendments. Composting is aerobic degradation of waste by microorganisms and fungi to produce compost that can be used as a soil improver and organic fertilizer. It is a natural process by which various organic wastes are converts into humus like materials by diverse microbial population under controlled conditions of temperature, moisture, and aeration. It is the feature of control that splits composting from natural decomposition or rotting processes which happens in sanitary landfill, open dump, and/or unmanaged waste pile (Roger et al., 1991). During composting, microorganisms alter organic materials, such as fruits, vegetables, leaves, sludge, and food wastes into beneficial product like soil humus (Reinikainen and Herranan, 1999; Rynk, 1992). During composting organic waste materials are rotten and converted into a product that may be used as organic fertilizer and/or soil conditioner (Ahmad, 2007).

Bacteria, fungi, and actinomycetes are prevalent in nature which acts as decomposers. These are indigenous to fruit and vegetable matters, soil, dust, and wastes of all kinds, so special organisms are not required (Rodrigues, 1996). Controlled decomposition of wastes occurs as a result of actions of these naturally occurring microorganisms. Composting also considered as microbial farming therefore, they require food, energy, and habitat. These microorganisms build protein from nitrogen and takes carbon as energy source. Bacteria break down into simpler forms by producing enzyme through their complex carbohydrates (Hamdy, 2005), and utilize them as food. Composting process goes on until the left over nutrients are consumed by the last microorganisms and the majority of the carbon is changed into carbon dioxide and water (Rynk, 1992). The nutrients are present during decomposition remain in the compost within the humus and dead bodies of microorganisms.

20.6.1 INOCULATION BY MICROBES

During the process of composting, easily degradable organic matter is utilized by microorganisms as a source of carbon and nitrogen. The final product is compost which consists of converted, slowly degradable compounds, the cell walls of dead microorganisms and intermediate breakdown products (Wei et al., 2007). A number of biological, microbiological, and physico-chemical techniques have been developed to describe the agrochemical properties and the maturity of compost. The nutrient status of wheat straw and sorghum stalk compost was enhanced after they inoculated with *Aspergillus niger* and *Penicillium* spp. (Gaur and Sadasivam, 1993). Singh and Sharma (2003) inoculated different types of wastes (MSW, mixed solid waste, and horticultural waste) with various microflora, namely *Pleurotus sajor, Trichoderma harzianum,* and *Azotobacter chroococcum* in different combinations. The waste was decomposed for different time periods and then subjected to successive composting for a fixed time period of one month. Ahmad (2007) prepared an organic fertilizer by the composting of FVW and augmented with nitrogen and used it as a carrier for plant growth promoting rhizobacteria strain, *Pseudomonas fluorescens* biotype G (N3) containing ACC-deaminase to prepare a biofertilizer. This bioorganic fertilizer applied @ 300 kg/ha was found to enhance the growth and nutrient uptake of maize. Studies by Diby et al. (2005) revealed the ability of *P. fluorescens* strains to improve nutrient mobilization in the rhizosphere of black pepper, which results in improved plant vigor. *Trichoderma viridae*, producing cellulase enzyme and having copper bioaccumulating ability (Anand et al., 2006), is also considered as continuously efficient inoculants for composting (Bhardwaj and Gaur, 1985).

20.6.2 BIOLOGICAL BENEFICIAL ORGANISMS

It includes the adding of earthworms and/or beneficial microbial inoculants. They feed on organic wastes, may have ability to consume more than their body weight, and only 5–10% of the feedstock are used for their growth and they ooze the mucus coated undigested matter as worm casts enriched in essential nutrients (Bhawalker and Bhawalker, 1993). These nutrients are also a rich source of enzymes, vitamins, antibiotics, growth hormones, and immobilized microflora (Bhawalker, 1991; Kale et al., 1982). Earthworms are often found in composting heaps, except on those materials which are disgusting to their growth. They work as pulverizers of organic materials so as to improve activity of microbes and provide aeration during composting. They have established significant attention as inputs for composting and the practice is known as vermicomposting (Edwards et al., 1985).

Many studies have shown the positive use of earthworms in the processing of organic wastes into compost (Kale and Bano, 1994; Tomati et al., 1983). It is a novel method for the discarding of nontoxic solid and liquid organic wastes. It helps in efficient and less costly recycling of agricultural residues, industrial wastes, and animal wastes using less energy (Jambhekar, 1992). Generally, vermicompost does not differ from wormless compost considerably in nutrient contents, but it differs only in its early processing and physical properties (Kale and Bano, 1994).

20.7 IMPROVEMENT OF HOUSEHOLD COMPOSTING RATE AND QUALITY

Composting of house hold wastes are usually carried out by aerobic methods in the attention of rapidity and quality of the product. Therefore, a number of mineral and biological additives in composting are needed for the rapid degradation of the wastes. Quality composts from the household wastes and other materials can also be obtained in a small duration of time by the use of microbial inocula as starter (Baheri and Meysami, 2002; Shin et al., 1999). Lei and Gheynst (2000) reported that mineral additive and other material inoculations can improve the microbial population, create valuable microbial communities, enhance microbiological quality and produce various preferred enzymes. These enhance the alteration of organics and reduce odorous gas emissions. Hamdy (2005) established the possibility of *Rhizopus orizae* microbe for the rapid degradation of orange peels with high cellulolytic and pectinolytic activities. Thus organic and mineral additives are explores the possible areas of research on composting process under different circumstances with admiration to climate, time, and space for the better organic fertilizer product from house hold wastes (Ahmad et al., 2007a). Mixing the compost with nutrient elements and preparing it quickly is necessary for intensive crop production, as well as for disposal of huge amount of accumulated organic waste material. For this purpose some chemical and biological additives have been found advantageous by different workers (Ahmad et al., 2007a). The quality of compost can be increased by the methods like mineral additives, plant hormones, inoculation by microbes, and biological beneficial organisms. These are briefly discussed below.

20.7.1 BY MINERAL ADDITIVES

Typically the compost additives are combinations of different concentrations of mineral nutrients and readily available forms of enzymes, carbon, and pH-balancing compounds that are used to increase microbial activity when the additive is in contact with the waste material (Himanen and Hänninen, 2009). Adding nitrogen and phosphorus mineral forms is eventually recommended to lower the C:N and C:P ratios of composting material, where required, and to enhance the microbial activity (Ahmad et al., 2007a). However, these inputs may significantly increase the nutrient content in the compost. Wong and Fang (2000) observed that the addition of lime to sewage sludge, prolonged the thermophilic stage. Koivula et al. (2004) reported 20°C higher temperatures in biowaste compost with added bottom ash. Gowda et al. (1992) also reported improved N and P content in the compost through amendment with rock phosphate. Considerable amount of mineral nitrogen is lost by volatilization during composting (Gaur and Singh, 1995). To beat this problem, pyrite minerals have been tested as nitrogen conserving additives. Banger et al. (1989) made nitrogen and phosphorus enhanced compost containing higher level of total nitrogen than usual and significantly high content of total phosphorus in the compost. Addition of NH_4NO_3 and NH_4Cl to the fermentation medium was found to increase the infusing potentiality due to boosting enzymatic levels (Hamdy, 2005).

20.7.2 PLANT HORMONES

Plant hormones are a number of chemicals that profoundly affect the growth and differentiation of plant cells, tissues, and organs. It is probably that enhancement of composted material with nutrients and/or plant hormones (kinetin, gibberellic acid, and indole acetic acid) can transform organic waste material into value-added organic fertilizers, for increasing crop yield. An organic fertilizer prepared by composting FVW and augmented with nitrogen and plant hormones was found effective at the minimum concentration of 300 kg/ha for enhancing growth and yield of wheat and maize (Ahmad et al., 2007b; Zahir et al., 2007).

20.8   CONCLUSION

Composting is one of the best options for treatment of solid waste. During the process of composting, organic matter breaks down by bacterial action resulting in the formation of humus like material called compost. The value of compost as manure depends on the quality and quantity of feed materials provided into the compost pit. Recycling organic materials in the home landscape makes sense, not only it helps to save our valuable landfill space but also it will actually improve our soils and growing conditions in our home environment. The utilization of the organic matter being continuously produced by most households everyday in the form of leaves, grass clippings, kitchen waste, and the like, without having to pay to have it carted away to land fill, is going to have a positive impact on the health of our environment as a whole.

KEYWORDS

- **biodegradable**
- **composting**
- **fertilizers**
- **microbial degradation**
- **organic waste**

REFERENCES

Adhikari, B. K.; Barrington, S.; Martinez, J.; King, S. Characterization of Food Waste and Bulking Agents for Composting. *Waste Manag.* **2008,** *28* (5), 795–804.

Ahmad, R. Use of Recycled Organic Waste for Sustainable Maize (*Zea mays* L.) Production. Ph.D. Thesis, Institute of Soil & Environmental Sciences, University of Agriculture, Faisalabad, Pakistan 2007.

Ahmad, R.; Jilani, G.; Arshad, M.; Zahir, Z. A.; Khalid, A. Bio-conversion of organic wastes for their recycling in agriculture: an overview of perspectives and prospects. *Ann. Microbiol.* **2007a,** *57* (4), 471–479.

Ahmad, R.; Shehzad, S. M.; Khalid, A.; Arshad, M.; Mahmood, M. H. Growth and Yield Response of Wheat (*Triticum aestivum* L.) and Maize (*Zea mays* L.) to Nitrogen and L-tryptophan Enriched Compost. *Pak. J. Bot.* **2007b,** *39* (2), 541–549.

Ahring, B. K.; Mladenovska, Z.; Iranpour, R.; Westermann, P. State of the Art and Future Perspectives of Thermophilic Anaerobic Digestion. *Water Sc. Technol.* **2002,** *45,* 298–308.

Albasha, M. O.; Gupta, P.; Ramteke, P. W. Management of Kitchen Waste by Vermicomposting Using Earthworm, *Eudrilus eugeniae*. International Conference on Advances in Agricultural, Biological & Environmental Sciences (AABES-2015), 2015, pp J22–J23.

Anand, P.; Isar, J.; Saran, S.; Saxena, R. K. Bioaccumulation of Copper by *Trichoderma viride. Biores. Technol.* **2006,** *97,* 1018–1025.

Araujo, A. S. F.; Monteiro, R. T. R. Plant Bioassays to Asses Toxicity of Textile Sludge Compost, Scientia Agricola (Piracicaba, Brazil) **2005,** *62* (3), pp 286–290.

Baheri H.; Meysami P. Feasibility of Fungi Bioaugmentation in Composting a Flare Pit Soil. *J. Hazard. Mat.* **2002,** *89* (2–3), 279–286.

Banger, K. C.; Shanker; S.; Kapoor, K. K.; Kukreja, K.; Mishra, M. M. Preparation of Nitrogen and Phosphorus-Enriched Paddy Straw Compost and Its Effect on Yield and Nutrient Uptake by Wheat (*Triticum aestivum* L.). *Biol. Fert. Soils* **1989,** *8,* 339–342.

Bhardwaj, K. K. R; Gaur, A. C. *Recycling of Organic Wastes*; ICAR: New Delhi, 1985.

Bhawalkar, U. S. Vermiculture Biotechnology for LEISA. Seminar on Low External Input Sustainable Agriculture: Amsterdam, Netherlands, 1991.

Bhawalkar, U. S.; Bhawalkar, U. V. Vermicultre Biotechnology. In *Organic in Soil Health and Crop Production,* Thampan P. K., Ed., Peekay Tree Crops Development Foundation: Cochin, 1993, pp. 69–85.

Bouallagui, H.; Ben Cheikh, R.; Marouani, L.; Hamdi, M. Mesophilic Biogas Production from Fruit and Vegetable Waste in Tubular Digester. *Biores. Technol.* **2003,** *86,* 85–9.

Breidenbach, A. W. Composting of Municipal Solid Wastes in the United States. Pub.SW-47r, U.S. Environmental Protection Agency, 1971.

Converti, A.; DelBorghi, A.; Zilli, M.; Arni, S.; DelBorghi, M. Anaerobic Digestion of the Vegetable Fraction of Municipal Refuses: Mesophilic Versus Thermophilic Conditions. *Bioproc. Eng.* **1999,** *21,* 371–376.

Diby, P.; Sarma, Y. R.; Srinivasan, V; Anandaraj, M. *Pseudomonas fluorescens* Mediated Vigour in Black Pepper (*Piper nigrum* L.) Under Green House Cultivation. *Ann. Microbiol.* **2005,** *55* (3), 171–174.

Dumitrescu, L.; Manciulea, I.; Sauciuc, A.; Zaha, C. Obtaining Biofertilizer by Composting Vegetable Waste, Sewage Sludge and Sawdust. Bulletin of the Transilvania, University of Braşov, **2009,** *2* (51) Series I.

Edwards, C. A.; Burrows, I.; Fletcher, K. E.; Jones, B. A. The Use of Earthworms for Composting Farm Wastes. In *Composting of Agricultural and Other Wastes*, Crasser, J. K. R., Ed., Elsevier Applied Science: Oxford, 1985; pp 229–241.

Epstein, E.; Willson, G. B.; Burg, W. D.; Mullen, D. C.; Enkiri, N. K. A Forced Aeration System for Composting Wastewater Sludge. *J. Water Pollut. Contr. Fed.* **1976,** *48,* 688–691.

Evans, T. D.; Boor, M.; MacBrayne, D. Biofertiliser Plant Design—Food Waste to Biofertiliser and Biogas. 12th European Biosolids and Organic Resources Conference, Nov. 2007, Aqua Enviro: Manchester, UK, 2007.

Furhacker, M.; Haberl, R. Composting of Sewage Sludge in a Rotating Vessel. *Water Sc. Tech.* **1995,** *32,* pp. 121–125.

Gaur, A. C.; Sadasivam, K. V. Theory and Practical Considerations of Composting Organic Wastes, In *Organics in Soil Health and Crop Production*; Thampan P. K., Ed., Peekay Tree Crops Development Foundation: Cochin, 1993; pp 1–22.

Gaur, A. C.; Singh, G. Recycling of Rural and Urban Wastes Through Conventional and Vermicomposting, In *Recycling of Crop, Animal, Human and Industrial Wastes in Agriculture* Tandon H. L. S., Ed., FDCO: New Delhi, India, 1995; pp 31–35.

Gaur, A. C.; Singh, G. Role of Integrated Plant Nutrient Systems in Sustainable and Environmentally Sound Agricultural Development. RAPA Publication: Bangkok, 1993; pp 110–130.

Gowda, T. K.; Radakrishna, S. D.; Balakrishna, A. N.; Sreenivas, K. N. Studies on the Manurial Value and Nutrient Enrichment of Municipal Waste Compost Produced from Banglore City Garbage. In: *Proceedings of National Seminar on Organic Farming*, MPKV: Pune, India, 1992; pp 39–41.

Gustavsson, J.; Cederberg, C.; Sonesson, U.; Otterdijk, R. V.; Meybeck, A. Global Food Losses and Food Waste, *Extent, Causes and Prevention*, Food and Agriculture Organization of the United Nations: Rome, Italy, 2011, p 38.

Hamdy, H. S. Purification and Characterization of the Pectin Lyase Produced by *Rhizopus oryzae* Grown on Orange Peels. *Ann. Microbiol.* **2005**, *55* (3), 205–211.

Haug, R. T. *The Practical Handbook of Compost Engineering*; CRC Press: Florida, USA, 1993.

Himanen, M.; Hänninen, K. Effect of Commercial Mineral-Based Additives on Composting and Compost Quality. *Waste Manag.* **2009**, *29*, 2265–2273.

Howard, A. An Agricultural Testament. Oxford University Press: London, 1943.

Ilaboya, I. R.; Asekhame, F. F.; Ezugwu, M. O.; Erameh, A. A.; Omofuma, F. E. Studies on Biogas Generation From Agricultural Waste; Analysis of the Effects of Alkaline on Gas Generation. *World Appl. Sci. J.* **2010**, *9* (5), 537–545.

Indumathi, D. Microbial Conversion of Vegetable Wastes for Bio fertilizer Production. *IOSR J. Biotech. Biochem.* **2017**, *3* (2), 43–47.

Ito, A.; Umita, T.; Aizawa, J.; Kitada, K. Effect of Inoculation of Iron Oxidizing Bacteria on Elution of Copper from Anaerobically Digested Sewage Sludge. *Water Sci. Tech.* **1998**, *38*, 63–70.

Jambhekar, H. K. Use of Earthworms as a Potential Source to Decompose Organic Waste. *Proceedings of National Seminar on Organic Farming*, MPKV: Pune, India, 1992; pp 52–53.

Kale, R. D.; Bano, K. Laboratory Studies on Age Specific Survival and Fecundity of Earthworm, *Eurilus eugeniae*. *Mitt. Hamb. Zool. Mus. Inst.* **1994**, *89*, 139–148.

Kale, R. D.; Bano, K.; Krishnamoorthy, R. V. Potential of *Perionyx excavatus* for Utilization of Organic Wastes. *Pedobiologia* **1982**, *23*, 419–425.

Kalpana, P.; Sai Bramari, G.; Anitha, L. Formulation of Potential Vegetable Waste Compost in Association with Microorganisms and *Spirulina platensis*. *Asian J. Plant Sci. Res.* **2011**, *1* (3), 49–57.

Kapanen, A.; Itavaara, M. Ecotoxicity Tests for Compost Applications, *Ecotox. Environ. Saf.* **2001**, *49*, 1–16.

Koivula, N.; Räikkönen, T.; Urpilainen, S.; Ranta, J.; Hänninen, K. Ash in Composting of Source-Separated Catering Waste. *Biores. Technol.* **2004**, *93*, 291–299.

Lei F.; Gheynst J. S. V. The Effect of Microbial Inoculation and pH on Microbial Community Structure Changes During Composting. *Proc. Biochem.* **2000**, *35*, 923–929.

Misi, S. N.; Forster, C. F. Semi-Continuous Anaerobic Co-Digestion of Agrowaste. *Environ. Technol.* **2002**, *23*, 445–451.

Ngumah, C; Ogbulie, J; Orji, J; Amadi, E. Potential of Organic Waste for Biogas and Biofertilizer Production in Nigeria. *Environ. Res. Engg. Manag.* **2013**, *63*, 60–66.

Novinsak, A.; Surette, C.; Allain, C.; Filion, M. Application of Molecular Technologies to Monitor the Microbial Content of Biosolids and Composted Biosolids. *Water Sci. Technol.* **2008**, *57*, 471–477.

Oncel, I.; Keles, Y.; Ustun, A. S. Interactive Effects of Temperature and Heavy Metal Stress on the Growth and Some Biochemical Compounds in Wheat Seedlings. *Environ. Polln.* **2000**, *107*, 315–320.

Park, J. I.; Yun, Y. S.; Park, J. M. Long-Term Operation of Slurry Bioreactor for Decomposition of Food Wastes. *Biores. Technol.* **2002**, *84*, 101–104.

Pleissner, D.; Chi, W.; Sun, Z.; Sze, C.; Lin, K. Biore source Technology Food Waste as Nutrient Source in Heterotrophic Microalgae Cultivation. *Biores. Tech.* **2013**, *137*, 139–146.

Pleissner, D.; Lin, C. S. Valorisation of Food Waste in Biotechnological Processes. *Sustain. Chem. Proc.* **2013**, *1* (1), 21.

Reinikainen, O.; Herranen, M. Different Methods for Measuring Compost Stability and Maturity. *Soil Sci. Soc. Amer. J.* **1999**, *55*, 1020–1025.

Rodrigues, M. S. Composted Societal Organic Wastes for Sustainable Wheat (*Triticum aestivum*) Production. Ph. D. Thesis, Wye College, University of London: London, 1996.

Roger, S. W.; Jokela, E. J.; Smith, W. H. Recycling Composted Organic Wastes on Florida Forest Lands. Dept. of Forest Resources and Conservation, Florida Cooperative Extension Services, University of Florida: Florida, USA, 1991.

Rynk, R. *On-farm Composting Handbook*; Northeast Regional Agricultural Engineering Service, Coop. Ext., NRAES-54: Ithaca, USA, 1992.

Sanmanee, N.; Panishkan, K.; Obsuwan, K.; Dharmvanij, S. Study of Compost Maturity During Humification Process Using UV Spectroscopy. *World Acad. Sci. Eng. Technol.* **2011**, *80*, 403–405.

Shin H. S.; Hwang E. J.; Park B. S.; Sakai T. The Effects of Seed Inoculation on the Rate of Garbage Composting. *Environ. Technol.* **1999**, *20*, 293–300.

Singh, A.; Sharma, S. Effect of Microbial Inocula on Mixed Solid Waste Composting, Vermicomposting and Plant Response. *Compost Sci. Util.* **2003**, *11*, 190–199.

Singh, C. P.; Amberger, A. Humic Substances in Straw Compost with Rock Phosphate. *Biol. Wastes* **1990**, *3*, 1165–1174.

Steger, K.; Sjogren, A. M.; Jarvis, A.; Jansson, J. K.; Sundh, I. Development of Compost Maturity and Actinobacteria Populations During Full-Scale Composting of Organic Household Waste. *J. Appl. Microbiol.* **2007**, *103*, 487–498.

Stenger, A. Experimental Valuation of Food Safety: Application to Sewage Sludge. *Food Policy* **2000**, *25*, 211–218.

Tchobanoglous, G.; Kreith, F. *Solid Waste Handbook*, 2nd ed., McGrawHill: New York, 2002.

Tomati, U.; Grappelli, A.; Galli, E. Fertility Factors in Earthworm Humus. *Proceedings of International Symposium on Agriculture and Environment: Prospects in Earthworm Farming*, Ministero della Ricerca Scientifica e Technologica: Rome, 1983; pp 49–56.

Tsai, S. H.; Liu, C. P.; Yang, S. S. Microbial Conversion of Food Wastes for Biofertilizer Production with Thermophilic Lipolytic Microbes. *Renew. Energy* **2007**, *32*, 904–915.

Umsakul, K.; Dissara, Y.; Srimuang, N. Chemical physical and Microbiological Changes During Composting of the Water Hyacinth. *Pak. J. Biol. Sci.* **2010**, *13*, 985–992.

Upadhyay, V. P.; Prasad, M. R.; Srivastav, A.; Singh, K. Eco Tools for Urban Waste Management in India. *J. Human Ecol.* **2005**, *18*, 253–269.

Verma, V. K.; Singh, Y. P.; Rai, J. P. N. Biogas Production From Plant Biomass Used for Phytore-mediation of Industrial Wastes. *Appl. Microbiol. Biotechnol.* **2007,** *98* (8), 1664–1669.

Vesilind, P. A.; Spinosa, L. Sludge Production and Characterisation, Production and Regula-tions, In *Sludge into Biosolids: Processing, Disposal, Utilisation* Spinosa, L.; Vesilind, A. Eds., IWA Publishing: London, UK, 2001; pp 3–8.

Wei, Z.; Xi, B.; Zhao, Y.; Wang, S.; Liu, H.; Jiang, Y. Effect of Inoculating Microbes in Municipal Solid Waste Composting on Characteristics of Humic Acid. *Chemosphere* **2007,** *68,* 368–374.

Wong, J. W. C.; Fang, M. Effects of Lime Addition on Sewage Sludge Composting Process. *Water Res.* **2000,** *34* (15), 3691–3698.

Wu, L.; Ma, L. Q. Relationship Between Compost Stability and Extractable Organic Carbon. *J. Environ. Qual.* **2002,** *31,* 1323–1328.

Zahir, Z. A.; Iqbal, M.; Arshad, M.; Naveed, M.; Khalid, M. Effectiveness of IAA, GA_3 and Kinetin Blended with Recycled Organic Waste for Improving Growth and Yield of Wheat (*Triticum aestivum* L.). *Pak. J. Bot.* **2007,** *39* (3), 761–768.

CHAPTER 21

Genetically Modified Microorganisms for Sustainable Soil Health Management: An Biotechnological Approach

MD. SHAMIM[1,*], MAHESH KUMAR[1], AWADHESH KUMAR PAL[2], RANJEET RANJAN KUMAR[3], and V. B. JHA[4]

[1]Department of Molecular Biology and Genetic Engineering, Dr. Kalam Agricultural College, Kishanganj, Bihar Agricultural University, Bhagalpur 813210, Bihar, India

[2]Department of Biochemistry and Crop Physiology, Bihar Agricultural University, Sabour, Bhagalpur 813210, Bihar, India

[3]Division of Biochemistry, ICAR—Indian Agricultural Research Institute, Pusa 110012, New Delhi, India

[4]Department of Plant Breeding and Genetics, Dr. Kalam Agricultural College, Kishanganj, Bihar Agricultural University, Sabour, Bhagalpur 813210, Bihar, India

*Corresponding author. E-mail: shamimnduat@gmail.com

ABSTRACT

There are several factors that cause loss of plant nutrient (macro and micro) from soil surface horizon. Plant nutrients from the soil are usually lost in different ways, among them big amounts are removed from the soil due to the harvest of crops, and weeds also remove a considerable quantity of plant nutrients from the soil. Leaching and erosion are also among the important causes that lead to the removal of nutrients from the soil. In the current global atmospheric condition, environmental pollution is increasing by agricultural chemicals, industrial wastes, and construction wastes etc. that results in the deterioration of the soil health. So, in order to cope up with the current emerging problem, it is important to exploit the microbial metabolic potential in the construction of several genetically engineered microorganisms (GEMs). This chapter deals with the various examples for the designing of safer GEMs for environmental release with specific emphasis on the use of plant growth-promoting rhizobacteria production, biological control of plant diseases, and biodegradation for agricultural production enhancement.

21.1 INTRODUCTION

To achieve healthy and productive plants, soil quality is of immense importance. Soil quality has been defined as the "capacity of a soil to function within ecosystem boundaries to sustain plant–animal productivity, maintain or enhance water and air quality, and support human health habitation" (Karlen et al., 1997; Chaparroeta et al., 2012). Improper use of chemical pesticides contributed in loss of soil health and productivity along with addition of salts to the soil (CCRC, 1991).To revive the soil health and micro fauna living on it has become an essential concept of biofertilizer that can be a good supplement for chemical fertilizers. The term biofertilizer includes selective microorganism like bacteria, fungi, and algae. Biofertilizers are organisms that enrich the nutrient quality of soil. The main sources of biofertilizers are bacteria, fungi, and cyanobacteria or blue-green algae. Biofertilizers are nutrient availability systems in which biological processes are involved, which are capable of fixing atmospheric nitrogen or convert soluble phosphate and potash in the soil, into the forms available to the plants. Biofertilizer is a cost-effective, eco-friendly, and renewable source of land nutrient which play a vital role in maintaining a long-term soil fertility and sustainability. They maintain the soil fertility cost by assuring yield and continuous use of biofertilizer makes the soil very fertile. The biofertilizer can be manufactured in solid form as well as in liquid form for spraying on the plants as Fertigation. Biopesticide and biofertilizer are need of modern agriculture since demand for safe and residue free food is increasing (Chang, 1987). The main and direct purposes of applying biofertilizers to soil are: to provide nutrient sources and good soil conditions for the growth of crops when used as a live body, to partially substitute and enhance the function of chemical fertilizer, and the capital used for making biofertilizers is cheaper than that of chemical fertilizers, and to lessen the negative effect aroused from applying chemical fertilizers to soil. The application of plant growth-promoting rhizobacteria (PGPR) inoculants in agriculture can be traced back to the beginning of the past century, when a *Rhizobium* based product named "Nitragin" was patented (Nobbe and Hiltner, 1896).

In view of overcoming this bottleneck, it will be necessary to undertake short-term, medium, and long-term research, in which soil microbiologists, agronomists, plant breeders, plant pathologists, and even nutritionists and economists must work together. According to the information from specific sources, the demand for agricultural production is supposed to increase by at least 70% by 2050. Actually, soil microorganisms play fundamental roles (microbial services) in agriculture mainly by improving plant nutrition and health as well as soil quality (Barea et al., 2013a; Lugtenberg, 2015). Accordingly, several strategies for a more effective exploitation of beneficial microbial services, as a low input biotechnology, to help sustain environmentally friendly agro-technological practices have been, and are being, proposed. The final goal is to optimize the role of the root-associated microbiome in nutrient supply and plant protection (Raaijmakers and Lugtenberg, 2013). Since the interactions between microbial communities and crops are influenced by diverse ecological factors and agronomic managements, the impact of environmental stress factors must be considered, particularly in the current scenario of global change, as they affect a proper management of the crop-microbiome interactions (Zolla et al., 2013).

Agriculture plays a major role in meeting the food demands of a continuous growing human population, which also led to an increasing dependence on chemical fertilizers and pesticides (Santos et al., 2012). In agriculture, promote alternate means of soil fertilization relies on organic inputs to improve nutrient supply and conserve the field management (Araujo et al., 2008). Chemical fertilizers are industrially manipulated, substances composed of known quantities of nitrogen, phosphorus, and potassium, and their exploitation causes air and ground water pollution by eutrophication of water bodies (Youssef et al., 2008). Organic farming is one of such strategies that not only ensures food safety but also adds to the biodiversity of soil (Megali et al., 2014). The additional advantages of biofertilizers embrace longer shelf life causing no undesirable effects to ecosystem (Sahoo et al., 2014). Organic farming is mostly dependent on the natural micro flora of the soil which constitutes all kinds of useful bacteria and fungi including the arbuscular mycorrhiza fungi (AMF) called PGPR. Biofertilizers keep the soil environment rich in all kinds of micro- and macro-nutrients via N_2-fixation, solubilization or mineralization of phosphate and potassium, release of plant growth regulating substances, production of antibiotics, and biodegradation of organic matter in the soil (Sinha et al., 2014). When biofertilizers are applied as seed or soil inoculants, they multiply and participate in nutrient cycling and benefit crop productivity

(Singh et al., 2011; Compant et al., 2005). In general, 60–90% of the total applied fertilizer is lost and the remaining 10–40% is taken up by plants. In this regard, microbial inoculants have vital significance in integrated nutrient management systems to maintain agricultural productivity and healthy environment (Adesemoye and Kloepper, 2009).

With the development of recombinant deoxyribonucleic acid (DNA) technology, the metabolic potentials of microorganisms are being explored and harnessed in a variety of new ways. Today, genetically modified microorganisms (GMMs) have found application in agriculture, human health, bioremediation, and in industries such as food, paper, and textiles. Genetic engineering offers the advantages over traditional methods of increasing molecular diversity and improving chemical selectivity. In addition, genetic engineering offers ample supplies of preferred products, cheaper production of product, and safe handling of otherwise dangerous agents. This chapter describes several molecular tools and strategies to engineer microorganisms; the advantages and limitations of the methods are addressed. Environmental releases of GMMs were proposed for the first time by Monsanto and Advanced Genetic Sciences, Inc. (AGS; Watrud et al., 1985; Lindow, 1985). AGS developed a product named Frost Ban®, a *Pseudomonas* syringae engineered such that a gene coding for a protein necessary for ice-nucleation had been deleted, and conducted a field release on strawberry fields in the University of California experimental plots under EPA and California Department of Food and Agriculture authority on April 24, 1987 (Smith, 1997). The danger posed by these GMMs is therefore related both to their dispersal into the environment and to their potential for adaptation to a new environment. In doing so, their development may, by altering the animal and plant microbial ecological balance, interrupt the environment to a greater or lesser extent.

Unfortunately, the use of GMMs on a wide scale is still hampered by a lack of knowledge of the possible ecological impacts that such organisms might have once they are released into the environment. One of the concerns is that GMMs may affect non-target microorganisms (Levin et al., 1987). The final part of this chapter reviews and evaluates several applications of GMMs currently employed in commercial ventures.

21.2 MOLECULAR TOOLS AND METHODS FOR GENETIC ENGINEERING OF MICROORGANISMS

Various different strategies have been applied to improve the competitiveness of a bioinoculant in the plant environment. These are either by promoting rhizobial multiplication in the plant environment, by inhibiting the growth of competing microorganisms, or by interfering with some of the signals perceived by the competing microbes provided these signals control (at least in part) the expression of functions central to microbial fitness (Savka et al., 2002). Because this is a triple interface (bacteria, plant, and soil) interaction, it is possible to modify one, two, or three of these factors to improve microbial colonization. An improvement of plant-microbe symbioses should involve the coordinated modifications in the partner's genotypes resulting in highly complementary combinations (Tikhonovich and Provorov, 2007).

21.2.1 GENE TRANSFER METHODS USED IN MICROBES

21.2.1.1 TRANSFORMATION

There are several methods used for the gene transfer into important microorganisms. The most frequently used method is transformation. In this process, uptake of plasmid DNA by recipient microorganisms is accomplished when they are in a physiological stage of competence, which usually occurs at a specific growth stage (Lorenz and Wackernagel, 1994). However, DNA uptake based on naturally occurring competence is usually inefficient. Competence can be induced by treating bacterial cells with chemicals to facilitate DNA uptake. For *Escherichia coli*, an organism used commonly as a cloning host and a "bioreactor" for the commercial production of numerous therapeutic proteins, the uptake of plasmid DNA is achieved when cells are first treated with calcium chloride or rubidium chloride (Sambrook et al., 1989). For many microorganisms, such as the antibiotic producing *Streptomyces*, transformation of plasmid DNA is a more difficult process. For these organisms, transformation involves preparation of protoplasts using lysozyme to remove most of the cell wall. Protoplasts are mixed with plasmid DNA in the presence of polyethylene glycol to increase the uptake of DNA. Growth medium, growth phase, ionic composition of transformation buffers, and polyethylene glycol molecular weight, concentration, and treatment time are variables that must be studied to identify the optimum conditions for protoplast formation and regeneration.

21.2.1.2 ELECTROPORATION

Electroporation is another method to transform DNA into microorganisms. This method, originally used to transform eukaryotic cells, relies on brief high-voltage pulses to make recipient cells electrocompetent (Neumann). Transient pores are formed in the cell membrane as a result of an electro-shock, thereby allowing DNA uptake. Growth phase, growth medium, cell density, and electroporation parameters must be optimized to achieve desir-able efficiency. The main advantage of this method is that it bypasses the need to develop provisions for protoplast formation and regeneration of cell wall. Electroporation is frequently used when the efficiency of protoplast trans-formation is insufficient or ineffective. Several reports have documented the application of this method to industrially important *Streptomyces* (Pigac and Schrempf, 1995; Tyurin et al., 1995; Mazy-Servais et al., 1997), *Corynebac-terium* (Wolf et al., 1989), and *Bacillus* (Brigidi et al., 1990; McDonald et al., 1995). Electroporation is also the primary method of choice for transferring DNA into lactic acid bacteria (von Wright and Sibakov, 1993; Kullen and Klaenhammer, 1999; Lindgren, 1999). In addition to using purified DNA for electroporation procedures, methods have been developed to transfer DNA directly from DNA-harboring cells into a recipient without DNA isolation (Kilbane and Bielaga, 1991).

21.2.1.3   CONJUGATION

Conjugation is another method used to introduce plasmid DNA into micro-organisms. This method involves a donor strain that contains both the gene of interest and the origin of transfer (oriT) on a plasmid and the genes encoding transfer functions on the chromosome (Grohmann et al., 2003). Upon short contact between donor and recipient, DNA transfer occurs. After conjugation takes place, donor cells are eradicated with an antibi-otic to which the recipient cells are resistant. Recipient cells containing the transferred plasmid are identified based on the selectable marker gene carried by the plasmid. One advantage of this method is that it does not rely on the development of procedures for protoplast formation and regen-eration of cell wall. In addition, this method suggests the possibility of bypassing restriction barriers by transferring single-stranded plasmid DNA (Matsushima et al., 1994). Introducing DNA by conjugation from donor *E. coli* has proven useful with *Streptomyces* and *Corynebacterium* (Compant et al., 2005; Schäfer et al., 1990; Mazodier et al., 1989; Bierman et al., 1992; Matsushima et al., 1994; Sun et al., 2002; Paranthaman and Dharmalingam, 2003).

21.2.1.4 VECTORS

Selection of a cloning vector to perform genetic modifications depends on the selection of the gene transfer method, the desired outcome of the modification, and the application of the modified microorganism. Several classes of vectors exist, and the choice of which to use must be made carefully. Replicating vectors of high or low copy numbers are commonly used to express the desired genes in heterologous hosts for manufacturing expressed proteins. Replicating vectors are also used to increase the dosage of the rate-limiting gene of a biosynthetic pathway, such as that used for an amino acid, to improve the production of the metabolite. Cosmid and bacterial artificial chromosome vectors, which accept DNA fragments as large as 100 kb, are necessary when cloning a large piece of DNA into a heterologous host for manipulation and high-level metabolite production (Rao et al., 1987; Sosio et al., 2000). Conjugal vectors facilitate gene transfer from an easily manipulated organism such as *E. coli* into a desired organism that is usually more difficult to transform. Gene replacement vectors allow stable integration of the gene of interest. Food-grade vectors differ from the conventional cloning vectors in that they do not carry antibiotic resistance marker genes. Special deliberation must be given when constructing GMMs for industrial applications. If a GMM is to be released into the environment as a biological control agent, conjugal vectors should be avoided to prevent the horizontal transfer of the vectors and the genes into indigenous microorganisms. If a GMM is used as a starter culture for food fermentation, conjugal vectors should also be avoided (Verrips et al., 1996), and food-grade vectors should be developed and used for genetic manipulation (Lindgren, 1999; Martin et al., 2000).

21.3 STRATEGIES FOR GENETIC ENGINEERING OF MICROORGANISMS

Various strategies have been developed to create GMMs for desired traits. They include (1) disruption or complete removal of the target gene or pathway; (2) over-expression of the target gene in its native host or in a heterologous host; and (3) alteration of gene sequence, and thereby the amino acid sequence of the corresponding protein.

21.3.1 DISRUPTION OF UNDESIRABLE GENE FUNCTIONS

Disruption of a gene function can be accomplished by cloning a DNA fragment internal to the target gene into a suitable vector. Upon introducing the recombinant plasmid into the host organism, the internal fragment of the gene, along with the vector, is integrated with the host chromosome via single-crossover recombination. The integration results in the formation of two incomplete copies of the same gene separated by the inserted vector sequence, thereby disrupting the function of the target gene. However, such integration is unstable because of the presence of identical DNA sequences on either side of the vector. The recombinant strain often undergoes a second recombination that will "loop" out the recombinant plasmid from the chromosome, thus restoring normal function of the target gene. To create a stable recombinant strain blocked in the unwanted gene function, a gene replacement plasmid carrying two selectable marker genes is required. The first selectable marker gene, originating from the cloning vector, is used to select the transformed cells, whereas the second selectable marker gene is inserted into the target gene. The recombinant plasmid is introduced into the host organism, followed by the selection of transformed cells based on the first selectable marker gene. Upon double-crossover recombination, the second selectable marker gene, now inserted into the target gene on the host chromosome, disrupts the sequence of the target gene and destroys gene function. The recombinant strain is selected based on its resistance to the second selectable marker gene product and its sensitivity to the first selectable marker gene product. Another approach to disrupting gene functions relies on antisense technology. The technology is based on antisense ribonucleic acid (RNA) or DNA sequences that are complementary to the messenger RNAs (mRNAs) of the target genes (Brantl, 2002). The binding of an antisense molecule to its complementary mRNA results in the formation of a duplex RNA structure. The activity of the target gene is inhibited by the duplex RNA structure because of (1) an inaccessible ribosomal-binding site that prevents translation; (2) rapid degradation of mRNA; or (3) premature termination that prevents transcription (Desai and Papoutsakis, 1999). Antisense technology has been used to downregulate target gene activities in bacteria (Desai and Papoutsakis, 1999). The chief advantages of this approach are quick implementation and simultaneous downregulation of multiple target genes. In addition, this method is ideal for downregulation of primary metabolic gene activities without creation of lethal events.

21.3.2 OVER-EXPRESSION OF DESIRED GENES

High-level expression of a target gene may be achieved by employing a high copy number vector. Eggeling et al. (1998) constructed several *Corynebacterium glutamicum* recombinant strains containing increased copy numbers of *dap*A, a gene encoding dihydrodipicolinate synthase at the branch point of the lysine and methionine/threonine pathway. Lysine titer was higher in the recombinant strain containing one extra copy of *dap*A than the wild-type strain and was highest in the recombinant strain containing the highest copy number of the same gene. However, gene expression systems based on high copy number vectors have a number of drawbacks. One is the segregational instability of recombinant plasmids, which results in the loss of recombinant plasmids and therefore loss of the desired traits. For example, expression of the *Bacillus thuringiensis (Bt)* toxin gene from a high copy number vector in *Pseudomonas fluorescens* was undetectable because of plasmid instability (Downing et al., 2000). Segregational instability of plasmids is usually resolved by maintaining recombinant strains under selective pressures, usually by means of antibiotics. However, concerns about the use, release, and horizontal transfer of antibiotic resistance marker genes suggest that other means of maintaining plasmid stability need to be developed. Baneyx (1999) outlined a few options to achieve this goal. One method relies on creating a mutation in a critical chromosomal gene that is complemented with a functional copy of the same gene on the plasmid. As long as the plasmid housing the critical gene is present, the recombinant strain will survive. Major disadvantages of this method are: the need to create a mutation in an essential gene of the host organism, the need to develop a specific growth medium, and the need to introduce an additional plasmid encoded gene to complement the deficiency (Baneyx, 1999). Another concern about the use of high copy number vectors for high-level protein production in bacterial cells, especially in *E. coli*, is the formation of insoluble protein aggregates known as *inclusion bodies*. Inclusion bodies are biologically inactive because of protein misfolding, which is a consequence of rapid intracellular protein accumulation (Baneyx, 1999; Le and Trotta, 1991). Although methods exist to isolate and renature inclusion bodies (Lilie et al., 1998; De Bernardez et al., 1999; Altamirano et al., 1999), these systems are often inefficient and add steps in the purification of active proteins. Also, in the process of renaturation of proteins, a significant percentage of the proteins remain denatured and inactive (Swartz, 2001). In most cases, the goal of protein production is to achieve acceptable levels of accumulation of desired proteins that retain biological activity. To achieve this goal, the rate of recombinant protein synthesis needs to be optimized (Makrides, 1996). An effective method to accomplish this is by lowering the growth temperature or by altering medium composition (Baneyx, 1999). A molecular approach to maximize active protein production is to co-express the genes that facilitate protein folding and improve transportation of the recombinant protein out of the cell to decrease the intracellular concentration of the protein (Makrides, 1996; Bergès, 1996; Miksch et al., 1999). Another molecular approach is to carefully select a vector or promoter system that does not overwhelm the cell's capacity to produce active proteins (Minas and Bailey, 1995). However, both methods result in GMMs that harbor plasmids. An another approach, which ensures expression of target genes at desired levels and avoids plasmid segregational instability and production of inclusion bodies, is to integrate target genes into the host's chromosome (Olson et al., 1998). Although integration of a single copy of target genes may not be enough to achieve the desired level of protein production, integration of multiple copies of target genes has yielded very encouraging results (Kiel et al., 1995; Peredelchuk and Bennett, 1997; Martinez-Morales, 1999).

21.4 APPLICATIONS OF GENETICALLY MODIFIED MICROORGANISMS

21.4.1 SOIL HEALTH IMPROVEMENTS

Genetic modifications to enhance soil fertility have also been developed on large scale. *Medicago sativa* (alfalfa), grown in soils with a high nitrogen concentration, has been shown to undergo improved root nodulation when exposed to a genetically modified *Sinorhizobium* (*Rhizobium*) *meliloti* expressing the *Klebsiella pneumonia nif*A gene than plants in the same environment exposed to wild-type *S. meliloti* (Vázquez et al., 2002). Another study showed that the recombinant *S. meliloti* significantly increased plant biomass when compared to the wild-type strain (Galleguillos et al., 2000).

21.4.2 FOR THE PLANT GROWTH-PROMOTING BACTERIA

Successful crop growth depends on the genetic make-up of the plant, adequate availability of nutrients, and presence of beneficial microbes in the soil, and the absence of phytopathogens. Plant beneficial bacteria, also known as PGPR positively influence plant growth promotion in two ways:

(2) directly by producing phytohormones such as auxins, lowering plant stress ethylene levels by 1-aminocyclopropane-1-carboxylate (ACC) deaminase activity or by helping plants acquire nutrients, for example, via nitrogen (N_2) fixation, phosphate solubilization or iron chelation (Spaepen et al., 2009) or (2) indirectly by preventing pathogen infections via release of antimicrobial agents, by outcompeting pathogens, or by establishing the plant's systemic resistance (Hardoim et al., 2008). Long et al. (2008) concluded that natural plant associated with plant growth-promoting bacteria traits does not have consistent and predictable effects on the growth and fitness of all host plants. These inconsistent and irreproducible results may reflect variations in inoculant-crop compatibility, soil composition, and moisture content, and, perhaps most importantly, an incomplete understanding of the mechanisms employed by plant growth-promoting bacteria to facilitate plant growth. Recently, biotechnology and genetic engineering have been used as tools for further understanding of the mechanisms behind plant growth promotion and to enhance their effects on plants and thereby on soil. RD64 strain, a *Sinorhizobium meliloti* 1021 strain engineered to overproduce indole-3-aceticacid (IAA), showed improved N_2-fixation ability compared to the wild-type 1021 strain. It also showed high effectiveness in mobilizing P from insoluble sources, such as phosphate rock (Bianco and Defez, 2010).

Expression in *E. coli* of the mineral phosphate solubilization (MPS) genes from *Ranella aquatilis* supported a much higher GA production and hydroxyapatite dissolution in comparison with the donor strain (Kim et al., 1998b). Genetic transfer of any isolated gene involved in MPS to induce or improve phosphate-dissolving capacity in plant growth-promoting bacteria (PGPB) strains, is an interesting approach. An attempt to improve MPS in PGPB strains, using this approach, was carried out (Rodriguez H and Fraga, 1999) with a pyrroloquinoline quinone (PQQ) synthetase gene from *Erwinia herbicola*. This gene, isolated by Goldstein and Liu (1987), was subcloned in a broadhost range vector (pKT230). The recombinant plasmid was expressed in *E. coli*, and transferred to PGPB strains of *Burkholderia cepacia* and *Pseudomonas aeruginosa*, using tri-parental conjugation. A genomic integration of the phosphoenol pyruvate carboxylase (pcc) gene of *Synechococcus* PCC in *P. fluorescens* 7942 allowed phosphate solubilization in the recipient strain (López-Bucio, 2003).

The transgenic *Azotobacter*, expressing *E. coli gcd*, showed improved biofertilizer potential in terms of mineral phosphate solubilization and plant growth promoting activity with a small reduction in N_2-fixation ability. *Azotobacter vinelandii* AvOP harboring pMMBEGS1 and pMMBEPS1, without supplementation of PQQ, showed pink coloration of the colony, solubilized TCP and also released inorganic phosphate in liquid media more than the wild type, suggesting that the *Azotobacter* was able to synthesize the cofactor PQQ. *Azotobacter vinelandii* AvOP genome sequencing project also revealed the presence of ORFs with homology to the PQQ cofactor biosynthetic genes (Sashidhar and Podile, 2009).

Many individual or *pqq* gene clusters from P solubilizing bacteria were cloned and expressed in *E. coli* which enabled them to produce GA leading to MPS phenotype (Table 21.1). Expression of *gab*Y (396 bp) gene in *E. coli* JM109 induced MPS ability and GA production. *gab*Y gene sequence was similar to membrane-bound histidine permease component which may be a PQQ transporter (Babu-Khan et al., 1998). *Serratia marcenses* genomic DNA fragment induces GA production in *E. coli* but it does not show any homology with either *gdh* or *pqq* genes (Krishnaraj et al., 2001). The DNA fragment did not confer GA secretion in either *E. coil gdh* mutant or in other *pqq* producing strain. Thus, it was postulated that the gene product could be an inducer of GA production. Similarly, other reports showed genes that are not directly involved in *gdh* or *pqq* biosynthesis could induce MPS ability. Genomic DNA fragment of *Enterobacter agglomerans* showed MPS ability in *E. coli* JM109 without any significant change in pH (Kim et al., 1997). MPS genes from *R. aquatilis* showed higher GA production and hydroxyapatite dissolution in *E. coli* compared to native strain (Kim et al., 1998). Over-expression of *P. putida* KT 2440 gluconate degydrogenase (*gad*) operon in *E. asburiae* PSI3 improved MPS phenotype by secretion of 2-ketogluconic acid along with GA35 (Kumar et al., 2013).

21.4.3 NITROGEN-FIXATION IN SOIL BY GENETICALLY MODIFIED MICROORGANISMS

Nitrogen-fixation is the ability to fix atmospheric nitrogen and supply it in a usable form to the host plant and is the most economically important plant growth-promotion trait. N_2-fixing bacteria can supply over 155 Kg nitrogen/ hectare in some agricultural regions that grow leguminous crops (Crews and Peoples, 2004) and can result in nitrogen residues remaining in the soil in sufficient quantities to supply the needs of non-leguminous crops in the same field. This rotation of legume and non-legume crops is particularly important for farmers in developing countries, where chemical fertilizers may be prohibitively expensive. The N_2-fixing bacteria that form symbiotic associations with these leguminous crops are collectively called rhizobia and currently the term is used to describe 44 species of plant nodulating bacteria dispersed among 11 genera of the α- and β-proteobacteria (Cummings et al., 2006). These bacteria colonize and infect the roots of legumes, induce nodule formation, form bacteroids (cell wall free bacteria), and fix N_2 in the oxygen limited environment of the nodule. The most commercially important N_2-fixing bacteria are *Rhizobium*, *Sinorhizobium*, and *Bradyrhizobium* which are sold as crop inoculants. The economic value due to increased crop yields and the reduction in fertilizer costs have led to significant research into the development of enhanced rhizobia inoculation systems, survival in soil, root colonization and nodule formation, and N_2-fixing activity.

ACC deaminase activity appears to be a widespread trait among *Rhizobia*, as a survey of 233 strains from Saskatchewan, Canada, yielded 27 isolates possessing this gene, mostly *R. leguminosarum* (Duan et al., 2009). Whether rhizobia possessing this enzyme is able to stimuli root growth directly has not been shown, but *Rhizobia* expressing ACC deaminase are more effective at forming root nodules on legumes. *R. leguminosarum bv. viciae* contains one copy of ACC deaminase, which when mutated reduces its ability to nodulate *Pisum sativum* L. cv. *sparkle* (pea; Ma et al., 2003). This ACC deaminase gene from *R. leguminosarum bv. viciae* was introduced into *S. meliloti* which does not have this enzyme; transgenic bacteria showed 35–40% greater efficiency in nodulating *Medicago sativa* (alfalfa; Ma et al., 2004).

Large number of endophyte genomes are now available, including the kallar grass endophyte Azoarcus sp. BH72 (Krause et al., 2006), sugarcane endophytes G. diazotrophicus Pal5 (Bertalan et al., 2009), and *H. seropedicae* Z67, corn endophyte *K. pneumonii* 342 (Fouts et al., 2008), rice endophyte *Azospirillum* sp. B510 (Kaneko et al., 2009) tall fescue endophytes *Epichloe festucae* N *coenophialum* and *N. lolii*, onion endophyte *B. phytofirmans* PsJN, AM G. intraradices (Martin et al., 2008), and several rhizobial and Frankia nodule forming bacteria. Several poplar endophytes have also been sequenced including *Enterobacter* sp. strain 638, *P. putida* W619, *S. proteamaculans* 568, and *Stenotrophomonas maltophilia* R551-3; the genome sequences suggest that these microbes possess the growth-promoting mechanisms of acetoin production, IAA synthesis, and gamma-aminobutyric acid (GABA) metabolism, but no functional ACC deaminase (Taghavi et al., 2009). As some endophytes live within plants in unculturable states, it may be important to undertake metagenomic approaches to acquire genomic information, and that is being done for rice in an ambitious project to understand the entire community of endophytes present by sequencing of 100 Mb of DNA extracted from inside rice plants. Isolation of the genes involved in these mechanisms may allow for their pyramiding within endophytes or their transfer into plants for enhanced NUE. Moreover, when a *S. meliloti* strain was genetically modified to express a *Rhizobium leguminosarum* ACC deaminase gene (Ma et al., 2003); the transformants successfully out competed the parental *S. meliloti* in its ability to nodulate alfalfa (Ma et al., 2004).

Genetic modifications carried on rhizobia have attempted to improve strains used as inoculum through enhancing (1) their ability to successfully compete with indigenous soil rhizobia for increased nodule formation and (2) their capacity to fix N_2 (Amarger, 2002). Transfer of ACC deaminase from *E. cloacae* into rhizospheric *Azospirrilum brasilensis* increased the root-elongation potential of this strain in tomato and canola (Holguin et al., 2001), suggesting that similar transgenic techniques may increase the root growth-promoting ability of endophytic strains. This has been shown to be an effective technique in nodule-forming rhizobia: ACC deaminase genes from *Sinorhizobium* sp. BL3 were introduced into *Rhizobium* sp. strain TAL1145, increasing its ACC deaminase activity, resulting in nodules with greater number and size, and producing higher root mass on the tree legume *Leucaena leucocephala* (Tittabutr et al., 2008). Root stimulation by transgenic auxin production in endophytes may also enhances root development: the entire tryptophan monooxygenase pathway was introduced into *P. fluorescencs* strain CHA0 elevating synthesis of IAA, and stimulating an increase in root fresh weight of cucumber by 17–36% in natural soil (Beyeler et al., 1999). Release of engineered *S. meliloti* with an extra copy of both *nif*A and *dct*ABD (a regulatory N_2-fixation gene and C4-dicarboxylic acid transport gene, respectively) into soil increased alfalfa (*Medicago sativa* L.) yield by 12.9% (Bosworth et al., 1994).

Constitutive expression of the N_2-fixing transcriptional regulator *nif*A was shown to significantly increase N_2-fixation by cornendophyte *E. gergoviae* 57-7 in planta and may be a useful trait to introduce into other diazotrophic endophytes (An et al., 2007). Another strategy to increase endophytic N_2-fixation has been added an additional copy of the *nif*HDK operon under a stronger *nif*H promoter, allowing *R. etli* to have increased nitrogenase activity up to 58%, and which increased *P. vulgaris* weight by 38%, increased plant nitrogen content by 15%, and increased seed yield by 36% (Peralta et al., 2004). Novel genes from endophytes may be used to make transgenic plants with improved NUE: tomato plants constitutively expressing bacterial ACC deaminase are able to better tolerate flooding and heavy metal stress (Grichko et al., 2000), while *N. tabacum* plants expressing a phytase gene from the soil fungus *A. niger* accumulated up to 52% more P than controls when grown in soils amended with either phytate or phosphate and lime (George et al., 2005). Many other transferable, genetic mechanisms to improve plant NUE by endophytes must exist in the countless undiscovered or understudied endophytes and will hopefully lead to genetically enhanced inoculants in the future. Despite much investment and some promising experiments, only one genetically modified endophyte has thus far been commercially released: strain RMBPC-2 of S. meliloti, sold by the American company, Research Seeds Inc., has been modified with genes to enhance C4-dicarboxylic acid uptake and N_2-fixation in symbiosis with alfalfa.

Robleto et al. (1998) introduced the genes for trifolitoxin production into Rhizobium etli. Trifolitoxin is an antibiotic to which many non-trifolitoxin producing strains are sensitive to. (Roberto et al., 1998) found that strains carrying the trifolitoxin genes were more competitive and persisted longer in soil than non-trifolitoxin producing strains. They showed that over 2 years, 20% more nodules contained the trifolitoxin producing strains. Van Dillewijin et al. (2001) introduced an over-expressed putA gene into *S. meliloti*. This is a metabolic gene, involved in root colonization. Once again, larger populations of the modified strain were initially identified in nodules. Orikasa et al. (2010) showed that insertion of a highly active catalase (vktA) gene in *Rhizobium leguminosarum* resulted in increased N_2-fixing activity of nodules, 1.7–2.3 times that of the wild type bacteria. Genetically, modified N_2-fixing strains have been commercialized both in the US and in Australia. *S. meliloti* strain RMBPC-2 has been genetically enhanced through the insertion of a *nif*A gene to increase N_2-fixing activity, the *dct*ABD genes which include the regulation and structural genes necessary for bacteroid uptake of C4-dicarboxylic acids from the plant and a spectinomycin resistance gene as a genetic marker. This strain was first commercialized in 1997 and has resulted in significant increases in the yield of alfalfa, although these increases were very much dependent on environmental conditions.

21.4.4 DEVELOPING PHOSPHOSPHATE SOLUBILIZING MICROORGANISMS/GENETIC ENGINEERING OF PHOSPHOSPHATE SOLUBILIZING MICROORGANISMS

Phosphorous is the most essential macronutrient required by plants in usable form for their growth. Inorganic phosphates occur in soil, mostly in insoluble mineral complexes, some of them appearing after the application of chemical fertilizers. These precipitated forms of P cannot be absorbed by plants. Therefore, the ability to convert insoluble phosphates to a form accessible to the plants, for example orthophosphate, is an important trait for a PGPR strain. There are numerous mechanisms through which bacteria can solubilize phosphates. These include the production of simple organic acids, such as gluconic or α-keto acids which simply redissolve insoluble phosphates and the production of phosphatases and phytases which liberate phosphates from organic matter (Rodriguez et al., 2006). Glucose dehydrogenase (GDH) is a membrane bound enzyme encoded by the gcd gene and converts glucose into gluconic acid (GA; Rodriquez et al., 2006). Recently, the gcd gene from *E. coli* was cloned under the control of phosphate transport gene promoters and then inserted into Azotobacter (Sashidhar and Podile, 2009). This modified strain showed enhanced phosphate solubilization and resulted in enhanced plant growth when inoculated into sorghum. Rodriguez et al. (2000) cloned the PQQ synthesis genes from *Erwinia herbicola*, and transferred them into two rhizobacterial strains, *B. cepacia IS-16* and *Pseudomonas* spp. *PSS* recipient cells. This trait is also involved in mineralization of phosphate and after this gene was transferred to both rhizobacterial strains, they produced larger zones with insoluble phosphate as the mineral phosphate source, in comparison with those of parental strains without PQQ.

Although knowledge of the genetics of phosphate solubilization is still very less, and the studies at the molecular level in order to understand how precisely the phosphosphate solubilizing microorganisms (PSM) brings out the solubilization of insoluble P are inconclusive (Rodriguez et al., 2006; Sharma et al., 2013). In comparison bacteria better perform than fungi in solubilization of phosphorus (Alam et al., 2002). Among all microbes population in soil, PSB having 1–50%, while PSF are ~0.1–0.5% in phosphorus solubilization (Chen et al., 2006). There are adequate colonies of PSB in rhizospheres. There are two types of bacterium aerobic and anaerobic, with prevalence of aerobic bacteria which are mostly found in submerged lands. A considerable higher population of PSB is mainly present in plant rhizosphere as compared to non-rhizosphere soil. Among soil bacteria, *Pseudomonas* and *Bacillus* are more common. The common species like *Pseudomonas*, *Bacillus*, *Penicillium*, and *Aspergillus* produced organic acids, which have the capacity to lower pH in their vicinity to bring solubilization of bound phosphates in soil. Inoculation of peat-based cultures of *Bacillus polymyxa* and *Pseudomonas striata* increased yields of wheat and potato. The available insoluble phosphorus in soil can be solubilized by PSB; they have the ability to transform inorganic unavailable phosphorus to soluble forms HPO_4^{2-} and $H_2PO_4^-$ by the process of producing organic acid, chelation, and ion-exchange reactions and make them available to plants (Kumar et al., 2014a; Kumar 2015a, b, Shivran et al., 2013). Hence, using PSB in crop production would not only decrease the high cost of manufacturing phosphatic fertilizers but also solubilize insoluble chemical fertilizers that are applied in soil (Chang and Yang, 2009; Banerjee et al., 2010, Kumar and Kumawat, 2014).

However, some genes involved in mineral and organic phosphate solubilization have been isolated and characterized. Initial achievements in the manipulation of these genes through genetic engineering and molecular biotechnology followed by their expression in selected rhizobacterial strains open a promising perspective for obtaining PSM strains with enhanced phosphate solubilizing capacity, and thus, a more effective use of these microbes as agricultural inoculants. The initial achievement in cloning of gene involved in P solubilization from the Gram-negative bacteria *Erwinia herbicola* was achieved by Goldstein and Liu (1987). Similarly, the napA phosphatase gene from the soil bacterium *Morganella morganii* was transferred to *Burkholderia cepacia IS-16*, a strain used as a biofertilizer, using the broad-host range vector pRK293 (Fraga et al., 2001). An increase in extracellular phosphatase activity of the recombinant strain was achieved. Introduction or over-expression of genes involved in soil phosphate solubilization (both organic and inorganic) in natural rhizosphere bacteria is a very attractive approach for improving the capacity of microorganisms to work as inoculants. Insertion of phosphate-solubilizing genes into microorganisms that do not have this capability may avoid the current need of mixing two populations of bacteria, when used as inoculants [N_2-fixers and phosphate-solubilizers (Bashan et al., 2000)]. There are several advantages of developing genetically-modified PSM over transgenic plants for improving plant performance and subsequently soil health: (1) with current technologies, it is far easier to modify a bacterium than complex higher organisms, (2) several plant growth-promoting traits can be combined in a single organism, and (3) instead of engineering crop by crop, a single, engineered inoculant can be used for several crops, especially when using a non-specific genus like *Azospirillum* (Rodriguez et al., 2006). Some barriers should be overcome first to achieve successful gene insertions using this approach, such as the dissimilarity of metabolic machinery and different regulating mechanism between the donor and recipient strains. Despite the difficulties, significant progress has been made in obtaining genetically engineered microorganisms (GEMs) for agricultural use (Armarger, 2002). Overall, further studies on this aspect of PSM will provide crucial information in future for better use of these PSMs in varied environmental conditions.

Over-expression of gene involvement in phosphorus mobilization in soil (both organic and inorganic) under rhizosphere bacteria is a very effective approach for improving the capacity of microorganisms for efficient absorption of phosphorous in soil. In addition, many PGPR may be combined in a single organism. Specific primers based on conserved regions of genomic region have been designed for various microorganisms associated with phosphate mobilization, including mycorrhizal fungi, *Penicillium* sp., and *Pseudomonas* spp. (Oliveira et al., 2009).

Introduction or over-expression of genes involved in soil phosphate solubilization (both organic and inorganic) in natural rhizosphere bacteria is a very attractive approach for improving the capacity of microorganisms to work as inoculants. Insertion of phosphate-solubilizing genes into microorganisms that do not have this capability may avoid the current need of mixing two populations of bacteria, when used as inoculants (N_2-fixers and phosphate-solubilizers; Bashan et al., 2000). Opportunity also exists for genetic manipulation of soil microorganisms. It is now a reality that gene technologies can be used to enhance specific traits that may increase an organism's capacity to mobilize soil P directly, enhance its ability to colonize the rhizosphere (i.e., rhizosphere competence, Lugtenberg et al., 2001), or perhaps to form specific associations with plant roots (Bowen and Rovira, 1999). Identifying microbial traits that are associated with P mobilization is an important step in this regard and subsequent isolation and manipulation of candidate genes is required. Over-expression of microbial phosphatase genes, MPS genes and genes that are directly associated with organic acid biosynthesis (e.g., citrate synthase, phosphoenol pyruvate carboxylase) are examples (Gyaneshwar et al., 2002). Several acid phosphatase genes from Gram-negative bacteria have been isolated and characterized (Rossolini et al., 1998). These cloned genes represent an important source of material for the genetic transfer of this trait to PGPB strains. Some of them code for acid phosphatase enzymes that are capable of performing well in soil. For example, the acpA gene isolated from *Francisella tularensis* expresses an acid phosphatase with optimum action at pH 6, with a wide range of substrate specificity (Reilly et al., 1996). Also, genes encoding non-specific acid phosphatases class A (PhoC) and class B (NapA) isolated from *Morganella morganii* are very promising, since the biophysical and functional properties of the encoded enzymes were extensively studied (Thaller et al., 1994; Thaller et al., 1995). Among rhizobacteria, a gene from Burkholderia cepacia that facilitates phosphatase activity was isolated (Rodríguez et al., 2000a). This gene codes for an outer membrane protein that enhances synthesis in the absence of soluble phosphates in the medium, and could be involved in P transport to the cell. Besides cloning of two non-specific periplasmic acid phosphatase genes (napD and napE) from *Rhizobium* (*Sinorhizobium*) *meliloti* was accomplished (Deng, et al., 1998, 2001).

Thermally stable phytase genes (phy) from *Bacillus* sp. DS11 (Kim et al., 1998a) and from *B. subtilis* VTT E-68013 (Kerovuo et al., 1998) have been cloned. Acid phosphatase/phytase genes from *E. coli* (appA and appA2 genes) have also been isolated and characterized (Golovan et al., 2000; Rodríguez et al., 1999). The bifunctionality of these enzymes makes them attractive for solubilization of organic P in soil. Also, neutral phytases have great potential for genetic improvement of PGPB. Neutral phytase genes have been recently cloned from B. subtilis and B. licheniformis (Tye et al., 2002); a phyA gene has been cloned from the FZB45 strain of B. amyloliquefaciens. In most bacteria, mineral phosphate-dissolving capacity has been shown to be related to the production of organic acid (Rodríguez and Fraga, 1999). Goldstein (1996) proposed direct glucose oxidation to GA as a major mechanism for MPS in Gram-negative bacteria. GA biosynthesis is carried out by the GDH enzyme and the co-factor, PQQ. Some genes involved in MPS in different species have been isolated (Table 21.1).

Expression in *E. coli* of the MPS genes from *Ranella aquatilis* supported a much higher GA production and hydroxyapatite dissolution in comparison with the donor strain (Kim et al., 1998b). The authors suggested that different genetic regulation of the MPS genes might occur in both species. MPS mutants of *Pseudomonas* spp. showed pleiotropic effects, with apparent involvement of regulatory mps loci in some of them (Krishnaraj et al., 1999). This suggests a complex regulation and various metabolic events related to this trait. Expression of a MPS gene in a different host could be influenced by the genetic background of the recipient strain, the copy number of plasmids present and metabolic interactions. Thus, genetic transfer of any isolated gene involved in MPS to induce or improve phosphate-dissolving capacity in PGPB strains, is an interesting approach. An attempt to improve MPS in PGPB strains, using this approach, was carried out (Rodríguez et al., 2000b) with a PQQ synthetase gene from *Erwinia herbicola*. This gene, isolated by Goldstein and Liu (1987), was subcloned in a broadhost range vector (pKT230). The recombinant plasmid was expressed in *E. coli*, and transferred to PGPB strains of Burkholderia cepacia and *Pseudomonas aeruginosa*, using tri-parental conjugation.

21.4.5 Fe-ENRICHMENT BY GENETICALLY MODIFIED MICROORGANISM FOR IRON NUTRITION IN RHIZOBIA

Although iron is the fourth most abundant element in the Earth's crust; it is essentially unavailable in aerobic environments at biological pH, as it tends to form insoluble Fe^{3+} oxyhydroxide (Guerinot and Yi, 1994). To combat iron deficiency, different organisms produce different types of siderophores. In nature the ability to utilize various sources of chelated Fe appears to be of much importance and that is the reason why microorganisms often employ more than one high affinity Fe transport system for acquiring iron. This broad transport capability may be achieved by two means: either like the *E. coli* model which involves many specific siderophore transporters for uptake of ferrisiderophores that it may encounter or like the *Streptomyces* model which involves a non-specific receptor/transporter for its own siderophore and other siderophores that has similar coordination of iron but different structures (Crowley et al., 1991). Iron acquisition by *Rhizobium* sp. is essential for N_2-fixation by the *Rhizobium* legume root nodule symbiosis. Root nodule bacteria, form a N_2-fixing symbiotic interaction along with their leguminous plant hosts and have a high demand for iron (Guerinot, 1991). Diverse types of siderophores are produced by the different rhizobial genera, for example, *Rhizobium leguminosarum bv. viciae*, the symbiont of peas, lentils, vetches, and some beans, synthesizes a cyclic tri-hydroxamate type siderophore called vicibactin, whereas *S. meliloti* under iron stress produces rhizobactin 1021 a dihydroxamate siderophore (Persmark et al., 1993). Catecholate siderophores are known to be produced by rhizobia from the cowpea group (Jadhav et al., 1992) and salicylic acid and dihydroxybenzoic acid are produced by *Rhizobium ciceri* isolated from chick pea nodules (Berraho et al., 1997). Citrate as a siderophore is produced by *Bradyrhizobium japonicum* (Guerinot et al., 1990) and anthranilate is produced by *Rhizobium leguminosarum*. Under natural conditions, the ability to produce siderophores has been demonstrated to confer a selective advantage to the producer organism (de Bellis and Ercolani, 2001). It has also been speculated that besides capturing iron quotas necessary for growth, siderophores are also a type of iron scavengers because they can mobilize iron from weaker ferric-siderophore complexes from other species and thus are mediators of competitive interaction among organisms. Microorganisms which themselves do not synthesize a particular type of siderophore may yet be proficient at the uptake of the iron-bound siderophore complex that they do not produce. The ability to cross-utilize heterologous siderophores may be accounted by the presence of multiple type of siderophore receptors that are expressed for the uptake of different types of siderophores (Brickman and Armstrong, 1999) or use of a low specificity (broad range) system that recognizes more than one type of siderophores (Crowley et al., 1991).

Most of the rhizobial biofertilizer strains are poor rhizospheric colonizers due to their inability to compete with the indigenous soil microflora for nutrients; iron being one of them because of the iron rich enzymes involved in N_2-fixation (Verma and Long, 1983). Utilization of foreign siderophores is considered to be an important mechanism to attain iron sufficiency.

Pseudomonades which are known for their rhizospheric stability have diverse iron uptake systems, and multiple receptor genes have been detected in their genomes (Dean and Poole, 1993; Crowley et al., 1987). From the above facts, it could be suggested that by increasing the number of outer membrane siderophore receptors rhizobial strains could be made more efficient with respect to iron acquisition, and hence colonizing the rhizosphere. Most of the hydroxamate siderophores present in soil are of the ferrichrome-type. Because ferrichrome is synthesized by a variety of soil fungi, it is likely an iron source in the rhizosphere where hydroxamate concentrations have been estimated to be as high as 10 µM (Crowley et al., 1987) and ferrichrome is found in nanomolar concentrations, as estimated by physicochemical (Holmstrom et al., 2004) as well as bioassay methods (Powell et al., 1983). As majority of soil bacteria are good utilizers of iron bound to hydroxamates (Jurkevitch et al., 1992), thus, the rhizobia isolates could be at a competitive disadvantage when residing free in soils. It is therefore pertinent to engineer these strains with a ferrichrome receptor to increase their iron acquisition property and hence survival. Thus, cloning of the ferrichrome receptor gene in rhizobial bio-inoculant strains and understanding the effect of ferrichrome utilization on rhizobial growth and survivability under conditions, wherein ferrichrome was made available by other producer species, were achieved by heterologously expressing the *E. coli fhu*A gene in *C. cajan* rhizobia. The *fhu*A was engineered under the control of the *lac* promoter and its expression was first confirmed in *E. coli* by the rescue of the phenotype of *fhu*A mutant and was subsequently introduced into the rhizobial strains and its expression monitored. The expression of *E. coli fhu*A in rhizobial strains imparted the associated phenotypes, namely, the ability to utilize iron complexed with ferrichrome and sensitivity to albomycin (Rajendran et al., 2007). Several studies have shown that the utilization of heterologous siderophores by genetically introducing the receptor gene provides growth advantage to the bacteria. Brickman and Armstrong (1999) found that the incorporation of alcaligin receptor gene alone could confer upon a siderophore deficient strain of *P. aeruginosa* the ability to utilize ferric alcaligin.

21.4.6 BIOFERTILIZER POTENTIAL AND POSSIBLE GENETIC ENGINEERING FOR POTASSIUM, ZINC, AND SILICON IN MICROBES

Out of the major plant nutrients, K is the most adequate nutrient in the soil. It is 7/8 common in ground, whose surface layer (lithosphere) on an average contains ~2.6 K (~3% K_2O). It is present in soil in four major pools according to the availability of potassium to plants (Zakaria, 2009). The availability of potassium depends on soil type and is influenced by physic and chemical properties of soil. Potassium in soil is generally categorized in four groups depending on the availability to plants, that is, water soluble, exchangeable, non-exchangeable, and structural forms. Water soluble potassium is directly absorbed by plants and microorganism and subjected to leaching exchangeable K which is electrostatically bound on outer sphere to clay surfaces of mineral (Barre et al., 2008). Potassium is most important for growth and development of plants. It is participated in adjustment of plant cellular osmotic pressure and the transportation of compounds in plants. It encourages activation of enzymes, the utilization of nitrogen, and the syntheses of sugar and protein. It also helps photosynthesis in plants (Zhang and Kong, 2014). Potassium is required for the cell metabolic mechanisms. Potassium is required in huge amount by the crop to obtain its highest yields. It is associated with movement of water, nutrients, and carbohydrates in plant tissues. A large range of rhizosphere microbes are isolated as potassium mobilizers like *Bacillus mucilaginosus*, *Bacillus edaphicus*, *Bacillus circulans*, and *Arthrobacter* sp. Potassium mobilization is mainly due to the produced organic acids. The use of microbial inoculants as biofertilizers can be an alternative tool of chemical fertilizer; the use of KSB by farmers can solubilize potassium available in their own fields and can save some requirement of potassic fertilizer. Among bacterium genera, these are *Pseudomonas*, *Bacillus*, *Achromobacter*, *Agrobacterium*, *Microccocus*, *Aereobacter*, *Flavobacterium*, and *Erwinia* (Zhao, 2008). Earlier, research findings have shown that potassium-solubilizing bacterium can stimulate the plant growth (Lin et al., 2002; Basak and Biswas, 2009). The studies have shown that inoculums can increase plant growth and increase nutrient concentration in plants (Basak and Biswas, 2010).

The process of potassium mobilization means insoluble potassium and structural unavailable forms of potassium compounds are mobilized and solubilized due to production of different organic acids. These organic acids are accompanied by acidolysis and complexolysis exchange reactions, and these are the main processes attributed to their conversion into soluble form. The organic and inorganic acids transform insoluble potassium (mica, muscovite, and biotite feldspar) to soluble form of potassium with the net result of increasing the availability of nutrients to crop plants. It is demonstrated that *Bacillus mucilaginosus* and *Bacillus edaphicus* can produce polysaccharide and carboxylic acids, such as tartaric acid and citric acid, which help to solubilize K compounds in soils (Lin et al., 2002).

Zinc (Zn) deficiency has received great attention in India because half of the Indian soils have poor availability of Zn (Cakmak, 2009). This is mainly due to crops grown in Zn deficient soils. In our country, ~50% of soils are low in Zn, and it remains most important nutritional disorder influencing crop production. The major causes of zinc deficiency are the use of chemical fertilizers, intensive agriculture, and poor irrigation facilities that lead to lowering of zinc content in soils (Das and Green, 2013). Zn deficiency is estimated to increase from 42% to 63% by 2025 due to continuous decreasing of soil health (Singh, 2009). Zinc has pivotal role in plant metabolism by affecting the activities of hydrogenase and carbonic anhydrase, stabilization of ribosomal fractions, and synthesis of cytochrome (Tisdale et al., 1984).

Zn plays a role in carbohydrate metabolism. The enzyme involved is carbonic anhydrase (CA). Zinc has a role in anaerobic root respiration in rice. The enzyme involved is alcohol dehydrogenase (ADH). Zinc is also necessary for the assimilation of tryptophan which is a precursor of IAA; it also plays an active role in the secretion of growth-promoting hormones and auxins (Alloway, 2004). Zinc also helps in plant diseases resistance, photosynthesis, protein metabolism, pollen development, and cell membrane integrity (Gurmani et al., 2012) and enhances level of antioxidant enzymes and chlorophyll within plant tissues (Sbartai et al., 2011). Numerous bacteria, mostly those associated with rhizosphere, have capacity to convert unavailable form of Zn into readily available form through mobilization processes (Cunninghan and Kuiack, 1992). The zinc-solubilizing Biofertilizer (ZSB) are potential alternative that could cater plant Zn essentiality by mobilizing complex Zn in the soil. Many genera of rhizobacteria related to *Thiobacillus thiooxidans*, *Acinetobacter*, *Bacillus*, *Pseudomonas*, and *Thiobacillus ferrooxidans* have been reported as zinc solubilizers (Saravanan et al., 2007). These microorganisms mobilized metal forms by protons, chelated ligands, and oxidoreductive systems present on cell surface and membranes (Table 21.1). This bacterium also demonstrated many traits beneficial to plants like production of phytohormones, antibiotics, siderophores, vitamins, antifungal substances, and hydrogen cyanide (Goteti et al., 2013).

Solubilization of zinc can be done by a range of processes, which are secretion of metabolites like organic acids and proton extrusion or production of chelating agents (Sayer and Gadd, 1997). In addition, producing inorganic acids like sulfuric acid, nitric acid, and carbonic acid could also facilitate solubilization (Seshadre et al., 2002). It is apparent from the zinc solubilization data that the solubilization potential varied with each isolate. Production of organic acid like gluconic acids (especially 2-ketogluconic acids) by microorganism isolate has been observed to be a major mechanism of solubilization (Fasim et al., 2002). This solubilization property is important in nutrient cycling. Fall in pH and acidifications of medium were noted in all cases. Higher solubilization of insoluble zinc sources was achieved in 72 h. The zinc-solubilizing potential also correlated with the zinc levels that are accumulated by plant leaves. The solubilization of zinc phosphate is done by a strain of *Pseudomonas fluorescens* (Simine et al., 1998). They found that gluconic acids are secreted in culture medium, which helps in solubilization of zinc salts. In their study also reported that acidic pH can solubilize bacterium due to production of organic acids and higher production of available zinc in the culture broth.

Pseudomonas aeruginosa has a potential to solubilize ZnO in liquid medium (Fasim et al., 2002). Bacterial inoculation has also the ability to increase bioavailable Zn in rhizosphere soil (Whiting et al., 2001) and Zn content in plants (Biari et al., 2008). PGPR produced siderophores (Saravanan et al., 2011); derivatives of gluconic acids, for example, 2-ketogluconic acid and 5-ketogluconic acid; and different other organic acids for the mobilization of Zn and iron (Tariq et al., 2007). These bacteria can be used to solubilize insoluble sources of Zn, such as ZnO and $ZnCO_3$ because most of the soils are rich in Zn content but less in soluble Zn. *Bacillus* and *Pseudomonas* spp. have much potential to solubilize these sources in soil system for taking economically efficient Zn (Saravanan et al., 2003). Rhizosphere microorganism may benefit plants through different mechanisms including mobilization of nutrients and also acts as a biocontrol agent (Khalid et al., 2009).

Silicon (Si) is an element which is useful for healthy growth and development of plants. There is plenty of total Si in soils, but most of them were unavailable to the plant. Silicate mineral-solubilizing bacteria (SSB) dissolve silicate minerals (such as feldspars and micas) and release elements of potassium and silicon. The SSB are of great interest in recent times because of their role in solubilization of silicate minerals rendering silica and potassium available for crop uptake thus reducing the potash fertilizer requirement (Sheng, 2005) and in desilication of ores like bauxite (Zhou et al., 2006). Studies have shown that these bacteria solubilized silica besides releasing phosphate, potassium, iron, and calcium from soil silicate mineral. Therefore, these microorganisms attracted attention of scientists to advocate these organisms as potassium-mobilizing biofertilizers. Silicon plays an important role in plants like accelerating growth and conferring rigidity to leaves, thus increasing leaf surface area for maximizing photosynthesis and mitigating effects of abiotic stresses such as drought, salt, and metal toxicity in many plants (Ma and Yamaji, 2006). Despite of its abundance in earth's crust, its major part occurs in insoluble forms that cannot be taken easily by

TABLE 21.1 Important Genetically Modified Microorganisms for Soil Health Enhancement.

Sr. no.	Gene name	Transfer from	Transfer to (host)	References
Nitrogen				
1.	ACC deaminase	*E. cloacae*	*Azospirillum brasilensis*	Holguin (2001)
2.	ACC deaminase	*Sinorhizobium* sp. BL3	*Rhizobium* sp. strain TAL1145	Tittabutr et al. (2008)
3.	*E. gergoviae*	*nifA*	Constitutive expression	An et al. (2007)
4.	*nifA* and *dctABD*	extra copy of both *nifA* and *dctABD* from *S. meliloti*	*S. meliloti*	Bosworth et al. (1994)
Iron				
1.	*fhuA*	*E. coli*	*fhuA*	Rajendran et al. (2007)
Phosphorus				
1.	napA phosphatase gene	*Morganella morganii*	*Burkholderia cepacia IS-16*	Fraga et al. (2001)
2.	PQQ	*Erwinia herbicola*	*B. cepacia IS-16* and *Pseudomonas* sp. PSS	Rodriguez et al. (2000)
3.	Gcd	*E. coli*	Azotobacter	Sashidhar and Podile (2009)
4.	PPC gene	*Anacystis nidulans*	*Pseudomonas fluorescens*	Srivastava et al. (2000)
5.	PQQ biosynthesis	*Serratia marcescens*	*E. coli*	Krishnaraj and Goldstein (2001)
6.	pqqE	*Erwinia herbicola*	*Azospirillum* sp.	Vikram et al. (2007)
7.	pqqD	*Rahnella aquatilis*	*E. coli*	Kim et al. (1998)
8.	Pqq/ABCDEF gene	*Enterobacter intermedium*	*E. coli DH5α*	Kim et al. (2003)
9.	Ppts-gcd, P gnl1A-gcd	*E. coli*	*Azotobacter vinelandii*	Sashidhar and Podile (2010)
10.	Gab Y Putative PQQ transporter	*Pseudomonas cepacia*	*E. coli HB101*	Babu-Khan et al. (1995)

TABLE 21.1 *(Continued)*

Sr. no.	Gene name	Transfer from	Transfer to (host)	References
11.	*gltA*/citrate synthase	*E. coli* K12	*Pseudomonas fluorescens* ATCC 13525	Buch et al. (2009)
12.	Gad/gluconate dehydrogenase	*P. putida* KT2440	*E. asburiae* PSI3	Kumar et al. (2013)
13.	mps	*Erwinia herbicola*	*E. coli* HB101	Goldstein and Liu (1987)
14.	*gabY*	*Pseudomonas cepacia*	*E. coli* JM109	Babu-Khan et al. (1995)
15.	pKKY	*Enterobacter agglomerans*	*E. coli* JM109	Kim et al. (1997)
16.	pK1M10	*Rahnella aquatilis*	*E. coli* DH5	Kim et al. (1998b)
17.	pKG3791	*Serratia marcescens*	*E. coli*	Krishnaraj and Goldstein (2001)
Potash				
1.	pcc gene	*Synechococcus* PCC	*P. fluorescens* 7942	López-Bucio et al. (2003)
PGPRs				
1.	ipdC [phytohormone IAA]	*Azospirillum brasilense* Sp245	*A. brasilense*	Baudoin et al. (2010)
2.	IAA,	*Sinorhizobium meliloti*	*Sinorhizobium meliloti* 1021	Bianco and Defez (2010)
3.	mps genes for GA	*Ranella aquatilis*	*E. coli*	Kim et al. (1998b)

ACC, 1-aminocyclopropane-1-carboxylate; IAA, indole-3-aceticacid; PQQ, pyrroloquinoline quinone.

roots (Vasanthi et al., 2012; Rodrignes and Datnoff, 2005). Si makes plant cell walls thick and stronge while enlarging the size of the vascular system of plant. The silicon cellulose membrane in epidermal tissue protects plants against excessive loss of water by transpiration (Meena et al., 2014a, b, c). Soil has a lot of various microorganisms, but fewer are able of solubilizing silicates such as *Bacillus caldolytyicus*, *Proteus mirabilis*, *Bacillus mucilaginosus* var. *siliceous*, and *Pseudomonas* was found most effective to release silica from natural silicates (Meena et al., 2014a, b, c).

These SSB have the ability to degrade silicate, especially aluminum silicates (Al_2SiO_5). These bacteria produced several organic acids during their growth, which can play a role in silicate weathering. These organisms help in releasing potassium from K-containing minerals. Silicon microorganisms released organic acids as part of its metabolism that plays dual role in silicate weathering. These are supplied H^+ ions to medium and stimulate hydrolysis and organic acids such as citric acid, keto acids, oxalic acid, and hydroxyl carbolic acids that make complexes with cations and made available to plants in readily form. Joseph et al. (2015) reported that some bacterial isolates can solubilize insoluble minerals like silicates, phosphates, and potash into soluble form by secretion of organic acids (2-ketogluconic acid, alkalis, and polysaccharides). Bacteria make available to silicates by producing proton, organic ligands, hydroxyl anion; extra cellular polysaccharides and enzymes (Barker et al., 1998).

21.5 CONCLUSIONS

Industrial fertilizer use has permitted the large increase in global agricultural production but this is not a sustainable solution to meet future food demands. Some barriers should be overcome first to achieve successful gene insertions using this approach, such as the dissimilarity of metabolic machinery and different regulating mechanism between the donor and recipient strains. Despite the difficulties, significant progress has been made in obtaining genetically engineering microorganisms for agricultural use (Armarger, 2002). However, studies carried out so far have shown that following appropriate regulations, genetically modified microorganisms can be applied safely in agriculture (Armarger, 2002; Morrissey et al., 2002). There are several advantages of developing genetically-modified PGPB over transgenic plants for improving plant performance: (1) with current technologies, it is far easier to modify a bacterium than complex higher organisms, (2) several plant growth-promoting traits can be combined in a single organism, and (3) instead of engineering crop by crop, a single, engineered inoculant can be used for several crops, especially when using a non-specific genus like *Azospirillum*.

KEYWORDS

- **biofertilizer**
- **genetically engineered microorganism**
- **PGR**
- **plant nutrients**
- **soil management**
- **sustainable soil health**

REFERENCES

Adesemoye, A. O.; Kloepper, J. W. Plant-microbes Interactions in Enhanced Fertilizer-use Efficiency. *Appl. Microbiol. Biotechnol.* **2009,** *85,* 1–12.

Alam, S.; Khalil, S.; Ayub, N.; Rashid, M. In Vitro Solubilization of Inorganic Phosphate by Phosphate Solubilizing Microorganism (PSM) from Maize Rhizosphere. *Int. J. Agric. Biol.* **2002,** *4,* 454–458

Alloway, B. J. In *Zinc in Soil and Crop Nutrition*; International Zinc Association: Brussels, 2004.

Altamirano, M. M.; Garcia, C.; Possani, L. D.; Fersht, A. R. Oxidative Refolding Chromatography: Folding of the Scorpion Toxin Cn5. *Nat. Biotechnol.* **1999,** *17,* 187–191.

Amarger, N. Genetically Modified Bacteria in Agriculture. *Biochimie.* **2002,** *84,* 1061–1072.

An, Q.; Dong, Y.; Wang, W. et al. Constitutive Expression of the *nif*A Gene Activates Associative Nitrogen Fixation of Enterobacter Gergoviae 57-7, an Opportunistic Endophytic Diazotroph. *J. Appl. Microbiol.* **2007,** *103* (3), 613–620.

Araujo, A. S. F.; Santos, V. B.; Monteiro, R. T. R. Responses of Soil Microbial Biomass and Activity for Practices of Organic and Conventional Farming Systems in Piauistate, Brazil. *Eur. J. Soil Biol.* **2008,** *44,* 225–230.

Ayres, M. D.; Howard, S. C.; Kuzio, K. J.; Lopez- Ferber, M.; Possee, R. D. The Complete Sequence of *Autographa californica* Nuclear Polyhedrosis Virus. *Virology* **1994,** *202,* 586–605.

Babu-Khan, S.; Yeo, T. C.; Martin, W. L.; Duron, M. R.; Rogers, R. D.; Goldstein, A. H. Cloning of a Mineral Phosphate Solubilizing Gene from *Pseudomonas cepacia. Appl. Environ. Microbiol.* **1995,** *61,* 972–978.

Banerjee, S.; Palit, R.; Sengupta, C.; Standing, D. Stress Induced Phosphate Solubilization by *Arthrobacter* spp. and *Bacillus* spp. Isolated from Tomato Rhizosphere. *Aust. J. Crop Sci.* **2010,** *4,* 378–383.

Baneyx, F. In vivo Folding of Recombinant Proteins in *Escherichia coli*. In *Manual of Industrial Microbiology and Biotechnology*, 2nd ed.; Demain, A. L.; Davies, J. E., Eds.; ASM Press: Washington, DC, 1999; pp 551–565.

Baneyx, F. Recombinant Protein Expression in *Escherichia coli. Curr. Opin. Biotechnol.* **1999,** *10,* 411–421.

Barker, W. W.; Welch, S. A.; Chu, S.; Baneld, J. F. Experimental Observations of the Effects of Bacteria on Aluminosilicate Weathering. *Am. Mineral.* **1998,** *83,* 1551–1563.

Barre, P.; Montagnier, C.; Chenu, C.; Abbadie, L.; Velde, B. Clay Minerals as a Soil Potassium Reservoir: Observation and Quantification Through X-ray Diffraction. *Plant Soil* **2008,** *302,* 213–220.

Basak, B. B.; Biswas, D. R. Influence of Potassium Solubilizing Microorganism (*Bacillus mucilaginosus*) and Waste Mica on Potassium Uptake Dynamics by Sudan Grass (*Sorghum vulgare* Pers.) Grown under Two Alfisols. *Plant Soil* **2009,** *317,* 235–255.

Basak, B. B.; Biswas, D. R. Co-inoculation of Potassium Solubilizing and Nitrogen Fixing Bacteria on Solubilization of Waste Mica and their Effect on Growth Promotion and Nutrient Acquisition by a Forage Crop. *Biol. Fertil. Soils* **2010,** *46,* 641–648.

Bashan, Y.; Moreno, M.; Troyo, E. Growth Promotion of the Seawater-irrigated Oil Seed Halophyte Salicornia Bigelovii Inoculated with Mangrove Rhizosphere Bacteria and Halotolerant *Azospirillum* spp. *Biol. Fertil. Soils* **2000,** *32,* 265–272.

Bergès, H.; Joseph-Liauzun, E.; Fayet, O. Combined Effects of the Signal Sequence and the Major Chaperone Proteins on the Export of Human Cytokines in *Escherichia coli. Appl. Environ. Microbiol.* **1996, 62,** 55–60.

Bertalan, M.; Albano, R.; De Pádua, V. et al. Complete Genome Sequence of the Sugarcane Nitrogen-fixing Endophyte Gluconacetobacter Diazotrophicus Pal 5. *BMC Genomics* **2009,** *10* (1), 450.

Beyeler, M.; Keel, C.; Michaux, P.; Haas, D. Enhanced Production of Indole-3-acetic Acid by a Genetically Modified Strain of Pseudomonas Fluorescens CHA0 Affects Root Growth of Cucumber, but Does Not Improve Protection of the Plant against Pythium Root rot. *FEMS Microbiol. Ecol.* **1999,** *28* (3), 225–233.

Bianco, C.; Defez, R. Improvement of Phosphate Solubilization and Medicago Plant Yield by an Indole-3-acetic Acid-overproducing Strain of *Sinorhizobium meliloti. Appl. Environ. Microbiol.* **2010,** *76,* 4626–4632.

Biari, A.; Gholami, A.; Rahmani, H. A. Growth Promotion and Enhanced Nutrient Uptake of Maize (*Zea mays* L.) by Application of Plant Growth Promoting Rhizobacteria in Arid Region of Iran. *J. Biol. Sci.* **2008,** *8,* 1015–1020.

Bierman, M.; Logan, R.; O'Brien, K.; Seno, E. T.; Rao, R. N.; Schoner, B. E. Plasmid Cloning Vectors for the Conjugal Transfer of DNA from *Escherichia coli* to *Streptomyces* spp. *Gene.* **1992,** *116,* 43–49.

Blouin Bankhead, S.; Thomashow, L. S.; Weller, D. M. Rhizosphere Competence of Wild-type and Genetically Engineered *Pseudomonas brassicacearum* Is Affected by the Crop Species. *Phytopathology* **2016,** *106,* 554–561.

Bosworth, A. H.; Williams, M. K.; Albrecht, K. A.; Kwiatkowski, R.; Beynon, J.; Hankinson, T. R.; Ronson, C. W.; Cannon, F.; Wacek, T. J.; Bowen, G. D.; Rovira, A. D. The Rhizosphere and its Management to Improve Plant Growth. *Adv. Agron.* **1999,** *66,* 1–102.

Brantl, S. Antisense RNAs in Plasmids: Control of Replication and Maintenance. *Plasmid* **2002,** *48,* 165–173.

Brigidi, P.; De Rossi, E.; Bertarini, M. L.; Riccardi, G.; Matteuzzi, D. Genetic Transformation of Intact Cells of *Bacillus subtilis* by Electroporation. *FEMS Microbiol. Lett.* **1990,** *55,* 135–138.

Buch, A. D.; Archana, G.; Naresh Kumar, G. Enhanced Citric Acid Biosynthesis in *Pseudomonas fluorescens* ATCC 13525 by Overexpression of the *Escherichia coli* Citrate Synthase Gene. *Microbiology* **2009,** *155,* 2620–2629.

Cakmak, I. Enrichment of Fertilizers with Zinc: An Excellent Investment for Humanity and Crop Production in India. *J. Trace Elem. Med. Biol.* **2009,** *23,* 281–289.

CCRC, Catalogue of Strains; Food Industry Research and Development Institute, Hsinchu, Taiwan, R.O.C., 1991.

Chang, D. C. N. Effect of Three Glomus Endomycorrhizal Fungi on the Growth of Citrus Rootstocks. *Proc. Int. Soc. Citriculture* **1987,** *1,* 173–176.

Chang, C. H.; Yang, S. S. Thermo-tolerant Phosphate-solubilizing Microbes for Multi-functional Biofertilizer Preparation. *Bioresour. Technol.* **2009,** *100,* 1648–1658.

Chaparro, J. M.; Sheflin, A. M.; Manter, D. K.; Vivanco, J. M. Manipulating the Soil Microbiome to Increase Soil Health and Plant Fertility. *Biol. Fertil. Soils* **2012,** *48,* 489–499.

Chen, Y. P.; Rekha, P. D.; Arun Shen, A. B.; Lai, W. A.; Young, C. C. Phosphate Solubilizing Bacteria from Subtropical Soil and their Tricalcium Phosphate Solubilizing Abilities. *Appl. Soil Ecol.* **2006,** *34,* 33–41.

Chilton, M. Agrobacterium. A Memoir. *Plant Physiol.* **2001,** *125,* 9–14.

Corsaro, B. G.; Di Renzo, J.; Fraser, M. J. Transfection of Cloned Heliothis Zea Nuclear Polyherosis Virus. *J. Virol. Methods* **1989,** *25,* 283–91.

Crews, T. E.; Peoples, M. B. Legume versus Fertilizer Sources of Nitrogen: Ecological Tradeoffs and Human Needs. *Agric. Ecosyst. Environ.* **2004,** *102,* 279–297.

Cummings, S. P.; Humphry, D. R.; Santos, S. R.; Andrews, M.; James, E. K. The Potential and Pitfalls of Exploiting Nitrogen Fixing Bacteria in Agricultural Soils as a Substitute for Inorganic Fertilizer. *Environ. Biotechnol.* **2006,** *2,* 1–10.

Cunninghan, J. E.; Kuiack, C. Production of Citric Acid and Oxalic Acid and Solubilization of Calcium Phosphate by *Penicillium billai. Appl. Environ. Microbiol.* **1992,** *58,* 1451–1458.

Das, S.; Green, A. Importance of Zinc in Crops and Human Health. *J. SAT. Agric. Res.* **2013,** *11,* 1–5.

De Bernardez Clark, E.; Schwarz, E.; Rudolph, R. Inhibition of Aggregation Side Reactions During In Vitro Protein Folding. *Methods Enzymol.* **1999,** *309,* 217–236.

Deng, S.; Summers, M. L.; Kahn, M. L.; McDermontt, T. R. Cloning and Characterization of a Rhizobium. Meliloti Nonspecific Acid Phosphatase. *Arch. Microbiol.* **1998,** *170,* 18–26.

Deng, S.; Elkins, J. G.; Da, L. H.; Botero, L. M.; McDermott, T. R. Cloning and Characterization of a Second Acid Phosphatase from *Sinorhizobium meliloti* Strain 104A14. *Arch. Microbiol.* **2001,** *176,* 255–263.

Desai, R. P.; Papoutsakis, E. T. Antisense RNA Strategies for Metabolic Engineering of *Clostridium acetobutylicum. Appl. Environ. Microbiol.* **1999,** *65,* 936–945.

Downing, K. J.; Leslie, G.; Thomson, J. A. Biocontrol of the Sugarcane Borer *Eldana saccharina* by Expression of the *Bacillus thuringiensis cry1Ac7* and *Serratia marcescens chiA* Genes in Sugarcane-associated Bacteria. *Appl. Environ. Microbiol.* **2000,** *66,* 2804–2810.

Doyle, J. D.; Short, K. A.; Stotzky, G.; King, R. J.; Seidler, R. J.; Olsen, R. H. Ecologically Significant Effects of *Pseudomonas putida* PPO301(pRO103), Genetically Engineered to Degrade 2,4-dichlorophenoxyacetate, on Microbial Populations and Process in Soil. *Can. J. Microbiol.* **1991,** *37,* 682–691.

Duan, J.; Müller, K.; Charles, T. et al. 1-Aminocyclopropane-1-Carboxylate (ACC) Deaminase Genes in Rhizobia from Southern Saskatchewan. *Microb. Ecol.* **2009,** *57* (3), 423–436.

Eggeling, L.; Oberle, S.; Sahm, H. Improved L-lysine Yield with *Corynebacterium glutamicum*: Use of *dapA* Resulting in Increased Flux Combined with Growth Limitation. *Appl. Microbiol. Biotechnol.* **1998**, *49*, 24–30.

Fasim, F.; Ahmed, N.; Parsons, R.; Gadd, G. M. Solubilization of Zinc Salts by Bacterium Isolated by the Air Environment of Tannery. *FEMS Microbiol. Lett.* **2002**, *213*, 1–6.

Fouts, D. E.; Tyler, H. L.; DeBoy, R. T. et al. Complete Genome Sequence of the N2-fixing Broad Host Range Endophyte Klebsiella Pneumoniae 342 and Virulence Predictions Verified in Mice. *PLoS Genetics* **2008**, *4* (7), e1000141.

Fraga, R.; Rodriguez, H.; Gonzalez, T. Transfer of the Gene Encoding the Nap A Acid Phosphatase from *Morganella morganii* to a *Burkholderia cepacia* Strain. *Acta Biotechnol.* **2001**, *21*, 359–369.

Fuller, M.; Hamed, F.; Wisniewski, M. E.; Glenn, D. M. Protection of Crops From Frost Using a Hydrophobic Particle Film and an Acrylic Polymer. Agricultural Research Service, US Department of Agriculture. 2001.

Galleguillos, C.; Aguirre, C.; Barea, J. M.; Azcón, R. Growth Promoting Effect of Two Strains (a Wild Type and its Genetically Modified Derivative) on a Non-legume Plant Species in Specific Interaction with Two Arbuscular Mycorrhizal Fungi. *Plant Sci.* **2000**, *159*, 57–63.

George, T. S.; Simpson, R. J.; Hadobas, P. A.; Richardson, A. E. Expression of a Fungal Phytase Gene in Nicotiana Tabacum Improves Phosphorus Nutrition of Plants Grown in Amended Soils. *Plant Biotechnol. J.* **2005**, *3* (1), 129–140.

Glandorf, D. C. M.; Verheggen, P.; Jansen, T. et al. Effect of Genetically Modified *Pseudomonas putida* WCS358r on the Fungal Rhizosphere Microflora of Field-grown Wheat. *Appl. Environ. Microbiol.* **2001**, *67*, 3371–3378.

Goldstein, A. H.; Liu, S. T. Molecular Cloning and Regulation of a Mineral Phosphate Solubilizing Gene from *Erwinia herbicola*. Biotechnology **1987**, *5*, 72–74.

Goldstein, A. H. Involvement of the Quinoprotein Glucose Dehydrogenase in the Solubilization of Exogenous Phosphates by Gram-negative Bacteria. In *Phosphate in Microorganisms: Cellular and Molecular Biology*. Torriani- Gorini, A.; Yagil, E.; Silver, S., Eds.; ASM Press: Washington, DC, **1996**, pp 197–203.

Golovan, S.; Wang, G.; Zhang, J.; Forsberg, C. W. Characterization and Overproduction of the *Escherichia coli* appA Encoded Bifunctional Enzyme That Exhibits Both Phytase and Acid Phosphatase Activities. *Can. J. Microbiol.* **2000**, *46*, 59–71.

Goteti, P. K.; Emmanuel, L. D. A.; Desai, S.; Shaik, M. H. A. Prospective Zinc Solubilizing Bacteria for Enhanced Nutrient Uptake and Growth Promotion in Maize (*Zea mays* L.) *Intern J. Microbiol.* **2013**, 1–7.

Grichko, V. P.; Filby, B.; Glick, B. R. Increased Ability of Transgenic Plants Expressing the Bacterial Enzyme ACC Deaminase to Accumulate Cd, Co, Cu, Ni, Pb, and Zn. *J. Biotechnol.* **2000**, *81* (1), 45–53.

Grohmann, E.; Muth, G.; Espinosa, M. Conjugative Plasmid Transfer in Gram Positive Bacteria. *Microbiol. Mol. Biol. Rev.* **2003**, *67*, 277–301.

Gurian-Sherman, D.; Lindow, S. E. Bacterial Ice Nucleation: Significance and Molecular Basis. *FASEB J.* **1993**, *7*, 1338–1343.

Gurmani, A. R.; Khan, S. U.; Andaleep, R. K.; Waseem, K. A. Soil Application of Zinc Improves Growth and Yield of Tomato. *Int. J Agric. Biol.* **2012**, *14*, 91–96.

Gutterson, N.; Howie, W.; Suslow, T. Enhancing Efficiencies of Biocontrol Agents By Use of Biotechnology. In New Directions in Biological Control. Alternatives for Suppressing Agricultural Pests and Diseases. Proceedings of a Ucla Colloquim Held at Frisco, Colorado, January 20–27, 1989. Baker, R. R.; Dunn, P. E., Eds.; New York, 1990, pp 749–765.

Gyaneshwar, P.; Naresh Kumar, G.; Parekh, L. J.; Poole, P. S. Role of Soil Microorganisms in Improving P Nutrition of Plants. *Plant Soil* **2002**, *245*, 83–93.

Hardoim, P. R.; van Overbeek, L. S.; van Elsas, J. D. Properties of Bacterial Endophytes and their Proposed Role in Plant Growth. *Trends Microbiol.* **2008**, *16*, 463–471.

Höfte, H.; Whiteley, H. R. Insecticidal Crystal Proteins of *Bacillus thuringiensis. Microbiol. Rev.* **1989**, *53*, 242–255.

Holguin, G.; Glick, B. R. Expression of the ACC Deaminase Gene from Enterobacter cloacae UW4 in Azospirillum brasilense. *Microbial Ecol.* **2001**, *41* (3), 281–288.

Jones, D. A.; Kerr, A. *Agrobacterium radiobacter* Strain K1026, a Genetically Engineered Derivative of Strain K84, for Biological Control of Crown Gall. *Plant Dis.* **1989**, *73*, 15–18.

Joseph, M. H.; Dhargave, T. S.; Deshpande, C. P.; Srivastava, A. K. Microbial Solubilization of Phosphate: *Pseudomonas* versus *Trichoderma. Ann. Plant Soil Res.* **2015**, *17*, 227–232.

Kaneko, T.; Minamisawa, K.; Isawa, T. et al. Complete Genomic Structure of the Cultivated Rice Endophyte *Azospirillum* spp. B510. *DNA Res.* **2009**, *17* (1), 37–50.

Karlen, D. L.; Mausbach, M. J.; Doran, J. W.; Cline, R. G.; Harris, R. F.; Schuman, G. E. Soil Quality: a Concept, Definition, and Framework for Evaluation (a Guest Editorial). *Soil Sci. Soc. Am J.* **1997**, *61*, 4–10.

Kerovuo, J.; Lauraeus, M.; Nurminen, P.; Kalkinen, N.; Apajalahti, J. Isolation, Characterization, Molecular Gene Cloning, and Sequencing of a Novel Phytase from Bacillus Subtilis. *Appl. Environ. Microbiol.* **1998**, *64*, 2079–2085.

Kerr, A. Biological Control of Crown Gall Through Production of Agrocin 84. *Plant Dis.* **1980**, *64*, 25–30.

Khalid, A.; Arshad, M.; Shaharoona, B.; Mahmood, T. Plant Growth Promoting Rhizobacteria (PGPR) and Sustainable Agriculture. In *Microbial Strategies for Crop Improvement;* Khan, M. S., Zaidi, A., Musarat, J., Eds; Springer-Verleg: Berlin/Heidelberg, 2009; pp 133–160.

Kiel, J. A.; ten Berge, A. M.; Borger, P.; Venema, G. A. General Method for the Consecutive Integration of Single Copies of a Heterologous Gene at Multiple Locations in the *Bacillus subtilis* Chromosome by Replacement Recombination. *Appl. Environ. Microbiol.* **1995**, *61*, 4244–4250.

Kilbane, J. J.; Bielaga, B. A. Instantaneous Gene Transfer from Donor to Recipient Microorganisms via Electroporation. *Biotechniques* **1991**, *10*, 354–365.

Kim, K. Y.; McDonald, G. A.; Jordan, D. Solubilization of Hydroxyapatite by *Enterobacter agglomerans* and Cloned *Escherichia coli* in Culture Mediu. *Biol. Fertil. Soils* **1997**, *24*, 347–352.

Kim, K. Y.; McDonald, G. A.; Jordan, D. Solubilization of Hydroxypatite by Enterobacter Agglomerans and Cloned *Escherichia coli* in Culture Medium. *Biol. Fert. Soils* **1997**, *24*, 347–352.

Kim, Y. O.; Lee, J. K.; Kim, H. K.; Yu, J. H.; Oh, T. K. Cloning of the Thermostable Phytase Gene (phy) from Bacillus sp. DS11 and Its Overexpression in *Escherichia coli. FEMS Microbiol. Lett.* **1998a**, *162*, 185–191.

Kim, K. Y.; Jordan, D.; Krishnan, H. B. Expression of Genes from *Rahnella aquatilis* That Are Necessary for Mineral Phosphate Solubilizaiton in *Escherichia coli. FEMS Microbiol. Lett.* **1998b**, *159*, 121–127.

Kim, C. H.; Han, S. H.; Kim, K. Y.; Cho, B. H.; Kim, Y. H.; Koo, B. S.; Kim, C. Y. Cloning and Expression of Pyrroloquinoline Quinone (PQQ) Genes from a Phosphate-solubilizing Bacterium *Enterobacter intermedium. Curr. Microbiol.* **2003**, *47*, 457–461.

Krause, A.; Ramakumar, A.; Bartels, D. et al. Complete Genome of the Mutualistic, N_2-fixing Grass Endophyte *Azoarcus* sp. Strain BH72. *Nature Biotechnol.* **2006**, *24* (4), 1384–1390.

Krishnaraj, P. U.; Sadasivam, K. V.; Khanuja, S. P. S. Mineral Phosphate Soil Defective Mutants of *Pseudomonas* sp. Express Pleiotropic Phenotypes. *Curr. Sci.* **1999**, *76*, 1032–1034.

Krishnaraj, P. U.; Goldstein, A. H. Cloning of a Serratia Marcescens DNA Fragment that Induces Quinoprotein Glucose Dehydrogenase-mediated Gluconic Acid Production in *Escherichia coli* in the Presence of Stationary Phase Serratia Marcescens. *FEMS Microbiol. Lett.* **2001**, *205*, 215–220.

Krishnaraj, P. U.; Goldstein, A. H. Cloning of a *Serratia marcescens* DNA Fragment that Induces Quinoprotein Glucose Dehydrogenase- mediated Gluconic Acid Production in *Escherichia coli* in the Presence of a Stationary Phase *Serratia marcescens*. *FEMS Microbiol. Lett.* **2001**, *205*, 215–220.

Kullen, M. J.; Klaenhammer, T. R. Genetic Modification of Intestinal Lactobacilli and Bifidobacteria. In *Probiotics: A Critical Review*; Tannock, G., Ed.; Horizon Scientific Press: Wymondham, UK, 1999; pp 63–83.

Kumar, C.; Yadav, K.; Archana, G.; Kumar, G. N. 2-Ketoglutaric Jacid Secretion by Incorporation of *Pseudomonas putida JAA*KT 2440 Gluconate Dehydrogenase (gad) Operon in Enterobacter Asburiae PS13 Improves Mineral Phosphate Solubilization. *Curr. Microbiol.* **2013**, *67*, 388–394.

Kumar, R.; Kumawat, N. Effect of Sowing Dates, Seed Rates and Integrated Nutrition on Productivity, Profitability and Nutrient Uptake of Summer Mungbean in Eastern Himalaya. *Arch. Agron. Soil Sci.* **2014**, *60* (9), 1207–1227.

Kumar, R.; Chatterjee, D.; Kumawat, N.; Pandey, A.; Roy, A.; Kumar, M. Productivity, Quality and Soil Health as Influenced by Lime in Ricebean Cultivars in Foothills of Northeastern India. *Crop J.* **2014**, *2*, 338–344.

Kumar, A.; Bahadur, I.; Maurya, B. R.; Raghuwanshi, R.; Meena, V. S.; Singh, D. K.; Dixit, J. Does a Plant Growth-promoting Rhizobacteria Enhance Agricultural Sustainability? *J. Pure Appl. Microbiol.* **2015a**, *9*, 715–724.

Kumar, R.; Deka, B. C.; Ngachan, S. V. Response of Summer Mungbean to Sowing Time, Seed Rates and Integrated Nutrient Management. *Legum Res.* **2015b**, *38* (3), 348–352.

Lampel, J. S.; Canter, G. L.; Dimock, M. B. et al. Integrative Cloning, Expression, and Stability of the *cryIA(c)* Gene from *Bacillus thuringiensis* Subsp. *kurstaki* in a Recombinant Strain of *Clavibacter xyli* Subsp. *cynodontis*. *Appl. Environ. Microbiol.* **1994**, *60*, 501–508.

Le, H. V.; Trotta, P. P. Purification of Secreted Recombinant Proteins from *Escherichia coli*. *Bioprocess Technol.* **1991**, *12*, 163–181.

Levin, M. A.; Seidler, R.; Bourquin, A. W.; Fowle, J. R.; Barkay, T. EPA Developing Methods to Assess Environmental Release. *Biotechnology* **1987**, *5*, 38–45.

Lilie, H.; Schwarz, E.; Rudolph, R. Advances in Refolding of Proteins Produced in *E. coli*. *Curr. Opin. Biotechnol.* **1998**, *9*, 497–501.

Lin, Q.; Rao, Z.; Sun, Y.; Yao, J.; Xing, L. Identification and Practical Application of Silicate-dissolving Bacteria. *Agric. Sci. China* **2002**, *1*, 81–85.

Lindgren, S. Biosafety Aspects of Genetically Modified Lactic Acid Bacteria in EU Legislation. *Int. Dairy J.* **1999**, *9*, 37–41.

Lindow, S. E. Methods of Preventing Frost Injury Caused by Epiphytic Ice Nucleation Active Bacteria. *Plant Dis.* **1983**, *67*, 327–333.

Lindow, S. E. Ecology of *Pseudomonas syringae* Relevant to the Field Use of Ice-deletion Mutants Constructed in vitro for Plant Frost Control. In *Engineered Organisms in the Environment Scientific Issues*. Halvorson, H. O., Pramer, D., Rogul, M., Eds.; ASM: Washington, DC, 1985; pp 23–35.

Lindow, S. E. Competitive Exclusion of Epiphytic Bacteria by Ice *Pseudomonas syringae* Mutants. *Appl. Environ. Microbiol.* **1987**, *53*, 2520–2527.

Lindow, S. E; Panopoulos, N. J. Field Tests of Recombinant Ice: *Pseudomonas syringae* for Biological Frost Control in Potato. In *Release of Genetically Engineered Microorganisms;* Sussman, M., Collins, C. H., Skinner, F. A., Eds.; Academic Press: London, UK, 1988; pp 121–138.

López-Bucio, J.; Cruz-Ramýrez, A.; Herrera-Estrella, L. The Role of Nutrient Availability in Regulating Root Architecture. *Curr. Opin. Plant Biol.* **2003**, *6,* 280–287.

Long, H. H.; Schmidt, D. D.; Baldwin. I. T. Native Bacterial Endophytes Promote Host Growth in a Species-specific Manner, Phytohormone Manipulations Do not Result in Common Growth Responses. *PLoS One* **2008**, *3* (7), e2702.

Lorenz, M. G.; Wackernagel, W. Bacterial Gene Transfer by Natural Genetic Transformation in the Environment. *Microbiol. Rev.* **1994**, *58*, 563–602.

Lugtenberg, B. J. J.; Dekkers, L.; Bloemberg, G. V. Molecular Determinants of Rhizosphere Colonization by Pseudomonas. *Ann. Rev. Phytopathol.* **2001**, *39*, 461–490.

Ma, W.; Guinel, F. C.; Glick, B. R. *Rhizobium leguminosarum* biovar *viciae* 1-aminocyclopropane-1-Carboxylate Deaminase Promotes Nodulation of Pea Plants. *Appl. Environ. Microbiol.* **2003**, *69*, 4396–4402.

Ma, W.; Charles, T. C.; Glick, B. R. Expression of an Exogenous 1-aminocyclopropane-1-carboxylate Deaminase Gene in *Sinorhizobium meliloti* Increases Its Ability to Nodulate Alfalfa. *Appl. Environ. Microbiol.* **2004**, *70*, 5891–5897.

Ma, J. F.; Yamaji, N. Silicon Uptake and Accumulation in Lower Plants. *Trends Plant Sci.* **2006**, *1*, 392–97.

Makrides, S. C. Strategies for Achieving High-level Expression of Genes in *Escherichia coli.* *Microbiol. Rev.* **1996**, *60*, 512–538.

Martin, M. C.; Alonso, J. C.; Suarez, J. E.; Alvarez, M. A. Generation of Food Grade Recombinant Lactic Acid Bacterium Strains by Site-specific Recombination. *Appl. Environ. Microbiol.* **2000**, *66*, 2599–2604.

Martin, F.; Gianinazzi-Pearson, V.; Hijri, M. et al. The Long Hard Road to a Completed Glomus Intraradices Genome. *New Phytologist.* **2008**, *180* (4), 747–750.

Martinez-Morales, F.; Borges, A. C.; Martinez, A.; Shanmugam, K. T.; Ingram, L. O. Chromosomal Integration of Heterologous DNA in *Escherichia coli* with Precise Removal of Markers and Replicons Used During Construction. *J. Bacteriol.* **1999**, *181*, 7143–7148.

Matsushima, P.; Broughton, M. C.; Turner, J. R.; Baltz, R. H. Conjugal Transfer of Cosmid DNA from *Escherichia coli* to *Saccharopolyspora spinosa*: Effects of Chromosomal Insertions on Macrolide A83543 Production. *Gene* **1994**, *146*, 39–45.

Mazodier, P.; Petter, R.; Thompson, C. J. Intergeneric Conjugation Between *Escherichia coli* and *Streptomyces species. J. Bacteriol.* **1989**, *171*, 3583–3585.

Mazy-Servais, C.; Baczkowski, D.; Dusart, J. Electroporation of Intact Cells of *Streptomyces parvulus* and *Streptomyces vinaceus. FEMS Microbiol. Lett.* **1997**, *151*, 135–138.

McDonald, I. R.; Riley, P. W.; Sharp, R. J.; McCarthy, A. J. Factors Affecting the Electroporation of *Bacillus subtilis. J. Appl. Bacteriol.* **1995**, *79*, 213–218.

Meena, V. D.; Dotaniya, M. L.; Coumar, V. In *A Case for Silicon Fertilization to Improve Crop Yields in Tropical Soils,* Proceedings of the National Academy of Sciences, India Section B: Biological Sciences, 2014a; p 505.

Meena, V. S.; Maurya, B. R.; Bahadur, I. Potassium Solubilization by Bacterial Strain in Waste Mica. *Bang. J. Bot.* **2014b**, *43*, 235–237.

Meena, V. S.; Maurya, B. R.; Verma, J. P. Does a Rhizospheric Microorganism Enhance K+ Availability in Agricultural Soils? *Microbiol. Res.* **2014c,** *169,* 337–347.

Megali, L.; Glauser, G.; Rasmann, S. Fertilization with Beneficial Microorganisms Decreases Tomato Defenses Against Insect Pests. *Agron. Sustain. Dev.* **2014,** *34* (3), 649–656.

Miksch, G.; Neitzel, R.; Friehs, K.; Flaschel, E. High-level Expression of a Recombinant Protein in *Klebsiella planticola* Owing to Induced Secretion into the Culture Medium. *Appl. Microbiol. Biotechnol.* **1999,** *51,* 627–632.

Minas, W.; Bailey, J. E. Co-overexpression of *prlF* Increases Cell Viability and Enzyme Yields in Recombinant *Escherichia coli* Expressing *Bacillus stearothermophilus* α-amylase. *Biotechnol. Prog.* **1995,** *11,* 403–411.

Naseby, D. C.; Lynch, J. M.; Impact of Wild Type and Genetically Modified *Pseudomonas fluorescens* on Soil Enzyme Activities and Microbial Population Structure in the Rhizosphere of Pea. *Mol. Ecol.* **1998,** *7,* 617–625.

Neumann, E.; Schaefer-Ridder. M.; Wang, Y.; Hofschneider, P. H. Gene Transfer into Mouse Lyoma Cells by Electroporation in High Electric Fields. *EMBO J.* **1982,** *1,* 841–845.

Oliveira, C. A.; Sa, N. M. H.; Gomes, E. A.; Marriel, I. E.; Scotti, M. R.; Guimaraes, C. T.; Schaffert, R. E.; Alves, V. M. C. Assessment of the Mycorrhizal Community in the Rhizosphere of Maize (*Zea mays* L.) Genotypes Contrasting for Phosphorus Efficiency in the Acid Savannas of Brazil Using Denaturing Gradient Gel Electrophoresis (DGGE). *Appl. Soil. Ecol.* **2009,** *41,* 249–258.

Olson, P.; Zhang, Y.; Olsen, D. et al. High-level Expression of Eukaryotic Polypeptides from Bacterial Chromosomes. *Protein Expr. Purif.* **1998,** *14,* 160–166.

Orikasa, Y.; Nodasaka, Y.; Ohyama, T.; Okuyama, H.; Ichise, N.; Yumoto, I.; Morita, N.; Wei, M.; Ohwada, T. Enhancement of the Nitrogen Fixation Eficiency of Genetically Engineered Rhizobium with High Catalase Activity. *J. Biosci. Bioeng.* **2010,** *110,* 397–402.

Paranthaman, S.; Dharmalingam, K. Intergeneric Conjugation in *Streptomyces peucetius* and *Streptomyces* sp. Strain C5: Chromosomal Integration and Expression of Recombinant Plasmids Carrying the *chiC* Gene. *J. Bacteriol.* **2003,** *69,* 84–91.

Peralta, H.; Mora, Y.; Salazar, E. et al. Engineering the nifH Promoter Region and Abolishing Poly-β-hydroxybutyrate Accumulation in *Rhizobium etli* Enhance Nitrogen Fixation in Symbiosis with *Phaseolus vulgaris*. *Appl. Environ. Microbiol.* **2004,** *70* (6), 3272–3281.

Peredelchuk, M. Y.; Bennett, G. N. A Method for Construction of *E. coli* Strains with Multiple DNA Insertions in the Chromosome. *Gene* **1997,** *187,* 231–238.

Pigac, J.; Schrempf, H. A Simple and Rapid Method of Transformation of *Streptomyces rimosus* R6 and Other Streptomycetes by Electroporation. *Appl. Environ. Microbiol.* **1995,** *61,* 352–356.

Plaza, G.; Ulfig, K.; Hazen, T. C.; Brigmon, R. L. Use of Molecular Techniques in Bioremediation. *Acta Microbiol. Pol.* **2001,** *50,* 205–218.

Rao, R. N.; Richardson, M. A.; Kuhstoss, S. Cosmid Shuttle Vectors for Cloning and Analysis of *Streptomyces* DNA. *Methods Enzymol.* **1987,** *153,* 166–198.

Reilly, T. J.; Baron, G. S.; Nano, F.; Kuhlenschmidt, M. S. Characterization and Sequencing of a Respiratory Burstinhibiting Acid Phosphatase from *Francisella tularensis*. *J. Biol. Chem.* **1996,** *271,* 10973–10983.

Robleto, E. A.; Kmiecik, K.; Oplinger, E. S.; Nienhuis, J.; Triplett. E. W. Trifolitoxin Production Increases Nodulation Competitiveness of Rhizobium etli CE3 Under Agricultural Conditions. *App. Environ. Microbial.* **1998,** *64,* 2630–2633.

Rodríguez, H.; Fraga, R. Phosphate Solubilizing Bacteria and Their Role in Plant Growth Promotion. *Biotechnol. Adv.* **1999**, *17*, 319–339.

Rodríguez, E.; Han, Y.; Lei, X. G. Cloning, Sequencing and Expression of an *Escherichia coli* Acid Phopshatase/Phytase Gene (appA2) Isolated from Pig Colon. *Biochem. Biophys. Res. Comm.* **1999**, *257*, 117–123.

Rodríguez, H.; Rossolini, G. M.; Gonzalez, T.; Jiping, L.; Glick, B. R. Isolation of a Gene from Burkholderia Cepacia IS-16 Encoding a Protein That Facilitates Phosphatase Activity. *Curr. Microbiol.* **2000a**, *40*, 362–366.

Rodríguez, H.; Gonzalez, T.; Selman, G. Expression of a Mineral Phosphate Solubilizing Gene from Erwinia Herbicola in two Rhizobacterial Strains. *J. Biotechnol.* **2000b**, *84*, 155–161.

Rodriguez, H.; Gonzalez, T.; Selman, G. Expression of A Mineral Phosphate Solubilizing Gene from *Erwinia herbicola* in Two Rhizobacterial Strains. *J. Biotechnol.* **2000**, *84*, 155–161.

Rodrignes, F. A.; Datnoff, L. E. Silicon and Rice Disease Management. *Fitopat. Brasileira* **2005**, *30*, 457–469.

Rodriguez-Mozaz, S.; Lopez de Alda, M. J.; Barcelo, D. Biosensors as Useful Tools for Environmental Analysis and Monitoring. *Anal. Bioanal. Chem.* **2006**, *386*, 1025–1041.

Rossolini, G. M.; Shipa, S.; Riccio, M. L.; Berlutti, F.; Macaskie, L. E.; Thaller, M. C. Bacterial Non-specific Acid Phosphatases: Physiology, Evolution, and Use as Tools in Microbial Biotechnology. *Cell Mol. Life Sci.* **1998**, *54*, 833–850.

Ryder, M. Key Issues in the Deliberate Release of Genetically-manipulated Bacteria. *FEMS Microbiol. Ecol.* **1994**, *15*, 139–146.

Sahoo, R. K.; Ansari, M. W.; Pradhan, M.; Dangar, T. K.; Mohanty, S.; Tuteja, N. Phenotypic and Molecular Characterization of Efficient Native *Azospirillum* Strains from Rice Fields for Crop Improvement. *Protoplasma* **2014**, *251* (4), 943–953.

Sambrook, J.; Fritsch, E. F.; Maniatis, T., Eds.; In *Molecular Cloning: A Laboratory Manual*; Cold Spring Harbor Press: Cold Spring Harbor, NY, 1989.

Santos, V. B.; Araujo, S. F.; Leite, L. F.; Nunes, L. A.; Melo, J. W. Soil Microbial Biomass and Organic Matter Fractions During Transition from Conventional to Organic Farming Systems. *Geoderma* **2012**, *170*, 227–231.

Saravanan, V. S.; Madhaiyan, M.; Thangaraju, M. Solubilization of Zinc Compounds by the Diazotrophic, Plant Growth Promoting Bacterium *Gluconacetobacter diazotrophicus*. *Chemosphere* **2007**, *66*, 1794–1798.

Sashidhar, B.; Podile, A. R. Transgenic Expression of Glucose Dehydrogenase in Azotobacter Vinelandii Enhances Mineral Phosphate Solubilization and Growth of Sorghum Seedlings. *Microbial. Biotechnol.* **2009**, *2*, 521–529.

Sashidhar, B.; Podile, A. R. Mineral Phosphate Solubilization by Rhizosphere Bacteria and Scope for Manipulation of the Direct Oxidation Pathway Involving Glucose Dehydrogenase. *J. Appl. Microbiol.* **2010**, *109*, 1–12.

Sayer, J. A.; Gadd, G. M. Solubilization and Transformation of Insoluble Inorganic Metal Compounds to Insoluble Metal Oxalates by *Aspergillus niger*. *Mycol. Res.* **1997**, *101*, 653–661.

Sayler, G. S.; Ripp, S. Field Applications of Genetically Engineered Microorganisms for Bioremediation Processes. *Curr. Opin. Biotechnol.* **2001**, *11*, 286–289.

Sbartai, H.; Djebar, M.; Rouabhi, R.; Sbartai, I.; Berrebbah, H. Antioxidative Response in Tomato Plants *Lycopersicon esculentum* L. Roots and Leaves to Zinc. *Am. Eurasian J. Toxicol. Sci.* **2011**, *3*, 41–46.

Schäfer, A.; Kalinowski, J.; Simon, R.; Seep-Feldhaus, A. H.; Pühler, A. High Frequency Conjugal Plasmid Transfer from Gram-negative *Escherichia coli* to Various Gram Positive Coryneform Bacteria. *J. Bacteriol.* **1990,** *172,* 1663–1666.

Seshadre, S.; Muthukumarasamy, R.; Lakshminarasimhan, C.; Ignaacimuthu, S. Solubilization of Inorganic Phosphates by *Azospirillum halopraeferans. Curr. Sci.* **2002,** *79,* 565–567.

Shanahan, P.; O'Sullivan, D. J.; Simpson, P.; Glennon, J. D.; O'Gara, F. Isolation of 2, 4-diacetylphloroglucinol from a Fluorescent Pseudomonad and Investigation of Physiological Parameters Influencing its Production. *Appl. Environ. Microbiol.* **1992,** *58,* 353–358.

Sharma, S. B.; Sayyed, R. Z.; Trivedi, M. H.; Gobi, T. A. Phosphate Solubilizing Microbes: Sustainable Approach for Managing Phosphorus Deficiency in Agricultural Soils. *Springer Plus* **2013,** *2,* 587.

Sheng, X. Growth Promotion and Increased Potassium Uptake of Cotton and Rape by a Potassium Releasing Strain of *Bacillus edaphicus. Soil Biol. Biochem.* **2005,** *37,* 1918–1922.

Shivran, R. K.; Kumar, R.; Kumari, A. Influence of Sulphur, Phosphorus and Farm Yard Manure on Yield Attributes and Productivity of Maize (*Zea mays* L.) in Humid South Eastern Plains of Rajasthan. *Agric. Sci. Dig.* **2013,** *33* (1), 9–14.

Simine, C. D. D.; Sayer, J. A.; Gadd. G. M. Solubilization of Zinc Phosphate by a Strain of *Pseudomonas fluorescens* Isolated from a Forest Soil. *Biol. Fertil. Soils* **1998,** *28,* 87–94.

Singh, M. V. Micronutrient Nutritional Problems in Soils in India and Improvement for Human and Animal Health. *Indian J. Fertil.* **2009,** *5,* 11–26.

Singh, J. S.; Pandey, V. C.; Singh, D. P. Efficient Soil Microorganisms: a New Dimension for Sustainable Agriculture and Environmental Development. *Agric. Ecosyst. Environ.* **2011,** *140,* 339–353.

Sinha, R. K.; Valani, D.; Chauhan, K.; Agarwal, S. Embarking on a Second Green Revolution for Sustainable Agriculture by Vermiculture Biotechnology Using Earthworms: Reviving the Dreams of Sir Charles Darwin. *Int. J. Agric. Health Saf.* **2014,** *1,* 50–64.

Sosio, M.; Guisino, F.; Cappellano, C.; Bossi, E.; Puglia, A. M.; Donadio, S. Artificial Chromosomes for Antibiotic-producing Actinomycetes. *Nat. Biotechnol.* **2000,** *18,* 343–345.

Spaepen, S.; Vanderleyden, J.; Okon, Y. Plant Growth-promoting Actions of Rhizobacteria. *Adv. Bot. Res.* **2009,** *51,* 283–320.

Srivastva, S.; Kulkarni, A.; Yadav, N.; Archana, G.; Naresh Kumar, G. International Conference on Managing Natural Resources for Sustainable Agricultural Production in 21st century, Feb 14–18, 2000, New Delhi, p 650.

Sun, Y.; Zhou, X.; Liu, J. et al. "*Streptomyces nanchangensis*," a Producer of the Insecticidal Polyether Antibiotic Nanchangmycin and the Antiparasitic Macrolide Meilingmycin, Contains Multiple Polyketide Gene Clusters. *Microbiology* **2002,** *148,* 361–371.

Sundheim, L.; Poplawski, A. R.; Ellingboe, A. H. Molecular Cloning of Two Chitinase Genes from *Serratia marcescens* and Their Expression in *Pseudomonas* Species. *Physiol. Mol. Plant Pathol.* **1988,** *33,* 483–491.

Swartz, J. R. Advances in *Escherichia coli* Production of Therapeutic Proteins. *Curr. Opin. Biotechnol.* **2001,** *12,* 195–201.

Taghavi, S.; Garafola, C.; Monchy, S. et al. Genome Survey and Characterization of Endophytic Bacteria Exhibiting a Beneficial Effect on Growth and Development of Poplar Trees. *Appl. Environ. Microbiol.* **2009,** *75* (3), 748–757.

Tariq, M.; Hameed, S.; Malik, K. A.; Hafeez, F. Y. Plant Root Associated Bacteria for Zinc Mobilization in Rice. *Pak. J. Bot.* **2007,** *39,* 245–253.

Thaller, M. C.; Berlutti, F.; Schippa, S.; Lombardi, G.; Rossolini, G. M. Characterization and Sequence of PhoC, the Principal Phosphate-irrepressible Acid Phosphatase of Morganella Morganii. *Microbiology* **1994**, *140*, 1341–1350.

Thaller, M. C.; Lombardi, G.; Berlutti, F.; Schippa, S.; Rossolini, G. M. Cloning and Characterization of the NapA Acid Phosphatase/Phosphotransferase of Morganella Morganii: Identification of a New Family of Bacterial Acid Phosphatase Encoding Genes. *Microbiology* **1995**, *140*, 147–151.

Tisdale, S. L.; Nelson, W. L.; Beaten, J. D. Zinc in Soil Fertility and Fertilizers, 4th ed; Macmillan Publishing Company: New York, 1984; pp 382–391.

Tittabutr, P.; Awaya, J. D.; Li, Q. X.; Borthakur, D. The Cloned 1-aminocyclopropane-1-carboxylate (ACC) Deaminase Gene from *Sinorhizobium* sp. Strain BL3 in Rhizobium sp. Strain TAL1145 Promotes Nodulation and Growth of Leucaena Leucocephala. *Syst. Appl. Microbiol.* **2008**, *31* (2), 141–150.

Tye, A. J.; Siu, F. K.; Leung, T. Y.; Lim, B. L. Molecular Cloning and the Biochemical Characterization of Two Novel Phytases from Bacillus Subtilis 168 and Bacillus Licheniformis. *Appl. Microbiol. Biotechnol.* **2002**, *59*, 190–197.

Tyurin, M.; Starodubtseva, L.; Kudryavtseva, H.; Voeykova, T.; Livshits, V. Electro Transformation of Germinating Spores of *Streptomyces* spp. *Biotech. Tech.* **1995**, *9*, 737–740.

Van Dillewijun, P.; Soto, M. J.; Villadas, P. J.; Toro. N. Construction and Environmental Release of a Sinorhizobium Meliloti Strain Genetically Modified to be More Competitive for Alfalfa Nodulation. *Appl. Environ. Microbiol.* **2001**, *67*, 3860–3865.

Vasanthi, N.; Saleena, L. M.; Raj, S. A. Silicon in Day To Day Life. *World Appl. Sci. J.* **2012**, *17*, 425–1440.

Vázquez, M. M.; Barea, J. M.; Azcón, R. Influence of Arbuscular Mycorrhizae and a Genetically Modified Strain of *Sinorhizobium* on Growth, Nitrate Reductase Activity and Protein Content in Shoots and Roots of *Medicago sativa* as Affected by Nitrogen Concentrations. *Soil Biol. Biotechnol.* **2002**, *34*, 899–905.

Verrips, C. T.; van den Berg, D. J. C. Barriers to Application of Genetically Modified Lactic Acid Bacteria. *Antonie Van Leeuwenhoek* **1996**, *70*, 299–316.

Vikram, A.; Alagawadi, A. R.; Krishnaraj, P. U.; Mahesh Kumar, K. S. Transconjugation Studies in *Azospirillum* sp. Negative to Mineral Phosphate Solubilization. *World J. Microbiol. Biotechnol.* **2007**, *23*, 1333–1337.

Von Wright, A.; Sibakov, M. Genetic Modification of Lactic Acid Bacteria. In *Lactic Acid Bacteria*; Salminen, S., von Wright, A., Eds.; Marcel Dekker: New York, NY, 1993; pp 161–198.

Whiting, S. N.; De Souza, M.; Terry, N.; Rhizosphere Bacteria Mobilize Zn for Hyper Accumulate or by *Thlaspi caerulescens*. *Environ. Sci. Technol.* **2001**, *35*, 3144–3150.

Wolf, H.; Pühler, A.; Neumann, E. Electro Transformation of Intact and Osmotically Sensitive Cells of *Corynebacterium glutamicum*. *Appl. Microbiol. Biotechnol.* **1989**, *30*, 283–289.

Yap, W. H.; Thanabalu, T.; Porter, A. G. Expression of Mosquitocidal Toxin Genes in a Gas-vacuolated Strain of *Ancylobacter aquaticus*. *Appl. Environ. Microbiol.* **1994**, *60*, 4199–4202.

Youssef, M. M. A.; Eissa, M. F. M. Biofertilizers and their Role in Management of Plant Parasitic Nematodes. *A Review E3 J. Biotechnol. Pharm Res.* **2014**, *5*, 1–6.

Yu, Z.; Podgwaite, J. D.; Wood, H. A. Genetic Engineering of a *Lymantria dispar* Nuclear Polyhedrosis Virus for Expression of Foreign Genes. *J. Gen. Virol.* **1992**, *73*, 1509–1514.

Zakaria, A. Growth Optimization of Potassium Solubilizing Bacteria Isolated from Biofertilizer. *Natural Resou. Eng. Univ.* **2009,** 40.

Zhang, C.; Kong, F. Isolation and Identification of Potassium-solubilizing Bacteria from Tobacco Rhizospheric Soil and Their Effect on Tobacco Plants. *Appl. Soil. Ecol.* **2014,** *82,* 18–25.

Zhao, B. G. Bacteria Carried by the Pine Wood Nematode and their Symbiotic Relationship with Nematode. In *Pine Wilt Disease*; Zhao, B. G.; Futai, K., Sutherland, J. R., Takeuchi, Y., Eds.; Springer: Tokyo, 2008; pp 264–274.

Index